REIGN OF THE BEAST

Reign of the Beast

The Atheist World of W. D. Saull
and his Museum of Evolution

Adrian Desmond

https://www.openbookpublishers.com
©2024 Adrian Desmond

This work is licensed under an Attribution-NonCommercial 4.0 International (CC BY-NC 4.0). This license allows you to share, copy, distribute and transmit the text; to adapt the text for non-commercial purposes of the text providing attribution is made to the author (but not in any way that suggests that they endorse you or your use of the work). Attribution should include the following information:

Adrian Desmond, *Reign of the Beast: The Atheist World of W. D. Saull and his Museum of Evolution*. Cambridge, UK: Open Book Publishers, 2024, https://doi.org/10.11647/OBP.0393

Further details about CC BY-NC licenses are available at
http://creativecommons.org/licenses/by-nc/4.0/

All external links were active at the time of publication unless otherwise stated and have been archived via the Internet Archive Wayback Machine at https://archive.org/web

Any digital material and resources associated with this volume will be available at
https://doi.org/10.11647/OBP.0393#resources

ISBN Paperback: 978-1-80511-239-6
ISBN Hardback: 978-1-80511-240-2
ISBN Digital (PDF): 978-1-80511-241-9
ISBN Digital eBook (EPUB): 978-1-80511-242-6
ISBN HTML: 978-1-80511-244-0
DOI: 10.11647/OBP.0393

Cover illustration: A spoof of the "Devil's Chaplain", the Rev. Robert Taylor (left, on the podium). His patron, the atheist Richard Carlile, is seen on the right, landing a punch. The wine merchant W. D. Saull funded both men and grounded his evolutionary talks in their dissident sciences. Such pastiches reinforced the prejudices of pious readers, by depicting the moral rot caused by irreligion. The wall posters on the left advertize contraception manuals and licentious memoirs, and a lecture by "Miss Sharples", Carlile's common-law "wife". Taylor's character is being impugned by portraying the mayhem caused by his infidel oratory. Beyond the brawling and debauchery, thieves are shown in the audience (bottom right) and a dagger-wielding agitator (centre). In reality, Taylor's congregations were respectable and attentive.
Etching, in the author's possession, entitled "The Triumph of Free Discussion" (the motto of Carlile's Fleet Street shop selling subversive prints). The caption reads, "A Sketch taken in the Westminster Cock Pit on Wednesday the 24th. of September 1834. Subject A Lecture by the Revd R. Taylor, A.B.M.R.C.S. 'On the importance of Character'."
Cover design by Jeevanjot Kaur Nagpal

Contents

About the Author	vii
Abbreviations	ix
Preface	1
1. Underground Evolution – Setting the Stage	13
2. Introducing Saull	59
Part I – 1820s: Dirty Dives and Subversive Origins	**75**
3. From Eternity to Here	77
4. From the Devil's Chaplain to That Dirty Little Jacobin	117
Part II – 1830s: The Shaven Ape at New Jerusalem's Gate	**157**
5. Perfectibility	159
6. Founding the Museum — June 1831	175
7. Monkey-Man —The Bristol Lecture 1833	201
8. The Antichrist and the Shaven Monkey	223
9. Damned Monkeys	239
10. An Appeal to the Revolutionary Enemy	247
11. Creation on the Cheap	257
12. Making Sense of the Museum	267
13. A Purpose-built Museum — 1835	277
14. Satires on Saull	289
15. Martyrs, Churches, and Vestries	309
16. Lease-holder of the New Moral World	321

Part III – 1840s: Atheists and Aborigines — **335**

17. Halls of Science — 337
18. The Atheist Breakaway — 357
19. Backlash — 385
20. Peace and Harmony — 397
21. Secularism and Salvage — 405
22. British Aborigines — 417
23. Reforming Scientific Society — 443
24. Museum and Pantheon for the Masses — 469
25. Celebrating the Dead — 489

Part IV – 1850s: Destruction — **501**

26. Provisions for the Afterlife — 503
27. Death and Dissolution — 531

Appendix 1 — 553
Appendix 2 — 555
Appendix 3 — 561
Appendix 4 — 565
Appendix 5 — 573
Appendix 6 — 581
Bibliography — 583
Index — 635

About the Author

Adrian Desmond is a science historian specializing in the social relations of evolution and palaeontology. His *Politics of Evolution: Morphology, Medicine, and Reform in Radical London* (1989) received the Pfizer Award of the History of Science Society. *Darwin* (1991, co-authored with James Moore) won the James Tait Black Memorial Prize for Biography, the Watson Davis Prize of the History of Science Society, the Dingle Prize of the British Society for the History of Science, and the Comisso Prize in Italy. He published a two-volume biography *Huxley* (1994, 1997), and collaborated with James Moore again on *Darwin's Sacred Cause: Race, Slavery and the Quest for Human Origins* (2009). *Reign of the Beast* is his tenth book.

Abbreviations

ANH	*Archives of Natural History*
AS	*Annals of Science*
BJHS	*British Journal for the History of Science*
CPG	*Cleave's Penny Gazette*
CGV	*Cleave's Gazette of Variety*
DPMC	*The "Destructive," and Poor Man's Conservative*
FTI	*Free-Thinker's Information for the People*
GM	*Gentleman's Magazine*
HO	Home Office Reports, National Archives, Kew
HS	*History of Science*
JBAA	*Journal of the British Archaeological Association*
JHB	*Journal of the History of Biology*
JHI	*Journal of the History of Ideas*
JSBNH	*Journal of the Society for the Bibliography of Natural History*
L	Letter
LI	*London Investigator*
LMR	*London Mechanics' Register*
LMI	London Mechanics' Institution
M	Memorial
MC	*Morning Chronicle*
MM	*Mechanics' Magazine*
MNH	*Magazine of Natural History*
NMW	*New Moral World*
NRRS	*Notes and Records of the Royal Society*
NUWC	National Union of the Working Classes
NS	*Northern Star*
ODNB	*Oxford Dictionary of National Biography*

OR	*Oracle of Reason*
PM	*Penny Mechanic*
PMG	*Poor Man's Guardian*
PP	Parliamentary Papers
PS	*Penny Satirist*
PSA	*Proceedings of the Society of Antiquaries*
TS	*True Sun*
UR	*Utilitarian Record*

Mr. Saull read an ingenious essay, to prove that the baboon is the original form of the human species, and expressed his hope that the day would arrive when the whole of the monkey species would be entitled to the elective franchise. Universal suffrage would not be complete without it. Lord Brougham said, that if this enfranchisement took place in Mr. Saull's day, he hoped that Mr. S. would be chosen as the first representative of the new elective body.

> *Penny Satirist*, 23 Sept. 1837, spoofing Saull's belief in
> democracy and a monkey origin for mankind

Preface

Was he joking? asked a co-operator in 1833, on learning that a founder of London's first labour exchange had lectured on men emerging from monkeys. Madness, surely, to think that such ribaldry could smooth our path to the socialist millennium. But the lecturer had been serious, and that is not the strangest part. The man had for the previous two years been running a museum of evolution. Imagine such a museum on a central London street in Darwin's younger day, almost three decades before the *Origin of Species* appeared. Impossible? After all, of the hundreds of Darwin biographies and histories of evolution, not one mentions it.

More intriguingly, this museum bucked the Victorian trend. It was free to men and women of all ranks, but artisans were especially invited, no embarrassing letter of introduction required. Just step in to understand how the present world had been produced and what promise fossil life held out for the future. Nor were these any old fossils. The museum held priceless treasures, expensive originals or 'type' specimens, some of which would become famous. Odder still, for a place expunged from the collective memory, it was lauded at the time as the biggest private geology museum in London, perhaps, some said, the country.

In its day the museum was difficult to miss. The two-storey, purpose-built edifice stood on Aldersgate Street, within view of London's magnificent new General Post Office. It was run by a proprietor who argued that life had 'evolved', and, more outrageously, that humans had ape origins. So how have historians and palaeontologists missed it?

True, it is easy for a myopic history to favour the scientific swells. They left their stories in expensive books and bequeathed a brilliant, accessible science to be reworked through the generations. The trouble is, switching the spotlight from the cut-glass crystal of the wealthy drawing room onto cut-price dives requires exhaustive work, even if the results are enriching. It could be argued that the blinkered emphasis on the mannered Charles Darwin and his urbane mentor Sir Charles Lyell, who now repose together in Westminster Abbey, actually acts to impoverish our cultural understanding. Not for our museum proprietor such a shrine. To find him you would have to search out an unconsecrated corner of Kensal Green Cemetery, a pilgrimage site where he is surrounded by radical heroes.

His name was William Devonshire Saull. Neither historians nor palaeontologists know much about him. And what image we do have of Saull was skewed by detractors, who thought him a misguided fool draped in satanic robes. Saull was a proselytizing socialist, atheist, and republican—a man who once outraged *Times* readers by reminding the monarch of the fate of Charles I. Saull was denigrated by decent society, which subsequently buried him with indecently obscurantist obituaries.

Saull's museum shared the same ignominious fate. It was destroyed and lost to posterity. But was it *really* of any consequence? Well, let us focus on the cultural impact of just one of its twenty thousand exhibits. The 'dinosaur', famously concocted by the upright Richard Owen in 1842, was based primarily on fossils from Saull's collection.[1] And this monstrous reptilian creature emerged from its furiously radical age to become one of the most iconic images of the media-obsessed twentieth century.

It was Owen's visit to the museum that piqued my original interest in Saull. The book might have begun in the 1990s, when Hugh Torrens allowed me to rewrite the "Saull" entry for the *Oxford Dictionary of National Biography*. But it had a much lengthier gestation. In mid-1970s, I became intrigued by the idea of the pious comparative anatomist Richard Owen finding an *Iguanodon* sacrum in a *socialist*'s museum, of all places. Why would a socialist have a fossil museum? What function did it serve in the 1830s and 1840s?—those violent decades of newly-established class warfare driven by Saull's friend Henry Hetherington in his illegal *Poor Man's Guardian*. (And who was the first man buried in Saull's funeral plot? Henry Hetherington himself.) How did the respectable Anglican Owen, the pet of the Tory nobility, a man who excoriated materialist transmutation as a moral and social poison, negotiate Saull

---

[1] Torrens 1997. The name "dinosaur" gained little vernacular traction to start (O'Connor 2012), and by the time it did Saull was dead.

and his radical and co-operative infidels? It was the question that led to *Reign of the Beast*.

Equally, the book could not have reached fruition without the vast digitization projects of the 2010s, especially of London's umpteen newspapers. It is only by getting back to these first drafts of history that we can make sense of Saull in his micro-context. They allow us to pin his activities down, almost to street level. And by this time a socially embedded history of science had become commonplace, which left formerly neglected actors—especially among the querulous working classes and their allies—crying out for study.

By far the greatest surprise to come out of this study was to find that, in Darwin's younger day, there *was* an open palaeontological museum, set up specifically to inform the great unwashed of their monkey ancestry and evolutionary destiny. That destiny was to realize the morally perfect man and woman, socialist of course (something Darwin would have abhorred). That Saull's artisan-friendly evolutionary warehouse had lain undetected under the noses of historians and palaeontologists has an explanation. The museum was shattered and fragmented on Saull's death, then lost as the traces were scattered to the four corners. In the same way, the evidence for Saull's evolutionary teaching was itself spread through hundreds of newspaper shards, which had to be pieced laboriously together. It was a gigantic job of cultural reclamation. My digitised database for Saull alone has over two thousand entries, mostly press snippets. And this is on top of six thousand cuttings of related street prints. Stitching it back together took decades.

My purpose, therefore, is not to discuss the history of evolution at the Victorian outset, or any of its bourgeois cultural cradles. This despite the popular appetite for such synoptic approaches, as shown by the wealth of books. Tellingly, most of these have the trigger word "Darwin" in the title, even if they now "try to avoid the tendency to see 1859 as 'year zero'".[2] By contrast, *Reign of the Beast* remains far removed from Darwin's gentrified world—our curious haunts are a dark Hades that would have horrified Darwin, not that he would have dreamed of

---

2   Conlin 2014, 5; Stott 2012; Quammen 2006. The "de-centring of Darwin" (E. Richards 2020, 9) encourages serious studies of alternative social and political contexts of Victorian evolution.

entering them.³ Nor does *Reign* make much contact with my *Politics of Evolution*. That dealt with the shabby-genteel bourgeois radicals and their deployment of the anatomies of self-development. These medical dissidents were looking first and foremost to career enhancements, and using materialistic sciences to attain that end. They were fighting a dirty war against the monopolistic medical baronets running the hospitals in the 1830s, and campaigning on behalf of the new order of lowly General Practitioners who ministered to the poor. The present study sinks a mine shaft much deeper into the social strata. There *were* points of contact between these medical democrats and Saull's street republicans—the fiery Thomas Wakley, founder of the *Lancet*, being one (he even teamed up with Saull to bring back the transported Tolpuddle Martyrs, Chapter 15)—but they were minimal. Some reforming physicians were known to frequent radical dives, John Epps being another case in point. But Epps, in his Quaker's hat, was a Christian who shared the democratic bent of the urban insurgents, not their vulgar atheism. Focusing on Saull allowed me to pursue the new sciences of palaeontology and evolution to a 'lower' level, right down to the socialist bedrock. *Reign* looks to the 'masses', not the 'classes'. It seeks to resurrect the street activists demanding complete emancipatory reform and to take seriously a previously-ignored ideological context. In this way, we can reassess the working-class threat that infidel 'evolution' (defined the old socialist way) could pose during the political upheavals of the 1830s. Not only was it a class threat to the conservative squirearchy; but Saull's monkey-stained materialism—and this is another theme of the book—equally frightened the wilder young millenarians inside the labour movement itself.

The real effrontery to them was that monkey. Saull was possibly the only lecturer in Britain in the 1830s to declare publicly that humans had ape forebears. And, arguably, it was his Grub Street milieu—infidel and socialist—that nurtured such a shocking view and sustained his public bravado.

---

3   Although Darwin did know of the "Devil's Chaplain", the indicted blasphemer (the Rev. Robert Taylor), who took his infidel mission to Cambridge and was hounded out of town by the students. Darwin also owned a cheap copy of *Lectures on Man* pirated by the notorious William Benbow, although I suspect in ignorance of its pornographic provenance: Desmond and Moore 1991, 81–84, 260.

How heinous was such a belief outside of the "blasphemy chapels" (dissenting chapels taken over by deists, atheists and co-operators in the later 1820s and early 1830s for their lectures, liturgical skits and political meetings[4])? Why was evolution so threatening in the 1830s? Two centuries later, it is hard to grasp how even the seemingly innocuous suggestion of one animal being able to turn into another could have caused such consternation. Yet it did. Evolution was abominated by many and left some hysterical: the Cambridge divine who saw total social collapse in its train; the British Museum grandee reduced to vulgarity in calling it vomit; the evangelical Christian who thought it heralded Satan's coming. The revolting prospect clearly raised deep social fears in an undemocratic, pious, conservative country. The fact that Cambridge catered to wealthy Anglicans, the British Museum feared admitting the uncouth classes, and that evangelical magazines were obsessed with artisan infidelity only reinforces the conclusion. The evolutionary spectre was a social threat, and *Reign of the Beast* looks to the weaponizing of such science by street deists, socialists, and radicals to underscore the roots of this dread.

We might look at Darwin, too, to understand his social fears. Saull and Darwin stood at opposite ends of the social spectrum and their diametric attitudes to evolution's social upshot are revealing. The wealthy, land-owning, would-be magistrate Darwin later confided that admitting that species could mutate was "like confessing a murder".[5] But for Saull, publicly advocating something far worse—that man was a transmuted ape—held no terror. He felt no qualms, no shame, in committing what Darwin feared would be seen as a capital offence by society. He even taunted young theology students on the subject. But then, for Saull, that society was a repressive, Anglican-dominated state, shielding massive inequalities. Undermining it was no crime at all, but morally justified and politically expedient.

Saull's question, 'What promise did fossils hold for the development of socialist man?' would have been unintelligible to Darwin. As the

---

4   The main ones we discuss were the former Congregationalist Salter's Hall Chapel on Cannon Street (taken over in 1826 by the Rev. Robert Taylor), the chapel in Grub Street, Cripplegate (set up for the Rev. Josiah Fitch in 1828), and the Optimist Chapel in Windmill Street (1829–31), re-branded the Philadelphian in 1831. Saull helped set up the former two, and lectured in the latter two.
5   Burkhardt et al. 3:2. Hodge 2009 on Darwin's landed-capitalist context.

literary historians would have it, Saull was imposing his own narrative 'plot' on fossil history. But where Darwin's evolutionary story privileged neither man nor his place,[6] millenarian co-operators gave their narrative real meaning for humanity. The fossils in Saull's emporium portrayed an evolution that was bursting with promise for the socialist New Jerusalem just over the horizon.

As a piece of idiosyncratic history of science, far from the mainstream, the book traipses along dark streets in the radical thirties and hungry forties to assess how even esoteric science could end up in disreputable rags. That it did so appalled Evangelical Christians. Such an unholy union of grubby atheism and abominations about the earth's long history and mankind's bestial ancestry proved Revelation's prophecy: Satan was abroad spreading his "filthy slime over Christendom" and the Second Coming was nigh.[7] The '*Beast*' of the title, in one aspect, was the levelling atheist's ancestor, the 'evolutionist's' monkey; but, to the outraged defender of tradition, it was the devil within, driving such blasphemous insanity. A biblical exegete investigating "evil" recalled in 1843 that the great Sir Isaac Newton considered "the reign of the beast to be the open avowal of infidelity".[8] That year, 1843, ten years after Saull went public, schismatic street atheists, sick of socialist quietism, were streaking past him and promoting mankind's bestial ancestry with a still greater vengeance as a stick to bloody the parson's nose.

This scenario—consciousness-raising working-class warriors using home-brewed astro-geologies to thrash the hated tithe-extracting Anglicans—is a world removed from the hackneyed 'warfare of science and theology' paradigm. That referred to *elite* gentlemanly thought, seemingly at war with itself as it tried to exclude an 'obscurantist' religion. For a century and a half, screeds have been written on how proper science ejected every tainting theological vestige. What started as polemical tirades by professionalizing scientists pushing out their boundaries to colonise new cosmological realms ended as a popular platitude.[9] *Reign of the Beast* adopts neither this military metaphor,

---

6 G. Beer 1985, 21–22; Zimmerman 2008, 2–3; A. Buckland, 2013, ch. 1. Throughout the text, the terminology of the day is used, which included non-inclusive gender language. 'Man' is taken to mean the whole of mankind.
7 Bickersteth 1843, 8–22; *Revelation* chs. 13, 16.
8 Bosanquet 1843, 115.
9 F. M. Turner 1978.

demolished in the 1970s,[10] nor does it engage with C. C. Gillispie's *Genesis and Geology*, whose problematic was also theology tainting high-blown theories of the earth.[11]

While most of these studies take a 'lateral' view, territorial essentially, looking to the frisson as middle-class professional boundaries jostled to-and-fro, here I add a missing dimension, the 'vertical', or class aspect. I look at a novel knowledge fashioned in situ to suit emerging socialist and infidel interests in an Anglican-privileged age. Our scurrilous social environment, back-alley dives and blasphemy chapels, and its grubby actors—anti-clerical deists, radicals, and co-operators—stand out of view of the 'higher' scientific echelons dealt with in today's demilitarised studies. These have demonstrated how religious affiliations affected scientists' attitudes, yet they avoid the 'lower' orders,[12] and how their views might have encouraged religious realignments in the dominant scientific class itself. Then again, when the grandees *are* seen to face threats, these are too often traced to the preceding century. True, the dons and divines were still "alarmed by the way that the Enlightenment of Diderot and Voltaire led to the French Revolution of 1789, the Reign of Terror, world war and Napoleon's military dictatorship".[13] But track this insurgent scourge forty years forwards, as Enlightenment ideas went 'underground', worming their way into insurgent 'pauper' Britain, and the more immediate threat becomes clear. How else to explain the often hysterical rejection of materialist evolution by don and divine? It was a *living* menace, shaking the ground under their feet. The grandees closed ranks against the rookery infidels for fear that their edifice-shaking sciences would topple the tiers of privilege.

Such a characterization also shows that *Reign* is not shaped by the old 'popularization of science' mould. That *noblesse oblige* model saw

---

10   J. R. Moore 1979, 19–122; J. H. Brooke 1991; P. Harrison 2015; Knight and Eddy 2016; Hardin, Numbers, and Binzley. 2018.
11   Rupke 1994b. The relations of Christianity and geology with its time and origins motif are understandably perennially interesting: Kölbl-Ebert 2009.
12   Using the terminology of the day. We have to be careful using this disparaging social-stratification language, for fear that it perpetuates rather than exposes the Victorian caste system stretching from royal 'highnesses' to 'low'-life. Conservative rags sneered at the latter, those who lived in some "filthy low street", and they execrated radicals for spreading their moral pestilence among "the low and ignorant": *The Age*, 28 Aug. 1842, 4; *Argus*, 28 Jan. 1843, 9.
13   D. Knight 2004, 53.

high-brow science being simplified and drip-fed to docile marginal audiences, a top-down activity with all the condescension that implied. Looking at the unexplored socialist and blasphemous forums actually exposes the poverty of this antiquated concept. But then this old diffusionist model has been heavily deconstructed by science historians recently.[14] *Reign of the Beast* is more an exercise in reclamation, in recovering an indigenous infidel science. Here, cannibalized scraps of subversive Enlightenment tomes were fused with upturned geological works to produce a blunderbuss science that was original, useful, and totally unacceptable to the establishment.

What we have in Saull's case was a dissident geology and astronomy, re-factored as munitions for new class interests, and shared with actively engaging audiences. These anti-clerical flocks were themselves of a new type. Their literacy was evidently not low; in fact, quite the opposite. Whether they bought into this subversive science at blasphemy chapels or socialist Halls of Science—which attracted the more "reflecting of the handicraftsmen"[15]—or in Saull's museum, they were obviously "periodical literate",[16] able to devour the co-operators' house journal, the *Crisis*, or the radical *Poor Man's Guardian*, and equally able to take in Saull's monkeying endeavours.

To sum up, *Reign of the Beast* focuses on illicit geology in infidel contexts. To say this is an unplumbed area would be an understatement. In 1990, Steven Shapin conceded that we knew pathetically little about the scientific beliefs of "lay members" of our own society.[17] If that is true of people today, imagine our ignorance of "lay" cultures in the 1830s. Agreed, the past few decades have seen an effort to amass Victorian plebeian autobiographies. But while these texts have been exploited, it is largely to illustrate hoi-polloi interest in literature, not science.[18]

If for Shapin's "laity" we read upstart urban groups, from fastidiously literate compositors to semi-literate but politically-articulate coalmen—the chaps, their wives and children, who haunted London's blasphemy chapels and socialist halls—then this area has remained the "cultural

---

14 Cooter and Pumfrey 1994; J. A. Secord 2004b; Topham 2009a, 2009b.
15 T. Coates to H. Brougham, 27 Sept. 1839, Brougham Correspondence 95, University College London; Coates 1841, 29.
16 Murphy 1994, 8.
17 Shapin 1990, 994.
18 Rose 2002.

wasteland" that Roger Cooter saw in 1984.[19] Only by filling in this missing class dimension will we achieve the necessary perspective on elite scientific authority. Traversing this formerly terra incognita will reveal the sciences of these subversive groups as locally-relevant constructions born of political necessity. Rejecting capitalist or anti-democratic authority, in society and science, the dissidents manufactured their own transformative knowledge on site, an indigenous production that they proclaimed as *really* useful.

Until we understand such a "contest between, rather than within, classes", say Roger Cooter and Stephen Pumfrey, and see how "ordinary men and women" tackled their own big scientific issues, we will not fully grasp how the grandees of science propped up the world of their paymasters by way of responding to the democratic threat.[20]

By focusing on a fashionable science, geology, utilized by the autodidacts among the "productive classes" (the co-operators' broader alternative to the radicals' emergent concept of the "working classes"), we can break up the old notion of a "common context" for all Victorian science. This idea of a "common context" has been prevalent since the 1970s.[21] Today, however, historians of science are no longer concentrating solely on the gentlemanly "intelligentsia", their shared ideas and ideals. Yet we still need to dig deeper, to further undercut the old paradigm by exposing the class bases of the rival 'pauper' sciences. In short, we need to get down to street level and ask really tricky questions: how did the science of the anti-union, capitalist 'blasphemers' differ from that of the anti-Malthusian co-operators or the democratic radicals? Only then will we understand how even the "scum"—as angry readers of the *Poor Man's Guardian* were branded[22]—made their own knowledges fit for purpose.

The resulting book probably takes too literally Jim Secord's injunction to view "science as a form of communicative action".[23] In *Reign of the Beast*, we see it as sloganeering shouts from behind the barricades. Nowhere better do we sense how political reform shaped the *elite* scientific culture

---

19   Cooter 1984, 2.
20   Cooter and Pumfrey 1994, 245, 249.
21   J. A. Secord 2021, 56–58.
22   *PMG*, 5 Nov. 1831.
23   J. A. Secord 2004b, 663.

of the 1830s than in Secord's *Visions of Science*. But his urban gentry with their expensive avocations producing ever-more pricey tomes were a far cry from our gutter-press infidels. Now we need to understand how political agitation shaped a responding working-class science, an important facet without which the whole cannot be understood.

The infidels and socialists left little in leather-bound form, nothing for the literary reviews even to sneer at. Their bleeding-edge politics created a much more jagged science. What artisans lapped up in Saull's museum was destined to serve distinct republican, democratic, and socialist ends. But penetrating this subterranean world through ephemeral squibs, illicit penny trash, and police informers' reports was a time-consuming labour of love, explaining why, as I say, the book was so long in the coming.

In truth, so many years have passed that many colleagues and correspondents are no longer with us to be thanked in person. In particular I am thinking of the late Mick Cooper and John Thackray, both of whom were encouraging and ever ready with information. Nellie Flexner read the manuscript many times over and suggested so many improvements. Bernie Lightman, too, acted beyond the call of duty and gave me his thoughts on the finished book. For fine reading of the text and stylistic suggestions I would also like to thank two anonymous referees. My heartfelt thanks also go to Hugh Torrens, Iain McCalman, Jim Moore, Roger Cooter, Jim and Anne Secord, Evelleen Richards, Ruth Barton, Steven Plunkett, and Frank James for plying me with offprints and coming to my aid over the years. I am also indebted to Angela Darwin for allowing me to read the T. H. Huxley family papers. Two of the greatest resources for radical literature are the Bishopsgate Institute, London, where David Webb was always enthusiastic and helpful in his searches; and the Co-Operative Heritage Trust Archive in Manchester, and here I must thank Jane Donaldson, Sophie Stewart, and Gillian Lonergan who have answered so many queries. At the Central Archive of the British Museum, Stephanie Clarke helped with the Trustees Minutes and Original Papers. I also received assistance from Valerie Hart at the Guildhall Library, Beverley Emery at the Royal Anthropological Institute, Rosie Jones, the Special Collections Librarian at the Natural History Museum, London, and, at the Geological Society of London Library, Caroline Lam and Wendy Cawthorne. To all these institutions I extend my thanks, which also go to the University College London archives, Birkbeck College, London, the Wellcome Institute for the History of Medicine Library, Imperial College, London, Archives, the Zoological Society of London, Library, the Linnean Society of London, the British Library, and last but not least The National Archives, Kew, whose preserved Home Office police spy reports proved so revealing.

Praise must finally go to Alessandra Tosi and her team, for their dedication in publishing open-access books. In particular, I would like to thank Jennifer Moriarty, who patiently accommodated my GNU/Linux manuscript submission. With open access everyone can share in knowledge, esoteric or otherwise, and not only those with deep pockets or privileged access to university libraries.

1. Underground Evolution – Setting the Stage

Extreme Geopolitics

> Let it not be forgotten that all proceedings with which the socialists desecrate the sabbath and outrage revelation, invariably open with a lecture on geology.[1]

So warned the appalled editor of the *Church of England Magazine* in 1840, after leaving a talk in a socialist hall by the London wine merchant and museum owner William Devonshire Saull (1783–1855). It was a reminder that the new science of the earth was not only startling and fashionable, but dangerous in dirty hands. Dissidents were harnessing geological armaments for use against the biblical props of priestly power. They were making the age of rocks undermine the Rock of Ages. An infidel geology was even being used to attack the top-down power structure of society, which denied the activists what they demanded: democracy for the radicals, and an anti-capitalist economy for the co-operators. In the wrong hands, seditious hands, the re-manufactured science could even serve the Antichrist.

Step in Saull with his filthy heresy of a monkey origin for man. Saull came tainted, having made his public debut in court, indicted on blasphemy charges. He was the financial backer of the jailed blasphemer Richard Carlile in the 1820s and of the socialist Robert Owen thereafter. His heresy was worse for being taught publicly, in London's largest private geology museum—*his* museum, which was dedicated to the evolving history of life. Astonishingly, this museum was founded early,

---

1   *Church of England Magazine* 9 (15 Aug. 1840): 120; *NMW* 8 (5 Sept. 1840): 159.

it was up and running in June 1831, only months after a young Charles Darwin had taken his degree at Cambridge.

How do we illuminate the back alleys where such strange views were fomenting? Trajectories are one way to throw light on mature views. The gentlemanly Darwin's path, his education, travels, materials, mentors, collections, political and religious convictions, have been meticulously dissected by scholars to plot his path to natural selection. Saull's background was the antithesis: untutored origins, trading status, socialist politics, atheism, and mentors whom Darwin would have detested. It is this peculiar set of circumstances that *Reign of the Beast* explores. With Saull leaving so little documentary evidence, we can only take a contextual approach, to show his very different trajectory through a series of underground dives. Where Secord's *Victorian Sensation* follows readers reacting to one pot-boiling book (the anonymous *Vestiges of the Natural History of Creation*), *Reign of the Beast* looks to a prior process in the making of knowledge: how the amalgam of plebeian science changed as it passed through successive blasphemous, radical, and co-operative furnaces.

Thus the following chapters show Saull moving from Richard Carlile's deistic clique with its eternalist geology, through the astro-theology of the "Devil's Chaplain",[2] the Rev. Robert Taylor, to the astro-geology of that "dirty little jacobin" Sir Richard Phillips.[3] Saull absorbed the new geology of fossil origins and progression along the way. It was a fit new science for a shadowy ideologue being watched by police spies as he moved to the centre of 'Social Father' Robert Owen's circle, with its emphasis on the perfectibility of man[4] (see Chapter 5). All this will help explain Saull's 'evolutionary' stance in the early 1830s—and his monkey-man, itself an outrageous provocation in a pulpit age.

Here, in the Introduction, I provide an overarching, non-chronological exploration of the historiographical conundrums of such a strange story.

Geology, the emerging account of the sequencing of the earth's strata and its fossil inhabitants, was the new flirtation of the emerging

---

2   Taylor's pride in the title can be seen in the police spy report, HO 40/25, f. 281 (15 Nov. 1830).
3   *Blackwood's Edinburgh Magazine*, 12 (Dec. 1822): 704.
4   The idea that, with right changed conditions—educational, religious, political—mankind could rise to moral heights in a socialist New Jerusalem.

British middle class through the 1830s, however troubling to the more conservative clergy. But it needed careful patrolling. This was a seething age of parliamentary and civic reform when young Turks "joined clubs of all sorts, heteroclitical [deviant], political, and Geological", as the *Herald to the Trades' Advocate* had it. Following the long Tory-dominated repression after the Napoleonic Wars, humble political activists now refused to be dismissed as "a grumbling swinish multitude", or to be cowed by "the haughty, domineering lordlings".⁵ A burst of reformist activism in the 1820s led to a rise in deism, unionism, and co-operation, with screeching demands for democracy and disestablishment. This could be well served by a new offensive science. The placarding and posting, blasphemy dens and radical agitation, the burning of jails and firing of the bishop's palace in Bristol during the Reform Bill riots, made this an age of fear for pious folk. Many believed "that 'the masses' were their natural enemies, and that they might have to fight ... for the safety of their property and the honour of their sisters".⁶

The threat of geology being co-opted by Satan's agents was spelled out by an Oxford Professor. He warned in 1834 that:

> the people every where are learning, and will learn, Geology. The first rudiments of the science bring them to successions of primaeval aeras totally different from the six days (whether natural days or longer periods) of the Book of Genesis. Next comes the emissary of infidelity. He points out the contradictions: the hearers cannot deny it: therefore he says you must reject the whole Bible and the whole of Christianity.⁷

There's the nub, the great attraction to the freethinker, the deist, the anti-clerical socialist. Geology opened up the subject of the age and development of the Earth. A new breed of pauper 'infidel' was being taught to associate a literal reading of Genesis with a tithe-rich, state-sanctioned "Priestcraft". What better way than this new upstart science to subvert the Anglican authority in the land?

Our entrée into this deistic netherworld is provided by one particular courtyard, full of wine caskets and brandy crates, shire horses, and heavy carts. It was a warehouse in one of London's main thoroughfares,

---

5   *Herald to the Trades Advocate* (11 Dec. 1830): 187 (9 Apr. 1831): 452–53.
6   Kingsley 1910, 3; Young 1960, 24.
7   Baden Powell 1834, 18.

Aldersgate Street, run by the wholesaler William Devonshire Saull. He was a fascinating, enigmatic, little-known trader: a hard core ideologue with a soft centre, an affable merchant who was deeply irreligious. Self-taught, and sensitive about it, he saw socialist schooling for the ignored young of the 'industrious' masses as a way to tackle poverty and raise awareness of social injustice. And his museum of evolution was to be central to this.

Saull put his money where his mouth was. As a self-made City merchant, he used his wealth to finance the movement. In the 1820s, he poured large sums into freethought venues, bailed prosecuted blasphemers (blasphemy being a crime), and defrayed defence costs. A teetotal advocate with a Robin Hood air, he plied the gentry with their favourite tipple and funnelled the profits to the poor. There was never any lack of seeming contradictions in his life. Come the thirties, he was the 'Utopian'[8] socialist Robert Owen's financier, putting up the money for institutes and halls of science. He was even owner of Owen's town house and mortgagee of Owen's home on the experimental co-operative estate of Harmony. Saull was a wealthy commercial gentleman who, somewhat incongruously, bankrolled co-operative equality. Always he was a facilitator, and there was hardly an infidel, Owenite, or radical pump that was not primed by his cash.

The City trader became not only the banker, but, in a strangely related way, the geologist to the cause. Most of all, Saull poured money into his museum. This raised his fossil emporium into one of London's top attractions by the end of the 1830s. It was hailed in the press as among "the most interesting and extensive geological collections" in the city, even "the largest private Geological collection in the United Kingdom".[9] By the 1850s, it contained over 20,000 exhibits, the lot said to be worth £2,000, equivalent to perhaps £200–300,000 today.[10]

How this courtyard museum inside his brandy depot *functioned* is the important thing. Arguably it was a 'radical' museum. The evidence

---

8   Though derided by Marx as "Utopian", labour exchanges, mutuals, and building societies were hardly utopian, even if Owen's followers did expect capitalists to voluntarily relinquish the means of production.

9   *Courier*, 27 Dec. 1841, 1; *Morning Post*, 31 Dec. 1841; *NS*, 31 Oct. 1846, 3. Karkeek 1841a, 73; 1841b, 175, too, called it "the largest private collection of fossil remains in the kingdom".

10 *UR*, 15 Sept. 1847, 83; *Mining Manual and Almanack for 1851*, 136; Timbs 1855, 542.

for this comes partly from its content but largely from the way it was pressed into service. In some museums, incoming exhibits, being serendipitously acquired, drove the exhibitions astray from their original goals and "disciplinary norms".[11] But Saull's from first to last was *designed* as an infidel Owenite cabinet. Even if the fossils were like those found in conventional museums, he used them for socialist educational purposes. And while some other institutions arranged their fossils stratigraphically, Saull did likewise, but for specific 'evolutionary' ends.

Unlike, say, the suffrage campaigner Henry 'Orator' Hunt's intended "radical museum"[12]—which was to illustrate 'loom and shuttle' lives of hard-done-by spinners—Saull's took a different tack on working-class problems. It was the connotations and context of Saull's exhibition that made it radical. First, as William Makepeace Thackeray's appreciative *National Standard* said, "his museum would be a sealed book to the many, were it not for his lectures".[13] The public, given free and unrestricted access once a week, were treated to a talk in which Saull welded the collection into an 'evolutionary' whole, whose progressive message was made the legitimation of social action. Second, the content spoke volumes. Henry Hetherington knew the best use of museums: to house the stuffed remains of the few remaining kings (as he laughed in the wake of the 1830 French revolution).[14] Hunt himself went beyond artisan 'manufacts' and included *memento mori* of the peaceful suffrage demonstrators killed by the Huzzars at Peterloo (or at least bits of skull hewn out by a yeoman's sword). But Saull went one better and gruesomely included the radical leaders themselves. His was the stuff of radical icons in a real, corporeal sense. It was not only a museum for radicals, but *of* radicals, as we will see.

Reign of the Beast thus straddles the line between labour studies and the history of geological culture. Thanks to studies of the 'underclass' in the last few decades, we have the potential for locating Saull in a way that previous generations found difficult. Thus, this work owes a huge debt to those pioneering investigators of dissident street culture, particularly

---

11 D. Porter 2019.
12 Huish 1836, 439–40.
13 *National Standard* 3 (18 Jan. 1834): 44–45.
14 *Republican* (Hetherington), 11 June 1831, 7.

Edward Royle's *Victorian Infidels*, I. J. Prothero's *Artisans and Politics*, and Iain McCalman's *Radical Underworld* on the blasphemy chapels and pornographic dives of Grub Street. (So named because it was "famed in former times/For half-starved poets and their doggrel [sic] rhymes".[15]) This ill-famed London road is more than a metaphorical reference point. Saull was financing a chapel-turned-infidel-forum on the actual Grub Street itself in 1828. This was to go "to further lengths in the abuse of Christianity" than any previous venue, as a police spy reported.[16] A generation of historians has built on these pioneering works, and they give us a framework to locate Saull, even if they themselves scarcely touch the man, except as a footnote.

Saull is equally a footnote in geological history, despite pioneering works such as Simon Knell's *Culture of English Geology*. Given Knell's sub-title, *A Science Revealed Through its Collecting*, the fact that Saull only figures tangentially in one note proves that his collecting spree is as little known as his cabinet, even though it was the "principal museum of geology in London", according to the press.[17] Nor, therefore, has there been any study of the ideology behind it. This is despite the fact that private museums and trading in natural-history artefacts have been studied from most sides, but rarely the political.[18] Here, then, we will see for the first time how differently structured a museum can appear when it was designed to fit an Owenite socialist agenda.

New Sources

This fractured footnote approach in studies of radical freethinkers and the material culture of geology adds to the difficulty of recovering the whole man. Indeed, the two camps have mutually exclusive toe-holds on Saull. Clearly, to break into the subject, we need new sources, in fact, new types of sources.

We can build on the "penny trash" literature familiar to labour historians—from Carlile's deist-cum-atheist rags in the 1820s to Owenite organs in the 1830s, and the plethora of illegal, unstamped weeklies,

---

15   *Lion* 2 (10 Oct. 1828): 471.
16   HO 64/11, ff. 43, 75, 77–78.
17   *Courier*, 12 Apr. 1841, 3.
18   Ville, Wright, and Philp 2020.

heroically churned out on hand presses by sharp compositors.[19] But there is an extra resource we have to examine more fully in Saull's case—the Home Office spy reports. These are essential, because Saull was the kingmaker who stood behind the scenes, and only these expose this shadowy activity. Police spies insinuated themselves into the infidel cadres, and the infiltration was deep: one agent even became Carlile's wife's confidant, allowing him to read personal letters. These surveillance snitches were thus privy to secret meetings. Their reports, however untrustworthy and hyperbolic, and full of garbled whispers of blasphemies and conspiracies, provide sensitive information available nowhere else.[20]

Why were spies tasked with tracking "blasphemous" outlets so assiduously? In the 1820s, blasphemy and sedition were often seen as two sides of the same coin. Christianity was routinely said by judges handing down harsh sentences to be the law of the land, although atheists in the dock disputed the legal basis for this.[21] Many deists in the Carlile camp were republicans; the King was head of the Anglican Church, so the lot was expected to topple as one. And with the exclusive Oxford and Cambridge seminaries catering largely to wealthy Anglicans, their ordinands often acted with magistrate and squire as policing agents in rural villages. These priests were paid out of state coffers, and this was the other major gripe of 'infidels', indeed of Dissenters generally: the huge sums raised in tithes and church rates to support the Anglican establishment. With the rising radical movement and working class warfare in the years around the Reform Bill (1832), this anti-clericalism became associated with democratic demands, linking still more closely blasphemy and sedition. Thus secret agents kept a close eye on the infidels, and these Home Office reports are a vital resource.

At the other end of the press spectrum, the respectable (that is, legally 'stamped' or taxed) London newspapers are equally little tapped. This is understandable, looking at the statistics. In 1837, the modern Babylon had fifty-one dailies. The papers catered to every party,

---

19   Hollis 1970; Wiener 1969.
20   Parsinnen and Prothero 1977 considered the spy reports under-used, and they remain so today. For a cautionary note on using Home Office spy material, see E. P. Thompson 1980, 532–38.
21   *Investigator* (1843): 71.

sect, class, and trade. There were morning and evening papers, papers published on two or three days a week, and, by 1831, nineteen 'Sunday Papers' alone.[22] To these could be added the fifty weekly periodicals on sale at news-stands. By the time of the first *Newspaper Press Directory* in 1846, well over 120 dailies and weeklies were for sale in London.[23] It was impossible in the past for historians to gain traction. But since the newspaper digitization projects of the early 2000s, access to this resource has become easier. It means we can not only trawl the ultra-radical *True Sun* but assess the reaction from *The Age, Atlas, Albion, Argus* (and that is just the *A*s) plus a dozen others as they frightened their gentle readers about 'geological infidelity' and the trampling of taboos by artisan demagogues. Now we can gauge the panoply of perspectives. No longer are we reliant on polished publications. We can read speeches, discussions, letters, and comments, all of which enable us to flesh out venues, audiences, and reactions.[24]

The dailies not only ensure immediacy but can provide the finer-grained sequence of events more commonly found in social than science history. Shorthand press reports, being the first draft of history, with their breathless on-the-spot coverage, can reveal the nuances of the moment. These get lost in the rose-tinted reminiscences written late in life, and in the romantic, filial, and often bowdlerized lives and letters so beloved of Victorians, the usual source of so much older history of science. Moreover, the built-in biases of partisan newspapers, rather than being a hindrance, can be a plus, helping us to understand the viewpoints of different sections of society. Their very diversity is an asset.

This leads us to our third new resource, one sort of press publication in particular. Satirical "mags", by the late thirties, were a news stand feature, putting the guffaws into working-class 'instruction'.[25] Nothing

---

22 *Penny Magazine* 6 (31 Dec. 1837): 507; *Political Magazine* (Carpenter), Nov. 1831, 98–101.
23 *Newspaper Press Directory* 1847, 63–74.
24 The value of such digitization has been well demonstrated by Pietro Corsi (2021). He has used mass scanning techniques to crack the Continental sources of anonymous (and long-disputed) snippets discussing faunal and geological change which appeared as cuttings in the *Edinburgh New Philosophical Journal* in the later 1820s, thus locating the original contexts and cultural meanings of these supposedly Lamarckian fragments.
25 J. F. C. Harrison 1961, 30; Maidment 2013 reappraises the visual comic caricature of the period. "Mags" was already contemporary slang, e.g. *Shepherd* 1 (11 July

escaped pastiche in the age of *Punch*. Later it would be Darwin who took the brunt of it (even his *Beagle* voyage was lampooned[26]). But a generation earlier it was Saull who was burlesqued. The working classes particularly came in for brutal mockery. And, since 'evolution' in the 1830s was a fringe obscenity promulgated by street radicals, socialists, and medical democrats,[27] which risked infecting the poor, we can appreciate why it was targeted too. Materialists were especially susceptible to satire, which always represented the cutting edge of conflict. The reductive power of caricature was used to laugh this disreputable ideology out of court (and, against infidels, it was often used *in* court as well). James Paradis describes it as an indispensable strategy to censor and ridicule, to prick the vanity of overblown, self-aggrandizing, materialist know-it-alls for their rigid and mechanical world lacking spirit and spontaneity.[28]

Satire runs close to abuse and exposes anxiety, and one sees it in attacks on Saull. In his case, it was inevitably a skit, as incredulous critics struggled with his "shaven ape". He was made a laughingstock to alert genteel readers about bestial transmutation long before Darwin was sketched as a hairy old ape, or the *Vestiges of Creation* (1844) was made the butt of jokes.[29] Most of the sarcasm on Saull's monkey-origin notion ironically came from a deviant Universalist preacher within the socialist movement itself, Saull's confidant, the Rev. J. E. Smith. Later, Smith took his pastiches out to new amusement-orientated middle-class weeklies, in particular to the *Penny Satirist*. The frequent foolery at the expense of Saull's "shaven ape" in the huge-circulation parlour publications of 1830s and 1840s brought Saull unexpectedly before a huge readership. Such drollery, in effect, took him mainstream. These weeklies ironically spread his name far beyond the confines of the back-street halls. Such mockery allows marvellous scope to follow Saull's trail from the tittering journalism of middle-class voyeurism right through to the mass-selling *Family Herald*.

1835): 366.
26 K. Anderson 2018; Browne 2001. Curtis 1997, on the development of ape satire later in the century, with different cultural targets.
27 Desmond 1987, 1989.
28 Paradis 1997.
29 J. A. Secord 2000, 318, 456.

If Saull's emporium is unplumbed by historians, Saull the collector is no better known.[30] Part of the reason is plain. Saull was a 'failure' because he had little presence in the gentlemanly journals and because he refused to obey the norms of elite society. The older histories of geology were compiled from these expensive journals and books, but we now see they only provided one class perspective. And Saull was ill-served after his death by gentlemen historians and professionalizing scientists, who had no interest in context and took their dismissive cue from high-brow reviews of his works. These had expressed shock at Saull's "peculiarities" (meaning atheistic politics). The trashing of his reputation left an image of an ignorant dilettante. We see it in the *Literary Gazette* obituary, which was perplexed by "this kind but crotchety philosopher". Here was "a man of excellent heart, and a great enthusiast in his pursuits, but his knowledge was rather superficial, and his views, in regard to politics and religion as well as science, were anything but orthodox".[31] Kindly but bizarre were the operatives. He was recollected as an oddball, always courteous, seemingly unruffled by the slings and arrows of outraged critics, hurled at him because of the "peculiarity" of his views.[32] Never were the 'peculiarities' explained, nor the politics, for they were too horrifying to be discussed. Saull's embarrassing socialist and blasphemous views were avoided, the context was stripped away and the museum's function was ignored in these obituaries.

The antagonistic anti-socialist, anti-infidel reviews and obituaries set the tone for his *Dictionary of National Biography* entry, which wildly missed the mark. This treated him as a "geologist" and "more enthusiastic than learned" (that is, a failure according to late nineteenth-century canons). In a positivist age, paying homage to professional science while reinforcing late Victorian conformity, such unrespectables from the radical thirties fared ill. Saull was branded a failed geologist,

---

30 Confusion compounds Saull's obscurity. He is often referred to as "Saul" or "Mr. Saul" in the press, even though he always signed himself "W. D. Saull" and his import company was "W. D. Saull & Co." To make matters worse, there was an unrelated shell collector, Miss Jane Saul (1807–1895), whose name was immortalized in G. B. Sowerby I's designation of a Pacific conch *Murex Saulii* (now *Chicoreus saulii*). Since Saull had bought Sowerby's father James' collection, all of this makes for laborious disentangling.
31 *Literary Gazette* 1998 (May 1855): 284.
32 *JBAA* 1st ser. 12 (1856), 186–87.

not a successful museum operator and deist, blasphemous, and socialist facilitator.

Saull spoke for the marginalized, and was himself marginalized from official history. Even would-be sympathizers misunderstood him. The 'official' scientific line rubbed off on the radical Joseph McCabe, in his biographical dictionary of freethinkers: Saull, the "Owenite Rationalist", was a "geologist" and "keen astronomer" (!) "though attached to somewhat fantastic theories".[33] The final indignity came at the hands of Saull's comrade-in-arms, that great survivor into the twentieth century, the secularist George Jacob Holyoake. The raconteur of early co-operation succumbed to the scientific put-down: praise for Saull's backing of the Harmony experiment was offset by his "enthusiasm for the suspected science", which he promoted "according to his knowledge".[34] With radical friends damning him with such faint praise, no wonder history took the dim view.

If ever there was an activist who has slipped through the historical net, it is Saull. This despite past attempts at resuscitation. Aleck Abrahams in *Notes and Queries* in 1922 pointed out the total disappearance of Saull's museum, both from the historical record and in real terms. Saull's bequest of his exhibition to a working man's institution after his death resulted in a complete shambles and its breakup and loss.[35] But nothing came of the query. As a result, Saull's fossil depository and its socialist *raison d'être*, his freethought financing and king-making are hardly known, never mind their inextricable relationship.

Thus Saull remains elusive, even though in his day he was a central figure in Robert Owen's circle. He was no less a prominent atheist, whose dissident activities led to public infamy. He was, after all, indicted for blasphemy, vilified in the *Times*, and lampooned by satirists. Yet within Owenite circles, he was ubiquitous in the 1830s: wherever a radical meeting needed a Chair or Treasurer, wherever a cause needed backing, a victim fund need financing or a radical institute funding, there he was. Like another pilloried atheist, the wealthy wag Julian Hibbert, a friend and fellow financier of radical causes (who became a 'donor' to the museum in a more ghoulish sense), Saull was a money man.

---

33   McCabe 1920, 708–709.
34   Holyoake 1906, 1: 190.
35   Abrahams 1922, s12–xi: 230.

The upshot is that the pitiably few secondary sources give little hint of Saull's freethought views, Owenite activity, or financing of dissident venues, or that his science and museum catered to an angry clientele after the disappointments following the Reform Bill. So, *Reign of the Beast* is an attempt to recover the radical milieus and rehabilitate Saull by taking him seriously. We have to 'de-peculiarize' his science by putting it back into context and to understand its propagation for contemporary ends.

Reconstructing his life is instructive, not merely as a pedantic exercise in recovery, but to illustrate a specific class activity in science. Working peoples' voices, excluded from science and politics in their own day, should not be silenced from histories of geology today. We need to spotlight them, not only, as Knell has done, as collectors, swappers, rock hunters, and fossil entrepreneurs,[36] but as participators in those vast political and social movements which rocked the 1830s and 1840s.

The advantage now is that we have modern digital resources in addition to traditional archival ones. With the scanning of more ephemeral literature, the daily papers, radical periodicals, street tracts and so on, a new contextual world for Saull is opening up. Indeed, a new arena for science is coming into view, populated by an unfamiliar cast. Given the growing availability of this esoteric literature, we can at last make strides in reconstructing the freethinking socialist sympathetically. We can shift the focus away from the failed 'professional' geologist. In its place comes an activist who ploughed his wine profits into a didactic museum for the masses—a facility for the propagation of a wilfully disruptive sort of fossil geology.

So much excellent work at the moment is devoted to science at the 'margins' (a term which needs total deconstruction). Science was made to fit needs, and needs varied across classes and cultures. As Prothero puts it, "The artisans were not passive recipients of ideas; they were a social group with certain ideals and interests according to which they moulded the ideas they met."[37] Reconstructing these unfamiliar milieus in science is finally showing up the vacuity of an older historiography which dismissed them and buried the clues, as not leading to "proper" science—that is, judged by a gentlemanly yardstick. Just how much the historical axis has shifted towards inclusivity is shown by Aileen Fyfe

---

36   Knell 2000.
37   Prothero 1979, 246.

and Bernard Lightman's *Science in the Marketplace*, as well as works on mesmerism, phrenology, electricity, and especially Anne Secord's study of artisan botanists clubbing together in the pub.[38]

Reign of the Beast is not so much science in the pub or market, as down the Labour Exchange—almost literally. One of the first such institutions of its kind Saull helped to set up in 1832, a co-operative exchange bazaar outside of the capitalist economy. The book's target is the ideologues here, agitators who thought science could supply republican and anti-clerical ammunition and underscore Robert Owen's perfectibilist and environmentalist socialism.

This is not to suggest that all Owenites used geology or astronomy or used them in this way. Some studiously avoided all science as politically and socially irrelevant or considered it suspect as an avocation of the rich. Others bought into the prevailing propaganda of its social neutrality put out by its gentlemen practitioners. Still more retreated completely, away from science *and* society. By the late 1830s, the 'sacred socialists' rejected the prevailing irreligious materialism of so many Owenites like Saull. They withdrew into 'aesthetic' institutions, where intuitive judgement replaced science as a source of knowledge, and the new morality of vegetarianism, teetotalism, pacifism, and celibacy became the human-perfecting instruments.[39] Still more treated bourgeois science with cynicism. An editorial inaugurating that "ferocious" illegal rag, *The Man*, talked of official science being tainted by "the cankering contamination of custom and pride", meaning it was poisoned by "prejudices".[40] This was shown by their Whig lordships' using science in socially-controlling, anti-radical ways—a subject worked up by Steven Shapin and Barry Barnes in the 1970s. An anodyne science cluttered up many mechanics' institutes, while innocuous articles about animals in 'improving' magazines were criticized as politically-useless pap. Working men were demanding emancipation, yet the Whig "thinks to stop our mouths with kangaroos."[41] Not that the strange kangaroos from the antipodes were uninteresting to mechanics,[42] more that they

---

38 A. Secord 1994, 1996; Fyfe and Lightman 2007; Winter 1998; Morus 2011.
39 Latham 1999, 20, 80, 168, 175.
40 *The Man* 1 (7 July 1833): 1; "ferocious": Noel 1835, 63.
41 Shapin and Barnes 1976, 243; 1977, 55–56.
42 Topham 1992, 1998, 2005, 2009a, 2009b, 2022, provide a more sympathetic reappraisal of the Society for the Diffusion of Useful Knowledge, Mechanics'

seemed like a distraction. Owenite house organs warned against these bourgeois-controlled institutes. They can teach you the 'abc' of practical science, said a Saull ally, but "only to make you better servants".[43]

Totally in step, Saull denounced any schooling run by the clergy and gentry, who simply want "to put your children in livery".[44] To be liberating, socialist science had to undermine such enslaving tactics. In this respect, Saull's views were typical of those of many radicals of the time, who saw the liberation of the mind accompany a liberation of governance. And, as a first step, new emancipatory sciences had to be developed at street level. For Saull, the moral of the French Revolution was that demolishing the old order without readying any replacement was ineffective—the forces of monarchy, church, and reaction would simply return.[45] Therefore new emancipationist sciences had to be developed in advance to replace the Creationist props of the *ancien régime*. They had to be fed into the educational system early, hence socialist junior schools countrywide by the early 1840s were training youngsters in progressive geology, or real, anti-Mosaic, earth history, as they saw it.[46]

This was proof, as Roger Cooter put it in *The Cultural Meaning of Popular Science*, that use of science as a "powerful tool in social and political debate ... need not necessarily have entailed endorsement of the dominant class's supposedly objective view of the structure of natural reality."[47] Saull's certainly did not. His 'evolutionary' lectures and museum promoted a different reality from the pulpit standard or geological norm. As he said at London's Rotunda building, just over the Thames on the Southwark Road, in its day the premier 'blasphemy' outlet in the metropolis, a new sort of materialist reasoning was needed to counteract such enslaving tactics upholding religious power, and he

---

Institutions, Bridgewater Treatises, and popular serials in general.
43 *Crisis* 2 (1 June 1833): 163; Johnson 1979, 85. This was Benjamin Warden speaking at Owen's institution. Warden was a master saddler in Marylebone. Warden, raised a Tory churchman, became a Unitarian and Freethinking Christian, finally renouncing all religion in the late 1820s. He and Saull worked in the British Association for Promoting Co-operative Knowledge (1830), the National Union of the Working Classes (1831), and the Labour Exchange. He was active at the Western Co-operative Institute, Poland Street, where Saull lectured on geology. Chase 1988, 150; Prothero 1979, 308–9, 306 n.18; Hollis 1970, 195.
44 *Crisis* 3 (4 Jan. 1834): 150.
45 *TS*, 28 Apr. 1835, 2.
46 *NMW* 11 (17 Sept. 1842): 99; (17 Dec. 1842): 203.
47 Cooter 1984, 203.

declared that his science as a "force would be fatal to that of tyranny and priestcraft. (Cheers.)".[48] His comrades equally sensed a subversive science's effectiveness against a repressive religious authority. Listen to George Petrie, a one-time insurrectionary and land reformer (and, another time, would-be assassinator of the hated Duke of Wellington),[49] whose contribution to Saull's museum would take a more grizzly form. Consult those "tutors which Nature has provided", the intellectual faculties, which "teach all sciences", Petrie told his ragged readers, and then ask "whether Religion has not in all ages, countries, and climes, produced the most debased slaves, the most demoralized people, and the most revolting carnage amongst mankind."[50] Saull's views precisely.

Atheism?

1830 was a pivotal time. The old onslaughts on "Kingcraft" and "Priestcraft" were slowly giving way to attacks on capital. Saull stood at an intersection, on the one hand slamming the old-style "tyranny and priestcraft" with Carlile's deists, while being about to set up the co-operators' Labour Exchange on the other. The exchange cut out the capitalist. It enabled the swapping of artisan manufactures, from bread to boots, or they could be switched for labour notes—Owenite 'bank' notes representing the hours of work a product entailed. The radicals were shifting targets from the "swaggering aristocrat", and were beginning to form "a labor theory of value that would make capital rather than hereditary privilege the antagonist of the 'useful and productive' classes."[51]

The 1820s–30s was also the time at which Saull becomes historically visible. Yet, however hazily he moved his sights from kings (under Carlile's influence) to capitalists (under Owen's), Saull saw religious authority as a root problem in both cases. He never stopped denouncing the Anglican undergirding of a political structure which he blamed for legally depriving the poor of their political rights. Saull's target

---

48 *Isis* 1 (3 Mar. 1832): 59–60.
49 Petrie [1841], 20–21; McCalman 1988, 197; Prothero 1979, 257–58, 289; Holyoake 1905, 102—05.
50 *The Man* 1 (4 Aug. 1833): 34.
51 Klancher 1987, 102.

remained the legitimating sanction of the state-supported and high-taxing Anglican authority. As he wrote (anonymously, see Appendix 2),

> *religion is a despotism*, reigning tyrannically over the human mind, blighting all its fair buddings, draining away or scorching up its proper nurture, misdirecting its energies, and making of human society one vast lazar-house, in which nothing but insanity is countenanced or encouraged...[52]

Of "all the evil genius that has ever existed", nothing was more guaranteed "to bring about the greatest amount of human misery". The screaming nature of such claims show their intensity. They were made time and again by Saull's comrades, both radical (those who sought enfranchisement first) and Owenite (those who looked to social regeneration as a prerequisite).

Take the radical Henry Hetherington, a republican democrat who is so often a counterpoint in our story. While Saull remained in the shadows, activists like Hetherington stood in the glare. Such men, proud and obstinate, refused to abase themselves before judges, let alone priests or kings, for they considered fighting for democracy and disestablishment neither immoral nor criminal. Hetherington was not at war with God (he was a Freethinking Christian) but with tithe-grabbing priests as a 'class', and the religions they peddled to retain their hegemony. Serving a term in Clerkenwell jail for publishing his *Poor Man's Guardian*, he wrote no less hysterically in 1832 about state-endowed clergymen fogging minds before emptying pockets. Religion was

> an artful scheme of robbers and tyrants to emasculate the mind of man—to rivet the fetters of slavery—to doom the honest and industrious portion of the community to the inextricable thraldom of ignorance and superstition—that they may ever remain an easy prey to their oppressors.[53]

Immersed in a sub-culture where such views were prevalent, Saull was in tune in seeing his museum's *raison d'être* as liberating, in kicking away the crutches of the Anglican regime.

---

52 [Saull] 1832a, 4, emphasis original here and throughout, except where noted.
53 Hetherington [1832], vi; Barker [1938], 15. Hetherington was eventually expelled from the brotherhood of Freethinking Christians for thinking too freely. The creed was often a halfway house for discontents on their way to deism or materialism.

With modern secularity viewed both as the jettisoning of theology and as "the fruit of newly-constructed self-understandings" embracing traditional moral values,[54] Saull's museum can be seen as a site that contributed to the shaping of the new secular man. In its challenging environment, new self-perceptions were being forged and complex class identities being reinforced. It was one of many emergent venues that were a seedbed for what would eventually come to be called "secularism" by Saull's close comrades.

But "secularism" was an endpoint in the 1850s. Saull passed through many earlier stages of unbelief, and these showed his irreligion progressing as the infidel milieu changed. Although I have used "atheist" in the subtitle, the word is shorthand and contentious. Saull never called himself an atheist. Probably, like Richard Carlile, he hated labels, and his changing standpoint can best be judged from the context. 'Atheist' shouted clerical (and thus class) antagonism in a radical age. But, then, "everything", said E. P. Thompson, "was turned into a battleground of class";[55] to which Joss Marsh, in *Word Crimes*, added aptly that an atheist was a person "who 'ignores God, just as a rude man might ignore the presence of his superior in rank'".[56] "Atheism" was never a stationary concept. As radicals deployed new vectors of attack on the gentry's sons dumped into the priesthood, so infidel positions adapted.

This gives us our trajectory from the 1820s to the 1850s. Saull was Richard Carlile's patron in the 1820s as Carlile, rejecting even Tom Paine's arguments as too superstitious, moved from deism to atheism (although he preferred the term "materialism"). On the last day of 1827, Saull could still write of the "goodness of the Supreme Being to all creatures" while denying the inspiration of the Bible.[57] Assuming he was not being facetious (this was a letter lambasting his vicar's position), he was still a deist or something more providential at this point. And, in the later 1820s, he sponsored the astro-theological theatre of the flamboyant dandy, the Rev. Robert Taylor, who taught that the Bible was a story-book personification of celestial events. Saull also financially underwrote another blasphemer, the deistical preacher, the

---

54 C. Taylor 2007, 22.
55 E. P. Thompson 1980, 914.
56 Marsh 1998, 21.
57 Saull 1828a, 4.

Rev. Josiah Fitch, in the twenties. By 1830, Saull himself was a committed materialist. London's blasphemy venues were infiltrated by police spies, so we have surveillance reports on Saull's speeches. One, on 22 November 1830, relayed rather breathlessly how Saull "ascended the pulpit" at the Optimist Chapel in Windmill Street

> and began a Lecture on Superstition in which he much abused the Ministers of all Religions and the Religions also and said he was glad to find that knowledge and Union of the people had begun to have some weight and pressed the Necessity of still further to unite for though slow they were sure in the end they would put down all Superstition and Tyranny. He also began to prove the eternal existence of all matter and contended that Materialism was the only true Religion ...[58]

By the 1830s, he was part of Robert Owen's co-operative movement and pinned his colours to its "rationalist" mast. This flew an Enlightenment flag proclaiming the sovereignty of the "laws of nature". Then, when a group of self-proclaimed 'atheists'-proper split from Owen around 1840, Saull supported them. Finally, he migrated to George Jacob Holyoake's catch-all "secularist" camp in the early 1850s. In short, a fine study of Saull shows him moving with the times, as so many did. However, behind the terminological facades, he probably shifted little from his 1830 denial of spirit, soul, and Christ's existence.[59]

The Missing Museum

Today's historiography tends to favour larger metropolitan and provincial public museums. These reflected national importance, regional assets, and civic pride. Fewer studies target difficult niche institutions, not least those with a radical working-class clientele, let alone tackle their politically transformative intent.[60] It is time to switch priorities from posh to poor, however hard it might be to penetrate this neglected class space, which left few archival traces. Saull's 'underworld' museum

---

58 HO 64/11, f. 167.
59 HO 64/11, f. 205 (1830); f. 462 (26 Dec. 1831).
60 Lundgren 2013 for a later nineteenth-century example (albeit non-emancipatory) of ambitions to transform the visitors' self-understanding in relation to social debates.

of 'evolution'[61], with its artisan clientele, emancipatory ideology, and palaeontological and pantheonic content is our entry point.

The uniqueness of Saull's endeavour was shown in the way it bucked trends. Generally "exhibitions rarely seek to explain their contents in terms of a broader social and political context",[62] being somewhat static, usually non-interactive, sometimes arranged aesthetically, and leaving visitors to bring meaning to often ill-labelled exhibits. But Saull's was completely the reverse. He was ever-present to point out why his fossils were chosen, how they fitted together, and what perfectibilist message they carried for the moral development of socialist man.

All of this suggests that the fossils might have been viewed somewhat uniquely in Aldersgate Street. At least, compared to the fossil cabinets being fitted up by dealers "in the first style of elegance" in fashionable drawing rooms,[63] the exhibits served a different purpose. Ralph O'Connor in the *Earth on Show* illustrates how fossils captured the imagination in polite society, invaded expensive literature and carried

---

61   The words "evolve" and "evolution", and "palaeontology", were on the cusp of use in the 1830s. "Palaeontology" was a neologism (*Report of the Third Meeting of the British Association for the Advancement of Science; Held in Cambridge in 1833* [1834]: 480), and, by 1837, the word was "becoming usual" (J. Phillips 1837, 1: 2). Although "evolve" generally meant a foetal unfolding, it was occasionally extended, even in the 1830s, to cover the emergence or unfurling of species through time. Sir Richard Phillips used it this way in a reprint republished by Saull. He said that secondary causes "must evolve … every thing that is possible" to leave a gradation of species (Phillips 1832a, 52). The word could also mean the emergence of latent capabilities. The Rev. Robert Taylor said in his (Saull-financed) pulpit in 1827 that the "purpose of nature to evolve and bring forth the moral capabilities of man, may be traced from the very first origination of animal life" (*Lion* 4 [9 Oct. 1829]: 462). Most often it referenced moral development, as in another Taylor sermon, when he claimed that, without struggle, the "latent faculties and capacities would never be evolved: man would seem to be born only to eat turtle, and to die like an alderman, choked in his own fat" (*Lion* 4 [6 Nov. 1829]: 607). Robert J. Richards 1992 maps the changes in meaning of the word "evolution" onto its underlying anatomical contexts. In Phillips's and Saull's use, the "multivalent discursive terrain of Romantic evolution—literary, scientific, aesthetic, philosophical, religious" (Faflak 2017, 14)—was being pinned down to the specific biological realm. In short, the word was transitioning to its more modern meaning, although it had yet to denote blindness in direction, for socialists and romantics still saw evolution in teleological terms, as aiming at human perfection.

62   S. Macdonald 1998, 2.

63   An early advert for this service can be seen in *Gardener's Magazine* 2 (May 1827): 356.

the new 'deep-time' message into the heart of a pious nation.[64] Here they could be displayed simply as curios, or for their beauty, to strike awe or spark curiosity about antediluvial times. While Saull might have used these aspects as lures, the meaning he extracted from the fossils was much more pointed.

In another aspect, too, he saw things differently. The learned were starting to suggest that *serious* museum collecting should result in the production of new knowledge—monographs, descriptions of new species, and specialist books.[65] In short, the fossils were there to be studied scientifically. Not so for Saull, who used them to sustain a new politics, not produce new knowledge. Anyway, despite growing demands that museums become knowledge-producing sites, it is clear, as Tony Bennett points out in *Birth of the Museum*, that they were never just places of knowledge acquisition. They always acted to regulate visitor conduct, marshal perceptions, reshape behaviour, and generally act to reform manners in such a way as to obviate more external coercive measures.[66] This appreciation makes Saull's venue, shaped by its Owenite ideology, particularly valuable as a sphere of study today.

Saull wove the fossils into his distinct narrative about the past to make a political point. The museum helped to empower an audience being made conscious of its dispossessed status by new class-awakening papers, particularly his friend Hetherington's *Poor Man's Guardian*. To this extent it served the same purpose—an assertion of power—as the unrealized museum projects of Henry Hunt and of the Grand National Consolidated Trades Union in 1833.[67] They show that Saull's was one of a number of possible museum tacks in the 1820s and 1830s, but the only one that took a geology-based turn. Saull's museum pre-eminently mated Carlilean irreligion and the social millennium. The fossil facility

---

64 O'Connor 2008.
65 Strasser 2012, 319.
66 T. Bennett 1995. There has been an avalanche of scholarship since the early 1990s on museums, warranting a "Focus" section in *Isis* (96 [Dec. 2005]: 559–608). Regarding natural history, many scholars have come in from the perspective of "popular science" (and on historicizing "popular science", see Topham 2009a, 2009b; O'Connor 2009). Audiences have been less studied in a historical context. Few historians have focused on exhibits designed to turn visitors into activists, despite the accounts of interactive displays, for example, Morus 1998, 2011.
67 *The Pioneer; or, Grand National Consolidated Trades' Union Magazine* 1 (2 Nov. 1833): 68.

was to help establish a new nature-based authority for an infidel Owenite society. Saull, in this quasi-millenarian[68] setting, was to show the great unwashed how, in fact, they had been invigorated by their 'evolutionary' bath, and how this evolutionary ascent guaranteed their progress to the promised land. Thus the museum exposes the use of geology in a naked class context, where it aids political campaigns to redress grievances and points to the inevitability of the coming Owenite man.

There are also more mundane reasons why we should be interested in Saull's lost museum—the number and nature of its exhibits. Let us start with the obvious: size. It was claimed to be *the* largest private geological collection in London. The press all agreed on this, from the Chartist *Northern Star* to the Tory *Morning Post*.[69] This point seemed uncontroversial. But size would count for little if visitors found the contents mediocre, meaning uninformative, unrelated to contemporary interests. The venue had to be exciting, disturbing or revealing, with star exhibits, something realized by all showmen. Studies have emphasized how exhibitors were looking for the exotic crowd pullers.[70] What drew the public were the ancient and marvellous—and what fitted the bill in Saull's case were those bizarre reptiles that would figure in his friend Gideon Mantell's double-decker *Wonders of Geology* (1838).

Possibly Saull's biggest coup was to bring in sea-rolled fossils of giant saurians from the Isle of Wight, which Hugh Torrens believes started arriving at the museum about 1836.[71] These gigantic creatures from the "Age of Reptiles"—as Mantell provocatively named it—were a sensation. The fossil bones of *Iguanodon* could be scaled up to suggest a living reptile seventy feet long, and the gigantic *Cetiosaurus* ("whale saurian") was even more colossal. Nothing like them had ever been

---

68 Rather than using the term "millennial", I follow J. F. C. Harrison 1979 in using "millenarian", since it refers to the newer, plebeian, and Southcottian prophetic tradition, which characterises some of Saull's fellow-travellers, notably the Rev. J. E. Smith. The term "millenarian" is also used for those infidels who anticipated a perfected socialist man in an eventual Heaven on Earth, the socialist New Jerusalem. Critics such as Henry Hetherington called this their "political millennium" (*PMG*, 14 Jan. 1832), to distinguish it from any religious expectation of Christ's Second Coming.
69 *NS*, 31 Oct. 1846, 3; *Morning Post*, 31 Dec. 1841.
70 E.g. Pearce 2008; Greenwood 1996, on William Bullock, a master of exotic crowd pullers.
71 Torrens 2014, 670.

seen before: disarticulated legs bigger than an elephant's, giant pelvises, eight-inch vertebrae, and a monstrous seven-inch claw, all of which left a huge amount to the visitors' imagination.[72] By 1839, Aldersgate Street had the greatest assemblage of *Iguanodon* bones from the Isle of Wight in Britain, and each new influx of exhibits swelled the ranks of visitors.[73]

These shipments naturally attracted the geological gentry as well. Saull gained personally from this. His stock rose with the museum's status. It provided his entrée and greased his otherwise difficult path through learned society, just as his provincial friend Gideon Mantell used his "Mantellian Museum" to garnish his profile as a fashionable doctor.[74] Saull's exhibits were a growing resource for desk-bound descriptive palaeontologists. And for none was this truer than the social-climbing young comparative anatomist, Richard Owen. Owen was the new Hunterian Professor at the Royal College of Surgeons in 1836, a pious man moving under wealthy patronage from anatomizing London Zoo's dead exotics[75] to the still more esoteric fossil reptiles of Britain's deep past.

Therefore we also should care about Saull's museum because it was exploited by the leading men of the day. Owen famously went on to make the *Iguanodon* sacrum (fused pelvic vertebrae) in Saull's collection the justification for his new 'Dinosaur',[76] a Brobdignagian creature which would become so iconic to future generations. That at least one major—and culturally crucial—taxonomic construction was based on Saull's specimens should underline the importance of his museum, at least with hindsight. Furthermore, these Aldersgate Street fossils became real bones of contention. Saull's museum was not only a site of political controversy, but palaeontological, as arch enemies Owen and Mantell tussled over Saull's prize *Iguanodon* sacrum, each figuring it and producing counter-reports.[77]

Though the leaders in their field, Owen and Mantell were far from the only elite visitors. Saull's collection was acknowledged and name-checked in the various fossil compendia and standard texts of the

---

72 Karkeek 1841a, 72; 1841b, 175; G. F. Richardson 1842, 402. On the scaling procedures, see Dawson 2016, 70–72.
73 *Morning Post*, 31 Dec. 1841; A. Booth 1839, 121.
74 Cleevely and Chapman 1992, 309.
75 Desmond 1985a, 235–41.
76 Torrens 1997, 2014; D. R. Dean 1999, 185; Dear 1986; Desmond 1979.
77 Cadbury 2000.

day.[78] It was recommended to students.[79] Thus it was widely known in the geological community, and was important enough to be routinely visited by specialists.[80] No less was it the stopping off point for visitors to London. Indeed, it was graced by the gamut of non-specialists, from foreign royals and ladies of leisure to Owenites and Chartist firebrands.

So we should care about the museum because Saull's contemporaries cared. Whether they loved or loathed him, they never objected to his museum's *contents*, which were usually lauded. Looking at the fossils alone, however visualized by the proprietor, they saw nothing to match, say, the "indecent" displays of anatomy museums,[81] which often engendered disgust in a puritanical nation. The reservations were solely about his Owenite explanations. And, because of these, the geological gentry might have found it uncomfortable to step inside an indicted blasphemer's private[82] museum. There was also a question of who they might meet there, the radical hot-heads who were specifically invited. This, too, raises questions. To what extent was the museum a place of mediation, where classes and masses, which might otherwise stand on opposite sides of the barricades,[83] could meet on common rocky ground?

The reason the gentlemen were here was to examine the unique exhibits, not least the 'type' specimens (the first found, named, and described fossil for any particular species, which set the standard). The museum was opened in 1831 after Saull bought the fossil collection of the late Lambeth mineralogist and natural history engraver James Sowerby. The Sowerbys have generated a huge literature—twenty papers in the *Archives of Natural History* alone, including a special issue

---

78 Dixon 1850, 55; Morris 1854, iv; G. F. Richardson 1855, 353, 379, 392.
79 G. F. Richardson 1842, 80.
80 Those known to have visited his establishment include Richard Owen, Gideon Mantell, Thomas Hawkins, Sir Richard Phillips, Thomas Rupert Jones (*Geologist* 6 [1863]: 312–13), Boucher de Perthes, and Edward Hitchcock. Identifying visitors is a chancy business, and certainly many more came but left no trace.
81 Bates 2008.
82 Swinney 2010 on changing meanings of "private" and "public" in relation to museums, and emerging attitudes to their access. The fast pace of early nineteenth-century palaeontology was partly dependent on the growing network of collectors and proliferation of private museums: M. Evans 2010; Knell 2000, 74; M. A. Taylor 1994.
83 Richard Owen certainly stood opposite Saull's barricade. Owen enlisted part-time in the Honourable Artillery Company in 1834, which backed the police and militia during the Chartist 'riots': Desmond 1989, 331–32.

on the family.⁸⁴ This is understandable because the Sowerbys produced expensive, beautifully illustrated monographs on shells, and were the 'trade' engravers for gentlemen's books. Their clientele was upmarket.⁸⁵ Yet, not a single paper has ever been published on Sowerby's museum after it passed into Saull's hands, where it functioned in a different class context. A study of this transition to Owenite territory is thus long overdue. The exhibits acquired by Saull were supposed to contain many 'type' specimens figured in Sowerby's multi-part *Mineral Conchology*.⁸⁶ But how many Saull inherited and then opened up to plebeian gaze has always been a matter of conjecture.⁸⁷

Through his publications, Richard Owen raised the status of some specimens, from dinosaurs to fossil whales. And Saull, through his contact with the French (he was, after all, a wine and brandy trader, not to mention supporter of the 1830 and 1848 revolutions), raised the profile of other fossils. He even had one tree fern from the Oldham coal seams christened *Sigillaria Saullii* after him by the great French fossil botanist Adolphe Brongniart. Saull was hardly a prophet in his own land, and this Parisian influence is another reason we should be interested.

The switch from Sowerby, a client of the gentry, to Saull, a patron of co-operators, provides one starting point to explain science changing with context. Sowerby and Saull used the same fossils in diametrically different ways, and this reflected in their different museum approaches.

Saull's Aldersgate Street museum was opened in the charged atmosphere of June 1831. The Reformers, having wiped out the Tories in the general election, were pushing the Reform Bill, which would lead to riots and incendiarism within months when the Lords tried to block it. As the museum was opening, the Whigs were contemplating swamping the Lords with new peers to ram the Bill through.⁸⁸ Even when passed, the Bill failed the working classes, Saull's target audience, resulting in

---

84 *Archives of Natural History* and its forerunner *Journal of the Society for the Bibliography of Natural History*. The special issue was *JSBNH* 6, iss. 6 (Feb. 1974). This is not to mention books on the Sowerbys, most recently Henderson 2015.
85 Dolan 1998.
86 Conklin 1995.
87 George Waterhouse thought a "large number": House of Commons, *Finance Accounts I.-VII of the United Kingdom of Great Britain and Ireland, for the Financial Year 1863–4*, Income and Expenditure of the British Museum, 24–6; while Anon. 1904, 322, had trouble identifying them.
88 M. Brock 1973, 234; Halévy 1950, 33–43.

two decades of mass action. While the Bill struggled, Saull was treasurer of the campaigning National Union of the Working Classes, collecting funds for jailed street vendors of pauper papers, and helping plan the Labour Exchange in the Gray's Inn Road (see Chapter 6). Although he supported Hetherington's radicals, who wanted democracy first, his base of operations would be the socialist institutions, where the onus was on re-modelling mankind for the social millennium. Even so, his sympathies lay with the republican deists and materialists. Whether they were the imprisoned publisher Richard Carlile in the 1820s, whose defence costs Saull paid, the flamboyant "Devil's Chaplain", the Rev. Robert Taylor, whom Saull sponsored, or the atheists of the *Oracle of Reason* jailed in the 1840s, Saull never failed in his financial duty. This was Aldersgate Street's wider context of resistance. The museum spanned the rise of the socialist Halls of Science, the atheist schisms, and the emergence of 'secularism' in the 1850s, flourishing until Saull's death in 1855, when it spectacularly vanished, just as Chartism and Owenism had done. Saull engaged at every radical level through a quarter of a century, and his lectures and museum artefacts, their arrangements and meaning, reflected this context.

 Looking at other museums shows how different Saull's was. What did working people get from fossils? One Tory in the later Museum of Practical Geology (that solid embodiment of industrial utilitarianism) told his fustian audience that collecting was more mercenary than moral—collectors could make money from selling their finds. This tacitly reduced the cliff-face poor to the status of suppliers. Sold on to experts, fossils helped identify strata and coal or mineral seams, which (it was left unsaid) would augment the wealth of mine barons and investors. And arranged in museums they gave "a deeper insight into the ... perfection of the Creator as exhibited in all his works".[89] There was often an underlying anti-radical, Christian message in such traditional views. In the ancient seminaries, Oxford and Cambridge, liberal Tories and Whig divines with a dual calling as "saurologists" and clergymen came to a consensus with the metropolitan gentry on safe science and

---

89 Edward Forbes, in *Working Man's Friend* n.s. 1 (28 Feb. 1852): 338–39; on selling fossils, e.g., Taylor and Torrens 1986.

ganged up to denounce the hated radicals.[90] They defended a descensive spiral of power from the Godhead reinforcing much of the hierarchical status quo, and turned fossils into a hymn to divine beneficence—coal, for example, being providentially arranged to ready Britain for industrial greatness.

Saull's anti-religious views, and thus the moral he drew from fossils, were a stark contrast. He went beyond attacking Anglican tithes and "other compulsory payments for the alleged support of religion"; beyond disestablishment of the Church (or stopping the "annexation of political power to episcopal rank"); beyond criticizing plural livings and the union of clerical and magisterial offices.[91] Even Dissenters would have agreed with much of this. Saull went on to assault Christianity itself, to slate the Bible as full of "contradictions, inconsistencies, and untruths", to consider all religions "nothing but insanity", perpetrated by a priestly caste in "the pursuit of wealth" at the expense of the industrious poor.[92] "What, then, is the course we should pursue, to counteract these direful effects?" he asked in 1833. The answer: contradict tradition, disrupt it through guerilla tactics, expose the astrological roots of Christian myth (a fashionable tactic in blasphemous back-street chapels), re-broadcast the anti-gravitational astronomy of "dirty little Jacobins" from the radical Enlightenment, use the new deep-time vistas of geology to refute Genesis, and surreally suggest our real simian origins. As a result, he used his fossil merchandise to conjure up disturbingly godless evolutionary images and opened the museum not only to mechanics but to coalmen, chimney sweeps, and charwomen, to blow away their religious "phantasies".[93]

This is the final reason why Saull's subversive science should be interesting. His museum was a site of political education, where geology was a tool to sharpen working men's ideals. It was also a site supporting a new sort of geology, fashioned for this purpose.

---

90 Morrell and Thackray 1981, 2; [Whewell] 1832, 117; descensive: Desmond 1989, 260 passim.
91 Calls made typically by the Society for the Extinction of Ecclesiastical Abuses, which he would chair: *TS*, 12 Oct. 1832, 1. On the society: *PMG*, 13 Oct. 1832; *The British Magazine and Monthly Register of Religious and Ecclesiastical Information* 2 (1832): 178–79.
92 Saull 1828a, 10; 1832a, 3–4.
93 Saull 1833a, 37; 1833b, 530.

What were the predisposing factors that brought an Owenite freethinker not merely to a love of fossils but to a singular understanding of their meaning—one that led to his heretical belief in monkey antecedents for mankind? The environmental determinism of his urban socialist milieu, which oozed republican, anti-clerical values and merged with democratic demands, provided a context where 'evolutionary' naturalism could flourish. Thus, *Reign of the Beast* is a contribution to the revisionist historiography of 'evolution'—in the sense of the self-emergence and unaided rise of life—in the early nineteenth century. It adds another enabling context. Our locus is outside of medical radical circles, the other context where transformist ideas could flourish,[94] and our time is long before Darwin published—Saull was dead by then. Indeed, his museum was thriving well before the blockbuster *Vestiges of the Natural History of Creation* (1844).

One can see why Saull's museum might have been different, given the angry agenda. And this is highlighted by a comparison to museums at the other end of the social spectrum. Look, for example, at two fossil fish aficionados, the Old Harrovian Lord Cole and Old Etonian Sir Philip Egerton. They ploughed money into elegant museum edifices—Cole converted a wing of his stately pile, Florence Court, in Fermanagh. There was a strange cachet to such fossils for these Grand-Touring grandees, and they were in a position to invite the Swiss Louis Agassiz—fossil fish expert *par excellence*—to visit and name their specimens. As patrons, they were taken seriously by the career geologists: their status, dedication, and duty to the nation, their cabinets exhibiting the intricacies of God's fishy works, all brought preferments, political and geological.[95] Their stately homes hosted Tory ministers no less than geological gentry. High rank, deep avocation, and deeper pockets paid dividends. Hence, these elite museum owners are better known. Saull's mercantile status counted for less, liquor money without rank was uncouth. His City museum spoke of neither career aspirations nor gentility, neither Christian humility nor political obedience, but the opposite in each case. Given this contrary ideology, the museum might have been expected to have had a different reception.

---

94 Desmond 1989; J. A. Secord 2000.
95 H. Woodward 1908, 301; James 1986.

In between Saull's City museum for ragamuffins and the squires' elegant edifice lay a whole host of private museums.[96] Most of these are better known than Saull's because they were catalogued in situ, and their content lists were printed. Gideon Mantell, for instance, published a self-promoting forty-four page guide to his Brighton museum.[97] Or the collections were auctioned off with descriptive brochures—the one for James Bowerbank's Highbury museum in 1865 ran to sixty-seven pages.[98] If there was an Aldersgate street catalogue, it has yet to be discovered, and the disastrous disposal of Saull's museum precluded any sales brochure. Saull's public-spirited gesture in bequeathing the museum to a new working man's institution ironically resulted in its breakup, with the riches cherry-picked and the rest hauled off in carts—the result of ignorant managers and unscrupulous predators. Other museums would be lost when owners or even curators[99] died but rarely in such a catastrophic series of circumstances as Saull's. With it went all systematic knowledge of its content. As with other lost private collections,[100] reconstructing its contents is a haphazard art. It involves scouring radical prints, tourist guides, press notices, monographs, museum repositories, and so on.

What was Saull's place in the geological order? The community was a vast assemblage, sorted by class, wealth, leisure, dedication, literacy, and commerce, with all the tangled patronage strings characterizing society at large. Historians have long dismissed the old 'amateur' and 'professional' categories back-projected onto the 1830s. There were no

---

96 These are becoming better known through pieces in the *Geological Curator*, Knell's *Culture of Geology*, and Hugh Torrens' indefatigable unearthing. Examples include Scarborough's William Bean, who specialised in molluscs, corals and sponges (McMillan and Greenwood 1972, 152–53); John Lee (1783–1866), whose museum was in Hartwell, Buckinghamshire (Delair 1985); Gideon Mantell (1790–1852), with a museum in Lewes, near Brighton (Cleevely and Chapman 1992); Thomas Hawkins (1810–1899), who sold his Glastonbury collection to the British Museum in 1835 (Carroll 2007; M. A. Taylor 1988–94, 112–14), and many more. Even in London there were competing collections: the James Baber (1817–1887) museum in Knightsbridge, built on oil-cloth manufacturing money (Anon. 1904, 242, 262); the Highbury museum of James Bowerbank (1797–1877), specialising in fossil fruits and sponges (Williams and Torrens 2016a; Robinson 2003); and that of the Strand mineral dealer James Tennant (1808–1881) (Tennant 1858).
97 Mantell 1836.
98 Anon. 1865.
99 K. Duffy 2017.
100 E.g., Fishburn 2020.

'professional' geologists at the time, no tiers of academically-trained, examined and accredited 'experts', the lab-coated men who would appear later in the century.[101] Quite the reverse, the actual professions—church, law and medicine—appeared as suspect, Latiny closed-shops, which circumscribed knowledge to retain their power and privilege. "They all live upon the ignorance of the people", was a typical radical rant. "They therefore think, if the 'mob' become too intelligent on one subject, they may grow too wise on others. Hence the 'Holy Alliance' amongst the professions" to keep the people subservient.[102]

It is better to talk in terms of cottage-industry fossil 'suppliers', the local beach-combers and flag-stone breakers, who traded their fossil finds and esoteric lore with exhibitors, academics, and gentlemen. To these unknowns at the source of the exchange chain, the fossils were often merely "trade goods".[103] The yokels would show the same sort of deferential attitude in their dealings with gentlemen that Anne Secord has revealed for Manchester's artisan botanists, as they gifted specimens to their 'betters' to ingratiate themselves or pique interest.[104] But these finders invariably go uncredited, as the buyers raise the fossils' status by making them 'specimens', and investing them with a scientific name to ratchet up their value. This is capitalist expropriation; being re-packaged—the fossil blocks are neatly trimmed and enclosed in mahogany cases—and publicized, the 'specimens' become bankable, or, as an old Tory said: "when once an animal subject is named and described, it becomes ... a possession for ever, and the value of every individual specimen of it, even in a mercantile view, is enhanced."[105] In short, bartered up the supply chain and shipped from province to metropolis, a fossil's intellectual and financial worth continually rises. Thus, as a first approximation, Hugh Torrens, in his study of the famous fossil finder Mary Anning—a Lyme Regis stall-holder of fossils—thinks that "collectors" (like Anning) and "gatherers" (the wealthy

---

101 Allen 2009; Desmond 2001.
102 *LI* 1 (June 1854): 41.
103 Lucas and Lucas 2014.
104 A. Secord 1994.
105 *Zoological Journal* 2 (Apr. 1825), 5. Simon Knell has begun drilling down to these local levels to snatch away the anonymity. For a broader view of the birth of the "specimen" in the natural history museum: Thiemeyer 2015: 401–03.

patrons and buyers, like Saull) is a better first-order breakdown.[106] In the end, though, the squirearchy controlling the prestigious learned bodies always got the kudos for fossil 'discoveries'. They were the ones authorized by the societies to produce published papers and take the credit.

Ultra-radicals were incensed by this sort of appropriation. The fossilists laboured, and the gentlemen capitalized. The activists were outraged by condescending patronage relations generally, and the poverty created by "respectable society". *The Poor Man's Guardian* railed not only against kings, priests, and "gentlemen"—"the real 'scum of the earth'"—but bankers and merchants, and only as a backer to the cause did Saull evade Hetherington's tarring brush. Plundering capitalists grew fat as workers were reduced "to the greatest possible misery, privation, and distress".[107] Mary Anning herself was near penury in 1836, her health "impaired from the hardships" of her lifestyle. Pressure from the geological gentry caused the Whig ministry to stump up cash for a £25 annuity, which ensured the survival of her fabulous fossil supply chain. But the ultras' evening rag, the *True Sun*, saw the petty sum demean the useful labourer when gigantic pensions were lavished on the mothers of Tory Dukes.[108]

---

106 Torrens 1995; Taylor and Torrens 1986. Torrens has done for the fossilists, what Anne Secord has done for the fustian botanists, opened up the province of the lost craftsman/woman and his/her patrons. Of course, even Anning was not at the base of the fossil chain, but required labourers to cut and transport her bulky rocks. On transmission up the hierarchy to "second-order collectors" and exploitation: Strasser 2012, 313–14. For a deconstruction of the derogatory "arm-chair collector" terminology, Barton 2022. On the provenance of specimens in supply chains: Lucas and Lucas 2014. See also Kohler 2007. On patronage in return for gifted or cheap fossils: Spary 2000, 77; and on field collecting, Endersby 2008, 54–83. Saull clearly collected some fossils, for example, his Hertfordshire hippopotamus molar (Mantell 1844, 2: 838–39). He also collected Eocene fossils in Bracklesham Bay, between Selsey Bill and Chichester Harbour (Mantell 1844, 2: 903). And he was a constant visitor to the Isle of Wight: *JBAA* 11 (1855): 66–67. Field collecting could be essential to establishing one's credentials, but I suspect in Saull's case it was desultory rather than systematic.

107 *PMG*, 30 July 1831.

108 Referring to the Duke of Newcastle's mother, whom Wellington had put up for a £1000 a year pension: *TS*, 3 Feb. 1836, 4; Torrens 1995, 269; *Cobbett's Weekly Political Register* 88 (4 Apr. 1835): 43. Poverty was the lot awaiting many old fossilists. Sandy M'Callum was a case in point. He was a "clever" Silurian collector in South Scotland who showed Sir Roderick Murchison the ropes but whose destitute wife had to be helped out after he died suddenly, by a fund to which Saull contributed: *Literary Gazette* 1984 (Jan. 1855), 49.

Saull, then, was a buyer, collator, and exhibitor. He bought the blocks, nodules, and slabs containing fossils rather than chiselling them out of the rock, which brackets him incongruously with those Tory patrons of palaeontology, Cole and Egerton. The brandy business put him on financial par with landed wealth and enabled him to vie with rich bidders at auctions or pay top price on site. The latter is probably how he obtained such a selection of choice Isle of Wight saurians.

But 'gatherers' also fed on one another. Thus, Richard Owen, who to my knowledge never scrambled over a rock-face in his life, thrived on museum specimens (Saull's included) and published on them extensively. This served both men well and they had a complex relationship, which probably remained icily formal. The poorer Owen's fast-publishing fame and elevated scientific status gave him a "Cuvierian rank without the means of doing it justice",[109] a sentiment echoed by fellow Tories requesting government help for him. Successfully so, for his Church-and-Queen traditionalism led to an offer of a knighthood, a civil list pension and trips to Buckingham Palace in short order. While the social-climbing Owen would have execrated Saull's blasphemous and socialist leanings, he needed access to his museum. On the other side, Saull, the richer merchant, had no scientific profile, so the elite exposure served his wine-depot museum well. And since a merchant amassing fossils risked being written off as a dilettante, no better than the hobbyists with their crazes for aquariums or ferns,[110] such imprimatur was crucial. It could help deflect conservative criticism. And if the museum was to be a site of political education, it was important to show that geological giants like Richard Owen had vouched for its contents.

Lectures and Venues

Saull was never one of the Geological Society's inner coterie of publishing specialists.[111] He remained a spare-time trader in fossil commodities. He

---

109 Richard Owen to C. Owen, 27 December [1841] (BL Add. MS 45,927, f. 38); Desmond 1989, 354–55; MacLeod 1970, 47–48.
110 Allen 1996.
111 Being a wine merchant did not preclude Saull's becoming one. Another City-based wine-trader, Joseph Prestwich (1812–1896) in Mark Lane, a fellow business visitor to France, showed that this was feasible. Where Saull's spare hours were spent in the Labour Exchange, Prestwich's were devoted to field geology. He descended the

collected fossil artefacts from all over Britain, blocks chiselled, sawn, and standardized—'manufacts' in effect. They were turned into an indoor representation of idealized progressive nature, by placing them in a single time-sequenced line to display life's inexorable ascent, pregnant with hope for the future. Most museums arranged their artefacts in a "relational" way and aimed for representational "completeness".[112] Saull's went further to make his museum the justification for political action. It carried the Enlightenment implication that, whatever the blundering, blocking, or bullying by obscurantist royals and religionists, this inexorably-rising nature was the guarantor of the coming social millennium. But an effort was needed to see it this way, or rather Saull's explanatory lecture. Without his talk, the museum to the uninitiated stood mute and uninformative, a jumble of rocks, a "sealed book",[113] the 'message' hidden. It needed Saull to open the book and read the narrative.

Saull's open-access Thursday lecture was one of many he gave on Owenism, geology, and (in the 1840s, as he crossed the porous border from deep-time geology to shallow-time archaeology) the rise of aboriginal Britons. He joined a growing band of independent lecturers at this time. The market for science talks was expanding in the 1830s and creating a host of itinerant speakers to exploit the new venues.[114] It was, said a magazine, "the rage of the present day to teach science to the people".[115] The political tumult pushed radical campaigners onto the boards—committed activists giving gratis talks, with entrance fees going to the cause, funding jailed news vendors or court defences. The *Brighton Herald* noted that:

> A new race of men has sprung up—full of energy, intelligence, and perseverance. They spread themselves in every direction; treat of every

---

Coalbrookdale coal pits to study the strata, and gained an FGS for it in 1833 (aged only 21), three years after Saull. Such dedication won him the Society's Wollaston Medal in 1849 for his work on the oldest Tertiary beds around London, despite being in full-time business (Prestwich 1899, chs. 2–3).
112 Strasser 2012, 321.
113 *National Standard* 3 (18 Jan. 1834): 44–45.
114 Hays 1983; Sheets-Pyenson 1985; Fyfe and Lightman 2007; Topham 2009a. 2009b; Huang 2016, 2017. Most of these concentrate on the entrepreneurial lecturing trade, rather than political propagandism, and thus they scarcely tap the underworld halls and grub-street venues.
115 *Shepherd* 2 (15 Feb. 1837): 33–35.

subject ... The platform is daily becoming a formidable rival to the pulpit, the theatre, and the concert or ball-room. These men are apostles of popular science ... sweeping away, wholesale, bigotry and superstition, enlarging men's minds, and compelling them to abandon those narrow and selfish prejudices which are the besetting sin of those who ... refuse to take common interest in the great family of mankind.[116]

But Saull was fairly unique in blending science and politics at a deep level: his geological lectures would end in a political harangue, and socialist talks would wind up with the crowd being invited to the museum.

Freethinkers called for their own social interpreters of science. The geological knights generated enormous respect but intense frustration. They were "party-writers", serving their class, producing content, at once interesting but socially worthless to the "productive" population, until it was dismembered and repurposed. Calls were for activists to interpret science themselves, to use geology to "contribute to the overthrow of every thing fabulous, vicious, or unreasonable".[117] And Saull was one of the few on the stump who could actually do this: turn geology to advantage. He rose to the call for radical and Owenite lecturers to fulfil social, religious, and scientific briefs.

Moreover Saull had a growing space to operate in. Not only had infidel theatres proliferated in the later 1820s, many of which he sponsored (see Chapters 3 and 4), but in the wake of the July Revolution in France (1830) and reform fever in Britain, a wealth of co-operative and radical halls sprang up (Chapter 5). These appeared in towns across the country, but London was the epicentre: chapels were converted, halls leased and assembly rooms were set up, thirty or more in the metropolis. These blasphemy dives, halls of science, and mutual instruction rooms stood outside the regular mechanics' institutions. Some were short-lived, a few became infamous in the bourgeois press. Since these venues are relatively unplumbed, I have drawn up an annotated list (Appendix 4) to show their geographic spread over the capital from the late 1820s. Usually they were set up by local cells and remained under working-class control, rather than being founded from philanthropic or socially-controlling motives by the clergy and gentry. They catered to deists,

---

116 Quoted in *NMW* 13 (19 Oct. 1844): 131.
117 *Republican* 14 (10 Nov. 1826): 561–65.

freethinkers, socialists, and radicals, and the surprising fact is that Saull is known to have financially backed or talked in at least half of them.

Because these halls sat far outside the social mainstream, they are almost ignored in modern studies of mechanics institutions, or of the respectable 'lits & phils'. But then they were equally eschewed by contemporary magazines in their listings of scientific venues.[118] Even the *Penny Mechanic* largely stuck with the expensive (one to two guineas per annum) Literary and Scientific Institutions "for the people". It rarely sank to the cut-price end of the market, although it did list a few of the better mutual instruction societies, including one of Saull's favourites, in Great Tower Street (at 1s a quarter, the cheapest on its books), which had a radical-Owenite cast.[119]

Despite this ostracism, some mechanics' institution managers still looked enviously at the socialist halls of science, which placed no bars on political, religious, or economic discussion. The halls hosted lively debates with clergymen on titillating topics such as "The Disadvantages of Christianity", or the "Genuineness, Authenticity, and Inspiration of the Bible".[120] These were real draws, yet such talks were taboo in mechanics' institutions. Nor could they match the free-for-all discussions after lectures, which made events spirited and participatory. This lured the more "reflecting" artisans.[121] And the convivial tea parties in socialist halls provided the kind of community feeling missing from more formal mechanics' institutions.

Learning from lectures was different from solitary book reading. Talks in social halls were entertaining and embracing—stump orators competed in crowd-pleasing rhetoric, cheered on or hissed, questioned and challenged. To work, talks had to be tailored to the local audience, so context was all. The halls were locked into local communities. As such, the lectures more resembled parish political rallies; they were a communal activity. Here was a more viscerally engaging way to learn of the new science and its community meaning. Visual excitement was often a key: hall walls were festooned with "splendid lithographic engravings", and

---

118 For example, *Magazine of Science, and School of Arts* 1 (1839): 320.
119 *PM* 2 (17 Mar. 1838): 279.
120 J. Baylee and F. Hollick 1839; *NMW* 11 (10 Sept. 1842): 90; both events chaired by Saull.
121 T. Coates to H. Brougham, 27 Sept. 1839, Brougham Correspondence 95, University College London; Coates 1841, 29.

tables covered in showy fossils, all handleable, or phrenological busts, or the latest electrical wizardry. And lectures often preceded *soirées*, again enhancing the joyous, community aspect.[122] They were cheap too, many of Saull's were free,[123] so they competed with circulating libraries and communal book clubs in penny-pinching terms.[124] In fact a penny bought you a night's entertainment and, in emancipatory Owenite circles, the wife or husband came too. Even then, any expense was made to seem worthwhile, for it was usually announced that profits would go to refurbishments or to bail out a celebrity activist. Saull's lecture profits invariably went to help jailed news vendors or finance tract distribution.[125]

Geo-Socialism

David Stack, in a bravura performance, has shifted the focus by concentrating on the "knowledge Chartist" William Lovett. Stack's claim is for what he calls the "isomorphic connections" between Lovett's political and scientific interests.[126] Science was not "coincidental" to Lovett's politics but actually helped shape his radicalism. In other words, the self-help sciences were not bourgeois imports which diluted political ideals, but were inextricable in their development, in Lovett's case moulding the fabric of his National Association. This Association was founded in 1841 to prepare the poor for their enfranchisement. Lovett's group set up a National Hall in Holborn (1842) for classes, lectures, and eventually a day school. Saull was Lovett's comrade-in-arms. Together they had sat lectures at the London Mechanics' Institution in the 1820s.[127] Then as fellow deists, radicals and co-operators they could be found in every London political union or co-operative association at the time of

---

122 E.g. *NMW* 3 (20 May 1837): 235.
123 When the National Political Union instituted lectures in 1832 they mooted a 2*d* entrance fee, but Saull proposed his be free: *MC*, 16 Feb. 1832.
124 A. Secord 1994, 278, on the Lancashire "weaver-botanists" and their pub-based book clubs.
125 *Crisis* 1 (6 Dec. 1832), 159; *PMG*, 24 Nov. 1832; *Lancashire and Yorkshire Co-Operator* ns no. 10 (n.d. [Oct. 1832]): 23; Hollis 1970, 194–202.
126 Stack 1999, 1029.
127 Lovett 1920, 1: vi, 36. No record now exists of Lovett in the LMI Members' Registers (1824–29) in Birkbeck College archives. Since registration was chaotic, occurring at multiple sites—at booksellers, the Crown and Anchor tavern, the secretary's office, and so on, resulting in *nine* collecting books in total—there was scope for confusion and loss during collation into one volume: Flexner 2014, 151.

the Reform Bill.[128] Saull had great sympathy for Lovett's "knowledge Chartists". But it is unknown whether he helped fund the Hall, although he promised Lovett his presence at the opening.[129]

Stack's work is suggestive but can we apply it to socialist geology? True, Saull's palaeontological progression was modified to meet Owenite needs, but one wonders how much it helped to re-shape Owenite political structure. Was there a dialectical relationship? One might have expected that an environmentally-determined rise of life would fundamentally re-ground the conditioning on which Owenism rested. In other words, within an 'evolutionary' scenario, human history became the history of the planet; therefore Owen's mantra, that "The Character of Man is Formed for Him, Not by Him"—which ran on the *New Moral World* masthead—could mean that geological forces must now be considered part of his character formation.

There certainly was recognition that the "social and moral world is subject to changes like those which geology points out in the physical world": both showed a progressive advance, but this only suggests a congruence.[130] And talk of basing "our new society on everlasting first principles, and to form society into a science in accordance with those first principles; first principles of the truth of which there shall be no more doubt than there is now respecting the sciences of mechanism, chemistry, or geology", again suggested no more than social adherence to the scientific gold standard.[131] Arguing that "human character is a formation, as obedient to fixed natural laws as any that have ever prevailed over the formation of geological strata" is simply invoking a naturalistic rationale.[132] Opponents argued that socialists demonstrated their social truths "by means of 'geology, chemistry, geometry, astronomy, and other modern onomies and ologies'",[133] but the protagonists were referring

---

128 The British Association for Promoting Co-operative Knowledge (1829–31), Metropolitan Political Union (1830–31), National Union of the Working Classes (1831–35), National Political Union (1831–34), and they sat together on many other committees. To make a fine distinction, Lovett was a radical with Owenite sympathies, where Saull was an Owenite with radical sympathies.
129 W. D. Saull to W. Lovett, 13 July 1842, British Library, Add. MSS., 78161 f. 162.
130 *NMW* 4 (16 June 1838): 268.
131 *NMW* 11 (9 July 1842): 9.
132 *NMW* 8 (29 Aug. 1840): 133.
133 *NMW* 7 (16 May 1840): 1205–1206; misquoting the *Quarterly Review*, 65 (Mar. 1840): 498, which said no such thing.

to socialism's infidel tendencies, using geology to deny Genesis. Many socialists did indeed make a religion of naturalism. In the *New Moral World*'s words: "As a false Geology" is "the basis of all imperfect systems of religion; true Geology ... will form the basis or fundamental principle of the improved religions of the Socialists".[134] This looks hopeful, and such programmatic statements inevitably suggest Saull's museum outlook. His work, uniting the evolutionary past to the socialist future, brought geology close to structurally enlarging Owenism.

Owenites certainly thought geology should be central. They talked of the laws of science stretching to society, and on every circumstance that goes to hone man being socialism's purview.[135] And historians have adverted to the Owenites' search for the causes forming man's character "before and from his birth".[136] Again, this largely points to Saull, given that his lecture titles typically invoked geology's influence in "Forming the Character of the Future Generations of Mankind".[137] But here Saull, the old infidel, largely seems to be arguing for the removal of religious impediments in order that man's character might develop its full socialist potential.

It appears that geology was deployed mostly in propaganda, disputation, and education as an arch-naturalistic science which de-sanctified and re-calibrated history; it propped up Owen's perfectibility stance, and promised a better future. In Saull's view, knowledge of geology would liberate and inspire man to the socialist heights. This was the meaning of his mid-1830s lectures "On Geology in

---

134 *NMW* 7 (6 June 1840): 1280.
135 *NMW* 11 (8 Oct. 1842): 117.
136 R. E. Davies 1907, 26. In truth, the Owenite literature concentrates largely on mankind's given organization at birth and the cultural forces shaping his upbringing. The atheists who split off in the 1840s came closest to discussing the pre-human conditioning of character as they explored the material ascent of life. For Charles Southwell, man was a "creature of circumstances" and "in every sense, a production of nature, no less than shrubs" (*Investigator* [1843]: 39–40). Nature had worked up to mankind, and Southwell opened his *Oracle of Reason* by invoking human progenitors, who were "not exactly either monkey or man". In short, "man could not have been always what he now is" (*OR* 1 [27 Nov. 1841]: 27). Southwell and the compositor William Chilton agreed that a person's character must partly reflect an inherited "organisation" at birth. This "original organisation", said Chilton, "is an effect", meaning it had prior causes (*Investigator* [1843]: 95).
137 *PM* 1 (29 July 1837): 322.

Reference to Human Nature", in Owen's Institution ("Admission free").[138] These talks, known by their titles, undoubtedly followed his line that an expansive geology would purge those "mischievous" religious "phantasies" which underwrote corrupt politics and corroded morals. No religious system has produced "sound morality, social happiness, or political elevation; on the contrary, they have all invariably tended to uphold the powers of the ruling few, at the expense of the welfare and happiness of the oppressed and deeply-injured many".[139] Socialist geology remained a cleansing agent, as it had been in his earlier infidel days, which would have a liberating effect on the human character.

This inflammatory new science of geology was one of the sensations of the age. As such it obsessed middle-class readers as well. Ralph O'Connor has shown how writers and poets co-opted traditional imagery—with the new fossil giants evoking Milton's fiends and Swift's Brobdingnagians—and sold otherwise unimaginable scenes of an extinct past to genteel folk, who needed a grab-handle on this alien science. By such "literary projection", they dampened fears and eased accommodation. But while the *Presbyterian Review*, as O'Connor noted, saw the respectable public placated by poetic narratives of a beguilingly exotic elsewhere, the *Review* did darkly comment that some took up geology "that they may consecrate it".[140] That is our infidels and their sanctification. They, by turn, revelled in the religious backlash. To social brethren, geology's ground was hallowed, for providing the deep historical contradiction of priestly phantasms in a visible, material form. So deep were they steeped in it that some radicals even turned the science into a career.

Two of Robert Owen's own sons became state geologists in America (the new mineral "Owenite" was named after one, David Dale Owen, in 1853[141]). This was largely as a result of the New Harmony community on the banks of the Wabash in Indiana, set up in the mid-1820s by Robert Owen and the enthusiastic geologist, social reformer and Pestalozzian educator William Maclure. David Dale made it the headquarters of the U.S. Geological Survey in the late 1830s. Robert Owen's youngest son

---

138 *NMW* 3 (12 Nov. 1836): 20; *PM* 2 (5 Aug. 1837): 8.
139 Saull 1833a, 37.
140 O'Connor 2008, 3, 6, 8.
141 Genth 1854, 297–99.

Richard was, by the 1850s, a professor at Nashville University with a geology textbook under his belt, and he was shortly to take over the Survey and become an expert on earthquakes.[142] So, it is no surprise that critics speculated that even Robert "Owen's religious opinions have received ... some material modifications from the geologist".[143]

In London, the American Henry Darwin Rogers—a future geologist of note[144]—was teaching geology at Owen's Institution in Gray's Inn Road over winter 1832–33. These were astonishing months: audiences at Gray's Inn Road could hear David Dale Owen on Chemistry, Robert Dale Owen on Geography, Robert Owen on the social system, Saull on geology (Tuesdays), and Rogers on geology (Thursdays).[145] Prothero notes that these lectures were "well attended" and introduced "by popular demand", adding they were "enormously successful because they avoided the mistakes of the mechanics' institutions and were pitched at the right level".[146] That they were, judging by Saull's.[147] His are the only ones we can reconstruct—but they also show something just as important. They were couched in infidel socialist terms and were integral to the Owenite agenda, which made them more communally relevant. This party aspect was a crucial factor. The season's success meant that science lectures would become a weekly feature at Owen's institutions in the 1830s.

Even if we move away from London, the case for geology's centrality is compelling. Social missionaries up and down the country took up geological arms just as passionately. These stump orators—and there were many of them (listed in Appendix 5)—were literally that, not accredited or career 'geologists', but political demagogues who often, like Saull, developed a real love of the science. Saull was far from the

---

142 Armytage 1951, 14, 18; Albjerg 1946, 21, 24–25; J. P. Moore 1947; Winchell 1890, 136–37; D. R. Dean 1989; Torrens 2000.
143 *NMW* 4 (21 July 1838): 306–07.
144 He was to become professor of geology at the University of Pennsylvania and carry out state surveys of Pennsylvania and Virginia: Gerstner 1994; S. P. Adams 1998.
145 *Crisis* 1 (29 Dec. 1832): 172; *PMG*, 22 Dec. 1832.
146 Prothero 1979, 253, taking his cue from Robert Dale Owen, who considered the language in mechanics' institutes often obscure and the scientific details arcane and "useless": *Crisis* 1 (15 Dec. 1832): 164.
147 Saull's were certainly pitched differently from the salt miner George Ogg's more technical, mineralogical, and experimental lectures (which started with Moses) at the London Mechanics' Institution, about which Saull (1826) was critical.

only geological orator on the circuit, even if he was the principal one. Everything was "Galileo, Geology, and Gaslights", said the droll young Holyoake in 1841. Punning awfully, he went on that only "fuddy-duddies" failed to march with the age, those defunct "Saurian remains of mankind", priests and nobles, whom "Geologists can tell" are on the road to extinction.[148] Nor was geology limited to talks. Spectacular fossils figured as *divertissements* at Owenite social festivals, while socialist children on school outings were taken to the strata themselves for inspiration. The Owenite Central Board advised branch lecturers to gen up on the subject, to send back fossil and rock specimens and to set up geology museums themselves.[149]

The infidels' obsession was held with almost religious reverence—but then, as Jim Moore once remarked, "Irreligion was never more variously religious than in Victorian Britain."[150] However, it produced the inevitable backlash: Saull's nemesis, the universalist preacher J. E. Smith, saw such "shrine"-like museums lead to an unhealthy "worship" of fossil relics. Smith, who could never let Saull's monkey-man drop, equally had no truck with his acquisitiveness. Smith ranked the geologists' "idolatry" alongside Catholic veneration of saintly remains, and Saull's own godless proprietorship could only have encouraged Smith's near accusation of ancestor-worship.[151] Even in socialist circles there were rumblings about this over-emphasis on geology, and "miniature geologists [school children] lisping out something about primary transition, secondary and tertiary!" when the educational goal was moral and social.[152] Some thought it took eyes off the political target. Others failed to see the science's remedial benefit in the depression. Learn all you want of coal seams, but you will not get coal any cheaper, sniped a social missionary one wintry February.[153] The reaction, if anything, proved the rule: deep within Owenite social seams, a stratal layer of geology was now firmly embedded.

---

148 *NMW* 10 (9 Oct. 1841): 114.
149 *NMW* 4 (6 Jan. 1838): 82–83; (25 Aug. 1838): 351–52; 5 (5 Jan. 1839): 170; 6 (24 Aug. 1839): 704; 8 (10 Oct. 1840): 240; 12 (22 July 1843): 32.
150 J. R. Moore 1988, 275.
151 J. E. Smith 1873 [1848], 1: 310.
152 *NMW* 6 (24 Aug. 1839): 697.
153 *NMW* 9 (6 Feb. 1841): 88.

Reaction, Prostitution, and Appropriation

Difficult debates among the Good and Great only made the science more attractive to socialists. Geology had become a trigger subject which polarized the press. Critics saw it tarnished by its trenching on Mosaic matters, some even thought it blasphemous and imbecilic. That it was upstart knowledge, awash in a sea of well-mannered Classicism, was shown by the reactions: geology is "to religion what ... foppery is to manners—silly, disgusting, and often injurious", said one protagonist.[154]

The consensus among geologists, by the 1820s, was of a sequence of strata laid down over aeons that housed the successive creations of life, but it seems to have caught many unawares. Hence the anguish among some sects about Charles Lyell's triple-decker *Principles of Geology* (1830–1833) and Oxford divine the Rev. William Buckland's Bridgewater Treatise on *Geology and Mineralogy Considered with Reference to Natural Theology* (1836). Liberal reviews could laud the works, but they invariably had to brush aside traditionalist worries and dismiss the "timidity" of religious souls who dreaded Moses being "compromised."[155]

The eight Bridgewater books were designed to 'rebaptise the sciences', in Jonathan Topham's memorable phrase, to mollify pious folk unsettled by the upstart sciences, strengthen faith, and prove God's plan. To show the providence of the existing social order had been the original intent of the louche Earl of Bridgewater, who bequeathed the cash to set up the series—that and suppressing the atheistic fallout from the French Revolution.[156] The sums were huge, £1,000 a book, the money being parcelled out by the President of the Royal Society, the Archbishop of Canterbury, and Bishop of London, which led one pundit in the *Mechanics' Magazine* to slate the lot as an "expensive hoax".[157] Given the political climate, conspiracy theories were rampant. The Owenite *Star in the East* even thought Buckland had suppressed his "sublime discoveries" of ancient life for years so as not to offend his patrons.[158] Letter wars flared up. Consider the infamous slanging match

---

154 *Freeman's Journal*, 17 July 1839.
155 *Monthly Review* 3 (Nov. 1836): 330–50.
156 Topham 2022, 3, 14, 26–28.
157 *MM* 21 (13 Sept. 1834): 412.
158 *NMW* 4 (28 Oct. 1837): 5. The *Star in the East* (Armytage 1961, 143) was owned by the agrarian reformer and Pestalozzian educationalist James Hill of Wisbech,

in the *Times* as late as 1845. Outraged correspondents thundered that Buckland's "disgusting nonsense" of yawning aeons spawning nothing but "Crocodiles and lizards!", uttered "without blush or shame", would yield poisonous atheistic fruits, not least a blasphemous belief in the natural ascent of life.[159] In reply, commentators laughed that Latiny letter writers even found "infidelity hiding in the mineral cases of the British Museum".[160] But the real fear for many was that some clever Voltaire would seize on these wrecked worlds to "spread evil".[161]

The infidel Owenites did nothing to assuage these fears. References to Buckland and Lyell pop up in their prints. The books were tooth-combed and cannibalized, regurgitated in epithets and snippets, or spewed out wholesale to prove the earth's antiquity and the unaided rise of life. Liberal *littérateurs* and co-operators alike were awed by the "grandeur" of Buckland's vision—his "vista of illimitable extension, filled with the multiplied consummations and colossal broods". But they baulked at the "theological requisitions, sophisms, and prevarications necessarily induced by the 'terms' of the Bridgewater Treatises".[162]

In his study of Bridgewater readers, Topham has shown Buckland walking a tightrope. The Oxford don and Canon of Christ Church ("£1000 per an.^{m} & no residence or duty required"[163]) was talking in his *Geology and Mineralogy* to an array of savvy, respectable, and religious audiences,[164] never to socialists. Yet they were talking back, and prostituting his sanctioned science in ways that would have appalled him.

Cambridgeshire. In 1845, Hill bought the *New Moral World* (Holyoake 1906, 1: 149–50).

159 *Times*, 23 June 1845, 6. This letter war ran from 23 June to 4 July 1845. Buckland's book had been contested by "scriptural geologists" from its publication (Topham 1998, 258).

160 *English Gentleman*, 5 July 1845, 10.

161 *Times*, 26 June 1845, 5.

162 *Monthly Repository* ns 11 (Jan.–June 1837): 269–78. So spoke Richard Henry Horne (1802–1884), fellow-traveller with the sacred socialists (Armytage 1961, 173) and editor of the *Monthly Repository*. This was shaking off its Unitarian roots to become a refined "ultra-Radical, if not Republican" literary organ, supporting the working classes by its "lofty eloquence". Unfortunately, its even loftier price, 1s 6d, put it out of their reach and made it a financial flop ([James Grant] 1837, 2: 327–28).

163 Wennerbom 1999, 104.

164 Topham 1998, 239, 249–61.

The Oxford-educated Whig Charles Lyell in *Principles of Geology* targeted a similar well-heeled audience. Lyell's expensive volumes oozed authorial gentility, just as Lyell himself oozed intellectual hauteur. He cultivated an apolitical air so as not to offend Tory reviews, arguing that the earth had been sculpted by a continuous stream of causes, no more violent in the past than they are now. There had been no catastrophic revolutions in nature. Lyell was urging what Secord calls a sort of slow "perceptual reform", non-violent, liberal. Lyell's aim was to raise the science above the sordid collecting, curating, and mapping level. But he sidestepped scripture, and in a "parson-ridden" age (Lyell's words[165]), this could smack of unrestrained naturalism. Worse, it could be seen as a snub to Moses. So Lyell, desperate not to offend his hail-fellow-well-met confrères, went to lengths to show the safety of *his* geology. He implied that this string of causation did *not* extend to animals and plants. To prove his point, he ratcheted up his attacks on the recently-deceased Parisian transmutationist Jean-Baptiste Lamarck. Lamarck was truly loathed, not least for his poisonous Jacobin philosophy which was thought to lie behind France's revolutions. The July 1830 outbreak in Paris only reinforced the disgust. In 1831, one evangelical Old Etonian at the British Museum railed

> against the abominable trash vomited forth by Lamarck and his disciples, who have rashly, and almost blasphemously, imputed a period of comparative imbecility to Omnipotence, when they babbled out their puerile conditions about a progression in nature.[166]

It was those *modern* disciples Lyell had to watch out for,[167] but instead he chose a softer target: Lamarck's near quarter-century-old musings on apes standing erect to be counted human. Lyell judiciously padded out his polemic, warning against accepting orangs as ancestors "with foreheads villanous [sic] low,"[168] so much so that his diatribe ended up as an entire volume of *Principles*. Lyell had a deep, aesthetic revulsion

---

165 J. A. Secord 1997, xiii–xxxiii.
166 J. G. Children to W. Swainson, 11 July 1831, William Swainson MSS, Linnean Society.
167 Corsi 1978, 2005, 2021; Desmond 1989.
168 Lyell 1830–33, 2: 2:60, paraphrasing Shakespeare.

at the bestialization implied by 'evolution',[169] but in so belabouring his attacks he massively raised Lamarck's profile.

Lyell's exegesis was a gift to the infidel socialists. Within weeks of this volume reaching the shops, the radicals' own "bricks and bludgeons" organ, the *True Sun*, dragooned "Monsieur Lamarck" into its pastiche. It turned to terrible doggerel, spoofing the evolution of lords and ladies:

> For what were Lords invented? Do you think
>   That Nature made them for no other uses
> Than just to talk about "destruction's brink,"
>   To plead for tithes, and to resist abuses?
> ...
> Oh! good Lamarck! how habit changes men!
>   How many plund'rers are there (we could score them)
> That ne'er had stolen, ne'er would steal again;
>   But that their fathers had been rogues before them![170]

The truckling geological gentry had long been upbraided in deist circles for their "false reasoning" "palmed [off], not only on the minds of the illiterate and the vulgar, but also on the ... better informed."[171] Now Lyell and his cronies were to be unceremoniously stood on their heads in Saull's museum lectures,[172] just as Saull's own monkey-man was making a debut.

The point is this: historians are starting to re-balance authors and readers, museums and museum-goers. If we want the view from below, we have to look beyond high-brow writers, whose works reinforced the cultural hegemony; beyond the Bucklands and Lyells, said the co-operators, who sought the "perversion of science" in order "to accumulate power and wealth in the hands of a few", instead of spreading its materialist "blessings" to the many.[173] These audiences are crying out for study. We need to probe their back-street halls, which stood far from Oxford's spires.

Hardly any attention has been paid to these subversive social groups, who scoured expensive geology books for their own

---

169 Bartholomew 1973.
170 *TS*, 9 Apr. 1832, 3; "bricks": [James Grant] 1837, 2: 105.
171 *Republican* 7 (28 Mar. 1823): 390.
172 *National Standard* 3 (18 Jan. 1834): 44–45.
173 *Crisis* 2 (13 July 1833): 222.

diametrically-opposed ends. They exacerbated the fault lines exposed by Lyell's and Buckland's books, glorying in the scripturalist discomfort. The result was that 'socialism and geology' were linked in many religious minds. 'Geological infidelity' became buzzwords. That giant 1s Sunday paper, the *Atlas*, advised the clergy to mug up on the upstart science "to guard it from this perversion".[174] Church of England primers were bolstered with geological rebuttals to arm ordinands.[175] Itinerant anti-socialist disputants took to the rounds to deny that fossils proved that creatures had "died before the creation of man", because God would not "have peopled this beautiful world with a race of beings who could neither return thanks for their blessings, nor who even knew the hand that made them."[176]

This riveting of socialist scepticism and geological chicanery explains the epigraph at the head of the chapter. In 1840 the editor of the *Church of England Magazine* (one of the largest circulation weeklies[177]) came away dispirited from a Saull Sunday lecture at London's socialist headquarters. Progression and the socialist Promised Land would have been Saull's theme. If true to form, he had illustrated it by monstrous *Iguanodon* bones from the museum to illustrate Britain's steamy Age of Reptiles. Outrageously, at its culmination, he would have mooted mankind's monkey forbears and rise from aboriginal savagery. It was too much for the editor. Such devilish events left him claiming that all attempts by socialists to "desecrate the sabbath and outrage revelation" started off like this. It was an overstatement, but it shows how inextricable the linkage between infidel socialism and geology now appeared to their enemies.

---

174 *Atlas*, 12 Nov. 1842, 730.
175 Johnson Grant 1840, xiii–xiv.
176 *NMW* 6 (12 Oct. 1839): 811. Also 7 (20 June 1840): 1326; *Courier*, 5 Jan. 1841, 3. This was the Owenites' *bête noire*, John Brindley, a former schoolteacher, and now a peripatetic socialist debunker, one of the "rabid maniacs" who hounded them, in the *New Moral World*'s words (showing the Owenites could match Brindley for personal abuse [J. F. C. Harrison 1969, 216]). His debates with Owenites could end in violence, with at least one broken jaw recorded (Buchanan 1840a, 142; Royle 1974, 64). Brindley was a government informant (Garnett 1972, 176; Hardy 1979, 58) and provocateur, whose lurid allegations left some Christians suspicious of dealing with him (Ainslie et al. 1840). He tried to persuade engineering bosses to sack infidel socialists, only to hear that this would entail dismissing most of their workforce (R. Cooper 1853, 76).
177 *Penny Magazine* 6 (31 Dec. 1837): 507.

2. Introducing Saull

> There are too few such men as Mr. Saull; men of great respectability, who are not content with holding Free-thought views, but lose no opportunity of avowing them, and impressing their importance upon their fellow-citizens ... His life affords fine example of public usefulness among a class most needing it—the middle and commercial. Rising above the sordid associations a competitive system is calculated to develope, he had an hour to spare for the instruction of the people, a purse ready to assist their cause, and a voice prompt to defend it.
>
> Atheist agitator Robert Cooper in
The London Investigator (1855).[1]

Respectability was a question of perspective. To the young firebrand Robert Cooper, the old lag Saull was wealthy and friendly, with a paternal attitude towards Cooper's atheistic *London Investigator*, which he helped distribute. Cooper never witnessed the younger Saull being dragged through the courts on blasphemy charges for supporting the Rev. Robert Taylor and his burlesque on Christianity. In the Reform Bill years—the early 1830s—respectability was far from a *Times* correspondent's mind as he damned Saull. Here, he was castigated as a rough trader, a "spirit-merchant in Aldersgate-street" who lectures to "mechanics at the Philadelphian-chapel" (a radical-blasphemous venue near Finsbury Square); "he assumes to be a great geologist" but "he is a very weak and conceited person,—a disciple of Mr. Owen".[2] That said it all. A mix of trade, blasphemy, and socialism spoke volumes to the *Times*' one and a half million buyers. Even then, Saull's appearing to onlookers as a mere "disciple" of infidels and co-operators was a pale shadow of the truth, as the Home Office knew from the tabs it was keeping on him.

---

1 *LI* 2 (June 1855): 46.
2 *Times*, 23 Jan. 1833, 2.

Next to nothing is known of Saull's personal life. Here we can only offer a series of glimpses through the political mist. Missing are almost all of the personal details. We do not even know what he looked like (there was a bust, an indication not least of his wealth, but it has vanished[3]). Equally obscure are his Northampton origins, and his relatives in that town. We know that one nephew there was a publican, suggesting that younger family members were in the trade. They also had freethinking leanings, and this nephew, John Saull, landlord of the "Admiral Rodney" pub near Northampton, refused to be intimidated by threats from civic leaders—the Anglican squirearchy was powerful in the provinces—and let his hall out to visiting freethinkers.[4] Press reports show another relative still fighting for universal suffrage after Saull's death.[5] That is pretty much the only political baseline we have, but it does suggest a freethinking radical family milieu.

William Devonshire himself was a generation older than young insurgent Cooper. When Saull arrived in London we do not know. Nor is his education documented, but it must have been minimal. Many ultra-radicals and soap-box co-operators were autodidacts, and he, too, appears to have been self-taught. Indeed, in a speech on Robert Owen's sixty-ninth birthday, he claimed that this was "the best education" available, being honed for purpose.[6] Still, he remained sensitive on the subject. Anecdotal (and undoubtedly apocryphal) evidence had him merely a "carman to a spirit dealer" at thirty, that is, in 1813, "barely able to do more than decipher the various addresses on the barrels".[7] This is extremely doubtful, for his younger brother Thomas, his partner in the wine trade, was obviously quite literate, judging by the 1813 ledgers and letters at Guildhall Library.[8] Moreover, their business, Saull & Saddington, "Wine and Brandy Merchts. 19 Aldersgate St", was already established by 1810, when Saull was 27, according to the *Post Office Directory*. Yet there is no doubt that he was self-made. Indeed

---

[3] Graves 1906, 5: 374–75.
[4] *UR*, 3 Feb. 1847, 20.
[5] *Daily News*, 21 May 1855, 3.
[6] *NMW* 7 (20 June 1840): 1319–25.
[7] *Preston Guardian*, 14 July 1855.
[8] Saull family of Aldersgate Street, papers, 19th century (Acc 2002/057), Ms 33957, Guildhall Library.

obituarists put down his lifelong interest in working-class education to the "defects" in his own.⁹

This in some part explains his strong support for Owenite 'rational' schooling. Not that such an emphasis on youth training was as obvious as it seems today. That doyen of labouring self-sufficiency William Cobbett, by contrast, was opposed to mentally restraining children. Education ("Heddekashun" as he laughed it off in his yokel-mimicking way) could thus be dangerous, and he also liked to cite examples of uneducated boys who later achieved brilliance.¹⁰ But Saull was soundly Owenite in his support for schooling beyond the clutches of clergy and gentry. He helped set up the "Rational School" in Owen's Institution of the Industrious Classes in Charlotte Street, London, in 1833. Chairing a patrons' meeting, he explained that

> you must not look to the gentry to commence a school on liberal principles, for if they did, the first thing they would do would be to put your children in livery, train them to be servants, to wait on them behind their carriages.

Having little schooling himself, "he would be always ready to assist [socialist training like this], as he was deeply interested in the education of youth; inasmuch, as he intends to leave his valuable museum for the purpose of education."¹¹ So, almost from the foundation of his museum, Saull was planning to bequeath it for Owenite educational purposes.

The cultural shaping of the young mind, as one believer put it, was like the geological sculpting of the landscape, and however questionable geology's role in the shaping of humanity, there was no doubt that, for socialists, geology was to be one of the fundamental axioms of this rational schooling.¹² That is how Saull saw it, as integral to a wider rational education—an education that had to be rigorous, comprehensive, and scientific to be effective.¹³ Geology was taught in the first co-operative school, set up at Salford in 1832, a democratic institution eschewing Owen's patriarchal approach, where the teachers

---

9 *JBAA*, 1st ser. 12 (1856): 186–87.
10 *Cobbett's Political Register* 88 (30 May 1835): 537, citing the case of Dr Adam Clarke, whom we will meet later in connection with apes and devils.
11 *Crisis* 3 (28 Dec. 1833): 144; (4 Jan. 1834): 150–51.
12 *NMW* 6 (5 Oct. 1839): 789–91; 1 (11 July 1835): 289.
13 Rigour was emphasized by Owenites: *Crisis*, 3 (14 Sept. 1833): 9–10.

were working men and the students had a say in its running. The school had a mineralogical museum at the outset.[14] By 1840, geology was part of even elementary instruction at the Owenites' Institution in London, while rational day-school boys were set exams in the subject all over the country.[15]

Saull's Owenite philanthropy and geological acquisitions depended on his business booming. And it did. He already had a wine and brandy warehouse at 19 Aldersgate Street by 1810.[16] In 1831, he moved the company to larger premises at No 15, a corner site a stone's throw away. This complex, with its bow-fronted shop, warehousing, stabling, and apartments, would be his home for life. Its large size, affirmed by £200 per annum rent, meant it could accommodate both his wine storehouse and museum.[17] Probably it was no coincidence that he bought out James Sowerby's fossil museum—making it the nucleus of his own—at this moment. We can assume that the new depot was actually acquired to accommodate the collection. Given that his entire business was re-located evidently to house the huge museum, his commitment was palpable.

Judging by his frequent trips across the Channel (on occasion accompanying Robert Owen to Paris), he specialized in French wines. Because high tariffs meant that only the finer wines were imported, we can be sure that "W. D. Saull & Co." was catering to the 'easy classes'. Charles Ludington, in *The Politics of Wine*, actually calls wine *the* demarcator of classes: favoured by the court and Church, it symbolized political power and social distinctiveness. And wine tastes reflected

---

14 *Lancashire and Yorkshire Co-operator*, No. 10 (1832): 6, 47; Yeo 1971, 91.
15 *NMW* 7 (30 May 1840): 1262–63; 11 (17 Sept. 1842): 99; (17 Dec. 1842): 203. See also *NMW* 4 (6 Jan. 1838): 82; 12 (22 July 1843): 32; Student in Realities [nd], Part 1: 254–55 on education beginning with the history of the earth; *Union* 1 (1 Dec 1842) 361–72.
16 "Saull and Saddington" traded until 1822, after which the company became "W. D. Saull & Co." and included at some point Thomas Saull and John Castle. The Castle partnership was dissolved in 1835, leaving the two Saull brothers: *London Gazette* 17857 (1 Oct. 1822): 1606; 19240 (13 Feb. 1835): 268.
17 House of Commons Parliamentary Papers, 19 pt. 1, 1840, Coms. of Inquiry into Charities in England and Wales: Thirty-second Report, Part VI. (City of London; General Charities, Essex), 20. He bought the property from a bankrupt leather cutter: *Perry's Bankrupt and Insolvent Gazette* 6 (1 Jan. 1831). Saull also owned a counting house with large wine cellarage in Burton Crescent, close to the house he let to Robert Owen: *MC*, 3 Oct. 1848, 1.

changing social mores. The port-swilling inebriety of the late Georgian age of aggressive masculinity was giving way to sherry sipping and more mannered ideals in the 1820s.[18] But even if Saull was now plying the pious, he was still doing a roaring trade. He seems to have been an astute manager, but tariffs also explain why business was booming. Wine duties were halved in 1825, after which French wine sales doubled or tripled. Then, in 1831, the tariffs were levelled, bringing French wines down on a par with the Portuguese and making them still more attractive. Foreign spirits were holding their own, despite swingeing government duties after the Napoleonic wars (a protectionist tax to favour British farmers and home-grown corn-spirit consumption).[19] By 1832, London, the sprawling, "monstrous smoke-hole" of a city, crammed with one and a half million residents, was consuming 10,000 gallons of spirits annually, and seven million gallons of wine.[20] Then a duty reduction in 1846 led to a fifty per cent rise in consumption over the following years.[21] All of this helps explain Saull's soaring profits. And the firm remained a success while Saull lived, but it crumbled into bankruptcy quickly after his death,[22] suggesting that he was the driving force. As a result, from the 1820s to the 1840s, he was comfortable enough to sink untold thousands into infidel chapels, Owenite halls, and court costs for prosecuted activists. He could shell out yearly subscriptions to numerous learned societies and think nothing of competing with the great institutions by bidding £40 (£3,500 in today's money) for a fossil.

Aldersgate Street was a well-known thoroughfare. It was home to the City of London Institution, with its newly inaugurated theatre in 1828.[23] This catered particularly to the sons of wealthy professionals. Up the street was the General Dispensary, an out-patient medical facility for the

---

18   Ludington 2013.
19   G. R. Porter 1843, 57–64; B. Harrison 1994, 65, on the massive rise in wine and sprit consumption in the 1820s and 1830s.
20   *Cosmopolite*, 19 May 1832, in HO 64/18, f. 657; W. A. Smith 1892, 89; *Lady's Magazine and Museum* 3 (Dec. 1833): 350.
21   G. R. Porter 1851, 559. Not only were Saull's relatives in the pub trade, but he himself can be located in the wider victualling business; he acted, for example, as an executor for London publicans: *County Herald and Weekly Advertiser*, 20 June 1835, 1.
22   It went bankrupt a year after his death: *The Law Journal Reports*, 1856, 53.
23   Denman 1828.

poor, to which Saull subscribed.[24] But its distinguishing landmark after 1829 was the General Post Office, London's new "pride and wonder" with its fifty-foot-ceilinged great hall supported by six Ionic columns. It gave the museum its instantly identifiable location, "a minute's walk from the General Post Office".[25]

Saull's house lay in one of the two parishes of Aldersgate ward, themselves marked by two ancient churches. The vicar of Saull's parish, St Botolph Without Aldersgate, clearly had issues with his recalcitrant parishioner, for he gave him Bishop Watson's *Apology for the Bible* (which had been written in reply to Tom Paine) in the hope that he would see the light. His Reverence must have been deflated to hear Watson slated as "deficient in reasoning" in Saull's privately-printed response. And the Bible itself Saull found wanting in the face of the latest "Astronomy, Geology, Geography, Ancient History".[26] Saull was never a profound thinker; he had none of the scurrilous Richard Carlile's deistical acumen (see Chapter 3), or the Rev. Robert Taylor's theatrical flourish (Chapter 4), nor the atheist compositor William Chilton's zoological stamina (Chapter 18), and certainly not the aggressive philosophical gall of the "Jew Book"-hater Charles Southwell (Chapter 18). Rather, Saull was an active, hurried business man with a freethinking passion and a long purse, ready for any infidel-Owenite eventuality. But he did share the others' Enlightenment belief in the omnipotence of science, and faith that science, rightly understood, could solve human problems. He naively echoed Richard Carlile's call in *Address to Men of Science* (1821) for the scientific clerisy to come clean about the anti-Christian implications of geology and astronomy. He demanded, in effect, that

---

24 Aldersgate's was the founding dispensary and a blueprint for others (Loudon 1981, 323). Dispensaries were financed by voluntary contributions and unique in that the doctors (including George Birkbeck, whom Saull would come to know well) would visit the poor at home. In 1845, the General Dispensary treated over ten thousand patients, including almost two thousand at home (*Daily News*, 30 Apr. 1846, 1). Saull attended yearly functions (e.g. *Times*, 12 May 1841, 2; *Daily News*, 30 Apr. 1846; *MC*, 11 May 1846, 1; 19 Oct. 1848, 7) and left the dispensary a bequest in his will. Ward meetings sometimes took place in the Dispensary theatre, so Saull might equally be found here on civic business (*Morning Post*, 22 Dec. 1832; *Examiner*, 29 Mar. 1845). Saull also subscribed to the Sanatorium founded in New Road for the middle classes: *MC*, 26 Mar. 1840; and he supported individual distressed medical men: *NMW* 12 (6 Jan. 1844): 224.
25 *Reasoner* 1 (6 Aug. 1846): 159; *NS*, 31 Oct. 1846, 3; Cruchley 1831, 43–44.
26 Saull 1828a.

they abandon their social base and act as fifth-columnists—impossible for the gentlemen of science because the Christian thread was woven so tightly into the social fabric that to unpick it would cause the whole cloth to shred. Saull demanded that materialist science lay its imperial claim to the realm of theology, believing, like all radical Enlightenment activists, that this would have profound social benefits. It was all neatly encapsulated in a book dedication to Saull by the Hackney Baptist and Bunyan expert George Offor, who saw Saull's work

> to draw mankind from the mad pursuit of phantoms, calculated only to injure or destroy human happiness, and to fix the mind upon realities most deeply interesting and valuable—to trace nature in her progressive developments from chaos towards perfection; these are researches calculated to check our baser, and elevate our nobler passions ...[27]

If infidelity marked Saull out in the parish, so did his politics, with the press pegging him as an extreme "Radical of the ward".[28] The London vestries had themselves become increasingly radical. Because they had many more skilled artisans on the electoral rolls, who allied themselves with the lower middle classes in their shared mistrust of "central authority", they were democratic hotbeds.[29] But Saull went further, and could even cause a public furore, most notably during a local Aldersgate election when he twitted the monarch over the fate of Charles I.[30] Being a republican, he also questioned the use of City of London funds for the King's domestic servants.[31] He would address wardmotes (meetings of merchants and citizens, chaired by the ward's alderman), urging municipal reform and support for the City's reform MPs.[32] And as a merchant, and thus an elector of delegates to the Common Council (which governed the City, with the Mayor and aldermen), he backed radicals who would push for "triennial Parliaments, universal suffrage, and vote by ballot".[33] Lobbying the Mayor with such ultra-radical demands prompted still more outraged letters to the *Times*, proving

---

27 Offor 1846, dedication, iii–iv.
28 *Baldwin's London Weekly Journal*, 24 Dec. 1836, 4.
29 Green 2010, 82–93. The exclusion of the vestries from the gagging Seditious Meetings Act of 1795 meant that radical expression could flourish here.
30 MC, 25 Dec. 1834; *Times*, 25 Dec. 1834, 2.
31 *Baldwin's London Weekly Journal*, 24 Dec. 1836, 4.
32 MC, 31 Dec. 1834; TS, 31 Dec. 1834, 2.
33 TS, 23 Dec. 1834, 8; 22 Dec. 1835, 8; MC, 23 Dec. 1834.

that Saull was actually far from inconspicuous.[34] Like many ultras and infidels, he saw both sides of the dock. When he was indicted for funding the Rev. Robert Taylor's blasphemous pulpit in 1828 (Chapter 4), it was the Court of Common Council in the City that he petitioned against the charge.[35]

Saull escaped prosecution, and such was the febrile political atmosphere in the Reform Bill years that the episode did not harm his City prospects. Many City aldermen were themselves reformers. As a wealthy merchant he was acceptable as an auditor of the City accounts only four years later, in 1832, a position he held through the decade.[36] With such visible bona fides, he was the obvious choice to audit, collect subscriptions, and act as banker to many of the radical and Owenite ventures. A City role was a guarantor of trustworthiness. Merchants were men "possessing public confidence", as important for committees collecting for Chartist widows as for the Guildhall.[37] Such credentials were even essential, given the horror stories of treasurers absconding with co-operators' savings or strike funds.[38] Wealth also allowed him to extend his financial dealings to deeds and promissory notes—Holyoake actually said he dealt in "bills and wine", reversing the priorities.[39] He even owned the deeds to Robert Owen's houses. Thus Saull became one of wealthiest Owenite backers, and he accepted whole-heartedly the socialist ideology: despite his huge business interests, he understood the need for individual regeneration, a non-capitalist labour exchange system, perhaps eventually the commonalty of property (see Chapter 6). He even went beyond Owen to demand a radical levelling via universal suffrage.

---

34 *Times*, 23 Jan. 1833, 2; *TS*, 17 Jan. 1833, 3.
35 *Trades Free Press*, 19 Jan. 1828, 206; *Times*, 18 Jan. 1828, 2.
36 *Courier*, 26 June 1832, 3; *Atlas*, 1 July 1832, 421; *TS*, 28 June 1834, 3; *Royal Kalendar*, 1838, 297.
37 *NS*, 27 Oct. 1849.
38 Chase 1988, 152; Chase 2000, 142; Goodway 1982, 47, 192; Rule 1986, 298, 319; G. Anderson 1976, 39.
39 Holyoake 1892, 2: 69. This is possibly an insider joke. 'Bills' were also the contemporary term for indictments, such as those handed to Carlile and Taylor, and on receiving one they would "immediately set about getting Bail", 'Bail' in this instance meaning the person who puts up the surety, and that was often Saull: e.g. HO 64/11, f. 200.

Actually, Saull's grand-sounding title, 'merchant', has to be treated cautiously, given that he was an uneducated self-made trader risen from humble origins. Although 'working class' and 'not of the working class' were the standard categorizations of the day, for example on committees at the London Mechanics' Institution, there was some fluidity, as individuals slipped effortlessly between categories,[40] and there was often mutual sympathy, especially before the Reform Bill. Before 1832, Gareth Stedman Jones reminds us, the class division was not "between employer and employed", but "between the represented and the unrepresented".[41] A tradesman was not as distinct as he might seem. Nor was Saull's position unique in combining commerce and radicalism. The wealthy George Rogers—a St Giles' tobacco and snuff manufacturer—did so too. He was another City parish reformer and radical benefactor, who joined Saull in the political unions; the two, for example, could be found co-operating to rescue the ailing radical paper, the *True Sun*,[42] or to pay off fines for indicted campaigners. As a result, Rogers was another slated by the Tory press as a politically suspect "low tradesman".[43] It was a time when these marginal mercantile men in a Church-and-Crown dominated society were flexing their muscles, as Steven Shapin has shown: they were changing the "boundaries of participation in science", deploying self-help phrenologies and anatomies to further their civic grip,[44] and now geology was equally being pressed into service.

Rogers became a Chartist, a physical-force one at that, and was a London delegate to the Chartist Convention. But Saull was never one for storming the citadel, nor was he a tub-thumper like Henry Hetherington, or quick with the repartee like Holyoake. Quite the reverse, Saull's were

---

40 Flexner 2014, 14, chap. 6.
41 G. S. Jones 1983, 106.
42 *TS*, 16 Oct. 1832, 1; 25 Oct. 1832, 1; HO 64/18, f. 702. Prothero 1979, 276, 311. George Rogers sat with Saull on the Metropolitan Political Union and National Political Union (NPU), and they worked on the condemned insurrectionary John Frost's defence fund (*CPG*, 21 Dec. 1839, 2). Rogers became the radical Thomas Wakley's election agent, running his 1835 parliamentary campaign in the new Finsbury constituency (Sprigge 1897, 239–52; Weinstein 2011, 50).
43 *The Age*, 28 Aug. 1842, 4.
44 Shapin 1983; Desmond 1989, ch. 4, on the sons-of-trade trained in London's back street medical schools adopting subversive approaches to science.

described as long-winded, "rigmarole" speeches, at least by opponents.[45] Perhaps that befitted the stolid, affable bank-manager of rational causes; certainly, it seemed proper for the dependable chairperson of radical committees. Always his cash dispersals showed his deep sympathy for the oppressed. This was illustrated by his first name-check in the newspapers: a guinea donated in 1825 to the cause of the Spanish and Italian refugees, who had fled persecution after failed rebellions and were exiled in poverty in Islington and Somers Town.[46] And nowhere was this sympathy more evident than in his role in the campaign to repatriate the transported Tolpuddle Martyrs.

While Saull's trading associations might have been detrimental in the eyes of elite geologists, there is no telling whether his fossil obsession, bizarre to some, was damaging to his business. It was, evidently, for another City merchant in the liquor trade, the Bishopsgate distiller James Bowerbank, whose collection, despite its emphasis on fossil fruits and seeds, rivalled Saull's own. It was said that Bowerbank's Highgate museum was amassed at a time when such a pursuit "was rather an opprobrium than a merit in a young commercial man".[47] On the other hand, a large museum could advance a merchant's reputation in learned society. Saull's wealth bought him rare fossils and, with them, access to geological high culture, including entrée to the geologist Charles Lyell's *soirées*.[48] Although uneducated, Saull was soon putting F.G.S (Fellow of the Geological Society), F.A.S (Fellow of the Astronomical Society), and F.S.A (Fellow of the Society of Antiquaries) after his name, while bandying around the Linnean binomials of ancient saurians and discussing runic inscriptions with the best of them.

It helped that Saull was clubbable, affable, and, as an Owenite, punctiliously moral (as he saw it), all of which gave the lie to the religious adage that materialists were evil people. So pervasive was this defamation that atheists constantly found it necessary to protest the

---

45 *Morning Post*, 17 Apr. 1838. Still, he could seem quite "affected", especially when eulogising Robert Owen: *Weekly Tribune*, 18 May 1850, 6.
46 *Courier*, 9 Feb. 1825, 1; *New Times*, 9 Feb. 1825, 1. J. White 2007, 140–41 on the refugees. But theirs was a *cause célèbre*, as likely to attract genteel ladies (Morgan 1862, 2: 147–48; Litchfield 1915, 1: 196) as sympathisers of the Carbonari revolutionaries.
47 Reeve 1863–64, 2: 133.
48 Morrell 2005, 137.

calumny. The *Christian Times* saw only "vice, and ignorance, and crime" accompany the "progress and power of infidel opinions". According to the *Patriot*, socialism was a malignant depravity and, for the *Christian Beacon*, its advocates were shiftless and profligate. This was a common perception in polite society, that socialists were "filthy fellows in their hearts".[49] The socialists' geology could be tarred as well. No one abhorred the materialists more than the Cromarty stone-mason-turned-editor of the evangelical *Witness* and influential author on Scottish fossil life, Hugh Miller, seemingly because of their disrespect for the cloth. No name calling was too foul; they were an "infestation", "vermin", a "slime"—castigations so severe that Miller's first biographer muted the barbs by calling them "half comic, half savage".[50] A later commentator wondered whether Miller had actually known anything of radical teachings.[51] Neither took Miller's hatred seriously enough. Nor was Miller alone in seeing an evolutionary geology pervert the "intelligent mechanics" of life. It was the rot that turned the infected into materialists, eating through their moral mooring and belief in salvation, and rendering them "turbulent subjects and bad men".[52] It was scarcely less hysterical at the other end of the social scale: the Regius Professor of Modern History in the University of Oxford, the Rev. Edward Nares, declared in his 1834 defence of Revelation, *Man, as Known to Us Theologically and Geologically*, that a bias against the Six Days of Creation was just another "vicious inclination" which pushes the "mind towards infidelity".[53] Had it not been an Owenite imperative, a holier-than-thou attitude would have been prudent anyway for Saull, given such mania. It greased the social wheels in hail-fellow, well-met geological society. As for Saull's suspect politics, the fossils themselves provided some deflection and diversion. With Tories marvelling at the beauty of his sea lilies and the rarity of his tree ferns, these artefacts could be seen mediating an otherwise deep

---

49 *London Magazine* 1 (1840): 105–11; *Patriot*, 28 Feb. 1839, 132; *Christian Beacon*, 2 (1840): 146–47; *Christian Times*, 30 Aug. 1851, 548. That materialism made them "bad men" was a common refrain: *Republican* 9 (9 Apr. 1824): 461.
50 Bayne 1871, 1: 271–72, 324, 328–29. Not that atheists could not respond in kind: Charles Southwell caned Miller and his *Witness* for their cant and fanaticism and defended the moral integrity of atheism: *Investigator* (1843): 185–86.
51 Mackenzie 1905, 185–86. All this despite Miller's own desire for social equality, a point made by early biographers, and explored by Lunan 2005.
52 H. Miller 1849, ix.
53 Nares 1834, 7.

political and religious divide.⁵⁴ They were the common rocky ground to facilitate discussion rather than dissension.

The republican banker was himself banking on his fossils. By the end they were worth some £2,000. This was a massive increase in intellectual capital: the dealer, in buying, collecting, and displaying, was indulging in a status-raising exercise. It enabled the parvenu to prise open intellectual doors. Trading in scientific commodities could be profitable in more ways than one. Unlike gold, hidden away as a hedge against inflation, fossil assets, like fine art, were kept visible and flouted to display one's affluence and learning. "Mr. Saull does not place his 'candle under a bushel,' nor, like a miser, lock up his stores", lauded Thackeray's *National Standard*. And perhaps we have to think in terms of fine art to understand why Thackeray's thrusting young blades would describe the museum merchandise doing "great credit to the taste, learning, and liberality, of its possessor". Liberality because it was opened to the downtrodden, learning because of its scientific pretension. But "taste"? Given the usual association of "taste" with class and character, we sense here an alternative aesthetic appreciation, for the museum's goal, "to elevate the moral character" and attack entitlement, which made Mr. Saull one of "the benefactors of humanity".⁵⁵

Saull made his fossil assets do work. They were didactic and often dramatic. Given the prevalence of infidel lectures on the "Antiquity and Duration of the World" to debunk biblical chronology,⁵⁶ his geological stockroom could capitalize in a visual way. But more, the rocks were said to talk direct, without religious intercession or obfuscation. They were thought to give an unmediated contact with 'reality' to testify directly against sacred texts. Saull's co-operative comrade William Lovett wrote that "In throwing open the stony records of geological science, the attentive student may read for himself without the aid of translators or commentators a true illustrated history of the various animal and vegetable tribes".⁵⁷ It was a common infidel belief, that Nature's Works

---

54 *Morning Post*, 31 Dec. 1841.
55 *National Standard* 3 (18 Jan. 1834): 44–45. On Thackeray and the palaeontologists, see Dawson 2016, 155–61.
56 *NMW* 1 (12 Sept. 1835): 364–66, for a typical case.
57 Lovett 1920, 2: 385–86, 417. Like Saull, Lovett was enamoured of geology. He himself wrote a geology book, but failed to get it published. Stack 1999 for a study of Lovett.

were truer than God's Word. As another Saull comrade, the former insurrectionary and land reformer George Petrie, put it in his influential poem "Equality" (1832), whose verses were pinned up on the walls in Saull's museum:

> Through boundless space new scenes of beauty rise,
> And Nature stands unveiled before his eyes;
> Her laws immutable he understands,
> Unmarr'd by vile translator's filthy hands.[58]

Nature was not bare rocks and fossils but pointed to something far more social: in infidel radical eyes, its truth and beauty exposed society's cruel deviance and suggested a remedy.[59]

The museum's seemingly unmediated contact with reality was enabled by Pestalozzian educational notions so enamoured of socialists.[60] This encouraged a direct understanding gleaned through contact with hand or eye, making fossils and models eminently suitable. While Dissenters argued that God talked in the Bible over the bishops' heads straight to them, radicals went further to see Nature talking over the heads of Dissenters and contacting the powerless directly.

> [G]eology is my subject ... and the book which is open to me, is not shut against the meanest of my readers ... Nature will ever display to those who pursue the path of her progress, not her secrets and mysteries, for she has none, but the powers of her action, and the method of her labours. These require not a variety of languages to understand or explain them, nor the imposing diligence of imposing schools and high-gifted seats of learning to comprehend them.[61]

For Saull, the truth of Nature was incontrovertible, and his display was designed to prove it. The stones do not lie: this was a leitmotiv of Owenite lectures through the 1830s and 1840s. All geology talks were therefore declared to be "free from assumptions and conjectures", as the rocks would in themselves expose religious obfuscations and "time-hallowed prejudices."[62] Direct instruction from the stones without priestly

---

58 *PMG*, 22 Sept. 1832; Petrie n.d. [1841], 5; on its publication: *PMG*, 11 Aug. 1832.
59 Murphy 1994, 113.
60 *NMW* 1 (1 Aug. 1835): 515; Greaves 1827; Silber 1965, 283; Armytage 1961, on the Pestalozzianism of William Maclure, Charles Lane, and J. P. Greaves.
61 *Republican* 14 (10 Nov. 1826): 561–65 (p. 562).
62 *NMW* 10 (13 Nov. 1841): 160.

intercession was what Owenites were offering, the fundamentalist belief in truth from immediate contact. Visiting Saull's museum, one lyricist penned a few verses for the Owenite house organ, the *New Moral World*, in 1840:

> Ye that would drink at learning's purest springs,
> Forget your books awhile, and study things;--
> See nature's volumes round you fair outspread,
> Cull'd from her library, too little read;—
> Each line from human pen may err or cheat,
> In her's alone, there cannot be deceit.[63]

Saull, like so many activists, ran the reform gamut. He was an infidel, co-operator, union sympathizer, campaigner against taxes on the pauper press, against church rates and tithes, an Anti-Corn-Law Leaguer to lower bread prices, parliamentary suffrage reformer, Aldersgate ward radical, republican, and so much more. He was often contemporaneously active in each sphere. His swift move from one campaign to another left them appearing as a blur, proving what Prothero says, that these were interlaced movements and cannot be artificially compartmentalized.[64] And he was financially committed to all of them.

Yet Saull figures only as an infrequent footnote in histories of artisan radicalism and Owenite socialism. We need to flesh out this skeleton, bring the bones back to life, and reorientate the story around his pride and joy, the Aldersgate Street museum. Resuscitating Saull requires us to be sensitive to his specific context, to appreciate how all the aspects of his cultic milieu, Carlilean materialism, 'Utopian' socialism, and rejection of Christianity, came together to produce an Owenite 'evolutionist' with a transformative museum that heralded the social Millennium. Agitators like Saull are often dismissed as of little consequence, and historians dealing with them are said to be walking on the wild side, as if what matters are only wealthy, expensive, official accounts of science. Such blasphemous, radical, and co-operative views, it is said, were heard by few, promulgated by fewer still, and, being advocated for political ends, were marginal to "real" science. This study suggests that, rather than a few promulgators, the critical factor might be how few are actually

---

63 *NMW* 8 (12 Sept. 1840): 175.
64 Prothero 1979, 4, 255–62.

known, because of the vagaries of historical preservation, rather than how few actually existed. W. D. Saull's case shows how these shadowy radical individuals, ignored in the histories of gentlemanly theoretical science, can be actively resuscitated and their lost worlds reconstituted.

What follows is a narrative and thus chronological approach to Saull's street-level science and its enabling and changing social context. It is the story of Saull's trajectory—in geological terms, from a world of eternal flux to one of origins and progress; in contextual terms, from a blasphemous theatrical astro-theology to a self-sustaining astro-geology. These transitions were nurtured by Saull's move from a culture of Malthusian liberal economics to a faith in social regeneration, progress, and socialist perfectibility. Those were the political shifts which edged him into tackling the ultimate question: how had humans originated on the Earth?

Our entry point is the Carlilean underworld of the 1820s. It is intended to show the dissident geologies and astronomies Saull first had to negotiate. Many of these, it turns out, were developed to grapple with, or circumvent, the problem of 'Creation'. That might have been expected, given that 'blasphemers' were chafing at the restraints of a law-backed Christian culture, and desperate to undermine the miraculous props of priestly power. Saull's unique solution unfolded as he negotiated the shifting underground movements at this time of political unrest.

PART I

1820s

DIRTY DIVES AND SUBVERSIVE ORIGINS

3. From Eternity to Here

Blasphemy, Eternalism, and the Emerging Question of Origins

> I challenge you Deists to say where but in the Bible can you discover any satisfactory account of the creation of the world, and of the "origin of man?"[1]

How Saull originally became radicalized is not known. But we can detect the company he kept. And it is this which allows us to trace the route he took through the dissident byways to arrive at his questions about mankind's origin in 1830.

Judging by police reports, Saull's first influential contact was with the recidivist, deist, and reviver of the blasphemy movement, the publisher Richard Carlile (1790–1843). This is confirmed by identification of Saull's first publication on geology (Appendix 1), a letter which appeared anonymously in Carlile's scurrilous sixpenny weekly, the *Republican*.

Carlile was the bogey-man of polite society. It was widely believed that never had such "*a scoffer at religion lived*" who so thoroughly "*merited the gallows or the pillory.*"[2] Indeed, an actual gallows orator once captured Carlile's reputation. Watching felons hang at a public execution in 1823, he exhorted the gawping onlookers never to "keep company with Deists" nor "to read any of Carlile's books", lest they share the same fate.[3] Not that such crude fairground demagoguery frightened Saull. The deistic *Republican* that printed his letter was an incendiary rag which ran from 1819 to 1826. This was Carlile's flagship, and it sold up

---

1 *Republican* 8 (18 July 1823): 52.
2 *Republican* 8 (17 Oct. 1823): 477–78.
3 *Republican* 8 (18 July 1823): 53.

to 15,000 copies some weeks, most notably during his sensational 1819 trial for blasphemy and sedition.[4] By the time of Saull's letter in 1826, on orbital wobbles explaining the geological strata fluctuating from polar to tropical and back, Carlile had only been out of Dorchester gaol a few months, having served a six-year term. And Saull's title spoke volumes, "Letter from A Friend: On Fossil Exuviae and Planetary Motion": Saull was already an intimate of the notorious Carlile.

Given that spies reported Saull financing Carlile's court costs and blasphemous publishing career, our starting point must be Carlile's shady Fleet Street premises in the 1820s. Even this building's appearance shouted its intent: in a window stood a statue of Tom Paine, and the front was placarded with badly written ads for the latest inflammatory pamphlet, while the first-floor windows were dominated by life-sized effigies of Old Nick and a Bishop.[5]

Carlile recast his own history in martyr's terms (religious language is inevitable when discussing Carlile's deism). He started as an itinerant tinplate worker and pamphlet hawker, who turned printer in 1817. His trenchant radicalism in Regency London was proved by his deliberately provoking the Tory government: he pirated publisher William Hone's "scandalous" (and funny) political parodies that year. These lampooned a repressive government through jests on the Lord's Prayer ("deliver us from the People. Amen.")—squibs that were tailored to alehouse readings, to further gall the Attorney-General.[6] Being charged with seditious libel and "blooded" by his first prison stay (he was locked up for months in the King's Bench) further radicalized Carlile.[7] He had Tom Paine's Enlightenment rationalist books smuggled into his cell. Then, on coming out, he republished them, bringing Paine back into popular view.

Paine's *Age of Reason* (1794–1807), execrated in the religious press, accelerated Carlile's conversion to deism. It left him brutally sceptical of scriptural truths. As a result, his *Republican* was marked by a rough handling of the "Holy Jew Book"[8]—an intentional slur designed to

---

4   *Republican* 1 (1819): xvi.
5   Vizetelly 1893, 68–72: the scene in 1831, when Vizetelly lived opposite.
6   Grimes 2000, 146–48; McCalman 1988, 122–23; Carlile 1832a, 342.
7   Wiener 1983, 17–23; McCalman 1975, v-vii.
8   *Republican* 6 (7 June 1822): 49. This racial imagery, playing on the anti-Semitic prejudices permeating society, was common among Carlile's circle. It can be seen

capitalize on wider social prejudices. Given the age's all-powerful sermon,[9] Carlile's onslaught on "Priestcraft" and "Kingcraft" were acts of disobedience, denying the church's authority by impugning Old Testament truths. Christianity was the law of the land and prop of polite society. Lack of this bulwark was unthinkable for most, and it was Carlile's brazen denial of the unthinkable that made him the notorious butt of scaffold moralizing. That he had a following made it even worse. But, in these times of social distress, there was real radical revulsion against the "bloated priesthood in the possession of those tithes which are the necessaries of life to the cottager".[10] Time and again, radicals echoed the sentiment, angry at the "plundering oppressors", the 18,000 privileged clergymen of the Established Church living comfortably off tithes (as Hetherington would say).[11] Because of Carlile, the word "infidelity" increased its currency in evangelical rebuttals from the 1820s, although, as he parried, it was a "mere word of cursing, abuse, and calumny".[12] Unlike a previous generation of radicals, Carlile had little time for labour demands and parliamentary reform, and certainly none for the growing unions and co-operation (quite unlike Saull). What he did was steer the radical cause to all-out blasphemy and put some fire in its veins.[13]

In this he *was* followed wholeheartedly by Saull. Not only did Saull castigate the clergy for emptying pockets but, even more so, for vacuuming brains. The local vicar of St Botolph was incredulous on learning from his parish reprobate that *"religion is a despotism*, reigning

---

    in his *Republican* and *Lion* (among both men and women, see Frow and Frow 1989, 49) and in the *Comet*. It became even more prominent in the 1840s with Charles Southwell. Yet, it is scarcely visible in Saull.

9   Young 1960, 12–13: "A young man brought up in a careful home might have heard ... a thousand sermons". These homilized and normalized every aspect of behaviour: Hilton 1988.

10   Carlile 1832a, 342.

11   Hetherington 1830; Hetherington [1832], vi. Another Saull associate, the Rev. Robert Taylor, would talk no less of the "Christian Priests and Bigots" robbing the hard working "of their reason and their substance which they did in this Country to the tune of Nine Millions Nine Hundred and Ninety Nine Thousands annually": HO 64/11, f. 167 (22 Nov. 1830).

12   *Republican* 7 (28 Mar. 1823): 397.

13   Epstein 1994, 68.

tyrannically over the human mind", and that no "evil genius that has ever existed" could have inflicted a greater "amount of human misery".[14]

That could have been Carlile, but it was Saull. Since so little is known of their relationship, all we can do is follow the money, as Saull converted his wine profits into heterodox capital. Ultimately, Carlile was imprisoned four times—in total he spent over nine years in jail, in "hell" as he put it, "preparing [society for] heaven".[15] This included his stretch in Dorchester gaol (1819–25), for publishing Paine's *Age of Reason* and the blind American deist Elihu Palmer's *Principles of Nature* (discussed below). Astonishingly, throughout these years he continued to edit the *Republican* weekly from his cell, which was some feat.[16] The Society for the Suppression of Vice tried in vain to shut his business down. His shop workers were incarcerated in Newgate. Here, they suffered shocking privations, being confined ten to a twenty-two-foot cell, forced to sleep on "*door mats*" and fed "*one pound of brown bread*" a day, plus a pint of gruel.[17] Carlile's wife and sister fared no better. They kept his 55 Fleet Street premises open and were themselves jailed. The government confiscated the shop's stock, in fact everything it "could put its harpy claws on". Being "reduced from comparative opulence to beggary" by the authorities,[18] Carlile desperately needed financing to keep the business afloat. How much Saull pumped in we do not know, only that he admitted to "many heavy pecuniary expenses" in aid of "free discussion" during this period. He later recalled an incident, though, during "one of the trials in the days of Carlile, in which he was responsible for the cost of the defence, in which the government withdrew a juryman at the last hour and involved him in the payment of costs".[19]

Saull was forking out frequently, given the succession of cases. Confirmation of this comes from a Secret Service source. A government spy had infiltrated Carlile's circle. His weekly reports to Bow Street Police Station give the feel of an old revolutionary who had been turned, but

---

14 [Saull] 1832a, 4.
15 Carlile 1832a, 342.
16 McCalman 1975, 78. Carlile had been jailed for publishing Paine and Palmer's works in his *Deist* in 1819.
17 *Republican* 9 (2 July 1824): 845; McCalman 1975, 76–78.
18 Carlile 1832a, 343–44; Wiener 1983, 70–72; Keane 2006.
19 *Reasoner* 16 (5 Feb. 1854) Supplement, 97–98.

his identity was cloaked in secrecy—he was referred to only as "*Him*" by his handler.[20] The mole was deeply embedded, clearly one of Carlile's assistants, always in his shop or at his meetings. He had Carlile's wife's confidence, to the extent that he was reading the letters Carlile wrote home while away.[21] Taken in by the family and by the shop assistants, the agent had intimate knowledge of their doings. He was vituperative in his reports, but these have to be treated carefully; they were, after all, designed to make his police handlers salivate. In selling information to the authorities, it pays to spice it up. Thus, the rheumatic, cane-supported Carlile he called "a very Wild and Extravagant Minded Person". "He is also Ferocious and I have no doubt but he is a <u>Calculating Bloodthirsty Person</u>".[22] With Carlile's imprimatur, the spy gained access to London's leading activists,[23] and from his briefings we get a deeper understanding of Saull's commitment. They show that Saull and his anti-Christian cadre would meet in Carlile's shop to plan strategies when blasphemy trials were pending.[24] And with Carlile again charged in 1830, Saull put up the bail to keep him out of prison until the trial.[25] Saull would frequently be heard talking at the Blackfriars Road Rotunda. Carlile had taken over this building in 1830 to make it the centre of London radicalism, where "a war to the death was to be waged against 'the aristocratical or clerical

---

20 HO 64/11, f. 350.
21 HO 64/11, ff. 63, 298. On the spies, see Parsinnen and Prothero 1977; Hollis 1970, 41–44. Sometimes the snitches were sussed, as on the occasion when one was fingered in the crowd at the Mechanics' Hall of Science in City Road: *Colonial Gazette*, 21 Apr. 1841, 252.
22 HO 64/11, ff. 3–4. The hyperbole was designed to impress the spy's handler, and it was probably the latter who underlined it. There were at least two spies operating. One was G. M. Ball, identified from later (1834) reports: HO 64/15, ff. 105, 107; HO 64/19, ff. 734–37. He was possibly a former Spencean revolutionary: see HO 64/11, f. 53 for his contacts going back to the Regency. Thomas Spence had been an agrarian reformer, who sought the expropriation of land and its hiring-out to small holders. An agrarian connection is also hinted at by Ball's membership of the Grand Lodge of Operative Gardeners in 1834. He recruited for the Lodge and was their delegate to meetings at Owen's Bazaar in April 1834: HO 64/15, f. 107; *Pioneer*, 26 Apr. 1834, 319; *Crisis* 4 (26 Apr. 1834): 23. Little is known about Ball: Oliver 1964, 83 n. 6. The other spy was Abel Hall, whose roots also went back to the Cato Street conspiracy: Parsinnen and Prothero 1977, 66–67. Hall was the vice-president of the local Tailors' Union lodge: HO 64/15, ff. 186, 198.
23 The Rev. Robert Taylor, Julian Hibbert, John Gale Jones, James Watson, Henry Hetherington, and many more: HO 64/11, f. 298.
24 HO 64/11, f. 75.
25 HO 64/11, f. 197.

despotism, corruption, and ignorance'". It was a venue that Saull helped keep afloat as a regular subscriber, even when others deserted it. In short, the nark reported, Saull had been one of Carlile's most consistent backers over the years.[26]

Another funder identified by the spy was Saull's ally, Julian Hibbert, a wealthy West Indies plantation heir expatiating his family's sins by bankrolling blasphemy causes. Saull and Hibbert were both, the spy reported, freethinking "men of property", which is what made them so dangerous in government eyes.[27] Hibbert was an avowed atheist, with a coruscating wit, always aimed at religious shibboleths. His was an austere life, involving temperance and vegetarianism, and he ended up in threadbare digs in Hampstead, all of which belied his wealth. The family's Jamaica plantation was run by 1,600 slaves and the profits had left them wealthy, genteel and landed, to the extent that they owned ships and quays, and financed the West India docks. Julian was duly Eton- and Cambridge-educated, picking up £10,000 on his father's death and as much again at the age of 25.[28] But, while his relatives became Church trustees, school governors, judges, and sheriffs, sharing the trappings of civic power, he worked with Saull on atheist propagandism. He would chip in with Saull to bail Carlile.[29] By 1831, the spy reckoned Hibbert had funded Carlile to the tune of £3000. Even that was probably an underestimate: Joel Wiener calculates that, by the end of Hibbert's life, £7000 might have been nearer the mark, showing the sort of sums needed to keep Carlile's Fleet Street press going in the face of state harassment.[30]

The epiphany moment for these 1820s' anti-clericals had often come on reading radical Enlightenment books, foremost among them one written by a deputy to the National Assembly during the 1789 revolution,

---

26 HO 64/11 f. 446 (29 Nov. 1831); "war": Wiener 1983, 164, 186 n. 2.
27 HO 64/11 f. 46 (Feb. 1828).
28 Donington 2014, 204, 224; E. Williams 1994, 88ff; Stange 1984, 48, 50, 170 on the family's Unitarianism. The family was massively compensated upon the abolition of West India slavery.
29 HO 64/11 f. 197.
30 HO 64/11, f. 446 (29 Nov. 1831); Wiener 1979. Hibbert would often lend Carlile money: HO 64/11, f. 67 (20 Sept. 1828). And when Carlile's house and furniture were sold in 1827 (HO 64/11, f. 17 [Sept. 1827]) and his books were being sold off cheap to recoup, Hibbert bought large stocks to give away to his friends: HO 64/11, f. 26.

Constantin François de Volney: *The Ruins: Or a Survey of the Revolutions of Empires*. From Saull's perspective, Volney's derivation of morality and virtue from nature's laws put science in a privileged position as part of the struggle, and he cheekily suggested theological novices read it.[31] Volney presented a typically radical Enlightenment view of nature's laws as beneficent and egalitarian. They were natural edicts constraining all, high and low, and the ground of ultimate authority, the highest court, beyond the jurisdiction of the clergy but available to everyman. A court sanctioned by Nature's immutable laws could overrule the capricious edicts of earthly tyrants. Carlile said *The Ruins* had started him on the road to deism, and "it has led thousands besides myself to search after truth."[32] He would pay back the debt by republishing it. If Enlightenment works had shaped Carlile, then Carlile equally reshaped Enlightenment works. He revamped them as the 'bibles' for a new deistic generation. The Word of Nature was spread with religious zeal by his followers—a group disparaged as "illiterate mechanics, silly fellows of weavers, beggarly lawyers" by detractors, but more sympathetically judged by McCalman to be "the respectable 'middling sort'—ambitious artisans, small shopkeepers and lesser professionals".[33] To this group, excluded from power and knowledge by an established culture, said E. P. Thompson, the "works of the Enlightenment came ... with the force of revelation."[34] They were liberating, leading one respectable 'middling sort' (Saull), snubbing his vicar's efforts at conversion, to announce in self-congratulatory style "I ... have a mind as free and unfettered [*sic*] as the air we breathe".[35] Such an exhilarating air surrounded all these anti-Christian cadres, as they cast off religious shackles and quoted chapter and verse from their Volneys, Paines, and Holbachs.

Unarguably, it was Paul-Henri Thiry Baron d'Holbach who was the greatest dissident inspiration. Being an extreme materialist, he had concealed his authorship in the eighteenth century, hence his works, including his monumental *System of Nature*, were usually attributed to "Mirabaud". Thus, Carlile was selling "Mirabaud"'s *System* in the

---

31 [Saull] 1832a, 13.
32 *Republican* 2 (18 Feb. 1820): 148. Palmer 1823.
33 McCalman 1988, 189–90; *Republican* 8 (18 July 1823): 52.
34 E. P. Thompson 1980, 798–99.
35 Saull 1828a, 21.

early 1820s,[36] and *Good Sense*, by "Curé Meslier", in 1826, although that, too, was by Holbach, his digest of the *System*. These libertarian pre-Revolutionary French books were seized on by Carlile and Saull as dissolvent and destabilizing in their own religiously backwards nation, fit to bring down Britain's *ancien régime*. Holbach demanded a re-grounding of social, political, and moral beliefs on 'rational' lines, and the subversive implications appealed to Carlile's anti-Church radicals. After all, if life for Holbach was an emergent property to be explained by matter acting deterministically, with no god needed to maintain it in motion, then the Church lost its authority, and man must "make one pious, simultaneous, mighty effort, and *overthrow the altars of Moloch and his priests*."[37] That was the nub for the anti-clerical Carlile and Saull: Holbach's admonition against waiting till the afterlife to redress the grievances in this one.

Saull, armed with Volney's and Holbach's "correct principles", and adopting a "fearless energy of mind", declared himself freed by "the complete eradication of all visionary fears, and superstitious ideas" to explore the more heretical scientific explanations of life.[38] And while social liberation for many marginal groups in the 1820s meant concentrating on the milksop self-help sciences, Saull would go to extreme lengths to develop a geo-astronomical explanation of life's ancestry. Such an approach was ultimately encouraged by his defiantly blasphemous context. Yet it was very far from the sort of science that he first encountered among the Carlile set, as we will now see.

Mankind Has Existed for All Eternity

Has the human species existed from all eternity, or is it only an instantaneous production of Nature? Have there been always men like ourselves? Will there always be such? Have there been in all times males

---

36 Thomas Davison's edition of "Mirabaud", *System of Nature* (1819), was bought up by Carlile in 1820, on Davison's imprisonment: McCalman 1975, 66, 219–21. For the century old lineage of Holbach's Enlightenment materialism and its social imputations, see Jacob 1981; Shapin 1980; Yolton 1983. Treuherz 2016 on Holbach's penetration of radical salons in eighteenth-century Britain, when his books reached a different audience from that aimed at by pirate presses in the early nineteenth.
37 Holbach [Mirabaud] 1 (1820), 185.
38 Saull 1828a, 23.

and females? ... Is this species without beginning? Will it also be without end? The species itself, is it indestructible, or does it pass away like its individuals? Has man always been what he now is, or has he, before he arrived at the state in which we see him, been obliged to pass under an infinity of successive developements? Can man at last flatter himself, with having arrived at a fixed being, or must the human species again change? If man is the production of nature, it will perhaps be asked, is this nature competent to the production of new beings to make the old species disappear?

Carlile's 1834 pocket edition of 'Mirabaud' (Holbach) *System of Nature*.[39]

Though men are seen to die ... the human species flourish in eternal being!

Carlile's 1824 edition of George Hoggart Toulmin's *The Antiquity and Duration of the World*.[40]

People in perpetuity was a strange concept to a pulpit age, an age steeped in stories of life being breathed into man, of corrupt birth, direction, hope, and redemption. And it was precisely this that attracted Carlile's materialists. It allowed them to sidestep the fundamental question of a divine genesis. That was a fable for "fanatics", said a *Republican* correspondent in 1823, who "swallow improbabilities ... wholesale" from the "Jew book": "a book composed of farce, fiction, and fanaticism, intermingled with tales of magic, morals and mystery", which "has enslaved all Europe" with the idea "of a God, making, contriving, or creating animal matter." Better to accept that humans and all macroscopic species were eternally existing, since they were composed of eternally-existing matter.

That correspondent was the outspoken and atheistical obstetrician James Watson. He derided talk of nature's 'design' by some incorporeal "manufacturer", and denounced the priesthood's effort to "defraud a credulous, puling, puerile, and idolatrous world". To him the key question was, "WHAT IS MAN and of what composed?" Since "the elements could never have had a beginning" and "man is a part of the elements", the solution to Holbach's riddle, and to the "difficulties thrown in his

---

39 Holbach [Mirabaud] 1 (1834), 75–76. This was touted as the "best translation": *PMG*, 20 Sept. 1834, 264.
40 Toulmin 1854 [1824], 46.

way by priests ... in assuming the creation of a *first* man", is to conclude "that, 'THERE WAS NO FIRST MAN'." Our species had no beginning.[41]

On this question, Carlile himself had initially been influenced by Palmer's *Principles of Nature* (1802, republished by Carlile in the *Deist* in 1819, and in book form in 1823). Palmer's solution was derived from common Enlightenment axioms. Since matter cannot appear *ex nihilo*, the earth must always have existed. But he did not entertain the idea of eternal *life*. For him the "vivifying influence of the sun" had originally produced the first animals from earthy matter. This raised the question, why is not it still doing so? We should "expect new beauties and wonders", but none are appearing. He thought that the earth's relation to the sun had reached an equilibrium (something Saull, pondering the same problem, was eventually to doubt), and, with this stabilization, the power to produce new life had dissipated. However, in earlier times, with the earth and sun in a different relationship, an "inconceivable exertion" must have occurred periodically to produce new life. For Palmer, "a graduated modification of physical energy has been exhibited through a past eternity" to generate the entire animal series "from man down to the lowest insect", all the life visible today[42]

Carlile played with the idea. In 1822, he argued that since "the power to produce anew would be equivalent to the annihilation of the existing species", it was no longer present, because species were not dying.[43] Initially, he, too, toed the Palmer line and assumed "that the first of all existing species of animals were organizations that resulted from some peculiar arrangement and compositions of matter". This was no advance, and, cajoled by Christian controversialists, Carlile could only plead ignorance "as to the origin of man or any other species". "The true Atheist", he affirmed, "holds no hypothesis about the origin of man;

---

41 *Republican* 8 (18 July 1823): 59; (28 Nov. 1823): 661, 666. Wickwar 1928, 225 incorrectly identifies him as Carlile's young shopman James Watson (1799–1874), who was at the time serving a one-year sentence (April 1823 to April 1824) in Coldbath-fields Prison for selling Palmer's *Principles of Nature* in Carlile's shop. The correspondent was actually a medical man, specializing in obstetrics, whose practice was in Brewer Street: *Republican* 8 (28 Nov. 1823): 655. Nor was he the Spencean apothecary "Dr" James Watson (who died in 1818), or his son, yet another James Watson, who allegedly once served as surgeon on a trawler (I thank Iain McCalman for the latter information).

42 Palmer, *Principles of Nature*, 53–55, appended to *Deist* 1 (1819); Palmer 1823, 53–55.

43 *Republican* 6 (11 Oct. 1822), 615–16.

nor is he ever in troubled doubt about that".[44] Constantly pushed, he constantly dodged, even as he argued what mankind was *not*: he was not conjured up by "an intelligent Almighty Power". That was only "an hypothesis to cover our ignorance". This was an answer straight out of Holbach. In *Good Sense*, which Carlile republished, Holbach answered his own question—"Whence comes man?"—in the negative: "I know not. Man appears to me ... a production of nature. I should be equally embarrassed to tell, whence came the first stones, the first trees, the first lions, the first elephants, the first ants, the first acorns, &c."[45] Not able to get anywhere, Carlile was left spewing out nihilistic *bon mots* to deflate mankind's spiritual majesty. As he declared from jail in 1823: animals (man included) "I look upon as a fungus springing out of the hot bed of change and corruption which exists on the whole surface of the earth".[46] Not so much an explanation as a materialist slap in the face.

But within months Carlile saw a better way to push beyond Palmer and Holbach. Mankind's eternity was largely predicated on the eternity of the earth (which, at least in an unchanged form, Holbach was not sure about[47]). The planet's everlasting existence was becoming a major arguing point. Palmer had accepted it. And Carlile was starting to dodge religionists' questions by affirming "that what you call the *world* never did come into existence, because it never was out of existence".[48] More poetic sources were pointing the same way. Grub Street materialists, armed with Carlile's pirated 1822 edition of Percy Bysshe Shelley's *Queen Mab*—itself indebted to Holbach and Volney—were encouraged by Shelley's concurrence on the "eternal duration of the earth".[49]

Further support was given to eternalism at this moment by the Norwich shoemaker Sampson Arnold Mackey's three-part *Mythological Astronomy of the Ancients Demonstrated* (1822–23), a book that came quaintly tied with cobbler's thread. Mackey was an extraordinary autodidact, influential in Saull's emerging understanding of planetary

---

44 *Republican* 8 (11 July 1823): 17; (18 July 1823): 52–53; (3 Oct. 1823): 397.
45 Holbach [Meslier] 1826, 17.
46 *Republican* 7 (27 June 1823), 822.
47 Holbach 1 (1820), 89–92, would not rule out the ongoing production of new beings. Nor was man exempted: new planetary conditions will require humans either to change or become extinct.
48 *Republican* 7 (28 Mar. 1823): 400.
49 *Republican* 8 (18 July 1823): 58.

history. Mackey had incarcerated himself in an attic to "penetrate the Mist", that is, decode the ancient Indian, Persian, Babylonian, and Hebrew texts for their astronomical indications, only to pay for such selfless dedication by ending up penniless in an almshouse. He had reasoned back from Hindu Scriptures only some seven or eight million years, but that was near-eternal enough in Grub Street.[50] Joscelyn Godwin's introduction to Mackey, in her *Theosophical Enlightenment*, reveals that Freemasons loved the book, but Mackey spurned them and stubbornly remained in Carlile's camp. And, while she has Mackey first in London in 1830,[51] secret service reports show that he actually made a trip to the capital in 1828, when Carlile was his constant companion. Carlile arranged for him to lecture in the City on astronomy "on a higher scale than as yet has been known",[52] and sold the *Mythological Astronomy* in his shop.[53] It was an extraordinary book, equally admired by Saull.

The eternity of the elements and earth was hotly debated while Carlile was incarcerated. A thousand turned up in one Leeds hall to thrash out the subject with Christian protagonists in 1823, only to have magistrates break up the "illegal" meeting.[54] So, when Carlile heard of an obscure book belabouring the point, not only of planetary eternity but of *human* eternity, he jumped at the chance of putting it back onto the street in cardboard covers. It had been penned forty years earlier by a hotheaded young doctor, George Hoggart Toulmin, in flaming Holbachian fashion. Dr Toulmin, in fact, had sent forth a string of re-vamped editions, starting with *The Antiquity and Duration of the World* (1780), hastily published the year following his medical doctorate at Edinburgh, when he was still only 26. In it, the world's discordant creation myths were shattered by the world itself: less from the untrustworthy records of ancient civilizations, more from time's immensity deduced from the rocks. Proofs were piled up: of the depths of cliff-face fossils, of petrified human remains in Gibraltar, the evidence of endless fluctuations of land and sea, of risings and fallings, of oceans as much as civilizations, of alternating warm and cold regions through time (shown by tropical

---

50 *Republican* 8 (12 Sept. 1823): 296; Mackey 1827 [1822–24], 33, also 201, 238, 263; J. Godwin 1994, 68.
51 J. Godwin 1994, 75.
52 HO 64/11, f. 92 (1828).
53 *Lion* 1 (27 June 1828), 804.
54 *Republican* 7 (25 Apr. 1823): 538.

elephants "transmuted into stone" in Flintshire and crocodile fossils in Derbyshire), and so on. Then there were the umpteen volcanic layers of lava around Etna, interspersed with layers of soil, each known to have taken hundreds of years to accumulate, and the depth of limestone beds, themselves the wrecks of endless empires of shellfish, compacted over the aeons by slow degrees. For Toulmin, climate and life on this age-old earth showed no overall directionality, no irreversibility, no extinction. Continuity, succession and fluctuation marked a self-sustaining system, the result of "laws fixed and immutable". And as an inseparable part of the planet humans must have had an equally "uniform and infinite existence". He stated it "without the shadow of hesitation": all life was timeless. During the planet's "endless periods of existence", life had persisted and all the while "the human species evidently must have been present".[55] Its population must have continually risen and fallen, as it ceaselessly recycled through phases of barbarism and civilization.

Such views might not have been uncommon in student Edinburgh. Toulmin had been studying in an Enlightenment hot-spot. The cosmopolitan university, the best in Britain, welcomed European students who infused their own Voltairian heresies. The democratic "Friends of the People", which met in Edinburgh in 1792, sometimes in Freemasons' halls, clearly had social roots in the 1780s, when mechanics could decry the *"purple and fine linen"* of a wealthy elite living off the back-broken "beast of burden".[56] The King's Birthday Riots of 1792 and plethora of inflammatory handbills speak the same. Even though modern work shows few Freemasons espousing radical views,[57] we do know that, for example, the 'Burke and Hare' anatomist Robert Knox's father, was a "leading Freemason" in 1780s' Edinburgh, a Holbachian and a supporter of the Revolution.[58] And European freemasonry, at least, as Margaret Jacob has shown, had a history of "pantheistic materialism" and contingent links to republicanism.[59] Anyway, from the heady

---

55 Toulmin 1854 [1824], 15, 37–38, 71.
56 Brydon 1988, 48, also 131–146 on the Friends and Edinburgh at the centre of Scottish radical activity in the 1790s.
57 M. C. Wallace 2007, 153; Jacob 2019, 124, 126, 134.
58 Lonsdale 1870, 3–4.
59 Jacob 1981, 225. Another fact suggesting that such geological views might not have been uncommon was the Edinburgh savant James Hutton's *Theory of the Earth* (1795), with its 'no traces of a beginning, no prospect of an end' theme. But, crucially, Hutton privileged mankind: as the chosen species, he had made a recent

Holbachian brew bubbling in the student underworld, soapmaker's son George Toulmin had distilled out *human* eternalism, a conclusion which made him so hated, as Roy Porter has shown. Defilers thought the "rack and gibbet" was too good for him.[60] Like Holbach, Toulmin saw the individual as an unprivileged production of nature, who "follows general and known laws". But, unlike Holbach, who thought the origin of humanity so unfathomable that "it cannot interest us",[61] Toulmin went straight for eternity. Less nuanced than Holbach, and more pointed, he seemed made for the 1820s' tub thumpers.

Antiquity and Duration was "an overt gesture of political radicalism" in geological dress.[62] It was subsequently modified as *The Antiquity of the World* (1783), then rebranded as *The Eternity of the World* (1785). One final, massive reworking left it as *The Eternity of the Universe* in the revolutionary year 1789, whereupon Toulmin dropped the subject, unsurprisingly, given events across the Channel and the British clampdown.[63] Nothing new existed under Toulmin's sun, and there was little new in this puffed-up edition, except that the "*unlimited* existence of the human species" was now foregrounded as if to suggest it had become *the* singular selling point.[64]

Four editions indicated an audience in febrile Enlightenment days, but, given the repressive years subsequently, the books had sunk into obscurity.[65] Damned as atheistical in their day, the books achieved pariah status among the geological gentry. Pious hammerers had rather stuck to empirical goals and shunned the cosmic question of origins. Such a blinkered attitude among chemists and astronomers had led Carlile to publish his blistering *Address to Men of Science* (1821). Here he called the scientific gentlemen cowards and demanded they come out

---

    appearance on an earth prepared for him, as "the apex of God's creation" (R. S. Porter 1978a, 345). Hutton's intent was diametrically opposed to Toulmin's assault on a "gloomy" theology's "Gothic barbarism and superstition" (Toulmin 1854 [1824], iv).

60 R. S. Porter 1978a, 439; R. S. Porter 1978b.
61 Holbach 1 (1820): 88.
62 R. S. Porter 1978a, 436.
63 Apparently, Toulmin recanted his 'atheism' after the Revolutionary Terror, not that Carlile knew it: R. S. Porter 1978a, 449.
64 Toulmin 1789, 53, also unpaginated second page of the Introduction, 9, 225, 229.
65 A fact commented on by G. F. Richardson 1842, 40, who put it down to Toulmin's scepticism.

as materialists,[66] an impossible request in a pulpit age, when, as Secord says, "The political authority of science was grounded not in doctrines of matter and natural law ... but in expert knowledge vouchsafed by an ideology of genius and divine inspiration."[67] Of course Carlile was whistling in the wind, or simply agitating the apothecaries and mechanics in their struggle against hospital consultants and work-place masters.[68] Mankind with his immortal soul was hived off as a special case by the scientific gentry (and many a pious fossilist). For them, humans were not amenable to physico-chemical explanations or eternally-operating geological ones. In a word, the devout damned Toulmin's books for the reason Carlile praised them—because belief in a human eternity on an uncreated earth would strip any divine rationale from the existing scientific and political hegemony.

Toulmin shocked genteel folk in the 1780s as much as geological gentlefolk in the 1820s. The Harley Street geologist Charles Lyell was the epitome of decorum. For him, like almost all his fellows, man was a moral being above geological explanation. Lyell, in the 1820s, knew about Toulmin but could never mention his name in print. Toulmin was a revolutionary wrecker, vandalizing cherished traditions, the geological equivalent of the British troops recently caught desecrating Burmese pagodas for trophies.[69] For looting Buddhist relics read smashing Christian idols, and it was Toulmin's brazenness that shocked Lyell. But such vandalism made sense to Carlile. Toulmin's pre-Revolution saleability and threatening posture meant that his books could have a flourishing afterlife in Carlile's urban underworld, which targeted such idols. Carlile needed this panegyric on Revolutionary geology for his list.[70]

From his prison cell, Carlile put out feelers for Toulmin's pantheistic books. They were rare, having been "in a suppressed state for nearly

---

66 Carlile 1821.
67 J. A. Secord 2000, 312–13.
68 Carlile's *Address* was penned in the wake of the William Lawrence case (see below), when one radical surgeon did 'come out', and suffer for it egregiously. The Lawrence episode could have been Carlile's catalyst.
69 J. A. Secord 2014, 159; 1997, xxvii; Wennerbom 1999, 43; Shortland and Yeo 1996, 23; K. M. Lyell 1 (1881), 174.
70 His list included Palmer's *Principles*, and by this point Carlile was also selling the jailed Thomas Davison's stock of Holbach's *System of Nature*.

forty years".⁷¹ By February 1824, he had *The Eternity of the Universe*, all but the opening, and correspondents supplied the missing pages after he put out a request for hand-written copies.⁷² In May, he sent his Fleet Street compositors the *Antiquity and Duration of the World*, which they compacted into fifty-four pages, and it still made a fast, titillating read. Carlile proudly put his preface, ostentatiously marked "Dorchester Gaol", to the finished *Antiquity* that September. What appealed was clearly Toulmin's leap, making planet and mankind coeval, which obviated any need for the "phantasmal aid of supernatural power". But it equally reinforced Carlile's steady-state, anti-origin mantra: all superficial changes on the earth's surface are balanced, with subsidence matching mountain building, with strata accumulating at one time and eroding at another, the sea invading here and retreating there, and this for all time. For Carlile, following Palmer and Toulmin, there was no progressive generation of species, no advancing sequence of productions. A balancing equilibrium became the apostle's creed. The earth's "self-regulating power" operated in perpetuity, and no designing God was needed "to superintend its changes".⁷³

The shilling *Antiquity* was already on sale in early October 1824.⁷⁴ The *Eternity of the Universe* was a larger book. It took longer to typeset and only appeared in August 1825. Although costing one and sixpence, it was still "as cheap a book, in point of worth, as was ever published," and recommended for young, uncorrupted minds.⁷⁵ Copies were eventually marked down to sixpence, and the title remained on Carlile's list for years among incendiary works by Paine, Volney, Palmer, and Holbach.⁷⁶ These pressings gave the obscure Toulmin a new exposure, and, for decades, all new printings were based on Carlile's editions, often carrying his preface. In the mid-twenties, they augmented the ideological armoury. Within months of publication, Carlile's imprisoned shopmen were quoting it in their deistic diatribes against anti-infidel

---

71 Carlile's preface to Toulmin 1854 [1824], vii.
72 *Republican* 9 (27 Feb. 1824): 259–60; 9 (7 May 1824): 605. Either Carlile or his compositor was still unsure of the obscure author, because the call initially went out for "Tailmin", only to be corrected later, ibid., 288.
73 Carlile's preface to Toulmin 1854 [1824], v–vii.
74 *Republican* 10 (8 Oct. 1824): 447.
75 *Republican* 12 (19 Aug. 1825), 224.
76 Carlile's list in *PMG*, 8 Dec. 1832, 640.

preachers.⁷⁷ A copy went into the London Mechanics' Institution library in 1826, not surprisingly, given the radical intake—Saull, Hetherington, and Carlile's shopman James Watson being active members at this time.⁷⁸ The darkling Toulmin suddenly found himself up with the heretical greats in the deists' pantheon. Even "Rule Britannia" was re-versed in radical chants to accommodate him:

> Nor British heroes lag behind—
>   Here Thomas Paine received his birth!
> Himself, an army of his kind,
>   And long may Britons boast his worth!
> ...
> Toulmin and Shelley lend their hand,
>   And many more deserve applause;
> Sages are rising in this land—
>   They rise to teach men Nature's laws!⁷⁹

In 1828, a spy reported that some two or three hundred men and women (Saull included) were still listening to Sunday sermons in one blasphemy chapel on "Toulmin's Duration and Antiquity of the world".⁸⁰ Indeed, into the thirties and beyond, Toulmin could be bought off-the-shelf at the usual radical and co-operative outlets.⁸¹ Not only did the books have a long shelf-life, but eternalism—of the earth, of species, of humans, with no birth and no cessation, no direction, no progression, and "no comprehender, much less Creator"—remained a

---

77   *Newgate Monthly Magazine* 1 (1 May 1825): 418.
78   *LMR* 4 (9 Sept. 1826): 313; Anon. 1833, 49. The latter also showed that the LMI held Volney's *Ruins*, (p.32) and works by Voltaire (pp. 16, 32). The scientifically well-stocked library of the LMI gives the lie to the notion that all mechanics' institutions libraries were "frivolous" and full of "fiction", an analysis also challenged by Walker 2013. On the radical strength inside the LMI, see Flexner 2014.
79   *Gauntlet* (23 June 1833): 319.
80   HO 64/11, f. 85.
81   At James Eamonson's shop in Chichester Place, John Cleave's in Shoe Lane, B. D. Cousins in Duke Street, and many more. It could also be had at the usual shops for seditious prints in other cities, at James Guest's in Birmingham, Joshua Hobson's in Leeds, Abel Heywood's in Manchester, and Thomas Paterson's in Edinburgh. Incredibly, Toulmin was still selling at the time of Saull's death, with Goddard's shop in London's John Street offering a combined edition of Toulmin's and the Rev. Robert Taylor's work for 2s: *Reasoner* 17 (1 Oct. 1854): 223.

potent trend. It was pushed as late as 1837 by Benjamin Powell in his *Bible of Reason*.[82]

But, as an anti-Creative *geological* stratagem, "eternalism" had become untenable no sooner than it had been published. Even though Carlile, in 1832, was still pushing Toulmin's books as "the best elementary treatises on this subject", for showing an "antiquity beyond calculation",[83] his equation of antiquity and eternity was no longer viable. This very year—1832—saw the death of the great Parisian palaeontologist Baron Georges Cuvier, and it was largely his fossil evidence for the rise of life that was already undercutting Toulmin's eternal balance, even at street level.

Blasphemy, Piracy and the New Science of Origins

> ...whence, in the first instance came man, and all the other superior animals? Now, most assuredly, they did not drop from the sky, and I need not say, they were not formed by the *Elohim*, or any other mythological gods; it follows, then, that they must have been ushered into existence, by the exalted generative powers of the earth...
>
> A letter writer to Carlile's new journal, the *Lion* (1828), offering a heretical exegesis of Cuvier's fossil geology.[84]

Geologists made breathtaking strides in the 1820s. In his definitive *Bursting the Limits of Time*, Martin Rudwick details how, between 1816 and 1825, the "Tertiary" era was established from the upper rocky layers of the earth's crust.[85] By then "Primary" and "Secondary" terminology was already common currency among the cognoscenti. The three eras were sequential and housed increasingly "higher" life forms, strange ones at times. The Secondary cliffs of Dorset and Somerset were revealing marine reptiles, ichthyosaurs and plesiosaurs, in the 1820s.[86] The first huge land living saurians were being disinterred. Nothing alive appeared remotely similar to some of these animals. The

---

82 B. F. Powell 1837, 2: 1. This started as twopenny numbers published by Hetherington and ended as a triple decker book in 1837–39. It was extracted in *NMW* 2 (16 July 1836): 298; 2 (3 Sept. 1836): 358.
83 Carlile 1832b, 371.
84 *Lion* 1 (6 June 1828): 734.
85 Rudwick 2005, 543.
86 M. A. Taylor 1994.

weirdest was what Baron Georges Cuvier in Paris would call a "ptérodactyle". Its structure was so unfamiliar that it had been a real bone of contention: was the animal bat-like, but with a wing on a single finger, a hairy cross between birds and mammals, or a flying reptile as Cuvier thought?[87] When another fossil skeleton turned up in Germany in 1824, Cuvier, finishing up the second edition of his magisterial seven-volume *Recherches sur les Ossemens Fossiles de Quadrupèdes* (1821–24), called it definitively a reptile.[88]

In these volumes, Cuvier famously described a hundred archaic fossil mammals from the Tertiary beds in Paris. Such a cavalcade astonished the popular press. No fewer than seventy represented "species most assuredly hitherto unknown to naturalists", and forty of those belonged to new "genera, or a different order of beings from any that now exist, which is quite a different thing!", said the *Cheap Magazine*'s editor George Miller in his *Popular Philosophy* (1826).[89] What "is more surprising still", added Miller, those lost had been replaced on the "busy stage of life". The directional progress of life, with "lower" forms departing forever and being replaced by "higher" ones, were major breaches in Toulmin's eternalist dam. The end result was a sequence of fossils that seemed to show a trend "upwards". Given romantic assumptions, unquestioned and uncontroversial, of man as the apotheosis, the "highest" type of life,[90] this series of forerunner animals could be seen progressing, aspiring, pushing or being pushed higher, until humans were created. Life had appeared as invertebrates and fish in the lower (and older) rocks, reptiles in the middle (Secondary) strata, and then archaic mammals as the Tertiary opened.

Of course, this new geology was arcane knowledge, buried away in learned journals or the chatter of expensive societies. Inaccessibility meant it made little impact in blasphemy circles to start with. Indeed, the fossil specifics were never of interest to most ideologues, except

---

87 As *Ornithocephalus*, it was already being described in the Edinburgh professor Robert Jameson's 1822 translation of Cuvier's introduction to *Ossemens Fossiles*, which Carlile's shopmen were known to be reading.
88 Taquet and Padian 2004.
89 G. Miller 1 (1826): 295. On the Victorians questioning whether extinction was still occurring, see Cowles 2013; Urry 2021.
90 For the tangled relationship between biological and social progress, and their morally-loaded effects in 'natural' ranking, manifesting in 'higher' and 'lower' beings, see Ruse 1996, Bowler 2021.

Saull. But the tenor of Cuvier's argument *did* permeate the underworld in the later 1820s, albeit arriving through indigenous, pirated, and more trusted channels.

That old doyen of radical studies, Simon Maccoby, once speculated that Cuvier's mention of a deluge as the destroyer of the old world was a reason "why the 'infidelity' of the streets was not yet using geological arguments in 1830".[91] Maccoby's reasoning was this. For Cuvier, the successive rock layers indicated the arrival of new environments—he called the turnover 'revolutions'—each complete with new species. And because Cuvier's last revolution, which he himself thought of as some local marine invasion, was associated so strongly in Britain with the biblical Flood,[92] it was unusable by radicals. But access to a wider range of subversive prints shows that some activists *were* using Cuvier as a cannonade. In fact, Carlile's imprisoned assistants in 1825 were quite able to extricate Cuvier's last revolution from Mosaic explanations. In contrast to the Flood of the "christian geologists", the Newgate-jailed infidels noted that Cuvier spoke "of *the small number of individuals of men and other animals that escaped from the effects of that great revolution*", so it was no universal inundation. Moreover, the strata were not jumbled higgledy-piggledy, as the detritus of a ravaging Flood, but lain in regular succession. This was evident, they said, quoting Toulmin, from the fossil creatures seeming "to be in the places where they have been generated, lived, and died". Fossil oysters and cockles were "deposited with as much regularity as beds of living shell-fish are in any part of the sea".[93]

But activists really only took up Cuvier when he was introduced by an accredited source, that is, a blasphemous hero. Trust played as big a role at the bottom of the heap as the top, where it has been brilliantly depicted in Steven Shapin's *Social History of Truth*. For the thinking dispossessed, credibility was a key issue, and Carlile's practical maxim, 'my enemy's enemy is my friend', served to assess it. Just as Shapin's gentlemen relocated "conventions, codes, and values of gentlemanly

---

91 Maccoby 1955, 459–60.
92 Because of Jameson's commentary on Cuvier's introduction to *Ossemens Fossiles*, which he titled "Theory of the Earth": Rudwick 2008, ch. 6; 2005, 556; 1972, 111–112, 133–35. On Jameson's intent on rendering his translation palatable to Edinburgh's Presbyterians: Dawson 2016, 48–54.
93 *Newgate Monthly Magazine* 1 (1 May 1825): 420–21.

conversation" into their philosophy,[94] so 1820s' infidels no less enhanced the dissident values in their swiped science. This took on the mantle of 'truth' in their tight-knit community because it was shared, soothing, and justified their rebellious action. Originally, Carlileans paid scant attention to the notices of bizarre reptiles and cave faunas that were beginning to figure in the proliferating trade journals. Many of these were, admittedly, second-hand press cuttings, no more. They might have been yearly round-ups of scientific snippets in the *Arcana of Science and Art* or an occasional report of the giant reptile *Iguanodon* from Tilgate Forest or of cave hyaenas found in Kirkdale.[95] These clippings focussed mostly on the odd and dramatic antediluvian finds guaranteed to awe. Typical were "Footsteps Before the Flood" (supposed tortoise tracks in ancient sandstone), vertebrae equal to the circumference of a human body, fossil lizards projected at 150 feet long, or fossil possums turning up in the Paris beds[96]. Tantalizing titbits, but their innocuous usage left them unnoticed by anti-Christian activists. What really swayed radicals was the imprimatur of the condemned. These were safe and sure sources, and the first to be exploited was Lord Byron.

Byron's poetic work *Cain* (1821) had been savaged by reviewers, who cried 'blasphemy'. As a result, the respectable publisher John Murray could not legally protect it from piracy, because blasphemous works were not copyrightable. The result was a plethora of pirate editions flooding the market. London's radical presses rushed to outdo one another. They were egged on by the knowledge that the King hated *Cain*, which Carlile thought a higher honour for Byron than a peerage.[97] Carlile, with his "dismally utilitarian" tastes,[98] cared little for poetry and less for Lord Byron, except as an irritant gnat on the Vice Society's rump. And the

---

94   Shapin 1994, xvii.
95   E.g. *Arcana of Science and Art* 1 (1828): 136–39; *Register of the Arts and Sciences* 2 (25 Dec. 1824): 142–43; *Gill's Technological Repository* 4 (1829): 189–90; *London Journal of Arts and Sciences* 5 (1823): 118; 9 (1825): 212–13; *LMR* 1 (18 Dec. 1824): 104–05; also 1 (12 Mar. 1825): 313; 4 (15 July 1826): 182, for cave fossils mentioned in George Ogg's lectures on geology at the LMI (which Saull attended in 1826, see Appendix 1).
96   *Arcana of Science and Art* 1 (1828): 105, 138; 2 (1829): 191–92; *MM* 9 (2 Feb. 1828): 15; *London Journal of Arts and Sciences* 2nd ser. 1 (1828): 53–54.
97   *Republican* 5 (8 Feb. 1822): 192. Wiener 1983, 62; Wickwar 1928, 269–70. Johns 2010 on piracy's long and unrespectable history.
98   Rose 2002, 35.

calls for Byron's prosecution were enough to have Carlile and the equally notorious blasphemy publisher William Benbow (who financed his own activism by selling pornographic prints[99]) competing to get copies out. The militant Benbow, a former shoemaker, was first away, as he always was with anything offensive to refined noses. His pirate edition in 1822 sold for 1s 6d. Pipped by his rival, Carlile undercut the price, pushing out a double-columned, small-type, sixpenny pamphlet later in April 1822.[100] By 1826, a fusillade of pirate editions had hit the stands. No fewer than five were jostling for place in the bookshops. "Poetry as cannon-shot", one press historian called it.[101] The piracy was designed to push their offended lordships into prosecuting one of their own, to widen the front. But what it actually did was push Cuvier to the fore.[102]

Cain had not merely mooted the immensity of time in Cuvier's lost worlds, it actually fingered Cuvier as the inspiration in the short preface. And, outrageously, it put the gory talk of the successively wrecked and remade planet, "before the creation of man", into the mouth of Lucifer. Satan flew back through misty time to expose a wide-eyed Cain to "The phantasm" of ancient worlds, "of which thy world Is but the wreck." Death had laid waste the empires of ancient life and rendered the "Mighty Pre-Adamites" so much mouldering clay, from which man might arise.[103] The "Mighty" included lost races of men no less than beasts—pushing poetic licence to its limit. It was as if Cuvier had spoken through Lucifer, and his beguiling portrayal appalled Byron's friends and foes alike. Cuvier's "desolating" conclusions caused a "deadly chill", infecting those who would otherwise "trouble their heads but little about Cuvier".[104] That alone raised Cuvier's stock among Carlile's circle, which happily endorsed *Cain*'s "ponderous blow at superstition".[105]

Byron had let the fossil cat out of the bag. *Cain* cast a rather glum eye on past immensities, as life fought through successive worlds on its way

---

99 McCalman 1984; 1988, 155–70, 205–12.
100 *Republican* 5 (15 Mar. 1822): 342–43; on sale: 5 (5 Apr. 1822): 448; reprinted, 6 (14 June 1822): 96.
101 Wickwar 1928, 259, 272; McCalman 1988, 211. The rival editions were Benbow 1822, Carlile 1822, H. Gray 1822, B. Johnson 1823, W. Dugdale 1826, while Benbow published a new edition in 1824.
102 O'Connor 2008, 104.
103 Byron 1822, vi–vii, 35, 49.
104 T. Moore 1854, 5: 321–22.
105 *Republican* 5 (8 Feb. 1822): 192.

to the present. No one could now avoid Cuvier, gloated a correspondent in Carlile's *Republican*: "Cuvier who hath re-engraved and illumed the illegible tablets of time, whose characters had been erased and darkened by the destructive hand and Cimmerian gloom of oblivion".[106]

Yet Byron was only a part of the piratical endorsement of Cuvier. The surgeon William Lawrence sat equally (and uncomfortably) in the radical spotlight. His was a carbon-copy case: a loss of copyright with a torrent of pirate editions, all appearing within weeks of *Cain*. Lawrence was the new professor of anatomy and surgery at the Royal College of Surgeons. But, in his first course, published as *Lectures on Physiology, Zoology, and the Natural History of Man* (1819), he had rashly excoriated his teachers for clinging to a belief that life depended on a divine vivifying power—that matter was animated by a vital principle, just as a soul animated man, and that such things must be publicly known to keep society "virtuous".[107] For Lawrence, life was a function of organization. It was an emergent property, appearing naturally, and needed no other explanation than the laws of physics and chemistry. But this was an inopportune moment to express materialist views, embedded in lectures which chafed at religious creeds and underscored republicanism. Within months, the Peterloo 'massacre' showed the authorities' intolerance, as 60,000 protesters were cut down by Huzzars in St Peter's Fields in Manchester, leaving eleven dead. Amid the heightened tension, Lawrence was slammed as socially irresponsible. With no soul, and no future rewards or punishments to keep the masses in check, what was to stop them from revolting? The Tory *Quarterly Review* revealed its cynical view of these "masses": unchecked, the effects of Lawrence's teachings would be "to break down the best and holiest sanctions of moral obligation, and to give a free rein to the worst passions of the human heart".[108] Lawrence found his motives questioned, by the imputation that he was removing social restraints. Even worse were his flippant protests, that the soul cannot be found "amid the blood and filth of the dissecting-room" and that no vital

---

106 *Republican* 11 (11 Feb. 1825): 163.
107 Lawrence 1822, 4–10.
108 [D'Oyly] 1819, 33.

spark can "impose a restraint upon vice stronger than Bow Street or the Old Bailey can apply."[109]

Lawrence was suspended from his post at the Bridewell and Bethlem hospitals and had to recant before the governor.[110] The *Quarterly* demanded his sacking from the College of Surgeons. It was all too much for Lawrence. He expediently withdrew his *Lectures* from sale and, in 1822, sought an injunction to stop the book being pirated. But the *Lectures*, being blasphemous, was refused copyright in the Court of Chancery.[111] With that, the pirate presses saw his compendious natural history of man as up for grabs.

Benbow again competed with Carlile. Lawrence is "coming out in all sizes and at all prices", said Carlile in April 1822, within days of his *Cain* appearing. His own octavo 3d sheets of the *Lectures* went on sale on 12 April 1822. They joined three other editions in 1822 alone: the J. Smith edition being run off down the road at 163 Strand; Benbow's smaller type octavo at 4d a sheet; and Griffin's in tiny duodecimo for 4d, word-for-word the best value. Faced with that, Carlile promised that "If the demand be so great as is expected", he would "print a very small edition in the cheapest and most compact form" to undercut the lot.[112] But Carlile's main selling point was that his edition also included Lawrence's 1816 lectures, besides those of 1817–1818 found in Smith's and Benbow's editions. Carlile's book was fatter.

With that, Lawrence's *Lectures* became another radical bible, henceforth to be found on every unrespectable bookshelf (and some respectable ones: even Charles Darwin owned a Benbow edition in boards[113]). Spies relayed how the book was occasionally read from an anti-Christian podium, "as the lesson for the evening", before

---

109 Lawrence 1822, 4–10; Jacyna 1983a explains Lawrence's College teachers' vested interest in upholding John Hunter's vitalist views.
110 W. Lawrence to Sir R. G. Glynn, 16 April "1832" [i.e. 1822], Royal College of Surgeons MS Add. 194. The letter was published alongside Galileo's recantation in *Republican* 6 (2 Aug. 1822): 317; Epstein 1994, 127–8; C. W. Brook 1943, 26–34. Lawrence's retraction, however, did not stop him from subsequently penning anonymous leaders in the *Lancet* denouncing the nepotistic elite at the College of Surgeons (Desmond 1989, 117–21).
111 *Republican* 5 (26 Apr. 1822): 538–39. Lawrence praised the "greater courage" of William Hone, himself prosecuted for his Lord's Prayer satires (Temkin 1977, 357).
112 *Republican* 5 (12 Apr. 1822): 465; (26 Apr. 1822): 538. Goodfield-Toulmin 1969, 307–08.
113 Desmond and Moore 1991, 260.

a blasphemy lecture.¹¹⁴ Lawrence's stature accordingly rose in the underworld as it sank above. He was treated as an unimpeachable authority. Saull, for example, when challenging his own vicar, quoted Lawrence on the absurdity of "Jewish Scriptures" in the light of modern geology or astronomy.¹¹⁵

Lawrence was one of those rare surgeons au fait with Continental science. French, German, and Italian sources littered his work, but it left him in the wake of Waterloo being portrayed as a turncoat. By not supporting Britannia's backwards view of immaterial vitalism and 'design' justifications of Creation, he was being unpatriotic. But then he *had* seen it as his mission to drag Britain into the modern world, and, as such, he had given the French savants their due. His *Lectures* acknowledged the turnover of fossil species, with new replacing old, of ancient rocks housing extinct types, of alternating strata laid down in fresh and salt water "indicating successive revolutions in the earth's surface". The image was one of advancing "approximation to our present species". The labours "of CUVIER, BRONGNIART, and LAMARCK, in France" had taken us beyond "the reach of history and tradition". Even more, they gave ground for "curious speculation respecting the extinct races of animals and the mode in which their place has been supplied by the actual species of living beings". In short, Cuvier was being "highly extolled" in a trustworthy source.¹¹⁶

French fossil zoology thus threw new grist into the freethought mill. After the mid-1820s, it was becoming difficult to admit that species were eternal. Long letters in Carlile's *Republican* took to arguing the point. In 1826, one saw Carlile's periodical as probably the last refuge of the "eternity of man ... advocates". The logic of eternal elements meaning eternal animals was finding fewer supporters, with geologists now showing that waves of rearrangement, extinction, and reconstruction had led to the progression of ancient life. Today's humans, one disputant said, would actually have found the earth uninhabitable when the early rocks were forming. And the new anti-Christian logic suggested that,

---

114 HO 64/12, f. 180 (27 Nov. 1832). This records a reading of Lawrence on humans existing "without the assistance of a first cause as the Superstitious Nonsense of the Clergy dictate", preceding J. E. Smith's "Antichrist" lecture at the Rotunda.
115 Saull 1828a, 10.
116 *Lion* 2 (4 July 1828): 30; Lawrence 1822, 5, 46, 48–49.

when man finally appeared, "he must have descended from some stock that had lately been formed from the energies of nature".[117]

The *Republican* folded in 1826, and by the time Carlile started up the *Lion* (January 1828) the tide had turned. Life's rise was the chatter in infidel chapels and coffee rooms before a Sabbath lecture. Orators would parade the growth from early "imperfect" life to today's creatures to show "that nature is progressive in the bodily as well as in the mental formation".[118] Cuvier's authority was accepted as recording life's gradual ascent. The first fishes were succeeded by "improved" amphibians and reptiles which had exclusive occupancy "for a considerable period", whence the dry land saw the birds and mammals emerge, "till, at last, the earth, by an effort or change, surpassing all his former ones, produced man". That said, these nihilistic republicans still believed that today's humans probably only constituted "a first and imperfect attempt towards the production of a class of rational beings."[119]

Quite technical matter entered the *Lion*'s pages. In 1829, Carlile cribbed a piece from one of the City's newer literary papers, the *London Weekly Review*, analyzing the findings of the rising 28-year-old French fossil botanist Adolphe Brongniart (who was to become the professor of botany at Paris Muséum d'Histoire Naturelle in 1833 and a Saull correspondent). It depicted in detail the earth's successive botanical "epochs" between the ancient ferns and modern flowing plants,[120] and correlated each with temperature changes and the peculiar animal life at the time. The latter was, of course, the discovery of Cuvier, Brongniart's colleague. What is telling is that most of the talk in Carlile's journal remained about Cuvier's mammal fossils. By and large, any discussion of English plesiosaurs and ichthyosaurs filtered back from these French sources. With Byron's and Lawrence's indicted imprimatur, only Cuvier was really trusted.

---

117 *Republican* 14 (11 Aug. 1826): 152.
118 *Lion* 2 (14 Nov. 1828): 614–15. This was a talk in the City Chapel in "wretched Grub-street", Cripplegate, delivered on 12 October 1828, possibly by Saull himself (see below).
119 *Lion* 1 (6 June 1828): 731–32. Carlile ran Roland Detrosier's address to the Banksian Society in Manchester (*Lion* 3 [23 Jan. 1829]: 103–12, esp. 109; Detrosier 1840 [1829]), which lauded Cuvier and his ninety fossil mammals unknown to modern naturalists, most of which seemed to have died out as "the result of constant but slowly operating causes," rather than by the biblical deluge.
120 *Lion* 3 (6 Feb. 1829): 171–73.

The Cuvierian muses not only affected Byron. By 1829, they influenced the other end of social scale, far from his Lordship's lofty heights. Street poets embraced extinction and emergence, particularly the first emergence of people to shape Heaven in earth's image. Carlile notoriously included verses of any quality in his journals. It did not matter how salt-of-the-earth they were, the operative was anti-Christian impact. Street poets now stretched out Cuvier's successive changes and saw drama in the ecological immensities. One portrayed the "myriads of years" as the earth was racked, ruined, and reformed, and at each turn "Its creatures, too, with ev'ry race / More comely-fashion'd grew".[121] Until, at last,

> My vision chang'd, I seem'd to stand
>   Amid a swarthy throng:
> Wond'ring they gaz'd on ev'ry hand,
> Upon themselves, the waves, the land,
>   But silence chain'd each tongue.
> Full long and ardently they view'd
>   Whatever met their ken,
> But on themselves with sighs subdued,
> They gaz'd in wonder's deepest mood,
>   They were the earth's first men!

Mute, instinct-driven, swarthy humans, the first of their kind, were entering the sacrilegious imagination, just as Saull was starting to ponder life's origins.

For many ideologues, geology's excitement lay in its confrontational value. Carlile was never particularly interested in the geological niceties, so much as their exposure of time's immensity. In the face of clerical hauteur, this could have a real nihilistic impact. As he typically put it, "the astounding revolutions, that, from time to time, occur on the earth's surface, [throw] the whole of human history into the shade of insignificancy, and [reduce] the conceit of man to animalcular importance".[122] If the clergy found this irksome, scriptural literalists could be goaded further by pointing out that death entered the world before Adam's sin. Cuvier's strange animals were obviously only known from their long dead and petrified remains. The deep strata

---

121 *Lion* 4 (20 Nov. 1829): 650–51.
122 Carlile 1832b.

interring them were ancient graveyards, the subterranean world itself a mausoleum. Such life, entombed before mankind's appearance, could be used to refute the biblical assumption that *"by one man sin entered into* THE WORLD" and that the wages of Adam's sin was death. With so many pulpits denying "that death was known till sin introduced it", Cuvier's fossils became attractive for the anti-clerical armoury. To biblical exegetes, the notion of suffering and death preceding man's fall was "inconsistent with all our views of the Divine perfections" and "would involve a dangerous concession ... as it implies that God was the author of natural evil in a world free from moral corruption." As such, an immensity of time when no "immortal" inhabitants existed to adore their Creator was incomprehensible to many.[123]

Carlile himself began half-heartedly exploiting Cuvier's evidence, twisting and turning it to his own anti-religious ends in a way that would have horrified *le baron* (Cuvier himself abhorred Lamarck's atheistic 'evolution'). For example, Carlile took a cutting on deep time from the fledgling but failing Tory paper, the *Representative*. The image it portrayed of the past was now pretty stock. First, a few plants of doubtful character, then tell-tale sea-shells and trilobites in beds just above; further up (and nearer us in time) came fishes, then lizard-like reptiles, and ultimately mammals. The fossils lay

> buried in beds that overlie each other, nearly in the order above detailed, and between beds or strata are generally found others which do not contain any fossil remains, and which mark the flux of considerable intervals of time in the process of their extinction.

Carlile realized that this defied any scriptural gloss on Cuvier: geology had revealed that each rock stratum

> was once its surface, and that one deluge [that is, the Biblical Flood] will not account for the great number of strata that are found. The succession of vegetables and animals explain the same conclusion, and all unite ... to overthrow that nonsense called religion.[124]

---

123 Biddulph 1 (1825): 126–29. Liberal exegetics might get round these geological conundrums. As a writer to the *Christian Observer* (24 Feb. 1829, 91–96) guessed, perhaps "the secondary strata may really have been deposited subsequently to the creation of man," or that death did not refer to any 'lower' creature, and only *mankind* was marked for death by Adam's transgression.
124 *Republican* 13 (24 Feb. 1826): 256.

Dragooned by the atheists, Cuvier's views, distorted and distended, proved so useful that he eventually acquired almost heroic status in blasphemy circles. How else are we to explain why, on his death in 1832, his French *éloge* was appropriated and run amid the anti-clerical and republican rants in the illegal rags of the day, the *Isis*, *Cosmopolite*, *Poor Man's Guardian*, and so on?[125]

It was becoming apparent that, even if the universe was uncaused and self-existent, fossil life was not. With Toulmin out of the way on this point, the question of origins became paramount. Clerical protagonists, of course, had a ready-made solution. One, moreover, that anti-infidel preachers were ready to throw in deist faces, by challenging them to say where but in the Bible could you find an explanation for the origin of man? It was all very well the blasphemy bards waxing lyrical about the "earth's yet open womb" producing "More comely-fashion'd creatures".[126] The question was, how?

* * *

For Carlile's deists and materialists, power lay inside nature, not outside in the hands of God. And gone was the notion of matter-in-motion causing all change; now matter itself was invested with immanent qualities. We see this already in Palmer's *Principles of Nature*. For Palmer, "dead matter" was an absurdity, "all is alive, all is active and energetic"— one could not "conceive of matter without power, or of power without matter." Saull would himself articulate this atheistic vision: there is "no power *superior* to that of matter", he would say.[127] Belief in such a nature provided Saull's Carlileans with their moral high ground. They claimed for it the status of true morality because it rested on non-idolatrous foundations.

Like their Enlightenment heroes, the 1820s' materialists were intent on liberating the mind from superstition, the body from clerical

---

125 *Isis* 1 (1 Sept. 1832): 455–56; *Cosmopolite*, 8 Sept. 1832; *PMG*, 7 Mar. 1835, 454. Cuvier was still being extolled in the *New Moral World* in 1838: *NMW* 4 (17 Feb. 1838): 129–30. Theirs was a vastly different image of Cuvier from that portrayed in respectable English journals, where he was co-opted as a conservative supporter of natural theology.
126 *Lion* 4 (20 Nov. 1829): 650–51; *Republican* 8 (18 July 1823): 52.
127 *Inventors' Advocate* 2 (11 Apr. 1840): 237; Palmer 1823, iv, 182, 184.

authority, and the people from kingly subservience. Not that their prime source, Holbach, was himself a real republican. He wanted wise, benevolent government and a monarch in harness, and argued that a materialist nature would teach princes that "they are men and not gods; that their power is only derived from the consent of other men."[128] But an innately-powered nature, driven from below, sat comfortably with the 1820s' radical ideal of people as sovereign atoms and the sole source of power, and it was equally useful to Carlile, himself no democrat, as a stick to beat the priests.

One corollary of this energetic nature was the resurgent notion of animate or living atoms. Not merely living but intelligent: thinking and awareness are widely manifested in nature, said a letter writer, so why not give matter "all those fantastical qualities" usually associated with souls and spirits?[129] Energetic matter, by assuming the old spiritual powers, was consequently believed to be capable of self-development. So, after Cuvier's revelations, the materialists were ready to look into nature itself to explain the production of new species through geological time.

Another letter writer in the *Lion* leapt even further: from animate atoms to an animate earth. This was specifically to accommodate Cuvier, for such an earth could intelligently arrange the generation of his successively 'higher' life forms. And Palmer's language, of the planet's *exertions*, an attribute of living things, only encouraged this sort of deduction. The earth possessed its own "exalted generative powers". The analogy between the rising perfection of life and advancing human mind showed that over-arching nature was no "blind power". The language was one of energetic consciousness. Nature would always "endeavour to improve, in consequence of former experience", and work "up its productions to current perfection". Such Schelling-like Romantic pantheism suggested that the earth itself "possesses the power of cogitation". "Every improvement or advancement" reflects the workings of "mind or experience". Intelligence was baked in at global level, life aiming, not at any godhead or attempting to become aware of itself, but at perfection. The object of mindful agency, 'design', was creeping in through the materialist back door. Nor would progress

---
128 Holbach 1 (1820): 109.
129 *Republican* 7 (25 Apr. 1823): 535.

end here. Perhaps the planet would see the obliquity of its axis become perpendicular (millenarian astronomy was never far away), when life would "co-order" itself to the changing winds and currents and reach a "more perfect harmony". Carlile had little time for the romantic twaddle of millenarian geology with its striving for perfection, and many readers had even less. But such efforts illustrate the accommodations being made by the later twenties to a life rising "through gradations of improvement".[130] The *Republican* and *Lion* were surely obligatory reading in Saull's house. He was, after all, Carlile's benefactor and published in the former (Appendix 1). Saull would have seen in these sixpenny street prints the subject of origins openly broached as an attractive part of an anti-Christian polemic. And, as Saull moved into Robert Owen's co-operative camp, with its emphasis on the perfectibility of man, he must have sensed how the new perfectible nature fitted his new political creed (Chapter 5).

Saull's comrades knew that Cuvier petrified some anti-infidel writers. And Carlile's scurrility and piracy was goading the less tolerant literalists into action. It is no coincidence that the scriptural extremists rose to prominence in the late twenties and early thirties, ignoring the more moderate evangelicals who "took a lively and on the whole constructive attitude to geology".[131] The *"great Armageddon* of infidelity seems rapidly to approach," heralded by this Satanic street geology, said one fulminating literalist. Cuvier and his "sorry warriors" were eroding our religious "mountain that standeth strong", and with it the faith that "is the pillar of our security". Such seditious science wants to carry us back beyond that "described in the Sacred History; and, with unauthorized effrontery, [it] presents us with a series of revolutions which have no foundation, whatever, in truth". The anti-Christian "warriors" in Cuvier's wake were pushing further, looking to the fossil strata for evidence of the "the progressive developement of organic life" and the rise of ever-more-perfect races.[132] But, the more anti-infidels

---

130 *Lion* 1 (6 June 1828): 731–34. 823; 2 (4 July 1828): 29–31.
131 Hilton 1988, 149. Fyfe 2004 on how cheap science and the suspect printing presses could threaten the faith of ordinary evangelicals, and how religious tracts responded to re-emphasize Revelation.
132 [Murray] 1831, xiii–xv, 22–23.

decried Cuvier's "nonsense", the more Saull with his Carlilean preconceptions and interest in geology saw the destructive potential.

Cuvier had vouched that there were no human fossils.[133] This suggested that "our species, comparatively speaking, is of a very recent origin", a fact now acknowledged by the infidels.[134] Man was the last, and most improved, animal to debut. Still, that mankind had a beginning, for Carlile, offered no support to the idea of "a revelation having been made to him"[135], or that he was a divine creation. But that finite origin, thrown up by the new temporal, sequenced science, was pushing Saull's deists to seek a rational explanation.

How Did Nature's Energetic Power Manifest?

Off-the-shelf solutions to the problem did exist but were not without their pitfalls. The Tory reviews of seditious trash revealed them with their pillory. Readers of the *Quarterly Review* loved to be incensed by the insane ravings of Enlightenment 'Frenchies'. English geologists were a gated community of sensible gentlemen who could be trusted not to rock the boat. They abjured all talk of origins as the abode of scoundrels. Not so the Gallic enemy who outrageously dabbled in such

---

133 In this, he was supported by reconcilers like F. J. Francis (1839, 156) at the Marylebone, Western, and Richmond Literary and Scientific Institutions, who denied nature's "self-origination" and scrubbed fossil geology clean using Thomas Chalmers's *On Natural Theology* (1835). However, new human skeletons were coming to light. The "Red Lady" of Paviland Cave, discovered by William Buckland in 1823, was assumed to be of recent origin, despite being associated with extinct mammals and chipped flints (Rudwick 2008, 77–79; Grayson 1983, 65–66; Riper 1993, 60). Another contender, the celebrated Guadaloupe skull-less skeleton embedded in a limestone block, had been placed in the British Museum as a Napoleonic war trophy. This, too, was thought of modern origin, though not by its original describer (Konig 1814), ironically. The literalist John Murray in *Truth of Revelation* (1831), who had examined it in the museum, thought the skeleton's fossilized nature should not be dismissed. By contrast, the *Christian Observer* insisted on its modernity (*Christian Observer* 34 [Aug. 1834]: 490; Rudwick 2005, 592; Grayson 1983, 95–97). Sir Richard Phillips calculated the chances against finding human fossils as astronomic. Given that the strata might be half a mile deep and the earth's surface 200 million square miles, he estimated statistically (he loved this sort of thing) that it would take 500 million bore holes to turn up another Guadaloupe "relique". Still, he thought that geologists should keep on looking for fossil humans (R. Phillips 1832a, 52–53).
134 *Lion* 2 (4 July 1828): 30; 1 (6 June 1828): 732; (27 June 1828): 806.
135 *Republican* 8 (11 July 1823): 17.

speculations. The *Review* pointed its finger at the "fooleries" of the French Consul General in Egypt a century earlier, Benoit de Maillet, who postulated our fish origins in a drying ocean. This had taken place over two billion years, a timespan so exorbitant that it was even toned down by de Maillet's editor to "millions", to make it more palatable.[136] The resulting posthumous and editorially-mangled *Telliamed*—De Maillet backwards—published in 1748, mixed sensible observations (the laying down of sedimentary rocks by the retreating sea) with what proved to be palpable absurdities (our fish ancestry being evidenced by mermen). The latter became the pretext for his scientific mauling. In an English gentleman's hands, science "lends no countenance to such insane and visionary 'theories'". The subversive notion of life's "'self-creating energies' [is] not less ridiculous than that of Demaillet and his mermaids".[137]

Portraying geology in the squire's hands as safe and De Maillet as deranged would have flagged him up to the deists, for whom a "self-creating" nature was now a given. Carlile admitted in 1824 that *Telliamed* was "the most interesting book I have read upon the subject".[138] De Maillet had actually been an astute observer and privy to esoteric Arab sources and legends, but merely mooting his fables of tailed mermen brought hoots of derision from readers of the *Republican*. Not "so much a fool as a Madman", wrote one. The critical reader continued: Carlile follows De Maillet and now makes us "the offspring of a Fish or some Amphibious Animal. I really pity the Man [Carlile], if his long Imprisonment has been the Cause of his Derangement."[139] Much of *Telliamed* would have appealed ordinarily—but for the mermaids. On the mermaids, at least, anti-infidels and anti-Christians could concur. That the human "began his career as a *fish*", or "for aught [De Maillet] defines to the contrary, an oyster or a cockle" was a "monstrous idea" to Christians, and obviously some *Republican* readers. To cap this, Carlile's warriors were now using Cuvier's and De Maillet's "arithmetic of infinites" to stretch the

---

136 Rudwick 2005, 129; Grayson 1983, 31.
137 *Quarterly Review* 27 (June 1822): 459–61.
138 *Republican* 10 (12 Nov. 1824): 592–93; Carlile's preface to Toulmin 1854 [1824], v.
139 *Republican* 11 (3 June 1825): 687; on De Maillet's life: [De Maillet] 1755, "Vie" 1–23. Mermaids fabricated in Japan were still being exhibited in London's tawdrier showrooms in the 1820s (Ritvo 1997, 178–80).

"six demiurgic days ... by the touch of this necromantic talisman" into millions of years.[140]

Not least is this interesting because it shows what was accessible on the street. While Pietro Corsi has singled out the "surprising" survival of De Maillet in continental geological literature into the nineteenth century,[141] it is no more of a surprise that his book surfaced on Britain's streets. While the French transformist Lamarck was relatively unknown in the 1820s, *Telliamed*, it seems, was accessible. It is not known whether Carlile was using an English edition or the American (1797) based on it. But that these were available is shown by old copies surfacing: for example, a second-hand *Telliamed* advertised for a shilling in a later *Reasoner*.[142]

Still the question had not been answered: *how* had prehistoric life appeared sequentially? *Telliamed* and the other Enlightenment authorities fixed on spontaneous generation.[143] This had an obvious attraction. Fundamental active, or living, particles provided a perfect democratic metaphor. They were self-organizing, self-willed, and in control of their own destiny—a natural legitimation of the right of 'social atoms' to better themselves through collective action. A shared social/biological lexicon reinforced the belief that nature was on the deists' side. The obstetrician James Watson mooted life rising through "the elements of matter in combination and by co-operative properties and powers".[144] Not for them the traditionalist argument that man and nature were subject to Divine edict, a sort of legislative command from 'above'. Kings might claim their authority from it, and priests their power, but an upstart nature was revolting. Power for the deists lay

---

140 Murray 1831, 22.
141 Corsi 2005, 75.
142 *Reasoner* 9 (1 May 1850): 47.
143 *Republican* 9 (28 May1824): 688–89 for Carlile's musings on the subject. Spontaneous or "equivocal generation" implied chance, and that tarred it in traditional eyes as materialistic and atheistic: Roe 1983, 171–72; Farley 1972, 1977; Desmond 1989, 70. This Enlightenment faith in species, indeed faunas, arising 'spontaneously' would ultimately settle into some sort of scientific respectability (in Germany anyway) as the theory of 'autochthons'—"sprung from the earth"—to explain the new ecosystems emerging after each geological revolution (Rupke 2005).
144 *Republican* 8 (15 Aug. 1823): 174. "I call myself a Social Atom—a small speck on the surface of society", an old foot soldier for democracy began his autobiography: W. E. Adams 1903, 1: xiii.

'below', with atoms as with people, uniting and co-operating. Rotting flesh served to show the regenerative properties and self-organizing ability of 'inanimate' matter as it brought forth tiny life. Carlile's jailed shop assistants rebutted Volney's religious critics, who made the atomic self-assembly of a human a laughingstock:

> how is it, that from a piece of putrified [sic] meat, thousands of animated, organized beings proceed? If the corruption of a piece of meat can do this before your eyes and you cannot account for it but by heat, acting on certain particles, why deny the power of unintelligent matter?[145]

But Carlile himself, discussing the 'spontaneous' appearance of intestinal worms, thought this example of little consequence: whether from egg or atoms, this only explained the appearance of individuals, not new races. Nevertheless, when it came to species, materialists could agree that, somehow, combinations of matter *had* originally made new ones, even humans, and would again under the same conditions.[146] But the question of *how* remained unanswered.

In the mechanics' literature, the 'vitality of matter' issue was heating up in the late 1820s. The question was whether the principle of life was a divine gift, or "whether each particle possesses inherent powers of life in its separate state, and thus spontaneously arises from decaying forms to engage in new scenes of activity."[147] Evidence for self-organizing vital particles was filtering in from France, where it was favoured by republican savants. They too saw life as an innate property of matter or organization. And they too abhorred a top-down spiritual "command structure", from which the king derived his warrant and matter its divine spark.[148]

---

145 *Newgate Monthly Magazine* 1 (1 Nov. 1825): 107.
146 *Republican* 7 (28 Mar. 1823): 401; 9 (28 Mar. 1824): 688–89.
147 *MM* 12 (12 Sept. 1839): 46, 88–91, debating Milne Edwards's work in France and the active molecules of the British Museum's Robert Brown (of 'Brownian motion' fame). The debate over "atomic atheism" had a long history among gentlemen philosophers, unknown to street propagandists (Goodrum 2002).
148 Jacyna 1983b, 325–26. French materialist sciences were coming in to Britain partly through press snippets discussing the republican and transformist Bory de St Vincent (Corsi 2021, 365). For example, "On the Tendency of Matter to become Organized" (*Edinburgh New Philosophical Journal* 4 [1827–8]: 194–96), and "Spontaneous Organization of Matter" (*Arcana of Science* 2 [1829]: 144). Jacyna 1987, on Bory's role in the 'immanentist' scientific tradition in Paris and its republican context.

Republican faith in self-generation was excoriated by religious critics. The church militant was stepping up; preachers denied any innate "tendency to a higher state of being". Life's adaptation was the supreme proof of "a creating Intelligence", meaning purpose was built in from the start. The Rev. Benjamin Godwin, rising to the challenge thrown down by local Bradford infidels in 1833, tore into Holbach's *System of Nature* and Carlile's *Deist*. No mind that "primitive man did, perhaps, at first, differ more from the actual man than the quadruped differs from the insect", he said, quoting Holbach. However much mankind had improved since his primordial production, he still had to start somewhere. But without the constraints of intelligence, a chance concurrence of atoms would have thrown up "thousands of monstrous shapes" of every useless combination, not organs designed for a purpose or animals adapted to niches.[149]

Favourable conditions or atomic intelligence guided this building process for infidels. While detractors laughed at their hocus-pocus of "mysterious chemistry",[150] the *Republican* materialists never lost faith in thinking matter steered by planetary conditions engendering life. Even Erasmus Darwin (who died 1802) was resurrected by deists in the 1820s, for his poetic attempts at a non-biblical production of man. He too had dramatically portrayed the primordial animation of a some simple "threadlet of matter", whence it hoisted itself on its upward path by striving for warmth, food, and moisture. A "pernicious" doctrine that would "infuse poison" into innocents, grumbled the humble *Magazine of Natural History*.[151] This was *Telliamed* updated and medically sober, with everything tracings its origins back to the "briny deep". Only this time the more sensible evidence came from the fact that "all quadrupeds and mankind in their *embryon* state are aquatic animals". They recapitulate their ancestral life and emerge from their embryonic fluids at birth. Darwin, as a doctor, had credence. He was seen putting reason above rhyme and was hailed as "the most philosophical, although not the

---

149 B. Godwin 1834, 168, 175, 180–81; Holbach [Mirabaud] 1834, 80; Morrell 1985, 11–13; Topham 2022, 359–62.
150 Rennie 1834, 51.
151 *MNH* 4 (Jan. 1831): 53–54, reviewing the new and orthodox King's College, London, professor James Rennie's *Insect Transformations*, published by the Society for the Diffusion of Useful Knowledge; [Rennie] 1830, 9. On Rennie's hack writing and short-lived King's career: Page 2008.

most perfect of poets". The technical magazines talked of his "beautiful" lines.[152] His poetry turned up everywhere in the mechanics' journals of the 1820s, as did his medical, technological, and scientific asides on nature.[153] The allusions only tail off in the 1830s, when they start to overlap with his grandson Charles's *Beagle* discoveries. In fact, you could still find him—"one of our finest poets"—railing against "the tyrant's power" in Julian Harney's *Democratic Review* as late as 1850:

> Hear nations hear, this truth sublime,
> He who allows oppression shares the crime.[154]

So, in the 1820s Erasmus Darwin still generated passion in circles high and low, with as much derision in one as veneration in the other. We find the radical co-ordinator Francis Place reading Darwin's *Zoonomia* in 1826.[155] And Carlileans believed "the beautiful speculations of a Darwin, throw much credit on modern philosophy," because he had stripped superstition out of life's equation.[156] Well almost. Watson (the obstetrician) thought him a genius but saw him pandering to patrons, with expensive poetry tomes "designed for the libraries of the higher and respectable classes as the wealthy people stile [sic] themselves."[157] It rankled that Darwin, despite praising the inherent properties of matter, could still in self-contradictory fashion assign it all to a "controuling [sic] power above nature":

> —And high in golden characters record
> The immense munificence of NATURE'S LORD.

This took the gloss off for some anti-clericals. Darwin was truckling to the "prevailing prejudices and cant of the day".[158] But his poetry sweetened

---

152 *LMR* 1 (22 Jan. 1825): 183; (29 Jan. 1825): 196; 2 (13 Aug. 1825): 265—all the praise coming from George Birkbeck at the LMI.
153 They can be found scattered through the *MM, LMR, London Journal of Arts and Sciences, Register of the Arts and Sciences*, and *Gill's Technological Repository*.
154 *Democratic Review* 1 (Apr. 1850): 418.
155 Jaffe 2007, 145.
156 *Republican* 10 (26 Nov. 1824): 666.
157 *Republican* 8 (15 Aug. 1823): 172; (12 Sept. 1823): 302. Browne 1989 on Erasmus Darwin's readership. Into the 1820s Darwin was still bandied around in high society. Recall Sheridan's put down of some "beautiful but far-fetched" idea of Darwin's being received "with great éclat" at Brookes's by the royal party around the Prince of Wales (*MC*, 2 Mar. 1827).
158 *Republican* 8 (12 Sept. 1823): 298, 303.

the bitter materialist pill,[159] and his radical influence and obnoxiousness to the authorities in the 1790s cannot be denied. It was best illustrated by one incident in that incendiary decade: the police raided a London Corresponding Society stalwart, John Thelwall—because the authorities were petrified that the LCS was about to call for a French Revolutionary-style Convention—and seized, among other things, a copy of Darwin's *Botanic Garden*![160]

Darwin's republican poetry was trashed after the French Revolution. Now it was trashed again in the 1820s, derided in 'higher' circles for its "fantastical dandisettism". "Sound was preferred to sense; high words to high thoughts," said a review in 1824. It debated whether his sort of dirty science was not killing the imagination. As the spiritual world is denied, mechanism is all that is left. "Frankenstein" is the most that imagination can inspire to—a magic spark animating dead flesh. But the Frankenstein monster is "a vile lump of earth, with nothing spiritual about him," just as Holbach's atomic man was an empty shell. Frankenstein stands in condemnation of what disreputable "philosophers have supposed possible". Dr Darwin and the deists, "who enquired how men were made" had so long talked up the issue that "they almost persuaded themselves that they had been in the manufactory" at the moment of production and had seen the atomic bodies rise from the dirt.[161]

* * * *

This gives a sense of how tantalizing the ideological question of origins had become in deist circles in the later 1820s. All the while, the incoming progressive geology, by ruling out the eternalist riposte to Christians, was forcing Saull's cadre to look for a new 'natural' solution to the recent emergence of humans. A new rhetorical strategy was needed, which

---

159 Goldstein 2017, 708–72.
160 Thelwall 1837, 164; Mee 2016, 181. Thelwall's interest in Darwin stemmed as much from his anatomical fascination. Thelwall had attended lectures at Guy's and St Thomas's hospitals, where he notoriously decried vitalism, as Lawrence would a generation later. But Thelwall's Jacobinism and medical materialism (related, as Solomonescu 2014 shows) were more long lasting than Lawrence's.
161 *Philomathic Journal* 1 (1824): 434; 4 (1826): 127. Fara 2012 on the political message behind Darwin's seemingly innocuous poems, which the pauper press and the authorities were equally attuned to.

would allow them to pick up the anti-infidel gauntlet, to say where, outside the Bible, you *could* find a satisfactory account of mankind's appearance. So far, the deist response had been inadequate. Watching the 'eternalist' argument evaporate, they could only feebly retort that, even

> though Materialists have not yet been able to prove the primary cause of the existence or origin of the *larger* animals, it does not follow that they are to despair of ever arriving at the great and mysterious secret; or that they are to jump at once, into the admission, of the existence of a supernatural almighty *designing* creative power or being; the existence of which, is as difficult or more difficult to be proved...[162]

By the late 1820s, the deists' Nature ran close to being personified as either the energetic Earth or the aggregate of its live atoms. And with the revelation of the *rise* of life, coupled with the assumption that *"lower"* forms were being pushed into *"higher"* ones—ranking remained unquestioned in biology, even as the new class warriors were starting to challenge it in society—deists saw successively greater power outputs needed to push life up the ladder. Increasingly greater pushes were needed to drive this emergent complexity. Thus it became commonplace in street propaganda to hear of Nature's power increasing through time to heave life ever "upwards".[163]

Simultaneously, with geology throwing up these new imperatives, a new breed of flamboyant deist was re-igniting astro-theological explanations to delegitimize the Jewish fables of Creation. Saull had come from Carlile's camp with a lot of baggage. Now he would take it to the chapels of these new provocateurs. So strong and financially extravagant was his support for the new blasphemy preachers that he would be indicted in court for it, in an episode marking his shocking public debut.

---

162 *Republican* 8 (15 Aug. 1823): 173. Interestingly, Lamarck was all but unknown among Carlile's cadre before Charles Lyell's exposé in his *Principles of Geology* in 1832. Therefore street deists in the 1820s had no recourse to Lamarck's escalating ladder of living species and his idea of needs causing bodily transformations.
163 Such beliefs ran right through to the 1840s (E. Martin [1844], 6).

4. From the Devil's Chaplain to That Dirty Little Jacobin

Robert Taylor ...William Devonshire Saull ... being persons of wicked, profane, and irreligious minds and dispositions, and disregarding the laws and religion of this realm ... did wickedly and impiously conspire, combine, confederate, and agree together, to blaspheme our Lord and Saviour Jesus Christ, and to bring into ridicule and contempt the Christian Religion and the Holy Scriptures. And ... did afterwards ... open a certain room, for the purpose, amongst other things, of delivering therein blasphemous and impious discourses, and did utter and deliver, and cause to be uttered and delivered, divers blasphemous and impious discourses, of and concerning our said Lord and Saviour Jesus Christ, and the Christian Religion, and the Holy Scriptures, in the presence and hearing of divers, to wit, five hundred persons ... and did also ... print and publish, and cause to be printed and published, a certain impious and blasphemous libel, in the form of an advertisement, in order to induce and persuade persons to be present at the said discourses, which said advertisement then and there contained therein the scandalous and impious matters following, of and concerning our said Lord and Saviour Jesus Christ and the Holy Scriptures, that is to say—"Christian Evidence Society. The 93d discussion will be held in the Areopagus on Tuesday, the 13th inst., at seven precisely. *Subject*—'The Character of Christ'....The rev. orator will deliver a *philippic* in exposure of the atrocious villanies [*sic*] that characterize the Jewish Vampire (meaning our said Lord and Saviour Jesus Christ) ... ["] ... And did also ... publish and cause to be published, divers other impious and blasphemous libels, of and concerning our said Lord and Saviour Jesus Christ, the Christian Religion, and the Holy Scriptures; to the high displeasure of Almighty God, to the great scandal of the Christian religion, to the evil example of all other persons, and against the peace of our said Lord the King, his crown and dignity.

The first count on the indictment of Taylor, Saull, and others at the Court of King's Bench, on Wednesday, 16 January 1828.[1]

---

1   *Times*, 17 Jan. 1828, 3.

Saull made his public debut in the dock. This court case is what first brought him to prominence, with the *Times* and leading dailies garishly reporting his indictment in 1828 at the King's Bench. A sensational trial was expected. He was charged, not merely as a disciple of the flamboyant infidel preacher, the Rev. Robert Taylor, shortly to be dubbed the "Devil's Chaplain", but as one of his financial backers. For Saull, this funding was to provide its own intellectual payoff, allowing him to use Taylor's astro-theology as the basis for his own scientific heresies. But it brought more immediate problems.

Prosecuting them both was the highest paid advocate in the land, Sir James Scarlett, newly knighted and appointed Attorney-General. It was he who had notoriously prosecuted the Peterloo protestors after the massacre. Scarlett, a Whig-turning-Tory, and turning more and more against parliamentary reform, specialized in sedition and libel cases. In his words, Saull was one of Taylor's "nest of vermin" to be cleaned out.[2] Scarlett intended this as a show trial, a warning to those who would contest the Christian law of the land. But the dandyish Rev. Robert Taylor intended it to be a show trial in quite another sense of the word.

Presiding on the bench was Lord Chief Justice Tenterden, who sat in awe of Sir James. He, too, had just been elevated. As Sir Charles Abbott, he had been a talent-less advocate, whose lack of eloquence was outweighed by his mastery of mercantile law. That specialism did not stop the Tory anti-reformer—who was made King's Bench chief justice in 1818—judging a Who's Who of insurrectionists, radicals, blasphemers, libellers, and seditious publishers. It was Abbott who had given Carlile his Dorchester sentence in 1819. He had gone on to jail Mrs (Jane) Carlile, while sending another Carlile shopworker Mrs Susannah Wright with her baby to Newgate.[3] One understands why infidels said

---

2   R. Taylor 1828a, 34. The words were actually used in Taylor's October 1827 trial by Scarlett (*Lion* 1 [8 Feb. 1828]: 167).

3   Frow and Frow 1989, 36, 40; Keane, 2006; Epstein 1994, 40–61, 107–08; Wiener 1983, 23–48; Marsh 1998, 68; Anon. 1821; Anon. 1822. Abbott had also sentenced Carlile's rival blasphemy publisher in 1820, Thomas Davison, leading Carlile to take over his stock. Up before Abbott at various times had been the revolutionary Arthur Thistlewood and the Cato Street conspirators, William Hone, William Cobbett, and Henry Hunt. As a staunch Tory, Abbott opposed the Corporation and Test Bill, Catholic emancipation, and shortly the Reform Bill.

they would rather face the black friars of the Inquisition than Abbott.[4] Carlile's "dingy and somewhat repellent-looking" Fleet Street shop was eventually arranged to thwart these continuing injunctions. Since prosecutions depended on an informant buying blasphemous prints from an assistant, whom the authorities could identify and charge, the shop was stripped bare, all stock being removed out of sight upstairs. A customer's request was answered by a disembodied voice from a hole in the ceiling above. A basket was lowered to collect the penny or so, and returned with the requested pamphlet.[5]

Theatrical defences provided a visitor spectacle, and Taylor's was expected to top the lot. Saull's "vermin" thus generated great excitement on their appearance at the Court of King's Bench, Guildhall, in the City of London on 16 January 1828. At Taylor's previous trial (he had been convicted but sentencing was deferred until this one, to take account of new charges), fashionable ladies had turned out en masse for the show. The *Morning Post* was relieved to see fewer this time. Still, crowds gathered even before the doors had opened, and court officers had trouble stopping them from flooding into the seats reserved for the Council. Taylor flounced in at 9.30 A.M. in "full canonicals, with white kid gloves, dress shoes and stockings, and all the attributes of modern dandyism."[6] When the "The King against Taylor" was called, his flock in the gallery rose up, causing Tenterden to threaten to clear the court. Taylor was in "high spirits", clutching rolls of paper, intending to defend himself. Nearby sat a supportive Carlile.

Coming to public prominence on a blasphemy charge might have been a badge of honour for rough and ready Carlileans, but for a "resident freeman and liveryman of this city" it could have consequences. Saull the merchant preferred to remain in the shadow, and, like the other backers, hired the prominent Whig 'civil rights' lawyer Henry Brougham for his defence. For Saull, particularly, as a City trader, this was the more necessary because it was two anti-infidel City aldermen who had brought the indictment, probably encouraged by the government. They paid for the case out of city funds (which

---

4   *Isis* 1 (25 Feb. 1832): 48.
5   Vizetelly 1 (1893): 68–69.
6   *Morning Post*, 17 Jan. 1828. MC 17 Jan. 1828, 3; *Times*, 17 Jan. 1828, 3; *New Times*, 17 Jan. 1828, 1.

was controlled by the aldermen) and had hired Scarlett as prosecutor.[7] Saull faced three counts, 1) a conspiracy to blaspheme and carry it out by setting up rooms as an infidel chapel—called the "Areopagus"—in Cannon Street, 2) to bring Christianity into disrepute by so doing, and 3) to continue to utter blasphemies up to the time of the trial. In fact, he was simply being indicted for funding the "Areopagus" venue, by way of warning other wealthy backers.[8]

As a City merchant, Saull acted the outraged innocent. A week before his court debut, he had already drafted a memorial to the Common Council of the City (the Mayor, Aldermen, and Commons). He indignantly protested his innocence, denying that he was "a person of an evil and wicked mind", or had caused "breaches of the peace". He disingenuously denied knowing "the said Rev. Robert Taylor or ... the other parties", even if he had "occasionally" attended "meetings, at which were assembled numerous and highly respectable persons of both sexes, and of all ages." And then he only entered because the posters had piqued his interest. He had been "falsely charged". Anyway, charging him was a misuse of City funds, and he appealed to the court's known tolerance of freedom of conscience, "which it is the boast of Englishmen". This charade of innocence, so necessary in these blasphemy cases, continued. "On the contrary," pleaded Saull, he was

> a person whose respectability of character has been long and well established and whose property has been acquired by his own industry; ... he has character to maintain and property to defend, and has, therefore, the strongest inducements to preserve, not to disturb the peace and good order of society.[9]

This was the crux. It was precisely because of his wealthy City status that Saull was being 'nailed', in the slang of the day. He was to be made an example. The government wanted the funding cut off to throttle the infidel chapels. The police had been keeping tabs on Taylor's group, and their reports hint at the motive for indicting Saull and the other backers. These financiers had kept Taylor afloat. The star performer

---

7 Saull 1828b; R. Taylor 1828a.
8 McCalman 1988, 190 on what little is known of the "Areopagus" hall up to this point.
9 Saull 1828b. This "Memorial" was written on 12 January and presented to the Common Council on the 17th.

could not have continued treading the boards to mock Christianity without them—they were the kingmakers, behind the scenes, "men of some property", said a spy, who "so far keep aloof from many of lower rate, but of more courageous or hardened principles".[10] That made them dangerous. There is some reason to suppose the spy's intelligence was being passed on. One anti-infidel alderman called Saull out when the Common Council debated his defence (Saull's "Memorial", see Appendix 3). Samuel Dixon, the longest serving councillor, said he

> had good ground for believing, that what Mr. Saull said was not true. It would be made evident in a Court of Law, that Mr. Saull did take part in the proceedings; that he held an office in [Taylor's] Society; and that he was a joint proprietor of the place of meeting.

That reeks of insider knowledge. Saull's wild canard, set flying, was now being shot down. A cabal of councillors was clearly out to expose him. One, Alderman Atkins, hated by radicals as 'Hell-fire Jack', was the prime instigator of the prosecution. Atkins, an anti-Catholic, anti-reformer—who was hissed on the streets and once had a brick thrown in his carriage—was a former Lord Mayor, and an "illiberal, peevish, ignorant bigot", in Taylor's words. The "Areopagus", being in his ward, was simply intolerable, and he caused a "tumult" at the council discussion by repeating Taylor's blasphemous crudities. Even though most councillors agreed that the prosecution (two called it persecution) was wrong, either not sanctioned by scripture or an infringement of freedom of conscience, and that Saull "was a respectable and very good man, whatever errors he may have adopted in speculative opinions", no action was taken to stop the case.[11]

Turning to the Rev. Robert Taylor, it will become apparent that Saull was more deeply embroiled in the "Areopagus" episode than ever he let on. And being enamoured of Taylor's "blasphemous" astro-theology, he would find good use for it as his scientific views matured.

---

10   HO 64/11, f. 46 (Feb. 1828).
11   The Common Council's deliberation of Saull's "Memorial" was widely reported: *MC*, 18 Jan. 1828, 1; *Times*, 18 Jan. 1828, 2; *Morning Post*, 18 Jan. 1828; *Courier*, 18 Jan. 1828, 3; *New Times*, 18 Jan. 1828, 2; *Trades Free Press*, 19 Jan. 1828, 206; *Atlas*, 20 Jan. 1828, 35; On Atkins: Spencer 2009; Welch 1896, 181; Beaven 2 (1913): lviii, 141. R. Taylor 1828a, 46, on Atkins the "bigot".

Blasphemy Chapels

The Rev. Robert Taylor was a meteoric phenomenon: an Anglican priest spouting deist blasphemies—a theatrical ranter of prodigious memory and encyclopaedic knowledge, whose debunking of Christianity using zodiacal esoterica would so dramatically influence Saull. His profanation was titillating, and it started pulling large audiences away from Carlile. His dandyism was beguiling,[12] hence the retinue of well-dressed ladies in his train. Of course, an exotic Thespian, puncturing the pious scripturalism of the age, simply infuriated the City fathers.

Taylor's restless trajectory had been extraordinary. From elite-trained surgeon at Guy's and St Thomas's hospitals to high-flying Cambridge graduate and ordinand, he seemed to have had it all. But a catastrophic collapse of faith, followed by recantations, more crises, and finally a lapse into deism, turned the "gay Lothario to a melancholy Jaques", as a biographer put it. His satires, pricking religious sensibilities, turned Swiftian, as his self-image became one of "champion, a martyr, a sufferer".[13] The more venomous the reaction, the more vehement his display. 1824 saw the itinerant hack traipsing round London's taverns peddling his theatrical sacrilege under the name of the "Christian Evidence Society". At first, it was biting dissections of biblical apologias. The act encouraged grog-house participation: more vaudeville than theatrical, with votes taken at the end, ayes and nays for the anti-Christian motion. As participatory profanation, it was even more subversive to the authorities. Nor did they make any bones about why they were going to 'nab' him (in another colloquialism of the day). With London a crime-ridden metropolis, this irresponsible apostate with his Christ-as-"Jewish vampire" wit would loosen the social restraints. In a sprawling city of massively unequal wealth, privilege, and power, the belief that only fear of other-worldly punishment would stop the masses rising up was widespread. So said the Lord Chief Justice on justifying Taylor's sentence: he might induce a convert to "commit crimes ...

---

12 Nor was this flamboyant dash incongruous in such a context. Elizabeth Amann, in *Dandyism in the Age of Revolution* (2015), has shown how the new sartorial cut was thrown up by revolutionary politics: it evoked a sartorial space that rejected the *sans-culotte* Terror but still endorsed an exuberant rational revolution.
13 Cutner n.d., 6, 8.

which, but for the removal of the restraints of religion, he never would have practised".[14]

The "Irreverend" took his show round the taverns: the Globe in Fleet Street, the Crown and Anchor in the Strand, the Crown and Rolls Room in Chancery Lane, and many more.[15] These were not small, some were huge auditoriums: the Crown and Anchor was a standard venue for political rallies and social celebrations and could accommodate 2,500. The radical Samuel Bamford, introduced to it by Benbow, was transfixed and thought it "wonderfully grand".[16] However grand, Taylor was fighting for elbow room in the city. The lecturing marketplace was crowded, but he gained a niche with his bleeding edge of bawd and blasphemy. Nor was there anything strange about picking up pennies in tavern venues. This was standard practice not only for political and religious orations, but, as we are coming to realize, scientific ones as well. Independent lecturing was a growth trade in London as orators and oracles selling the latest science took to the stump.[17] And Taylor's particular dramas would eventually compete with the London stage—reaching a peak later with his popular character, the "Archbishop of Cant". All of this got him huge audiences, with many followers poached from Carlile. However much Carlile approved of Taylor spoofing the Christian liturgy, he hated it when Taylor developed his own mock liturgy. "Such trash", he said, not mincing his words.[18] But even he recognized that Taylor had carved out a unique blasphemous corner in London's lecturing empire.

Success led to his apostles looking for a permanent venue. They had to vie with Dissenters for these halls, and there was sweet satisfaction at

---

14 *Times*, 8 Feb. 1828, 4; R. Taylor 1828a, 45. Details of Taylor's life from Cutner n.d.; *Comet* 1 (3 May 1832): 35–37; R. Taylor [Talasiphron] 1833.

15 Cutner n.d. Co-operators also met in the Crown and Rolls Room: *Co-Operative Magazine* 1 (Feb. 1826): 56. Taylor held court in smaller dives as well, favouring Lunt's Coffee House in Clerkenwell Green, where John Gale Jones was a regular (*Gentleman's Magazine* [Nov. 1844]: 550–51). "Irreverend" was a common joke, for example, *Republican* 14 (1 Dec. 1826): 669.

16 Bamford 1893, 18; Timbs 1866, 179–80; Parolin 2010, ch. 4, for a modern study.

17 Science 'marketplace' studies have taken off recently: see the informative essays in Fyfe and Lightman, 2007. Besides venues, the period began to see a rise in publishers' hacks—cheap science popularizers—trying to create a new type of authorial vocation (Fyfe 2005; Lightman 2007). Venues and theatricality seem to be a lesser explored topic, but see Morus 1993, 1998, 2010; Huang 2016, 2017; Hays 1983.

18 *Republican* 14 (11 Aug. 1826): 130. Marsh 1998, 348 n.81; McCalman 1992, 57.

taking a chapel from their nemesis, the Congregationalist preacher Dr Bengo Collyer, the impressive Salter's Hall in Swithin's Lane, Cannon Street.[19] They put the deposit down in 1826, but something prevented them from gaining access, and they had to set up first in the run-down Independents' Founder's Hall Chapel, in Lothbury, near the Bank of England. Here Taylor cut "a very extraordinary figure ... with a reverend hat and a glass suspended from his neck by a broad blue riband". The hall lacked the dash of the man, but, from July 1826, Taylor ran his Sunday "Divine Service" and "sacred dramas" here for a few months and attracted large congregations, "chiefly mild, sober, respectable and moral people". These were middling sorts who could afford the sixpence entrance fee for some titillating Sunday morning excitement and were prepared to run the gauntlet of Christian saboteurs trying to break up the proceedings.[20] The talks were exciting, disputants would rise from the audience, and a show of hands at the end would decide the Bible's verisimilitude. Late in 1826, the Taylor ensemble finally took over Collyer's splendid Salter's Hall Chapel and properly inaugurated the "Areopagus". No dive this: designed by the architect of St Paul's School, it was "handsome and very elaborate", with its impressive four-pillared portico entrance and huge interior, lighted by "semicircular headed windows, over which are tablets beautifully sculptured with the Grecian honey-suckle". Nor was the neighbourhood down and out, for the hall was "prettily situated in a planted garden". Deism was going upmarket. After "fumigating it well, in consequence of its late occupation by Dr. Collyer", as Carlile quipped, Divine Services were resumed. Each Sunday would see Taylor in full canonicals, with the public seated in rows and fashionable ladies in the side boxes. But it was a church service mocked. As word spread, his sacrilegious services became ever more popular, with shopkeepers jostling with mechanics for a seat in the pews.[21]

---

19   GM [Nov. 1844]: 550–51.
20   *Republican* 14 (28 July 1826): 73; (11 Aug. 1826): 129–35; (8 Sept. 1826): 263–64; (29 Sept. 1826): 353–61; (6 Oct. 1826): 401. The rent was £60 a year and expenses the same, so Taylor had to recoup £120 in sixpences. *Reasoner* 5 (16 Aug. 1848): 188–90, recalled the scene, with the chairman in the pulpit, and Taylor on the rostrum below him, combatting arguments from an audience that included the future M.P. for Tower Hamlets George Thompson.
21   McCalman, 1988 189–90. Prothero 1979, 260; *Republican* 14 (1 Dec. 1826):669; *Comet* 1 (3 May 1832): 35–36; Shepherd 1827, 152.

Taylor's astro-theology was already in evidence. We know it from Carlile—a grumbling regular in the pews—because he wanted Taylor to do away with the "nonsense about the sun being our father, the earth our mother, and the homage of one star to another..."[22] But the straight-talking Carlile was now sidelined. Astro-theology would become more and more central to Taylor's drama.

How far Saull was responsible for financing we do not know. The evidence is fragmentary for this early date. But he had clearly swung into Taylor's camp. The group had bought the Salter's Hall for £1,850 in autumn 1826, paying for it by issuing £5 shares and borrowing money. That the wealthy Saull was paying out is suggested by the fact that, only months later, a nark reported secretly that Saull was not only part of Taylor's "Committee", but one of the hard-core who remained faithful through thick and thin.[23] By now, Taylor was calling Saull "my kind friend",[24] and this "kind friend" was to stand bail for Taylor barely six weeks after the "Areopagus" opened. In February 1827, when Taylor was first arrested, Saull put up the £100 bail. Even then, he knew he was liable to forfeit it, because it depended on Taylor's good behaviour while free—and Taylor had no intention of discontinuing his blasphemous liturgy.[25]

This was Taylor's first court appearance on a blasphemy charge, with a hearing on 21 February 1827. The case was brought by the Lord Mayor and Alderman Atkins, who were determined to detoxify this "moral poison". Taylor was charged with "having wickedly, maliciously, unlawfully, scandalously, and blasphemously" impugned "in a loud voice" Our "Lord and Saviour of the World, Jesus Christ". But Taylor was not one to be intimidated at this Mansion House hearing. He stood in the dock, dressed in an embroidered blue cloak,[26] and outraged the proceedings by invoking parables about the Gadarene swine as the first

---

22 *Republican* 14 (1 Dec. 1826): 670.
23 HO 64/11, f. 6 (13 Aug. 1827). The spy was Abel Hall. On the costs: Royle 1979, 468; *Comet* 1 (3 May 1832): 37.
24 R. Taylor 1828a, 35; *Lion* 1 (8 Feb. 1828): 168.
25 *New Times*, 22 Feb. 1827, 4; *Times*, 22 Feb. 1827, 4; *Examiner*, 25 Feb. 1827; *Bell's Life in London and Sporting Chronicle*, 25 Feb. 1827; *Atlas*, 25 Feb. 1827, 119. Saull spoke for the other bails, who were Charles Grimwood, potato dealer; Samuel Purnell, fishmonger; and Christopher Scales, butcher.
26 *Atlas*, 25 Feb. 1827, 119; *Examiner*, 25 Feb. 1827; *Comet* 1 (3 May 1832): 37; *Bell's Life in London and Sporting Chronicle*, 25 Feb. 1827.

martyrs of Christianity (to the cheers of three hundred followers) and ended his court performance with the City Solicitor publicly dubbing him "The Irreverend!". Saull and others each stumped up £100 for Taylor's bail. But free again, pending the trial proper, Taylor kept up his mockery of Divine Services.

Then, in the Spring of 1827, Atkins added two new counts, and now included Saull and five others on the indictment sheet for "a conspiracy to overthrow the Christian religion"—and "conspiracy" is the operative, because they were the financial facilitators. Taylor was nonplussed that Atkins was "actually involving Mr. Saull himself in the meshes of law; for the alleged crime of conspiring with me and five other persons, to bring the Christian Religion into contempt; because, (my Lord), and there were really no better grounds of presumption against him—because 'he was my friend, faithful and kind to me'."[27]

All of this proves that, whatever Saull's protest that he barely knew Taylor, Dixon was right: Saull was one of the Committee and partly responsible for the chapel. To complicate matters, Taylor was conned by some swindlers and jailed for debt in the King's Bench in June 1827 for some months,[28] causing a management crisis at the Areopagus. The spy staked out King's Bench to identify Taylor's visitors. And it was Saull, one of the few stalwarts supporting him in jail, who was caught advising Taylor in August that, in light of events, he should give up the hall and return the money.[29] Accordingly, the chapel was sold at a £150 loss, with the shareholders taking a hit, although the loans were paid off fully. The winners in all this were the Dissenters, who avenged themselves by

---

27 R. Taylor 1828a, 35; *Lion* 1 (8 Feb. 1828): 168. Those indicted with Saull were two labourers (William Freeman and John Hanger); the radical printer John Brooks; Thomas Brushfield, an oil-man; and a "gentleman", John Roome, who "was the principal in trust for the Salter's Hall Chapel": *Comet* 1 (3 May 1832): 37; *MC*, 17 Jan. 1828. The only one fairly well known is Brooks, on whom see *Lion* 2 (10 Oct. 1828): 451–53 (for his refusing to take the oath in court); and *PMG*, 21 Nov. 1835 (having his property seized for refusing to pay the church rates); and Brooks 2009, which shows that Brooks's wife was a friend of Harriet Robinson, who married Taylor in 1834.

28 This was a sobering experience. Taylor called it a "hideous dungeon" and compared the prison marshal to "the triple-headed dog of hell": R. Taylor 1828a, 40; *Lion* 1 (8 Feb. 1828): 165–76; (29 Aug. 1828): 273–81.

29 HO 64/11, f. 6 (13 Aug. 1827); f. 28. Abel Hall was the spy. Taylor was not released until about December: HO 64/11, f. 30.

buying the hall back cheap, so "for ever precluding the possibility of this chapel again becoming the Areopagus of Infidelity".[30]

From this point on, Taylor's troubles only compounded. His blasphemy trial came up on 24 October 1827 at the Guildhall. Huge numbers turned up, including his fan-base of "well-dressed and youthful females". Taylor in flowing gown was escorted by friends in mock episcopal procession: "his neat clerical hat was conspicuously borne in his hand, an eye-glass depended from his neck, and the little finger of either hand was ornamented with a sumptuous ring".[31] Despite the show and his own three-hour defence, he was found guilty. But sentencing was deferred, because the aldermen intended to bring more charges and rope in Saull and the others. Therefore, a further trial was planned, the one on 16 January 1828, referred to in the epigraph.

The conviction deflated Taylor's backers, and many split off. But not Saull. Undeterred, in December 1827, he and a couple of other diehard supporters, reported the spy, secretly managed to lease a new hall in Hanover Street, Long Acre, without divulging Taylor's name; but the landlord "found out their real intentions" and cancelled the contract.[32] Saull, above all, remained loyal and at Taylor's house the two men continued their scheming, according to the spy. Nothing better shows how deeply embedded Saull was in London's small but noisy anti-Christian community, whatever his public protestations. How much he now owed to Carlile and Taylor can be seen from his flat rejection of the Bible as inspired. By the end of 1827, he was openly cavilling at the book's flat-earth incongruities, the absurdities of the sun standing still and other apparent suspensions of the "unalterable laws of nature". He leant heavily on William Lawrence, quoting him on the "ridiculous" Ark and impossible repopulation fantasies. Like a generation of deists, Saull saw these Pentateuchal legends as pale appropriations from the Chaldeans. And, pointing to his real nascent interest, he contrasted the

---

30 *Lion* 1 (9 May 1828): 605–06. Sale: *Comet* 1 (3 May 1832) 37.
31 R. Taylor 1828a, 3, for details of the "sermons" that got him convicted in 1828, one of which described Christ as the "Jewish Vampire".
32 HO 64/11, ff. 33, 41–42. Scheming on this with Saull was his long-time associate, the teacher F. A. Augero, who would become Secretary of the Radical Reform Association, and active, like Saull, in the Metropolitan Political Union and National Political Union.

immensity of astro-geological time and space with the myopic scriptural image of the earth's centrality and age.[33]

Saull's own trial for blasphemous conspiracy, on 16 January 1828, ran over to the next day. But Brougham's defence was not needed. On the 17th, the trial was suspended, because some of the jurymen failed to turn up.[34] In fact, it never resumed against Saull and his co-conspirators, even though Taylor himself was prosecuted. Still, the trial had figured prominently in all the papers, sometimes on the front page. Such show trials were sensational but controversial. They were obviously a threatening tactic, but they were coming to be seen as ineffective. They were even counterproductive: Carlile's well-publicized trials bumped up his sales, but when the government stopped incarcerating the "disgusting" man, the "sale of trash in that person's shop fell 50 to 1".[35] Still, so many deists had ended up at His Majesty's Pleasure that a socialist wag later turned the scales to suggest that the parsons were lucky that geologists were not in charge and able

> to proclaim that their opinions alone were correct, and that all who presumed to differ from them, were blasphemers, who would be sent to prison, and visited, in addition, with heavy fines, if they dared to promulgate their heterodox notions.[36]

Though Saull was never prosecuted, the indictment hung over him like a sword of Damocles. His annoyance was shown by the fact that, of the very few letters he published, three refer to this deferred prosecution. His vicar sent him Watson's *Apology for the Bible*, hoping to convert the parish reprobate. But all he got back on Christmas Day 1827, was a printed twenty-three-page tirade praising geology, astronomy, and their spread via the new printing presses, for debunking Old Testament absurdities. It ended up: "and although bigotry and fanaticism seem to be forging their chains, ready to fetter and manacle the bodies and minds of myself and others, yet I will resolutely proceed in the path I have chosen ... whether in prison, or enjoying the sweets of liberty".[37]

---

33 Saull 1828a.
34 *Comet* 1 (3 May 1832) 37.
35 *Trades Free Press*, 19 Jan. 1828, 206.
36 *NMW* 8 (26 Dec. 1840): 409.
37 Saull 1828a, 23.

Over the months he appealed to three Chief Justices to quash or resume the case, but none complied.

Ironically, seven months after the postponement, and with the threat of indictment *still* hanging over his head, Saull was summonsed for jury service. This led to a tortured letter to the Lord Chief Justice, published in the *Examiner*. He noted the incongruity in being called to sit in judgement on his fellows, because "at the present time I am actually a prisoner on bail". While "this charge hangs over me", he said, he could hardly be considered an impartial juror. And then there was the "deep mental degradation and pain" Saull felt on being forced to swear on the Bible as a juryman, when this book, being thought divinely-inspired, "is declared to be 'part and parcel of the law of the land'", and was responsible for him being in the dock in the first place.[38] Being a stout believer in trial by jury, like so many freethinking radicals, he objected to swearing on the Bible. Saull insisted on "solemn affirmation" for those who demurred from "moral motives". That is, he wanted atheists and deists to be treated like Quakers.[39]

While Saull's trial was postponed, Taylor's was not. It was slated for 7 February 1828. Optimism at first reigned, as Saull's group expected him to "get off". Accordingly, they started arranging for a new chapel, to replace Salter's Hall.[40] But as the time drew nearer, pessimism set in, with many predicting "he would get Three Years". The informer was now paying close attention to how the "men of property" would react to the verdict. On the trial day, Taylor presented his usual spectacle,

---

38 The justices he appealed to were Tenterden, Sir Stephen Gaselee, and Sir William Draper Best. Saull 1828c. Taylor, in Oakham jail, saw Saull's "excellent letter" on oath-taking in the *Examiner*. Oaths sworn on the Bible for him, too, were "an insult to our honour, and an offense [*sic*] to our reason": *Lion* 1828 2 (28 Nov. 1828): 689. Saull (1828c) also complained that he was relegated to trivial Guildhall trials, rather than grand juries, despite having paid considerable taxes as a City merchant. He saw his moral integrity being impugned: he was being barred from important trials because of his beliefs. Marsh, 1998, 49–50, on the legal deprivations suffered by blasphemers.

39 Saull 1828c. For Saull on Aldersgate wardmote oaths: *TS*, 24 Dec. 1835, 2; 26 Dec. 1835, 4. Another who notoriously refused to swear on the Bible was his friend Julian Hibbert. Juries acquitted so many London Corresponding Society heroes in the 1790s that the "Instauration Of Trial By Jury" was celebrated yearly by radicals, including Saull, well into the nineteenth-century: *MC*, 3 Nov. 1846, 1; *The Era*, 8 Nov. 1846; *Morning Post*, 7 Nov. 1846, 2; *Nonconformist*, 10 Nov. 1847, 799; *Daily News*, 6 Nov. 1852.

40 HO 64/11, f. 45.

with clerical attire and sumptuous adornments, including the bevy of young ladies in his train. But despite a two-hour plea for mitigation, his scripture-scoffing and "exposure of the atrocious villanies [sic] that characterise the Jewish Vampire" at Salter's Hall got him a year in Oakham Gaol. After this sentencing, more supporters peeled away, leaving only Saull's "aloof" and wealthy hard-core to rally round.[41]

Oakham was a 100-mile, two-day coach ride north, and, as such, designed to isolate Taylor physically. But it was never the "hideous dungeon" that was King's Bench. It was salubrious by comparison, surprisingly tiny, with five or six inmates at most. Fortunately, too, the jailer and his family were accommodating, and, for the exorbitant sum of 14s a week, Taylor was allowed a "very snug and decently furnished parlour, which, together with the bed-room, and a servant's attendance", befitted a gentleman of the cloth. This made it more like a hotel, which was just as well, because Taylor made a bad martyr and preferred his creature comforts. He was even allowed to stroll round the extensive gardens.[42] And wander further afield, it appears. For he posted back "SERMONS IN STONES" to his "geological friends", describing local fossils as so many more "Christian Evidences" of "the falsehood of the Mosaic account of the creation" and proofs "of the earth's having undergone changes, that could have been brought about only in the revolution of millions of ages." Taylor was obviously rambling in the Rutland hills. "If our geological and stone-analyzing friend [Saull] cares to pay the carriage", wrote Taylor, "I can send him a hundred weight of philosophical dirt." Included would be the "shells of fishes that were inhabitants of the County of Rutland, when Rutland's hills and vales were the deep unfathomed caves of ocean". He even jokingly offered the "vertebrae of men, that have waited for the resurrection, till the archangel's trumpet itself is oxydised."[43]

---

41 HO 64/11, f. 46. The spy named this rump as "Saul [sic], Augero, Pummell and the three others who are indicted with him for conspiracy". Sentence: *Times*, 8 Feb. 1828, 4; R. Taylor 1828a, 45; *Lion* 1 (8 Feb. 1828): 165–76.
42 *Lion* 1 (15 Feb. 1828): 196; Wiener 1983, 147; *Comet* 1 (3 May 1832) 37. Here, too, he found time to dig himself deeper into trouble by writing the *Syntagma*—based on the propositions that Christ never existed—in answer to the Rev. John Pye Smith (R. Taylor 1828b).
43 *Lion* 1 (21 Mar. 1828): 372.

Distance meant isolation—"Oakham MONASTERY", Taylor called it. For a clerical *roué* with an eye for the ladies, this was the hardest part. "My only punishment here, will be solitude," he sighed, but Saull managed to get round even that. Rushed off to Oakham on the night of his trial, Taylor had no time to pack linen and the necessities. So Saull used a pub contact at the George Inn in Oakham to supply Taylor with bed sheets and engaged this intermediary to pass on letters.[44] Back in London, Saull's support group moved up a gear. The spy watched closely. They "set about immediately to make the case public. They met at Carlile's the same evening [as the trial] and he placarded his Windows with the Sentence". Day after day, Carlile's shop was the focus of campaign meetings, and "it was settled by Saul [sic] Carlile and others at these Meetings that the Lion [Carlile's new weekly] is to ... feature all correspondence [from Taylor]". They were to broadcast his case through the press. They considered calling a public meeting, but that fell through. A fund was opened with £4 8s in subscriptions immediately raised, "and Saul [sic] who has Two persons in Oakham who deal with him in his business ... wrote that they are to pay to Taylor between them One Pound per Week and a Bottle of Wine every Sunday he remains there."[45] So the accommodation and servant were being funded by Saull's campaigners.

After his postponed trial and Taylor's conviction, Saull remained unbowed, but he moved further into the shadows. It means that we now have to dig deeper into police records and identify anonymous publications in order to trace his continuing anti-Christian activities. That he was not cowed is shown by the fact that, before Taylor's trial, as we have seen, Saull was trying to get him a new chapel in Hanover Street, in the expectation of his acquittal.[46] Now, barely days after Taylor's incarceration, we find the group putting in a tender for yet another infidel chapel.

Competition among the sectaries for accommodation and congregations was fierce. Deists and materialists were vastly outnumbered, of course, a miniscule Leonidas force facing the Persian might of Christian preachers. But, deist or Christian, all knew that the

---

44 *Lion* 1 (15 Feb. 1828): 195–97; (27 June 1828): 815. *Comet* 1 (3 May 1832) 37.
45 HO 64/11, f. 75.
46 HO 64/11, ff. 41–42, 45.

best chapels and orators brought in the biggest cash sums. Yet again, Saull's activists avenged themselves with the acquisition of a splendid Grub Street Chapel. They took it from vacating Presbyterians, led by the Scottish 'prince of preachers', the Rev. Alexander Fletcher, a fierce anti-infidel who had been challenged by Taylor at his trial.[47] Fletcher was a wealthy preacher whose book sales were as big as his congregations, and it was his "very large" chapel that Saull's group now snapped up, while Fletcher moved on to build himself the largest temple in London, in Finsbury, at colossal cost, complete with theatrical interior and Grecian Ionic pulpit[48]—the sort of construction which could only be dreamed about by deists. No love was lost between these congregations. Three years earlier, when an uncomfortable Fletcher found himself embroiled in a breach of promise suit, his chapel trustees were desperate to keep the news from "that villain Carlile", lest he exploit it. The churches were not only vying for space, but personnel as well, and at least one of Fletcher's congregation surprisingly came over to Carlile.[49]

The spy in 1828 reported this Grub Street front opening up:

> There is intended to be a New society of Deists and is got up entirely at the expense and under the sole direction of Saul [sic] ... and some others of Taylor's Committee. I do not find they have at present any other motives in view than that of going if possible to further lengths in the abuse of Christianity than Taylor did...

It was all hush hush, as Saull's cadre moved carefully behind the scenes, ignorant of the watching spy—who reported that even Carlile was caught unawares by the move. Saull's confreres put up £400 to buy the lease, while Saull himself took care of the £100 per annum rent. The deal was signed in February, with Taylor barely settled into his cell. The converted chapel on Grub Street, Cripplegate, was opened on 2 March 1828.[50]

---

47 Fletcher 1815; R. Taylor 1828a, 9; Fletcher *ODNB*.
48 Fletcher *ODNB*; Shepherd 1827, 163; H. G. Clarke 1851a, 73; *Lion* 1 (29 Feb. 1828): 273.
49 *Republican* 1 (17 Sept. 1819): 57; 11 (4 Mar. 1825): 258.
50 HO 64/11, ff. 4, 50, 75, 78. The Secret Service report fingered another of Saull's close friends, the Paineite Edward Henman, who contributed to the £400. Henman had helped keep Carlile's shop open, collecting funds to pay off Carlile's massive £2000 fine (Royle 1976, 26). For a long letter of Henman's denying the existence of the soul, see *Republican* 8 (14 Nov. 1823): 593.

Where Fletcher had told his congregation "he did not preach Reason to them, but Religion",[51] it was hoped that the new infidel orator would preach reason, not religion. Advertisements were placed in the *Sunday Times* and the city placarded to announce that the former school master and Taylor associate, the Rev. Josiah Fitch, would begin "Divine Service" in Cripplegate on Sundays. The opening saw "at least 300 of Both Sexes and many of them the same who attended Taylors Lectures and were very respectable in appearance".[52] Carlile and his shop workers (including the spy) turned up. There, too, was what McCalman calls the bevy of "socially frustrated 'gentlewomen' from middle-class backgrounds".[53] Carlile, his nose out of joint, was appalled by the liturgical charade and singing of deistical hymns. He preferred plain materialistic sermons. But that was missing the point, the entertainment value. By taking something so familiar, the solemn liturgy, and spoofing it, the very sacrilegious act, surreal and edgy, could draw crowds from the music hall, even as it pandered to the more knowing, doubting, anti-establishment theatre-goers. And being so risqué, the act turned its preachers into deistic matinee idols. A po-faced Carlile wanted straight-talking disquisitions; what he got was a vaudeville parody pricking pomposity and attacking the religiosity of the tight-laced age. For Carlile, the service threatened to be another "Punch and Judy kind of burlesque of religious worship", and he set about rubbishing it in the *Lion*, "as being as much 'Superstitious as the Christians'".[54]

Saull might have been more sympathetic. There is evidence that he was still a deist at this time, or at least prepared to give lip-service to the existence of a Creator when talking to his vicar.[55] However, and perhaps heeding Carlile, Saull's infidel elite in July 1828 did set up a sober mutual-instruction group inside this Cripplegate Chapel, calling it "THE ATHENAEUM, or SCHOOL OF MORALS AND SCIENCE." It was specifically for devotees (5s a quarter), and restricted to fifty in number. Keeping

---

51　*Lion* 1 (29 Feb. 1828): 273.
52　HO 64/11. f. 78: this is the most substantial source. *Lion* 1 (29 Feb. 1828): 273. Prothero 1979, 260; McCalman 1988, 190.
53　McCalman 1988, 189–90. One was the ex-actress Eliza Macauley, who would lecture in Grub Street, and, in a few years, help Saull's co-operators open their labour exchanges.
54　HO 64/11, f. 85. *Lion* 1 (14 Mar. 1828): 348–49; (21 Mar. 1828): 359; (4 Apr. 1828): 438–39.
55　Saull 1828a, 4, 16.

it select meant they could push their scientific heresies and Christian critiques to the limit without fear of prosecution. Here infidel scientific topics were aired more seriously, away from Fitch's flocking crowds.[56]

No doubt some dissident science was threaded through the Divine Service. On one occasion, for example, Fitch preached to two or three hundred (including Saull) on Toulmin's *Antiquity of the World*.[57] But it was Saull's "Athenaeum" cadre in their members-only Sabbath talks on geology, astronomy, and infidelity that drew Carlile's praise. They made it the "one chapel in the metropolis devoted on the Sunday to useful purposes". Street bards were no less rhapsodic about the "Athenaeum":

> IN classic Grub-street, famed in former times
> For half-starved poets and their doggrel rhymes,
> The "City Chapel" stands in humble state,
> Without allurements to attract the great.
> No playhouse singers, organ, or divine,
> No splendid silver for the bread and wine;
> No paintings, gildings, or a grand Te Deum,
> But free discussion, like the Athenaeum,
> And such 'tis named, for here no bigot raves
> Of hellish torments for his listening slaves,
> Who sigh and groan and trembling kiss the rod,
> And think *his dogmas* are "the word of God."[58]

One Athenaeum speaker (Saull?) eked out the subversive implications of astronomy, chemistry, and geology. He ridiculed a biblical deluge as any sort of sensible explanation, and insisted that the rise of life from the more "imperfect" in the lower strata to today's complex creatures shows a self-developing progressive pattern. It also bespeaks a staggering antiquity: "Who that explores the stratification of the crust of the earth" can doubt "that more millions of years have elapsed than the Bibleists will allow thousands?" And he emphasized that only planetary orbits and tilts can explain why fossils of tropical animals

---

56 See the printed two-page flyer, "The Athenaeum", dated 20 July 1828, enclosed in W. D. Saull to Robert Owen, n.d., ROC/18/6/1, Co-Operative Heritage Trust Archive, Manchester. Previous historians have confused Saull's private "Athenaeum" society with Fitch's public services: Carlile praised the former and condemned the latter.
57 HO 64/11, f. 85.
58 *Lion* 2 (10 Oct. 1828), 471.

are now found in temperate regions[59]—a subject that would become a lifelong obsession for Saull.

Saull had a deep and continuing commitment to practical anti-Christianity and science. But the spy makes clear that secrecy was now paramount. Saull became more circumspect in his publications. Whereas his printed letter to his vicar (dated 25 December 1827, *before* the trial) was signed, the next time he printed a letter with a frontal assault on Christianity (in 1832), it ran under the *nom de guerre* "D." (see Appendix 2).

For a merchant on bail, caution was now the order of the day. The need for it was continually apparent. When he published a seemingly innocuous letter in February 1829 in the *Morning Chronicle* (on a boy turning up at a dissenter's chapel to sign a petition against Catholic emancipation, because, the boy explained, Catholics *"don't believe in Jesus Christ"*), the *Morning Journal* ran a diatribe against Saull. It reminded readers that this "wiseacre" was the "warm patron" of Taylor, indeed that he had stood beside him in the dock. In an age of "infidel, sectarian, and Popish attack upon the church and state of England it is desirable to let the public see" that it is the "doughty champion of Deists and Papists" who would assail the constitution.[60] This public pillorying shows how tarred Saull had become by the Taylor episode.

Saull might not be caught out again, but that did not stop him delivering uncompromising lectures. Most notably, he started weekly talks in his strange friend Pierre Baume's Optimist Chapel in Windmill Street, Finsbury. 'Suspect' might be a better word for Baume; the French *émigré* had a murky past and a future that would be overshadowed by tittle tattle. Many considered Baume a bit "doubtful"—so said the spy, who had been ordered to keep an eye on him.[61] His republican deism barely disguised a dubious history. He had been Secretary to the Neapolitan ambassador in Paris, allowing him to amass a fortune, many thought through spying. Saull had known the chimaerical character since 1828, and Baume and his half-sister Charlotte often came to dinner.[62] Whatever

---

59   *Lion* 2 (14 Nov. 1828): 615–16.
60   Saull 1829. Response titled "mr. saull" by Vigil in *Morning Journal*, 2 Mar. 1829, 7.
61   HO 64/16, ff. 127–28 (Oct. 1830).
62   Roger Cooter, pers. comm. Baume *ODNB*. I would also like to thank Roger Cooter for sharing his transcriptions of the Baume-Saull correspondence, particularly Manx Museum MM 9950 uncatalogued: Baume to Saull, 9 June 1837.

Saull thought of Baume, he later let him use his brandy warehouse to store goods, and Baume even listed 15 Aldersgate Street as his mailing address.[63]. Baume had opened up a print shop in Windmill Street and, in early 1830, had bought this dilapidated chapel. It was transformed into a freethinking, bible-bashing venue, where republican talks attracted sizeable audiences after the July Revolution in 1830.[64]

Saull spoke here weekly, but we only know it from the informer's reports. On one Sabbath in November 1830, the spy related (rather breathlessly), Saull ascended the pulpit and

> began a Lecture on Superstition in which he much abused the Ministers of all Religions and the Religions also and said he was glad to find that knowledge and Union of the people had begun to have some weight and pressed the Necessity of still further to unite for though slow they were sure in the end they would put down all Superstition and Tyranny. He also began to prove the eternal existence of all matter and contended that Materialism was the only true Religion which would in time be known.[65]

There might, or might not, have been a God, but matter was all that mattered on earth. Whether or not Saull had moved on from deism, Baume himself could still gaze with "gratitude towards the *First Cause*", even if that distant being was glimpsed only through nature, not revelation. But they could agree that belief in the devil, sin, atonement, and hell was itself a blasphemy, and, far from a fallen being, man is "constantly advancing"—hence Baume called his own radical rag *The Optimist* (1829). Whatever is thought "the BEST, at this instant" will be "BETTER", he said, introducing his paper. The motto was to provide a natural legitimation of political action. As creatures of circumstance, we yield to the "NECESSARY or omnipotent influence" of nature's law, which drives life and society onwards.[66] Nature's writ, a law seen as a sort of judicial order, ensured progress. It meant that the "lower orders" of society, like those of nature, could expect a "higher" and brighter future.

The soul was another absurdity that Saull and Baume agreed on. It was an "impossibility", Saull told his Optimist audience.[67] But Saull's view

---

63 *The Sessional Papers Printed by Order of The House of Lords*, 1846, vol. 12, 32.
64 Prothero 1979, 259–61.
65 HO 64/11, f. 167 (22 Nov. 1830).
66 *Optimist* 1 ([no. 1] Dec. 1829).
67 HO 64/11, f. 205 (c.1830).

of Baume himself remains a mystery. It was not a match made in heaven. They might have concurred on Christianity and republicanism (Baume went so far as to placard his house with posters and fly a tricolour flag from his second floor[68]). They both might have been abstemious, with Baume going teetotal and becoming gaunt from his frugal diet. But, as Roger Cooter says, many thought Baume "dangerously mad". And this was not for his eccentricity, although that was shocking enough. In an age lacking cadavers for dissection, and with grave robbery endemic, only Baume could suggest, not merely that deists bequeath their bodies for research (many materialists agreed with this), but failing that—and rather than the sacrilege of letting corpses rot in "holy" ground—the skin should be tanned for chair covers, the skull donated to phrenologists, and the bones be whittled as knife handles.[69] No, what really marked him was the tragic death of his half-sister in childbirth in 1832—a child actually thought to be his. Baume donated both their bodies to University College Hospital and was mistakenly arrested for murder. The papers now had him pegged as the 'Islington Monster'.[70] Saull's morally-upright self-image is difficult to square with so politically fickle and sexually delinquent a friend. It may explain why Saull's letters were addressed formally "Dear Sir", with none of the intimacy of the day, never "Dear Baume". That, perhaps, was the most telling.

The Devil's Pulpit

By now out of prison, the Rev. Robert Taylor trod new boards in 1830. In May, Carlile had taken over the huge Rotunda venue, close to Blackfriars Bridge. It was a massive gamble, with the lease, taxes, and refurbishment running to £1300.[71] There was a decayed opulence to this huge building. It was more a complex than a venue, with billiard rooms, apartments, bar, coffee room, library, and two theatres. The smaller circular theatre had once been a museum, with a gallery supported by marble pillars, and

---

68 HO 64/16, ff. 127–28 (Oct. 1830).
69 *Lion* 3 (27 Mar. 1829): 397–98; Baume 1829, 4–5; R. Richardson 1989, 168–71, 236–37. Baume *ODNB*.
70 Baume *ODNB*; Cooter 2006, 3–5. Baume also adopted an orphan, whom he had the Social Father, Robert Owen, re-christen "Julian Hibbert Baume": *Crisis* 3 (22 Feb. 1834): 214–15.
71 *Prompter* 1 (2 July 1831), 555.

on its dome Carlile had painted the signs of the zodiac in readiness for Taylor's sermons. Even the spy thought Carlile had fitted up the smaller theatre in a "very handsome manner".[72] With its Ionic portico crowned by a statue of Contemplation, the whole place resembled a crumbling Grecian temple, which somehow seemed made for Taylor's dramas on the Mithraic sun-worshipping origins of Christian myth. Coupled with this was a larger theatre, so large it was once used for horse shows, and able to accommodate 2000 at political rallies.[73]

The radical nature of the institution was immediately visible: two tricolour flags hung on poles in the entrance.[74] Nothing was more guaranteed to sting traditionalists: the hated French flag, much revived after the July Revolution, "the Symbol of Treason", as one loyalist put it.[75] From the start, informants were tipping off the authorities. One was horrified to see William Cobbett, after an inflammatory lecture, hand the baton over to Taylor—"thus shocking to relate *Blasphemy* followed closely on the heels of *Sedition*",[76] he told the iron Duke of Wellington. The weekly pattern of fixtures at the Rotunda cemented this flip from one to the other: the "Reverend Blasphemer"[77] took the rostrum each Sabbath, complementing the Monday meetings by the new radical National Union of the Working Classes (NUWC, founded May 1831). These were simply the two faces of Janus. Many in the NUWC were themselves anti-clerical. Carlile emphasized from the podium that "Religion and Politics ... were intimately connected", with the church no more than a dumping ground for the younger sons of the aristocracy.[78] Then, with Taylor denouncing the political bishops, deploring the government funding of the Established Church, and decrying Christianity's part in the law of the land, his "blasphemies", like Saull's, were themselves radical and seditious. The context made them so. They were attacks on the state. All of this ensured that the Rotunda was of special interest to the Home Office.

---

72   HO 40/25, f. 262.
73   Parolin 2010, chs. 6–8; McCalman 1992, 52; *Prompter* 1 (13 Nov. 1830): 8; *Crisis* 2 (30 Mar. 1833): 89; Carnall 1853–54; Brayley 1850, 5: 319–20.
74   HO 40/25, f. 211.
75   HO 40/25, f. 258 (11 Nov. 1830).
76   HO 40/25, ff. 157–58 (9 Nov. 1830).
77   HO 40/25, f. 218.
78   HO 40/25, f. 235.

Saull was in the thick of it. He might even have helped fund the Rotunda.[79] He certainly contributed to the events. The ubiquitous treasurer, he took over the NUWC's finances,[80] but he had little input into their Monday meetings. Sunday's was a different matter. He could be seen in the coffee shop with Hibbert, the fishmonger John Pummell (another indicted alongside Saull), Carlile, and Taylor before an astro-theological drama[81] or taking to the stage to talk on astronomy before Taylor's main event. In fact, the nightly sequence perfectly captured the context of his developing science. On each Sabbath, Carlile would kick off, warming up the audience (which could reach 1000) with a lesson from Volney's *Ruins*. Then Saull might talk, followed by the main event, Taylor's drama, the lot topped off occasionally by a Hibbert skit on Church services, to "much laughter".[82] Another who often topped off the evening was the veteran Jacobin of the 1790s, the golden-voiced John Gale Jones, an orator with form who was to become close to Saull.[83] He would add an "abusive" onslaught on miracles, as priestly devices "to Gull and Rob the People".[84] Sometimes Saull followed Taylor, entering

---

79   Wiener 1983, 164–66. On p. 186, n. 2, Wiener cites HO 64/11 f. 446 (the spy's report of 29 Nov. 1831) as evidence for Saull's help in paying for the Rotunda. But this only states that Saull was one of two hundred subscribers (tickets were 10s a quarter for a box, or 5s for the gallery), and he remained one when the number dropped to twenty after Taylor was jailed. The spy added that, besides Julian Hibbert, Carlile was funded by several "individuals who are known only to himself and also from Saul [sic] and Pummell." This seems to be referring to past ventures, not contemporary Rotunda financing. The ambiguous phrasing means we cannot say for certain that Saull helped defray the cost, although it is likely.
80   *PMG*, 16 July 1831; 30 July 1831.
81   HO 64/11, f. 212.
82   HO 64/11, f. 445.
83   John Gale Jones had been an apothecary-turned-activist. He had possibly met Saull by the early 1820s, when both were supporters of the LMI (Hudson 1851, 49; Claeys 2000, 160). Newgate had hosted him in 1810 for publishing a "scandalous" attack on the Tory M.P. Charles Yorke, that haughty remnant of 'Old Corruption' (Harling 1996, 120; Kent 1898, 259). Then came another 12 months for a libel on the detested Lord Castlereagh (Miles 1988, 73; Kent 1898, 258–59; Maccoby 1955, 259). His republicanism and hostility to the Church had only strengthened. Blind assent was being demanded to state-blessed Christian dogmas, he argued, after Paine; and, if we dissent, "we immediately feel the chain pressing heavily upon our necks, reminding us of our wretched thraldom". Were Jesus alive in London, he notoriously suggested, he too would be in the dock for denouncing these "pernicious doctrines", never mind his sympathy for hovel-dwellers and prostitutes (J. G. Jones 1819, 5, 16–17; Epstein 1994, 107; Parolin 2010, 1–4).
84   HO 64/11, f. 445. For Taylor's huge audiences in November–December 1830, reaching a thousand: HO 64/11, ff. 212, 213. The spy reported that Volney's *Ruins*

into the free-for-all that ensued. Typically, Sabbath meetings would end with Hibbert, Saull, and Gale Jones addressing the crowds with "their usual abuse of religion and Political Government". While, on one Christmas Day, said the spy, "Hibbert, Saul [sic] and Jones gave their usual long speeches against all Religions and insisted that no such person as Jesus Christ was born on that day or that he ever existed at all".[85]

Amid all of this came Taylor's astro-theology sermons, which were inspirational to Saull. But their heightened radical context was now itself practically apocalyptic. As Taylor started his astro-theology dramas on 7 November 1830, the political roof seemed to be falling in—and taking the church spire with it. The King's speech at the opening of Parliament on 2 November had caused uproar for ignoring parliamentary reform. With the King due to visit the City on the 9th, radicals prepared a warm welcome: tricolour flags came out, stickers went up advising the populace to arm, and the authorities anticipated a "Riot", as extremists spoiled for a fight with the hated new police. Taylor stirred things up. He threatened that if the King did not stand with reform "he should take care that he was not served on that day in the same manner his predecessor Charley was served" (the beheading of Charles I). Nor was Carlile more temperate: he told listeners to "prepare themselves to fight for their Liberty" and, were the government to bring in emergency legislation, he would "immediately call on the people to take up Arms".[86] "Loyal subjects" sent hysterical notes to Wellington, relaying overheard titbits or copies of "seditious" flyers, or relating how the "lower orders" were being fired up. All damned the "abominable proceedings taking place at the Rotunda"—"His Majesty, His Majesties Ministers, and all that is great, and good, are there denounced as the vilest of the dregs of humanity." Conspirators intended "to surround the Royal Carriage" and "demand a pledge from his Majesty on the subject of Reform".[87] One informer saw "fury and desperation" written on the faces of every

---

was read as a preamble every week right into 1832, for example, HO 64/11, ff. 212, 227, 289, 317, 445, 454, 458, 462; HO 40/25, f. 386; HO 64/12, f. 49.
85 HO 64/11, ff. 458, 462. By 1832, audience numbers were dwindling, and "Hibbert, Jones and Saul [sic]", despite their "usual abuse" of Christianity, were being little heeded, according to the spy: HO 64/12, f. 2.
86 HO 40/25, ff. 154, 211.
87 HO 40/25, f. 258 (11 Nov. 1830).

man in the "mob". It was "horrible to hear the expressions of vengeance against the Government in general and against the 'The b[l]oody old Duke of Wellington' in particular". Real venom was reserved for the Iron Duke. If he comes to the City, a plotter was heard saying, "we will take care he shant come home alive".[88]

These November nights the Rotunda was overflowing. With up to six thousand unable to get in, orators delivered their speeches from the top of the portico outside.[89] The crowds, said an informer, "seem ripe for any species of revolutionary crime and threaten vengeance to the King, and his Ministers".[90] The Rotundanists were "seedsmen of sedition"; they "ought to be brought to justice", and the "riotous & desperate characters" milling around Blackfriars be forced to disperse.[91] Panic set in: local shops shut, and special constables were housed in stables opposite the Rotunda. Magistrates were called up and the police put on alert, as the government heard of "plans to cut the gas-pipes, rip up street stones, fire the town and kill Wellington (the 'English Polignac')".[92] The Reform Bill agitation had begun.

Astro-Theology

At this moment, and at the epicentre of insurrection, Taylor started his astro-theology dramas. Throngs greeted him, and he revelled in the adulation. Feeling "great pride" in being branded the 'Devil's Chaplain', he announced at the start "that he meant to play Hell and the Devil too with [the] whole System of Religion as that was the greatest Radical Reform the people stood in need of."[93] He started with an audience of 700. By Christmas 1830, 1000 were paying their threepences (or sixpences for a circle seat).[94] In the small theatre the rakish Taylor in

---

88 HO 40/25, f. 157 (9 Nov. 1830).
89 HO 40/25, ff. 153, 199, 214.
90 HO 40/25, f. 55 (5 Nov. 1830).
91 HO 40/25, f. 33 (4 Nov. 1830).
92 Quoted by Prothero 1979, 277–79; specials: HO 40/25, f. 209 (10 Nov. 1830). HO 40/25, f. 218, suggests they were planning to cut off the water mains as well, so that the expected fires could not be doused. Orders went out to guard the gasometers and pipes: HO 40/25, f. 115.
93 HO 40/25, f. 281 (15 Nov. 1830).
94 Audience numbers were staggering: 700 were reported by the spy on 15 Nov., HO 40/25, f. 281; 800 a week later: HO 64/11, f. 167 (22 Nov. 1830); 1000 by Christmas:

full canonicals, tricolour ribbon draped across his shoulders, supping wine, was roared on by the crowd, taking their threepenny delight in his Bacchanalian parody of Christian rites. Many were young women, mesmerized by the priest, dapper in his bright gown and flashy rings, the ensemble set off by gold buckles on his shoes. Not everyone was enthralled. So titillating was blasphemy that youngsters would sneak in to the show. One later recalled that he found Taylor "a vain, conceited fop" who flourished "a scented cambric handkerchief ... at every pause in his discourse."[95] But the audience was generally well-heeled and appreciative, despite the occasional rowdies drifting in off the street. To bounce them, Carlile ordered "constable staves" to be made and used on his authority.[96] At other times informants claimed that the audience was mostly of the "lower orders", with the well-dressed in the boxes.[97]

Taylor knew he was flirting with danger. As crowds were raging outside at their lack of power, Taylor was raging inside at the priests' usurpation of it. No such person as Jesus had lived, he said, Christ was just a poetic incarnation. The "Astronomical senses of the Words God, Jesus and Christ [were] nothing more than the hieroglyphs of the old ages deduced from the signs of the Zodiac." Weekly reports on such arcana went off to the Home Office. Christianity Taylor "called the Bloodiest System ever yet known." Turning these astrological images into flesh was "a Barbarous species of fraud by which the Clergymen and priests of all ages" robbed the "human race of their senses and substances."[98] And, by "substances", Taylor meant robbing "to the tune of Nine Millions Nine Hundred and Ninety Nine Thousands annually out of the pockets of the hard working people."[99]

The Bible was nothing but a celestial *"picture in words"*, according to Taylor, its actors portraying the sun's annual trajectory through the zodiac. Prophets and apostles were personifications. They poetically depicted the rising and falling of constellations through the year, as originally envisaged by the Babylonian and Chaldean astronomer-priests,

---

HO 64/11, ff. 212, 213.
95 Vizetelly 1893, 98–99. Taylor was now drinking too much and often intoxicated: HO 64/11, f. 209.
96 HO 64/11, f. 207 [1830].
97 *Prompter* (23 July 1831): 643.
98 HO 40/25, f. 281 (15 Nov. 1830).
99 HO 64/11, f. 167 (22 Nov. 1830).

who had incarnated the dramas of the sun god's journey and turned them into stirring parables for the masses. The folk tales were absorbed and adapted by 'seers', or masonic initiates across the Middle East, one group of whom would come to be called Hebrews.[100] Others in Persia and India would canonize these astronomical allegories in their own sacred texts. Taylor summed up his first talk, on "The Star of Bethlehem", the spy reported, by claiming "that he had found the key of the stable unlocked the door and found little Jesus and had swept the stable of all its Christian filth and Superstition". The apostles were dispensed with in turn as poetic figments of celestial events. What astonished the spy was the rapt attention of the well-heeled audience to all of this, and how each new twist "appeared to give great satisfaction."[101] So it went on weekly in the galleried auditorium, Taylor surrounded by orreries and a giant crucifix, flamboyantly pointing with his lace handkerchief to walls covered with astrological charts.

But the audience's rapturous applause suggests more than seductive fascination with a dangerous deism. It reflects the captivation with all things oriental. This was to be seen city-wide. The new London University had just established chairs of Hebrew, Sanskrit, and Hindustani,[102] not only defying the Anglican seminary norms of Cam and Isis but setting up the Dissenters' sons for service in the East India Company. The London Oriental Institution was also newly founded. Egyptomania was in full swing, and the theatrics of mummy-unrolling all the rage.[103] When it came to the exotic, nothing beat Giovanni Belzoni's spectacular Valley of the Kings exhibition in the Egyptian Hall in Piccadilly, opened in 1821 with a public unwrapping of a mummy.[104] A mock-up burial chamber with its hieroglyphed walls whetted the public's appetite and set the scene for Taylor's oriental decoding. By calling the 'sacred' words used to describe Jesus's birth "a direct plagiarism from the Sanscreet text of the Bhagavat Pourana (that is, in English, the *Book of God*) of the Hindoos", Taylor was tapping into a modern vein. Here was familiar

---

100 R. Taylor 1831, 34, 55, 195, 247–52.
101 HO 64/11, f. 207. If the spy was accurately recalling the spoken lecture, then the subsequent printed version was toned down: R. Taylor 1831, 15, 30. Wiener 1983, 165–66.
102 Bellot 1929, 37–44.
103 Moshenska 2014.
104 Tromp 2008, 184.

biblical imagery exposed in Egyptian chronicles, in "the Mythriacs of Persia, and in the fabulous writings ascribed to Zoroaster",[105] and all traceable to the same astronomical events. This eclectic mix of oriental erudition, astronomical pizazz, and biblical exegesis gave him the competitive edge in a crowded market place. There was even something of an "Astronomical Mania" developing at the time,[106] with commercial astronomy lecturing just beginning to take off. The Devil's Chaplain was riding a crest. But his astronomy had a more intimate and threatening cultural depth. His foppish extravaganzas in Carlile's theatre set him widely apart from conventional lecturers. If, in Altick's words, astronomy was an exhibition "exploring infinite space in a little room",[107] Taylor's was giving it radical depth as the resort of the gods.

His exegesis of Persian, Indian, Hebrew, and Greek texts, presented in its final form at the Rotunda (November 1830–July 1831), might have been some years maturing. Saull, too, had long been fascinated. In 1827, he had already evoked astrological images lifted from Volney's *Ruins*—familiar fare for every radical and a major Taylor source.[108] Parrying his vicar in 1827, Saull took one episode—the Garden of Eden—to decipher astro-theologically. He presented it in short-hand, rather cryptically, and his epistle itself required some decoding. The vicar must have been perplexed to hear from his wine merchant that all biblical characters were "purely astronomical" fictions and that "The language of religion

---
105 R. Taylor 1831, 23, 34, 44.
106 Huang 2016, 2017.
107 Altick 1978, 80.
108 With Taylor's *Devil's Pulpit*, backwards Britain was finally catching up with Europe on solar mythology. Besides Volney, Taylor was indebted to Charles François Dupuis (1742–1809), a member of the National Convention after the revolution, who in the *Origin of All Religious Worship* (7 vols. 1794–95) had traced the mythic elements in the world's religions to a common root (Epstein 1994, 140ff; Cutner n.d., 30ff. On Dupuis, Butler 1981, 78–82). Another of Taylor's sources was the rare Macon Reghellini de Schio, *Freemasonry Considered as the Result of Egyptian, Jewish and Christian Religions* (1829). Taylor (*Comet* 1 [23 Dec. 1832]: 326) said that he had one of the only two copies in England. Only further research will show actually how original Taylor was. Others shared Taylor's interests. Carlile's shoemaker friend, Sampson Arnold Mackey, who knew his Volney and Dupuis, had looked to the astronomical roots of ancient names in *Mythological Astronomy*. This was designed to prove that the ancients knew, and kept alive by their parables, the fact that the pole of the earth had once swung down to become parallel to the ecliptic, "with all the fiery consequences that must arise from such a state of the heavens" (Mackey 1827, 62). The knowledge, he thought, would be useful in geological speculation, which made it of interest to Saull.

throughout, is the language of the skies." As "proof" he pointed to a celestial globe for an astro-exegesis of Eden:

> Boötes [the Herdsman] the Osiris [a founding god, pushed into the under-world] of the Egyptians—the Adam or first man of the Persians and Chaldeans, who, by setting heliacally at the Autumnal Equinox, delivered the world over to the wintry constellations, and in falling below the horizon introduced into the world the Genius of Evil, Ahrimanes [the Zoroastrians' principle of evil, formed out of darkness, on which, according to Dupuis, the Devil of Genesis was based], represented by the constellation of the Serpent. Here is the woman [Constellation Virgo] who gave her husband the fruit of the tree, and by setting first, seems to draw him after her; and when the Virgin and the Herdsman fall beneath the western horizon: Perseus the Cherub with the flaming sword rises on the other side, and drives them out of the garden. And here again, at the opposite or vernal Equinox, we behold the Lamb [Constellation Aries] that taketh away the sins of the world, typical of the Christian religion:— The Sun appearing in the sign of Aries, the Ram, brings back the reign of the summer months, and appears triumphant over the Serpent, who disappears from the skies![109]

Taylor managed to get through his sacrilegious fare until Good Friday, 1 April 1831, when the Society for the Suppression of Vice acted. Paid informants sat among the audience that night watching Taylor, dressed "like the Archbishop of Canterbury", make a pantomime of "The Crucifixion of Christ".[110] It was his yahoo-mockery as much as the subject matter that got him arrested.[111] From his shorthand, the informant quoted Taylor on Christ's crucifixion:

> The Everlasting ceased to be. The Eternal God was no more. The great I AM was not. The living God was dead. There was a Radical Reform in the Kingdom of Heaven. The boroughmongers were turned out. God, over all, was put under. The blessed, for ever more, was no more blessed. And the Holy, Holy, Holy, was WHOLLY kicked out.[112]

---

109 Saull 1828a, 19–21. This is cobbled together from Volney 1819, 103–04, 132.
110 *Prompter* (23 July 1831): 641–48.
111 The more sober but equally sacrilegious Mackey could just as easily explain virgin births, devils, the Christian imagery of fishes and why the Ram or lamb was the lead in Christian dramas, without being prosecuted (Mackey 1827, 208, 216–24).
112 HO 64/17, f. 48. The wording, taken from 22-year-old informer Joseph Stevens' shorthand, differs from the *Devil's Pulpit* printed version (*Comet* 1 (11 Nov. 1832): 226. This has the wrong Rotunda date, the Good Friday Sermon "The Crucifixion of Christ" was preached on 1 Apr. 1831).

A "bill" (indictment) was quickly issued at the Surrey Sessions, charging Taylor on seven counts. Again, Saull stood bail. Keeping Taylor out of jail would keep the Devil in his Pulpit, at least until the trial. But despondency set in among the brethren, as they feared, in Hetherington's words, "that Taylor is 'nailed'".[113] So it was to be at the Surrey Sessions in Lincoln's Inn on 4 July 1831. Saull had been lined up as a character witness in this, Taylor's last judgement, but judge and jury had little time for the Christ-denier and cut the trial short. It meant two years' incarceration, not this time in the cushy surroundings of Oakham, but the "disgusting Horsemonger-lane gaol", just across the river in London.[114]

This jail was a huge, intimidating, brickwall-enclosed building, with its own gallows. Here capital offenders awaited execution, and Taylor must have felt like one from his treatment. A clampdown by the justices meant that supporters including Saull and Hibbert were barred from seeing him, "even to shake hands" or to provide "refreshment or news". Complaints were made to the governor, petitions got up, and the Rotundanists met to discuss the cruelty, but to no avail.[115] The treatment was clearly designed to break the infidel but also to warn his followers, like Saull. Persevering, in September 1831, Saull was finally given exception to visit, for an hour a week.[116] Carlile's new rag, the *Prompter*, painted the conditions as extreme: Taylor was treated like the worst felon, because (so the justices said) his crime was of the highest "moral degradation"—hence the apparent restrictions, privations, and solitary confinement at the whim of a "cruel gaoler".[117] This jail term was what finally extinguished Taylor's own shooting star. His infidel mission was effectively over.

---

113 HO 64/11, f. 229. Bail: HO 64/11, ff. 200, 296.
114 *Prompter* (23 July 1831): 641–48. On Saull, Hibbert and others supporting Taylor in court: HO 64/11, f. 337. On the "disgusting" jail: *Prompter* (23 July 1831): 641–48.
115 HO 64/11, f. 337 (7 July 1831). Carlile's claim (reported by the spy) that Taylor was being "slowly Murdered" was stretching the point.
116 *House of Commons Papers; Accounts and Papers: Reports and Schedules pursuant to Gaol Acts*, vol. 33, 1831–32, pp. 224–25.
117 *Prompter* (13 Aug. 1831): 713; (26 Aug. 1831): 727; (24 Sept. 1831): 811; (15 Oct. 1831): 860; (29 Oct. 1851): 886; (12 Nov. 1831): 920; Cutner n.d., 29. For an alternative view of Taylor's confinement, mentioning brandy, porter, meals from the local inn, and visits once a week by a woman who calls herself his "wife", see the antagonistic *Spectator*, 23 Jul. 1831, 706.

Astro-Geology

Taylor's fate, and Saull's friendship, help to explain what happened next. Saull effectively switched his political and scientific stratagem at this moment, 1831—the year he set up his geology museum. As Taylor's patron, he saw his irreverence being tortured for high crimes. Carlile, too, was in jail again, for sedition (1831–33).[118] Paranoia began to grip the community. They believed "they should now be well watched" by secret agents (as one agent ironically reported); indeed, that Taylor's sentence was a signal and "was intended to 'floor the Rotunda'".[119] For Saull, the writing was on the wall. As Taylor's chief disciple, he had been fingered publicly in the papers and would be so again.[120] His actions would now become more circumspect. The next time he broached astro-theology in print would be the last time, and, tellingly, he would only do so anonymously. The result was a published letter on 1 January 1832, *From a Student in the Sciences to a Student of Theology*, whose authorship has long caused confusion (see Appendix 2 for proof that it was Saull's).[121] It is perhaps no coincidence that Saull's first documented public lecture on geology (that I can find) was itself on that very day, 1 January 1832, at the Western Co-Operative Institute in Poland Street.[122] So his first geological speech coincided with his last frontal assault on Christianity. Although Saull remained a materialist for the rest of his life, from this point on he would mostly attack theology from behind the protective shield of geology. Scientific subterfuge was to provide the smokescreen he needed to assault parsondom while evading prosecution. And almost as a corollary of this, Taylor's suspect astro-theology was displaced as Saull shifted to the old Jacobin Sir Richard Phillips's unorthodox but unindictable astronomical explanations of events in earth history. Thus was started another of Saull's life-long obsessions, astro-geology.

Conceivably, Taylor's celestial allegories had piqued Saull's interest in astronomy proper. For months to come, Saull would still join in Gale

---

118 Wiener 1983, 177; Saull and Hibbert, as usual, had offered bail: HO 64/11, f. 197.
119 HO 64/11, f. 229.
120 *MC*, 26 Feb. 1829; *Times*, 23 Jan. 1833.
121 [Saull] 1832a, 15 for dating.
122 *PMG*, 31 Dec. 31 1831.

Jones's and Hibbert's anti-Christian harangues after Rotunda lectures.[123] But when he did ascend the Devil's Pulpit himself to talk now, it would be on astronomy ("of which he says he is a Master", reported the spy contemptuously in November 1831[124]). Saull was already a member of the Astronomical Society of London by June 1831.[125] Such bona fides provided sanction and status, and his credentials would be exploited for public (and public-house) speaking on science and education. Even if he preferred infidel chapels and co-operative halls to the corridors of science, his Fellowships of the Geological and Astronomical Societies still validated his competence when speaking in public. For example, the *Quarterly Journal of Education* (who clearly did not know who he was) vested him with an importance based on these magical credentials when running a story in 1831. Saull was chairing the radical MP Joseph Hume's meeting to promote children's education, the state having failed miserably in this regard:

> EDUCATION OF THE WORKING CLASSES.—A very numerous meeting of the working classes residing in the Tower Hamlets, took place on the 12th of July, in the grounds of the Ben Jonson public house, at Stepney, to consider the best means of establishing 'Societies for the Promotion of Public Instruction.' Mr. D. Saull, Fellow of the Geological and Astronomical Societies of London, was called to the chair. Mr. Hume, M.P. addressed the meeting at some length, expressing his hope to see the day when the state, like America and other countries, would make a proper provision for educating every child...[126]

Conversely, because these affiliations were pushed to the fore in a pulpit-age uneasy about scientific authority, conservatives could use them as targets. An example occurred January 1833, when Saull was trashed in the *Times*. It started on the 16th when he and other City electors requisitioned the Mayor in the wake of the Reform Bill. They presented a petition with a thousand signatures—"the most numerously signed that was ever presented at the Mansion-house"—which requested that

---

123 HO 64/11, ff. 458, 462; HO 64/12, f. 2.
124 HO 64/11, f. 445.
125 *Memoirs of the Astronomical Society of London* 4 (1831): 683. Saull might have been a frequent attendee, but the only reference I can find is in the *Morning Post* (9 Jan. 1836, 3), where he is listed among the "distinguished members" in attendance. He was probably more embedded in the Uranian Society.
126 *Quarterly Journal of Education* 1 (1831), 391; *Examiner*, 17 July 17 1831.

the Mayor approach Parliament with their democratic demands.[127] The City's elite artisans and shopocrats were taking the lead: they wanted triennial parliaments, secret ballots (to thwart intimidation and bribing), and abolition of assessed taxes (direct taxes, including the hated house and window duties which could put urban tenement rooms in the same band as country houses). The Guildhall meeting to discuss it took place on the 21st.[128] The next day the *Times* rubbished it in a leader (as a "lamentable failure" convened by "busybodies") and ran a letter on the defeat of the "'destructive' clique". Then, having inquired about the ringleaders, let another letter writer ("Verax") vent his spleen on these "poor insignificant creatures" the day after. Each of the conveners was smeared in turn through character assassination based on appearance and employment. It became highly personal and intrusive: "Mr. Nicholson is a coarse, stout, vulgar man ... He was formerly a tea-dealer in New Bond-street"; "Mr. Williams is ... [a] warehouse man in Watling-street; he is very conceited of his own opinions, but his knowledge is confined to the extracts he may make from his weekly reading of the [Cobbett's] *Political Register*"; "Mr. Newell is a cabinet-maker in Whitecross street, and considers himself a great politician, because he interferes in the politics of his ward, Cripplegate-without, but he is a person of no cultivation of mind", and so on. It set the tone for the damning appraisal of Saull, which took in his geological credentials:

> Mr. Saul [*sic*] is a spirit-merchant in Aldersgate-street, and lectures to wondering mechanics at the Philadelphian-chapel, near Finsbury square; he assumes to be a great geologist, having some smattering of the terms employed in that science, and has got his name on the list of the Geological Society: he is a very weak and conceited person, —a disciple of Mr. Owen, and a supporter, I have understood, of Mr. Robert Taylor.[129]

The spiteful effrontery led to a slanging match, the ultra-radical *True Sun* lashing "the writer of the contemptible twaddle" in the *Times*.

> The speakers at the Guildhall meeting are described as coarse, stout, weak, conceited, ignorant, uncultivated, and vulgar men. One of them,

---

127 *Cobbett's Weekly Political Register* 79 (19 Jan. 1833): 155; *TS*, 17 Jan. 1833, 3; 22 Jan. 1833, 2; *Atlas*, 20 Jan. 1833, 31. Maccoby 1935, 65, 84.
128 *Atlas*, 27 Jan. 1833, 47–48; *Bell's Life in London and Sporting Chronicle*, 27 Jan. 1833; *Examiner*, 27 Jan. 1833.
129 *Times*, 22 Jan. 1833, 2; 23 Jan. 1833, 2.

> Heaven forefend! is actually convicted of being a hatter; another commits the unpardonable crime of selling tea...and a Mr. Saul [sic] is charged with being a "disciple of Mr. Owen!!"

"It is amusing enough to observe this insolence towards tradesmen...."[130], but, in truth, the attack left the City traders smarting, and they extracted a niggardly half-apology in the *Times*.[131] By now, the morning (and evening) papers were all chipping in, on the Whigs' "aristocratical airs" or the tradesmen's impertinence.[132] It all goes to demonstrate that Saull's democratic activity could leave his geological qualifications being questioned. Ultimately, fellowship of a scientific society was not an impenetrable shield for a radical.

It is at this point, the dawn of the 1830s, that Sir Richard Phillips becomes visible in Saull's circle. As like minds, Saull and Phillips were intimately acquainted. Phillips, the veteran radical, had himself been written off in his day as "a dirty little jacobin".[133] He was another who had done an obligatory eighteen months inside for selling Paine's *Rights of Man*. Saull and Phillips shared anti-clerical, republican views, and Phillips advocated free universal education and public libraries,[134] endorsing an ideology of open accession that would define Saull's museum. Phillips, like the Owenites, sought a reformation of the individual by the rooting out of prejudicial customs, and one in particular: meat-eating, a dehumanizing custom that was morally debilitating. In his day, Phillips's "Pythagorean diet" (the original name of vegetarianism)[135] made him infamous, a laughing stock to the John Bull brigade. But, recently, Sky Duthie has cast it in a much more sympathetic light. He shows how such dietary dissidence rested on views of the liberation of all life from injustice and how it underlay a broader critique of societal customs which desensitized humans and sustained tyrannies. Indeed, by emphasizing the suffering common to all animal life, Phillips might well have eased Saull into his 'evolutionary' views, as we will see. Phillips had a mania for publishing cheap encyclopaedic

---

130 *TS*, 24 Jan. 1833, 4.
131 *Times*, 28 Jan. 1833, 2; 25 Jan. 1833, 3.
132 *British Traveller And Commercial And Law Gazette*, 24 Jan. 1833, 1.
133 *Blackwood's Edinburgh Magazine*, 12 (Dec. 1822): 704.
134 Duthie 2019, 86.
135 R. Phillips *ODNB*.

texts, dictionaries, and factual compilations, scores of them under a variety of pseudonyms.[136] Here was another man after Saull's heart, whose radicalism embraced science and politics. The old Jacobin could be found taking tea at Saull's Aldersgate Street depot. And once Saull's museum was up and running, Phillips, in his most famous compendium, *A Million of Facts* (1835), would extol its "ten thousand" exhibits.[137]

Saull and Phillips shared political platforms, which spoke volumes in these months. Notably, the two were together in the Metropolitan Political Union, a middle- and working-class pressure group pushing for parliamentary reform. Founded in March 1830,[138] it had a short life, like so many of these volatile unions before the Reform Act.[139] Saull was prominent, as were many of his colleagues. But they were seen as fanatics in the union: complete suffrage ultra-radicals, republicans, Church disestablishers and de-funders, wanting to ditch all church rates and tithes. They were too extreme for the middle-class moderates urging only household suffrage. This was particularly the case after the July Revolution in 1830. The French uprising put the fire in radical veins,[140] and the ultra-radicals' jubilation and Gales Jones's and Taylor's republican rhetoric in the union was seen to border on sedition, causing a ruction. Saull spoke up when they were expelled in August 1830, undoubtedly to defend his friends.[141] But many extremists thought it better to jump than be pushed; they seceded from the union and set up the more radical National Union of the Working Classes.

---

136 Topham 2007, 144ff.
137 R. Phillips 1835, 293; tea: J. A. Cooper 2010, 50.
138 *Spectator*, 13 Mar. 1830, 4. Other members included Brooks, Baume, Cleave, Hetherington, Lovett, Gale Jones, and Carlile's erstwhile shopman James Watson.
139 LoPatin 1999 on the rise of political unions running up to the 1832 Reform Act, when the working and middle classes were still in tandem.
140 HO 64/11, f. 161.
141 Saull's speech was reported in the (unobtainable) *Reformer's Register*, Part I (for July, August, and September 1830), as announced in an advertisement in Carpenter 1830–31, 16. On the expulsions: *Weekly Free Press*, 14, 21 August 1830. The ultras retrenched back into the Radical Reform Association and eventually formed the nucleus of the NUWC, with Saull as treasurer. Lovett 1920, 1: 57–58; Wiener 1989, 21–22; Belchem 1985, 200–05; Prothero 1979, 276–77. Saull was also in the Southern Metropolitan Political Union, founded in the Hercules Tavern, Lambeth, in October 1832. At the inaugural meeting, he seconded the resolution demanding "universal suffrage, vote by ballot, and triennial parliaments", as well as no property qualifications, and immediate abolition of the newspaper tax: *Examiner*, 21 Oct. 1832.

Phillips's science was celebrated as even more extreme. He had ditched the great Sir Isaac Newton's gravitation, which, for many, had been the defining moment of English science. As a radical Enlightenment ideologue, Phillips substituted matter and motion for attraction and repulsion. He simply rejected Newton's gravity, like Christian grace, as an impossibly occult action at a distance. No radical who trashed Newton was going to escape conservative censure, and Phillips was doubly-hated as a "filthy jacobinical dog".[142] Detractors laughed that he lacked gravitas himself: his roaming mind, "unincumbered by knowledge", said a scathing *Quarterly Review*, had led to him freewheeling into his own "secret" space. But deists were attracted, in equal parts by the simplicity and iconoclasm of his challenge.[143] Attraction and repulsion were "mystical terms", agreed Saull, so much "dust...thrown into the eyes of the world".[144] For Lovett, another seceding Metropolitan Political Union hard-liner, Phillips ratcheted up the heresy by walking him round St Paul's Churchyard one moonlit night, explaining his anti-gravitational theory, sacrilegiously chalking up diagrams on the cathedral walls as he went.[145]

Nor was Phillips's innovative technique to carry audiences with him better liked by the ranking elite. What could you say about a publisher who pleaded for a consensual approach to science, wanting, of all things, the working-class reading public to become part of the forum for establishing truth?[146] But in inviting this "public Jury" approach—democratic participation in the knowledge-making process—he was enthusiastically emulated by Saull in his open museum.

Phillips's orbital causes of environmental changes were equally trumpeted by Saull. Plebeian interest in planetary motion affecting

---

142 *Blackwood's Edinburgh Magazine*, 12 (Dec. 1822): 704.
143 *Republican* 8 (15 Aug. 1823): 169; 14 (8 Sept. 1826): 274–75. *Quarterly Review* 19 (July 1818): 375–79. Fellow travellers approached Newton from another side: an Owenite and lecturer on the Saull circuit, Thomas Simmons Mackintosh, in his "Electrical Theory of the Universe" (1837) rejected Newton's use of an initial Divine push to move the planets and ran the celestial machine by electrical fluids (Morus 1998, 135). Nor was it only radical deists who resisted the Newtonian consensus; plebeian Muggletonian Protestants could also produce rival astronomies (Reid 2005).
144 Saull 1832b, iii-v.
145 Lovett 1920, 1: 37.
146 Wallbank 2012, 165–70.

ancient ecologies had been growing through the 1820s.[147] What partly made the science so attractive for Phillips and Saull was its huge time frames, those "millions of years or ages" to confound "rabbinical or monkish commentators".[148] Saull had been trying to get behind the fashionable mythological astronomies to find real mechanical explanations, the sort Phillips now proffered. Saull's search had started at least by 1826, judging by the pseudonymous "Letter From A Friend: On Fossil Exuviae and Planetary Motion" published in Carlile's *Republican* (Appendix 1).

Here we have Saull's first-known published letter. It suggested that repetitive planetary wobbles could account for both the regularity of strata and the periodic switching of "torrid and frigid zones" (shown by animals from hot countries being found as fossils in what are now cold regions). Long-term planetary cycles would change the earth's tilt and bring "the north and south poles eventually into the position originally occupied by the equator".[149] This dramatic image of the swinging obliquity of the ecliptic and the resulting Armageddon scenario as the earth's polar axis lay flat on the orbital plane seems to have come out of Mackey's *Mythological Astronomy of the Ancients*.[150] The melting Arctic or Antarctic ice sheets, when faced with the baking overhead sun, had caused the massive periodical inundations required by Cuvier. But evidence existed that humans had survived the last such "age of horror". For Mackey, the "stupendous" walls of Babylon, 300 feet high, were proof that they had been built to withstand the onslaught. In 1826, shortly after Mackey's book, Saull used a 50,000-year precessional cycle to explain a deeper geology: the often-alternating geological sediments of marine and terrestrial origin, and hot-house animals and plants turning up in British rocks. Then, two years later, in the 1828 letter to his vicar, he repeated this near-apocalyptic scenario. But now he made it a 25,000-year cycle (like Mackey) and put it down to the precession of the equinoxes, which must have continued without limit to totally flip the north and south poles.

---

147 R. Phillips 1821, 100–11; Mackey 1823, 1825; Byerley 1831.
148 R. Phillips 1832a, 2.
149 [Saull] 1826.
150 Mackey 1823, pt. 2, 81, 94, 115–16; 1827, 75–76, 80, 85–86, 90–93, 214. As J. Godwin 1994, 68, explains, the precession of the equinoxes for Mackey describe a spiral, with each cycle altering the earth's tilt by four degrees.

What was absent from this letter was any talk of progress or perfectibility. The probable reason is that Saull had yet to swing into the co-operative camp,[151] or join up with Robert Owen. At this stage, he was simply seeking mechanical explanations for geological periodicity. He was still searching for them when he came under Phillips's wing. The old Jacobin was a different cut altogether. Phillips hated the "antiquarian" mystics—naming Dupuis and Volney, but he meant his nemesis Mackey.[152] "Stellarizing all ancient history and poetry is exactly akin to the spiritualizing of John Bunyan"[153], and he would have none of it. He was a mechanist, whose anti-occultism and anti-scripturalism had appeal. Saull now ditched the allegorical pyrotechnics which landed Taylor in gaol and looked to orbital astronomy to deepen his emerging geological persona. Henceforth, he would become a scientific cypher for Phillips's planetary explanations of cyclical geological events. In 1832, Saull persuaded him to flesh out his views more thoroughly. The result was the eighty-page *Essay on the Physico-Astronomical Causes of the Geological Changes on the Earth's Surface .... By Sir Richard Phillips. Re-published, with a Preface. by William Devonshire Saull*.[154] In it, Phillips split out the twin causes as they affected geology: first, the diminishing or rising obliquity, or tilt, of the Earth's axis, which was responsible for Britain having alternating hotter and colder climes.[155] And, secondly, a complicated gyration which affected the precession of the Earth's axis.[156] This explained the alternating marine and terrestrial sediments, as the

---

151 The earliest sign we have so far of his co-operative commitment is his appearance at a shareholders' meeting of the London Co-Operative Trading Fund Association in Red Lion Square in 1827: *London Co-operative Trading Fund Association meeting of the shareholders held...11th Dec. 1827* (1827).
152 The feeling was mutual: Mackey 1825; *Lion* 1 (27 June 1828): 804.
153 R. Phillips 1832a, 37.
154 Saull's preface was dated May 1832. The essay was published on 29 June 1832 (dated from *TS*, 29 June 1832, 3; *Courier*, 29 June 1832, 1).
155 R. Phillips 1812, 122–23; 1821, 80, 109; 1832a, 22–26, 33.
156 This was not a straight precession of the equinoxes but complex motions involving the earth's orbital changes due to perihelion forces. When Sir John Byerley suggested that Phillips thought the precession of the equinoxes was enough to bring about the geological changes, Phillips corrected him by insisting that "I taught that the geological changes arise from the advance of the line of apsides around the ecliptic in about 20,930 years" (R. Phillips 1832b; 1832a, 36–37; 1821, 104–05).

oceans accumulated in either the southern or northern hemispheres.[157] But the critical point for Saull was that he saw Phillips use these changes of terrestrial climates to "work changes of species". It was an environmental necessity, with the slow ecological changes wrought by a gyrating planet through tens of thousands of years inevitably producing "the wonderful gradation of being which we witness".[158] Phillips's planetary views were thus a major factor facilitating Saull's evolutionary direction of travel. This final component would be strengthened by the emerging co-operative movement.

---

157 R. Phillips 1812, 118–20; 1821, 103–04, 109; 1832a, 5–12, 20, 31. Mackey (1832) attacked Phillips's explanation of shifting oceans between hemispheres, in a lecture in Dean Street, Soho, in which he also censured Saull for supporting Phillips's anti-gravitational views. Another denunciation of Phillips's orbital explanations and anti-gravitation appeared in the *MNH* 6 (July 1833), 361–62.

158 R. Phillips 1832a, 48, 52.

PART II

1830s

THE SHAVEN APE AT NEW JERUSALEM'S GATE

5. Perfectibility

The Whigs came into office in 1830, for the first time that century. This switch in political fortunes refocused minds—radical debates now raged over suffrage, the secret ballot, and annual parliaments in the run up to the Reform Act of 1832. These months saw renewed labour activism, the growth of political unions, and the rise of anti-capitalist alternatives. All of these served to expose deep divisions among Carlile's fellow-travellers. While the emphasis had been on debunking Christianity and de-funding the clergy, Saull had stayed in line. But the new political imperatives were forcing his re-evaluation of allegiances.

Many of Carlile's supporters, fired by these new concerns, drifted away. The poverty and powerlessness of the increasing numbers of urban workers meant that the radicals looked to ever more democratic solutions. But others—including Saull—while supporting this radical move, urgently began to seek co-operative alternatives to the capitalist economic system. Saull's social interests as a City merchant and exponent of the new Cuvierian palaeontology of progress made co-operation and, ultimately, socialism an apposite choice. The root of both progressive palaeontology and politics lay in nature's power, delegated from below, which pushed life ever upwards. This unaided climb, life pulling itself up by its own bootstraps, was a powerful democratic image. Push and power came from below, not from God's fiat passed down via a priesthood. Some street activists already saw in the new palaeontology an inbuilt perfectibility principle. With Toulmin's eternalism out of the way, this upwards ascent of life could provide a scientific rationale for the social doctrine of human perfectibility. Man was not depraved and fallen; as an animal he carried nature's principle on through his social ascent. It legitimated the utopian drive towards the perfected man, a socialist, while, dialectically, socialist belief in perfectibility reinforced the image of nature striving linearly 'upwards'.[1]

More than political exigencies were causing fallout. Personal ones, too, were forcing Saull's re-orientation. His shift was made easier by his growing circumspection in the face of custodial threats. But Carlile's behaviour pushed him further. Carlile's star had waned through the late 1820s. By 1831, he was off the scene, in prison, and his shop was in a parlous state. In 1832, they were selling off stock cheap. A "wreck", the spy called the business, as Carlile's house in Fleet Street was let in a last desperate measure. There had always been grumblings about Carlile's extravagance and brusqueness, too, with the spy reporting that "most of Taylor's and Fitch's friends" thought him "too Rash" and that he was no longer "respected".[2]

But what ultimately cost Carlile so much support was "the way he treats his wife".[3] This came to a head in 1830. Carlile started an affair with a young evangelical-apostate Eliza Sharples, fresh from the mill town of Bolton.[4] He moved her, pregnant, into his house and his long-suffering wife and children out. Jane, who had kept his shop open through thick and thin, and gone to prison for him, was booted out. This was too much for many: "moral delinquency", Hetherington called it. He added a few years later, on looking back, that "nearly all your best friends were ashamed of you—they had entirely abandoned you".[5] Hetherington and Carlile now loathed one another. But then Hetherington was devoting himself fully to working-class agitation. The hounded editor of the illegal *Poor Man's Guardian* (founded 1831),[6] still republican of course, and as anti-clerical as ever, was emerging pre-eminently as a class warrior, something Carlile never was. Hetherington's demands

---

1. Bowler 2021 on the interdependence of utopianism's pre-determined social goal and a *linear* view of 'evolution'.
2. HO 64/11, f. 85; HO 64/18, ff. 602, 736; *Cosmopolite* 1 (5 May 1832); 2 (26 Jan. 1833).
3. HO 64/11, f. 7.
4. HO 64/12 f. 38. The "Lady" is from Liverpool, the spy reported erroneously; she was from Bolton (Frow and Frow 1989, 38).
5. *PMG*, 1 Nov. 1834. 308; also 15 Nov. 1834. 326; 6 Dec. 1834. 347–49; Wiener 1983, 81, ch. 10; Keane 2006; Frow and Frow 1989, 36–38.
6. It was illegal because Hetherington refused to pay the government stamp duty (which was itself designed to wipe out the inflammatory street press): Wiener 1969; Hollis 1970. The *PMG* got everywhere; distributors even impishly left copies on the Duke of Bedford's doorstep: HO 64/12, f. 165.

for workers' rights reflected the heated political rhetoric inside the Rotunda, Optimist, and other radical venues: full representation and fair wages to end poverty and oppression, which meant suffrage, the ballot, and abolition of property qualifications.[7] None of these were Carlile's priorities.

They were, however, Saull's, who proved the point by taking on the treasurer's role in the Hetherington-inspired National Union of the Working Classes (the spy's main target for surveillance). Saull began dissociating himself from Carlile; in truth, their differences became irreconcilable as the reform crisis loomed. How he reacted to Carlile's "moral marriage" we do not know, although he was later to help Jane Carlile and her children.[8] But, for Carlile himself, the cash was drying up. It now appeared as if the two men shared nothing but a disgust of Christianity.

While Saull backed the new political unions urging universal suffrage, Carlile called them "contemptibly devoid of intellect and useful purpose".[9] For him, they were all resolution and no action. Carlile, the strident individualist, was moving in an opposite direction, playing the prima donna, sounding more and more the bourgeois liberal overlord, insisting that these "dastardly associations, contemptible, frivolous, paltry nothings" should stop posturing and build on *his* infidel framework.[10] He accepted no need for any further reforming foundations

---

7   For Hetherington these were the prerequisites before schemes like co-operation could be considered, as he insisted time and again up to the passing of the Reform Bill: *PMG*, 14 Jan. 1832, 245–46; 28 Jan. 1832, 254; 2 Jun. 1832, 407; 1 Sept. 1832, 513; 8 Sept. 1832, 528; 22 Sept. 1832, 541; 22 Sept. 1832, 537; 29 Sept. 1832, 548; 29 Sept. 1832, 551; 3 Nov. 1832, 588; 1 Dec. 1832, 631. Among many incubators of Hetherington's emerging class consciousness might be considered the LMI, which both Hetherington and Saull attended (Flexner 2014). Here self-teaching groups formed, strengthening self-reliance, and management was divided into "working class" and "not of the working class", emphasising the distinction.

8   Saull never neglected Carlile's family, adding to funds to make sure they were provisioned later in life: *NS*, 22 Nov. 1851; *Reasoner* 12 (10 Dec. 1851): 64; (21 Apr. 1852): 367. In this he worked in conjunction with his closest friends (see Appendix 6). These included the apothecary Thomas Prout, another Carlile bankroller who sat with Saull in every political union; Dr Arthur Helsham; and the Paineite Edward Henman, who had also funded Carlile.

9   *Lion* 4 (9 Oct. 1829): 449–52; Belchem 1985, 198; Wiener 1983,171.

10 *Prompter* 1 (3 Sept. 1831): 753; Wiener 1983, 172. Admitting that the more moderate National Political Union was the "best thing of the kind that had been attempted" was simply damning with faint praise: *Prompter* 1 (3 Sept. 1831): 754.

than his anti-Christianity, all else was hot air. As a result, he was expelled from the Radical Reform Association, while the Metropolitan Political Union members actually hissed the "comical blade" for his reactionary views before kicking him out.[11]

As for London's co-operative experiments favoured by Saull, Carlile took still graver exception. From the first, he branded them a "retrogression", believing that, without competition, society would level all down to the "mediocre". He saw co-operative efforts as stifling the "dynamic motives of human action", leading to a "diminution in production", in McCalman's paraphrase. The more rejection Carlile suffered, the more aggrieved and opinionated he became as he built bulwarks against the trend. "I hate the co-operative system that would monotonously tie down the talent or utility of mankind, so as to make the ingenuity of the genius subservient to the dulness of the dolt". Not for him the "new millenium" [sic], as he lashed all such schemes as "Utopian".[12]

Carlile had never really advanced labour's claims, now he rejected more revolutionary action. He was no less vehement against the trades' unions and their "Tom-fool tricks". He founded his *Gauntlet* (1833) to take on the unions, who blinded their adherents with "secresy [sic] and nonsense".[13] Everything about Carlile now smacked of betrayal. The last straw for Saull's radical friends was Carlile's acceptance of the Whigs' £10 household franchise as the basis of the Reform Bill, which would give democratic power to the middle classes while cutting out labour. The 'base Whigs' seemed to have got him. It was confirmed when, on top of endorsing a classic capitalist economy, he approved its Malthusian base, the ultimate horror. Unlike almost all ultra-radicals, he had accepted Thomas Malthus's dictum that population outstripped food supply, making struggle, despair, and death the norm in the fight for resources. In this, he appalled Saull, and even Taylor berated Carlile's "Anti-social"

---

11 Wiener 1983, 171–72; *PMG*, 1 Nov. 1834, 309.
12 *Lion* 1 (29 Feb. 1828): 258–62; McCalman 1975, 150.
13 *Gauntlet* 1 (1833): iii-iv. Even the anti-Owenite Trades' Union journal, *The Agitator, and Political Anatomist* (Dec. 1831: 8, in HO 64/19, f. 138), criticized Carlile for demanding unions give up secrecy, without which their members could be picked off by the government.

views on this score.¹⁴ Saull said that Malthusians should "blush with conscious shame". Such pessimism rested on ignorance of the earth's true "productive powers" and a failure to appreciate that a proper technical education would push up productivity and put mankind on "the correct path of improvement".¹⁵

Owenism, Geology, and the Social Millennium

For Saull, the bridge was burnt. He never wavered from Carlile's anti-Creation and anti-clerical materialism, he simply carried it into the co-operative camp as he worked up his palaeontology. By late 1827, he was already a shareholder in the London Co-operative Trading Fund Association, which planned to buy or rent land on which labourers could make and sell goods at their full value (with no middle men).¹⁶ But, as with so many nominally-agrarian and co-operative goals of the London-based activists, it ended up promoting education and sending out speakers to local groups ("missionary work", in Prothero's words). Finally, as Malcolm Chase says, it reflected this "growing didactic function" by changing its name in 1829 to the equally ponderous "British Association for Promoting Co-operative Knowledge" (BAPCK).¹⁷ By 1831, this was the London lynchpin of some five hundred local co-operative societies and hosted a galaxy of activists—all now straddling the radical/co-operative line. The radical aspect was evident as they targeted the "rapacious aristocracy" for appropriating the land, turning labourers into "slaves" and "making their labour a marketable commodity".¹⁸ Reclaiming the land remained the agrarian

---

14 *Lion* 1 (28 Mar. 1828): 372; Wiener 1983, 172. Huzel 2006 on the near unanimous detestation of Malthus in the post-Carlile pauper press and the widespread belief among radicals that social inequality was to blame for pauperism, not profligacy.

15 Saull 1853, vii. He would also shortly attack the Whigs' Malthusian-inspired New Poor Law and the workhouses. Hale (2014), focussing on later Victorian times, rightly emphasizes the politically-constitutive dimension of antagonistic radical anti-Malthusian and capitalist Malthusian attitudes to the study of human origins and the ordering of society.

16 *London Co-operative Trading Fund Association meeting of the shareholders held...11th Dec.1827* (1827), 3pp.

17 Chase 1998, 148–51; Prothero 1979, 243. Its activists included many Saull associates: William Lovett, James Watson (the former Carlile shopworker), George Petrie, John Cleave, and Henry Hetherington. Claeys 2002, 175–82.

18 Chase 1988, 150.

goal, but it was soon overtaken by more ambitious urban concerns. And many of these were to become central to the later movement—indeed the blueprint for Saull's agenda—including establishing schools, dispatching missionaries, and opening halls for "lecturing on co-operation and the sciences."[19]

Much of this—the sciences and schooling—was dear to Saull's heart. He backed the BAPCK, which, in 1831, would take up the fight against the government clampdown on the unstamped press and Hetherington's jailing for publishing the *Poor Man's Guardian*.[20] 'Associations' for advancing causes were in the air at the time. The British Association for the Advancement of Science held its first meeting this summer (1831). And Saull was on the working committee of the co-operative "Association for Removing the Causes of Ignorance" (founded 1831). He guaranteed £20 yearly for seven years to this particular institution, which was dedicated to buying land and starting an infant school based on rational lines and instigating programmes for educating the "unemployed and uneducated", women and men alike.[21] This was another grand scheme that was better in the planning than the execution. In truth, hardly a radical/co-operative society or rational/educational scheme passed by that Saull did not support.

Co-operators had taken matters into their own hands to start collective endeavours through the late 1820s. Their paternal inspiration might have come from the philanthropist Robert Owen, but he was away in America at the time, and the speed of events took him by surprise. Owen was a man of humble origin, enormous energy, and good people-management skills. He was known mainly for his model village and innovative school at his New Lanark mill, which had drawn worldwide interest. He was back in London in 1830, when the spy tipped off the Home Office that Saull was "one of his best friends and supporters".[22] Owen encouraged many of the co-operative schemes (and was eventually honoured as the 'social father'), even though he was radically outflanked by the young guns. Every bit the cultural determinist, he made social and cultural

---

19   *The Co-Operative Miscellany; or, Magazine of Useful Knowledge* 1 (Feb. 1830): 25–26.
20   *British Co-operator* 1 (5 Aug. 1830); *PMG*, 30 July 1831, 30–31; *Cobbett's Weekly Political Register* 73 (27 Aug. 1831): 562–65.
21   *Morning Post*, 21 Dec. 1831, 1; *Radical Reformer*, 24 Dec. 1831, in HO 64/18, f. 706; *PMG*, 25 Dec. 1831; *Examiner*, Dec. 1831, 826, 831; 22 Jan. 1832; *MC*, 18 Jan. 1832.
22   HO 64/11, f. 238.

conditions the nurturing agent: change the home environment, and a child's moral and ethical growth can be steered. Saull applied this *mutatis mutandis* to ancient history. Taking his cue from Phillips, Saull accepted that a change in the ecological conditions brought about by planetary movement could direct the change of species. Owen's environmental necessitarianism might have been a sticking point for critics, from Carlileans to Owen's Christian fellow-travellers,[23] but it never was for Saull. It simply sharpened his approach to the development of life.

Owen, with his unshakable faith in human perfectibility, became Saull's icon. Indeed, many idolized Owen at the moment, excepting of course Carlile, who thought him a "fame-seeking opinionate" who "far exceeds all other fanatics".[24] When Saull and Owen first made contact we do not know, but it was before Owen left for America. While Owen was away, Saull sent him a copy of his stinging *Letter to the Vicar* in 1828.[25] With Owen's homecoming, philanthropist and financier began working together. Owen's "New Religion", the subject of his February 1831 lectures in town, was the old religion that "all Religion was in error and that the only one necessary was that of Nature which caused Man and all other animals to act in all they did because it could not do otherwise".[26] So said the undercover agent, reporting to the police. Nothing would have struck Saull more than that Owen hit the ground running in London with an anti-religious message. Published by Saull's co-conspirator John Brooks as *The New Religion; or, Religion Founded on the Immutable Laws of the Universe* (1830), Owen's talks demanded that, as any first step to social change, the religious warping of the infant mind must cease. As John Hedley Brooke has said, such secular religion was pursued with all the fervour of the sacred,[27] and the rapture was evident in Owen and Saull.

Since character was shaped by circumstance, all delusional input must be removed. Religious dogmas, often held by hypocrites or imposed for socially-controlling motives, were harmful to the moral

---

23 For example, the Freethinking Christian and anti-priestcraft Owenite T. Simmons Mackintosh ([1840]), who was in later years to lecture in tandem with Saull.
24 Wiener 1983, 24.
25 W. D. Saull to Robert Owen, n.d., ROC/18/6/1, Co-Operative Heritage Trust Archive, Manchester.
26 HO 64/11, f. 237; Robert Owen [1830].
27 J. H. Brooke 1991, 205.

development. It is hardly surprising that Saull immediately gravitated to Owen, these were to be his guiding precepts for life. And, of course, Saull immediately put his finances at Owen's disposal. The spy now targeted a new venue, Albion Hall. This stood behind Albion Chapel, a "pleasing", domed building on the corner of London Wall and Moorgate. It had been erected at huge cost by the infidels' nemesis, the Rev. Alexander Fletcher. For some years, the hall had been the home of the City of London Literary and Scientific Institution, which trained the merchants' sons and bankers' clerks, while the Cecilian Society practised its sacred music there every Tuesday evening.[28] The surveillance records show that, by February 1831, Saull had acquired it for Owen's lectures:

> This place was originally built for a School to Albion Chapel, but Saul [sic] has become a Leasee and the Society of Co-operatives of whom he is one of the strongest have had it made higher and have altered it as a Lecture Room, or a Concert room having also had a small Organ built there for that purpose.[29]

This was the start of Saull's lifelong financial commitment to Owenism. This very transaction, in fact, was the template for a succession of acquisitions to house Owenite lectures and social festivals, hence the obligatory organ. Seventy turned up to hear Owen's inaugural speech, "but many left before it was over disgusted", reported the spy, smugly. Yet, a hundred were there in subsequent weeks, with Saull in the audience.[30]

Many were now moving over to Owen's party or splitting their loyalties between the co-operators and radical unions. One could see it in the Optimist Chapel. It was still delivering blistering broadsides against Christianity in 1831, but increasingly the talk was Owenism. For example, another intelligence target, the Thames dockworker, leader of the shipwrights' union and erstwhile Cato Street conspirator John Gast, was reported lecturing here in the Spring. He was not seen so much

---

28 *LMR* 2 (24 Sept. 1825): 362; *Register of Arts and Journal of Patent Inventions* ns 2 (10 Mar. 1828) 45; Cruchley [1831], 141; Shepherd 1827, 170.
29 HO 64/11, f. 237.
30 HO 64/11, ff. 204, 237, 238. For a flyer announcing these talks on the "New Religion of the Science of Society" at Albion Hall see f. 216. These venues rarely lasted long, this one persisted for three months. Owen on 7 April 1831 started at a new chapel near Brunswick Square: HO 64/11, f. 249.

among the "political parties" these days, the spy reported, "having joined the Co-operative Society on Owens Plan and is chiefly among those who meet here [at the Optimist] and in the [Owenite] Tea Parties of Men and Women who now and then meet about London."[31] So Saull was only one among many activists gravitating to Owen, or splitting their time between co-operators and the working-class unions.

Whatever the crossover to co-operation, the activists remained radical, in that they still agitated against the aristocracy, the state church, and government oppression. Oppression took many forms. For example, the fourpenny newspaper stamp duty was designed to gag the street presses and put them out of business. Hence any paper with a penny cover price was illegal, because it had not paid the duty and passed it on. The beleaguered printers, led by Hetherington, became a *cause célèbre* on the street. The 'liberty' of the press, one unencumbered by taxes, was, like trial by jury, hailed by all radicals as a guarantor of British freedom[32]—and the hand-cranked press in the commoner's hand was now heralded as the saviour of a corrupt society. The way the law was selectively applied proved it was targeting the agitators. The "soporific" *Penny Magazine* was left alone because it was "harmless", whereas the "obnoxious" rags pedalling blasphemy and sedition were singled out.[33] The activists cleverly branded it a "Tax on Knowledge", and the catch-phrase caught on. Opposition to it became a rallying point as editors stuck to a penny and went to prison. Not merely editors, mostly it was the street sellers who were picked up with tricolour placards and bundles of the *Poor Man's Guardian*. Over a couple of years, possibly 200 were given three months' detention (despite pleas that they were *lending* papers for unlimited periods at a penny a piece!).[34] A "Victim's Fund" was set up, with Saull as Treasurer, and subscriptions poured in to Saull's

---

31 HO 64/11 f.209; Prothero 1979, 259–61. Gale Jones was another crossing the floor. On his sympathy for Owen: Claeys 2002, 64.
32 Epstein 1994, 62ff.
33 *Church Examiner, and Ecclesiastical Record*, 15 Sept. 1832, in HO 64/18, f. 384; "soporific": *The Thief*, 5 May 1832, in HO 64/18, f. 568.
34 *Republican* (Hetherington), 13 Aug. 1831, 5. Hollis 1970, vii, reports that from 1830 to 1836 740 men, women and children went to prison for selling the 'unstamped'. Hetherington was caught by the Bow Street runners and jailed in 1831, and again in 1832 (Barker [1938], 15).

wine depot.³⁵ No matter whether Saull had his co-operative BAPCK or radical NUWC hat on, he collected funds for the jailed vendors.³⁶ The pot paid out 5s a week to those incarcerated, a good going rate which explains the mock heroics in court of otherwise destitute sellers. One defiant vendor retorted to a magistrate: "imprisonment, I care nothing about it, as long as I am supported by the 'National Union.'" The fund also supported their wives and encouraged new sellers despite the mass arrests. Of course, "citizen Saull" chipped in, putting guineas into the pot where others put in pennies.³⁷

Owen's return from America had reinforced the flagging message of man's moral and physical perfectibility. His "New Religion" and "New State of Society" rammed home the point as he took to Saull's Albion stage. Only a change in "circumstances" could "produce a superior physical, mental, and moral character", and this required a new secular and scientific schooling for children, whose plastic minds provided the substrate. Man was "no more a free or responsible agent" than any other creature.³⁸ He was the product of his environment: tweak that, and he could be moulded and perfected. A messianic belief in adaptability swept the Owenite communities. In this secular theology of deism and religion of nature, "Science was the new providence, education was to be the redeemer of mankind; for by understanding and controlling circumstances, man could shape the human clay."³⁹ As Stedman Jones says, Owen's "historically unencumbered language" inspired huge numbers, "clearing the ground for a belief in natural and universal equality, human perfectibility, the malleability of social and political institutions".⁴⁰

---

35 For the opening subscription lists, see *PMG*, 6 Aug. 1831; *Republican* (Hetherington), 20 Aug. 1831, p. 8; thereafter Saull's name (with Lovett as his assistant) appeared in every subscription list published in the *PMG*, *Political Register*, *Republican*, *Radical*, or *Cosmopolite*. Hollis 1970, esp. 194–202; Wiener 1969, 89, 203.

36 *PMG*, 16 July 1831, for Saull on the NUWC committee on subscriptions; *PMG*, 30 July 1831, for both Saull's BAPCK and NUWC subscription work. Hollis 1970, and Wiener 1969, on the radical 'war of the unstamped', and Hewitt 2014 on the wider effects of the stamp duty, paper excise tax, and advertising tax on the newspaper industry.

37 *Republican* (Hetherington), 13 Aug. 1831, 5. Saull's guineas: *PMG*, 17 Sept. 1831.

38 Robert Owen 1830, 45, 60.

39 Royle 1974, 23.

40 G. S. Jones 1983, 126–27.

'Malleability' was a key concept. Saull's Optimist comrade Pierre Baume even announced in messianic eugenical fashion that he intended to leave trustees his wealth "in order to encourage experiments on PERFECTIBILITY, which have been tried succesfully [sic] upon almost every kind of vegetables and animals, except upon the HUMAN SPECIES; to find out whether we may or not form characters of an extraordinary superiority above every one now in existence!"[41]

But fellow travellers often fell out over priorities—many radicals argued that political and economic equality was a prerequisite to social regeneration. Some, just over the fence, in the NUWC, got so fed up with the talk that they "despised those who wrapt themselves in the perfectibility" jargon.[42] Few doubted that humans could be improved, but the population had to be "morally and politically free" before the experiment could begin.[43] Others would struggle with Malthus's attack on such optimism—his belief that stress was inevitable given population growth, despite Owen's counter argument that man could produce more than he could consume. One young surveyor with a passion for wild life, the future 'Darwinian' evolutionist Alfred Russel Wallace, whose "first love" was Robert Owen, would shortly wrestle with these contradictions.[44]

This search for Heaven on earth inevitably affronted religious sensibilities. It was the damnable dream of the "licentious, or the profligate", in short, the proud, who "would concede to no higher tribunal" and would deny "the necessary infirmities of our fallen nature."[45] And at least one geological don at the exclusive Anglican seminary of Cambridge University, the Rev. Adam Sedgwick, hysterically saw it raise the spectre of the French Revolutionary Terror. Genuine fear was struck into some hearts by Robert Owen's "*moral fanatics*" spreading their pernicious panaceas about earthly perfection: "no human system can bring the rebellious faculties of man under the law of obedience; and ... no external change of government whatsoever can make him

---

41 *Lion* 3 (27 Mar. 1829): 396; Baume 1829, 4.
42 *PMG*, 18 Feb. 1832.
43 Detrosier 1831.
44 G. Jones 2002, 74, 86–95; J. R. Moore 1997; Durant 1979, 35; R. Smith 1972, 191–96.
45 Rennell 1819, 25.

even approach toward a state of moral perfection—an idle dream of false philosophy ... and directly opposed to the word of God."[46]

The reaction shows how frightening a reinvigorated perfectibilist faith could seem. Nor did it need another Cambridge ordinand reviewing for the Tory *Quarterly* to point out where such nonsense must end:

> Dr. Darwin, indeed, carried the hypothesis still farther—for it was a favourite part of his creed that man, when he first sprang by chance into being, was an oyster, and nothing more; and that by time alone, (a lapse of some chiliads or myriads of ages, for he has not given his chronology very particularly,) and the perfectibility of his ostraceous nature, he became first an amphibious, and then a terrestrial animal![47]

That shaft was aimed at the pirates' favourite, William Lawrence. In his street-saturating *Lectures on Man*, he saw both the individual *and* the human species being perfected.[48] Man might be unique in his moral perfectibility, but there were no species limitations to the concept. The prospect was opening up of the improvement of all life—the spectrum from the oyster to the infant. Just as a child's mind was malleable, so, as Baume pointed out, domestic breeds were equally pliable. Saull's client, the Rev. Robert Taylor, portrayed it as a case of releasing latent potential. This was the "purpose of nature", he had announced at the Areopagus, and nature's effort "to evolve and bring forth the moral capabilities of man, may be traced from the very first origination of animal life".[49] Taylor, perhaps in talking to Saull, had crossed the line. In an Owenite world where circumstances shaped development, uncontrolled by a capricious deity, a certain symmetry prevailed. The "immutable laws of nature" applied to all; therefore man, being an animal, "is equally subjected to these laws with all earthly animal and vegetable existences".[50] And, while humans were "generated by nature" and could be *re*generated by a social and economic realigning with the "immutable laws of nature", it was short step to regenerating species into more perfect or 'higher'

---

46 Sedgwick 1833, 76–77. This was the Sedgwick who just as vehemently damned books on transmutation as a "paradise of fools": Adam Sedgwick to Richard Owen, 30 March (no year), British Museum (Natural History), Owen Collection, 23: f. 298; Desmond 1982, 189.
47 [D'Oyly] 1819, 14.
48 Lawrence 1822, 202.
49 *Lion* 4 (9 Oct. 1829), 462.
50 Robert Owen 1830, 152.

forms as ecological conditions changed. Improving circumstances could lead to improved species: Nature could act like a super-Owenite. If impediments had only to be removed to achieve human advancement, perhaps through prehistory ecological impediments had been removed for the species in each era to be improved.[51]

Some wealthy, well-read co-operators were already flirting with dangerous ideas. If removing impediments was the way to social change, then obstacles to women's education should be the first to go. It was the pre-eminent call co-operators had learned from the emancipist William Thompson. His Mary Wollstonecraft-homaging *Appeal of One Half the Human Race, Women, Against the Pretensions of the Other half, Men, to Retain them in Political, and thence in Civil and Domestic, Slavery* (1825) was aimed squarely at the "backsliding" utilitarian James Mill, who shockingly saw women's interests represented by their husbands and fathers. An improving Cork estate-owner—so improving that he was dubbed the "Red Republican"—Thompson was the movement's foremost anti-capitalist theorist. His *Inquiry into the Principles of the Distribution of Wealth* (1824) was a staple in radical libraries (including the NUWC's).[52] He disputed that capital should flow to the middle classes, giving them the leisure to indulge in intellectual activities. With labour fairly rewarded, mechanics would invade the scientific realm and give it new objectives and class allegiances. Mechanics' institutions, he urged, should be run by the workers themselves, and they should be equally open to women.[53] For Saull, with his emphasis on artisan education, this would have had a sweet sound. Thompson's work would have been well known to Saull, perhaps even the man himself. For when the wealthy philanthropist died in 1833, and his will leaving £10,000 to the co-operators was contested by relatives, Saull was part of the committee set up to back the executors.[54]

---

51 Robert Owen 1830, 89, 245. The *NMW* (1 [31 Jan. 1835]: 110) was still stressing "primitive man, generated by nature".
52 Pankhurst 1991, 57, 145.
53 W. Thompson 1826a, 46–47; 1824, x–xvi, 274–76.
54 *People's Conservative (Destructive)* 1 (28 Dec. 1833): 380. Saull and Anna Wheeler were among those deputed to raise fighting funds to settle it in the courts. But Thompson's writings on the despotism of marriage—which the relatives read into their testimony to suggest that the bequest was to further an immoral onslaught on the sacrament—did not dispose the Irish court to the co-operator's case: Pankhurst 1991, 130–36.

When it came to women's education, Thompson's arguments were intriguing. Perfecting the species relied on unblocking potential, particularly that of the oppressed sex, women.[55] Not only that, but any positive gain had to be passed on through the generations. It had to be cumulative. The cosmopolitan Thompson was well travelled and au fait with French thought. He had digested Jean-Baptiste Lamarck's "valuable" *Histoire Naturelle des Animaux sans Vertèbres* (7 tomes, 1815–1822). Thompson never mentioned that this book, by the professor of "insects and worms" (or invertebrates, in Lamarck's later neologism) at the Muséum d'Histoire Naturelle, advocated the transformism of the living chains of life.[56] Thompson, rather, was interested in Lamarck's laws of inheritance—that characters acquired in life were passed on, by some irreversible process. Without such an inheritance, Lamarck believed (and Thompson quoted him in the *Co-operative Magazine*), "nature would never have been able to diversify animals as it has done, and to establish amongst them a progression". The crux for Thompson was that transmission to the offspring was only possible if *"the changes acquired are common to the two sexes"*. There he had it: the justification for co-education—"if the females do not partake of every improvement equally with the males" the effort would be futile: "all our labors at improvement, as concerns the progression of the race, will be rendered abortive". Denying the downtrodden women schooling led to a lose-lose scenario: mothers could not pass on any improvements, nor, being uneducated, could they train their children.[57] Saull shared all these views; equal education, for him, became a mantra, and, tacitly, he adopted a 'social Lamarckian' outlook. Still, even such a tangential mention of the 'evolutionist' Lamarck in co-operative literature was rare, and he seems only to have been exploited for social ends.

It was obvious with the rise of co-operation and Owen's return to London why Toulmin's eternal nature was a dead letter. It had negated any directional, progressive trend, and with it any hope of society advancing to Utopia. For Toulmin, like other Enlightenment

---

55 B. Taylor 1983, 24–27, 68–69.
56 For the correct interpretation of Lamarck's transformism before Charles Lyell's re-imagining, see Hodge 1971, Sloan 1997, and Corsi 1988, on Lamarck's cultural context.
57 W. Thompson 1826b, 250, 253, 254.

philosophers, the masses were an "unmoveable and unimproveable threat".[58] But the "masses" now had their champions in the articulate class warriors of the NUWC, nature added its fossil backing to the calls for progress, and Saull sat astride the new Cuvierian geology while praising Owen's faith in perfectibility.

While 'Cuvier' was just a name to infidels, who twisted his views alarmingly to fit their own needs, he was actually well known to the elite philanthropists. In 1818, Georges Cuvier, accumulating posts alarmingly, and now mooted as Minister of State, had visited Britain to study British administration and scientific bodies. Owen, famed for his New Lanark school and community, hosted Cuvier, his wife, and step-daughter and returned to Paris with them on a specially-dispatched French frigate. Although Owen could not speak a word of French, the Genevan savant and diplomat Charles Pictet, who had himself studied Owen's New Lanark methods, acted as his companion and interpreter, as Saull would do later. Cuvier and Owen shared a carriage to Paris, where, for six weeks, Owen was introduced to all the leading lights, including the biogeographer Alexander von Humboldt and Pierre-Simon Laplace, known for his nebular hypothesis of solar system development. Owen's own self-aggrandizing account in his *Life* has him sitting "in the celebrated French Academy, of which my constant friend, Cuvier, was secretary".[59]

But how infidels and Saull's Owenites interpreted Cuvier's fossil geology was highly contentious. Some saw life on earth as a self-propelling, endless climb, and controversially claimed that it was "proved by the researches of Cuvier". Moreover, life was "directed towards some *end* or *final purpose*". The "exalted generative powers of the earth" had ushered in a succession of creatures treading a path towards the production of man. Cuvier's fossil progression was the proof of perfectibility. The earth's final "effort" was an "imperfect attempt towards the production of a class of rational beings." Still imperfect, mankind had a way to go as social regeneration succeeded

---

58   R. S. Porter 1978a, 445.
59   Robert Owen 1857, 1:166–70; R. D. Owen 1874, 121–22; Outram 1984, 103, on the Ministry offering. Owen's Christian acolyte John Minter Morgan, author of the Owenite allegory *Revolt of the Bees* (1826), was another who hosted Cuvier. This was probably in 1830. At the time, Cuvier was on his second visit to London, just as the July Revolution broke out: Morgan 1834, 1: 127–28.

physical generation. But the height of mechanical millennialism was reached by one letter writer, tipping his hat to Mackey's *Mythological Astronomy*. The obliquity of the earth's axis, he suggested, had yet to tilt so much that upended conditions would usher the planet to "the acme of its perfectibility".[60]

Such was the political and millennial maelstrom in which Saull was developing his geology. And by this point he was an ace away from stretching 'perfectibility' to its limit, in a way that even some socialists found horrifying.

---

60 *Lion* 1 (6 June 1828): 731–34.

6. Founding the Museum — June 1831

Historians have long complained that too little attention is paid to the content of museums.¹ But simply enumerating items is insufficient. We need to understand the way they functioned in discrete contexts, how the contents were presented, how they were viewed, and what social message they carried. On the extreme fringes it is often easier to gauge the underlying intention, which is too easily masked in genteel bourgeois settings. Particularly in times of crisis, in the aftermath, say, of the French Revolution, the ideology can become overt as controlling or liberating factors become visible.² The year 1831 was one of those stressful times, with angry demands reaching a crescendo in the run up to parliamentary reform.

Saull opened his museum in June 1831 at the beginning of a long, hot summer, a summer which saw three months' debate in the Commons over the Reform Bill. Radicals grew ever angrier at events. Hetherington started his *Poor Man's Guardian* on 9 July 1831, days after the museum opened, with the clarion call, so redolent with multiple meanings, "we … deny the authority of our 'lords' to enclose the *common* against us". For his suffrage campaigners the bill was a "deceit" perpetrated by the "'*liberal*' (Ha! ha!) WHIGS".³ And Hetherington knew just what to do with museums, stuff them full of dethroned kings and defrocked priests.⁴ So heightened were tensions that when the Lord's threw out the Reform Bill on its second reading that October—with twenty one bishops

---

1   Torrens, 1995, 282. Only recently has this begun to be rectified: Knell 1997, 2000; Taylor and Anderson 2017; Berkowitz and Lightman 2017.
2   Morrell 1971, 43.
3   *PMG*, 9 July 1831.
4   *Republican* (Hetherington), 11 June 1831, 7.

voting against it—riots broke out, some church congregations walked out in disgust; and, in Bristol, despite three cavalry troops arriving in the city, the bishop's palace was burned down.[5] Saull's whirlwind of activities—those outside of his regular wine and fossil trading—was astonishing in these months. He was simultaneously operating in multiple radical, infidel, and co-operative spheres. While negotiating for the museum in April, he was trying to bail Taylor for his Easter sermon on the Crucifixion, with its call for a "Radical Reform in the Kingdom of Heaven".[6] Fearful, like Hetherington, that they were all at this point under intelligence scrutiny, Saull nevertheless secretly helped keep the Rotunda afloat.[7] He was attending its Sabbath blasphemy extravaganzas and its new inflammatory Monday NUWC meetings, which started up in May, within days of his finalizing his museum purchase. He was talking at the Optimist Chapel, looking for new venues for Owen to succeed Albion Hall, and fund-raising at the BAPCK for the jailed news vendors. So many irons were being forged in the political fire of the moment. If, however, we pull focus, we can see that, in simple strategic terms, the museum was founded at the junction between the end of Saull's 'blasphemy' phase and beginning of his Owenite one.

This was also a fleeting, forlorn moment of revolutionary optimism. Within days of the museum's founding, Saull was organizing the first anniversary celebrations for the July Revolution in France and its "victory over kingly despotism". Here with Carlile's erstwhile shop assistant James Watson (1799–1874) and others, he sang the *Marseilles* in French (*de rigueur* at such events).[8] Henry Weisser has even claimed that this public anniversary meeting was a "turning point" as an all-working-class affair, a symbolic moment when class consciousness became incarnate.[9] Spirits were high and expectations still higher: "N.B.—If another Revolution should occur in the mean time, they will both be celebrated at the same time," ran Hetherington's advert for the

---

5   M. Brock 1973, 244–55; Halévy 1950, 42.
6   HO 64/17, f. 48; HO 64/11, ff. 200, 296.
7   HO 64/11, ff. 229, 446 (29 Nov. 1831).
8   *Republican* (Hetherington), 25 June 1831, 8; 6 Aug. 1831, 6; *PMG*, 6 Aug. 1831. Robin Eagles' thesis in *Francophilia in English Society 1748–1815* (2000) might easily be extended to this period, at least so far as many in the radical working classes were concerned.
9   Weisser 1975, 35.

meeting, three days after the papers announced Saull's museum open. With the self-identifying group hailing one another "Citizen Watson" and so on in the euphoria, Saull became, for a fleeting moment in 1831, "Citizen Saul" [sic].[10] It was a propitious moment to announce to the *sans culottes* his own geological Temple of Reason.

No wonder the year saw new church militants crusading against the infidels, those "sorry warriors" whose pernicious and "illegitimate" geology threatened the "great Armageddon".[11] These friends of the French Revolution had made "Omnipotence" impotent and "babbled out their puerile conditions about a progression in nature".[12] But whether geology threatened or fascinated, there was no doubting its draw. While science could serve many masters, some apprentices wanted it to go further. Again, in April 1831, as Saull was preparing to negotiate for his exhibits, the *Herald to the Trades' Advocate* heard from its readers that it featured too little science.[13] Even the young Hetherington was warmly sympathetic to science. The *Poor Man's Guardian* would, admittedly, become famous for its distraction-free advocacy of workers' rights, with reform first, science education second. And its radical correspondents attacked the Whigs' milk-sop mechanics' institutions, which diverted the workers with so much pap, and featured complaints that artisans were "saturated" with science.[14] Saull's printer friend John Cleave would equally lash the Whig institutes, accusing them of diverting the mechanics from more threatening economic studies with "zoological and geological sciences, and all the other ologicals".[15] The message was 'emancipation first'. But this socially-controlling, fodder-stuffing image[16] did not apply to all mechanics' institutions, and notably not to London's. Helen Flexner's study, by contrast, has shown that it allowed partial worker control, worker self-instruction, women's participation (at least on occasions), and in one respect it went to

---

10 *Republican* (Hetherington), 25 June 1831, 8; 29 Nov. 1831, 192.
11 [Murray] 1831, xiii–xv.
12 John George Children to William Swainson, 11 July 1831, William Swainson Correspondence, Linnean Society; Desmond 1989, 147.
13 *Herald to the Trades' Advocate*, 9 Apr. 1831.
14 *PMG*, 6 June 1835, also 1, 8 Sept. 1832.
15 *TS*, 31 Dec. 1835, 4. Cleave was talking in John Savage's radical Mechanics' Hall of Science in Marylebone, and was referring to a conventional Creationist geology and innocuous zoology.
16 Shapin and Barnes 1977; cf. Topham 1992.

extremes, presenting "science as negotiable rather than given".[17] And, surprisingly, many of our subsequent radical activists cut their teeth here, including Saull, Hetherington, Lovett, and Watson. Hetherington, active in a management capacity, actually planned to publish his own "Monthly Journal of Philosophy, Science and the Arts" in 1828, three years before he started the *Poor Man's Guardian*.[18]

Many ultra-radicals recognized that geology, rightly cast in materialist mould, could be liberating. So long as the god of the Anglican dons could be portrayed as miraculously creating new species through history, then a self-sufficient alternative could help kick away the church's Creationist crutch. Geology thus became part of the anti-clerical chorus, now reaching its crescendo. This made the science more than suspect for many in the pews, with its long ages and succession of ancient worlds, supposedly tenanted by repellent crocodiles and "disgusting" lizards,[19] long before the advent of man: the very idea was "silly, disgusting, and ... injurious".[20] To suggest that grotesque reptiles had the earth to themselves for untold aeons was daft, for they could neither have adored nor given thanks to their creator.[21] Even to moot such times without "immortal" humans was worrying, despite the reassurance of apologists on the providence of Britain's coal fields, which proved that man was in God's mind from the start.[22] Anti-infidel preachers warned of geology in Jacobin hands, because of the bastardized anti-Christian deductions being drawn from it. Making it accessible to the masses meant that "hundreds of sciolists can shoot off some philosophical popgun against the rock of ages".[23]

Saull Puts His Money Where His Mouth Is

All the while Saull had been collecting fossils. His out-of-pocket expenses were now split between Owenite stumps and fossil auctions. In 1839, when Abraham Booth published his literary and scientific

---

17 Flexner 2014, 189–90.
18 LMI management minutes, 29 Dec. 1828: Birkbeck College, London University.
19 *Times*, 27 June 1845, 6.
20 *Christian Advocate*, 29 Dec. 1834, 415; *Freeman's Journal* (Dublin), 17 July 1839.
21 NMW 6 (12 Oct. 1839): 811.
22 J. H. Brooke 1979, 40.
23 *British Critic* 1 (Jan. 1827), 200.

compendium, *The Stranger's Intellectual Guide to London*, he said Saull had been collecting for ten years, and the results "may vie with any private Museum of a similar nature in the kingdom".[24] That would put his start date around 1829. We know that, by this time, Saull was visiting the huge museum in Lewes, near Brighton, built up by the surgeon and self-publicizing antiquarian and fossilist, Gideon Mantell. By 1830, Saull and Mantell were sending one another parcels and swapping specimens.[25] Late in life, Saull put the start date for his collection at around 1828. But, in fact, his interest can be traced back further. In the *Letter* to his vicar, explaining how the changing obliquity of the ecliptic could explain Britain's previous torrid climes, Saull mentions as proof "the innumerable fossil remains of plants and animals found in the higher Northern latitudes, which could exist only in tropical climates, many specimens of which, I am possessed of."[26] That printed letter was dated Christmas Day 1827, so we know that by then his collecting had begun.

The fact that he was elected a Fellow of the Geological Society in June 1830 is circumstantial evidence that the collection was already sizeable. After all, it was presumably the reason he was nominated, for there is no sign he had started his geology lectures by that point. The Geological Society was embracing wealthy buyer-collectors as much as rock-face hammerers and aristocrats mindful of their civic duty. But how did an indicted deist, Carlile supporter, "Devil's Chaplain" backer, and Owen acolyte become a Fellow? Being warm-hearted, wealthy, and easy among old money helped, and having huge fossil assets helped more. But it was notably the reform lobby that got him in.[27] The body

---

24 A. Booth 1839, 121.
25 J. A. Cooper 2010, 38, 43, 47. Mantell's museum concentrated on fossils from the South-East of England. For descriptions of it at this time: Bakewell 1830; *American Journal of Science and Arts* 28 (1835): 194–97; Mantell 1836; and radical Thomas Wakley's appraisal in the *Lancet*—keen to play up the "philosophical" accomplishments of GPs in his campaign against the medical baronets: *Lancet* 2 (29 June 1839): 506–07; Cleevely and Chapman 1992; A. Brook 2002.
26 Saull 1828a; 1853, viii.
27 Saull was elected on 4 June 1830. I should like to thank Wendy Cawthorne, Geological Society Library, for the information on Saull's backers, who included George Birkbeck, more radical than is generally supposed at the LMI (Flexner 2014). Saull had been an LMI member from 1824, had stood (unsuccessfully) as a committee member in 1825 and had donated numerous books to its library, including Jean Louis de Lolme's *Constitution of England*, which advocated an extension of the franchise. Birkbeck was also a physician in the General Dispensary in Aldersgate Street, a charity supported by Saull. Another backer was

geologic, like the body politic, had its bourgeois radical contingent, but they were a small minority; and, even then, an out-and-outer like Saull sat on the fringe. Those who were initiated with him that June prove the point. Of the five inducted into the society, three were Cambridge divines, including a future Dean of Hereford and Archbishop of York.[28] Anglican priests were more a force in the gentlemanly body, on both the front and back benches, than co-operative collectors. The fiercely anti-clerical Saull, who would shortly chair meetings of the "Society for the Extinction of Ecclesiastical Abuses" (that is, the radical reform and disestablishment of the Church), was far from a typical candidate.[29] It shows how much a museum counted. While the divines were keeping up with the challenging science as part of their calling, Saull was admitted because of his enthusiastic collecting.

In 1831, he moved his business a few doors up the road, from 19 to No. 15 Aldersgate Street. This was a more substantial corner site, with entrances on both Aldersgate Street and Falcon Square, allowing for warehousing, stables, and the new museum. When a bankrupt hatter sold the lot a decade earlier it was advertised as a

> capital and very extensive PREMISES, most eligibly situated ... comprising a spacious and very attractive shop, of considerable depth, and with double bowed front, light counting-houses, extensive manufactory, including bowing-rooms, making-shops, dye-house, stiffening-shop, finishing-rooms, warehouses, large reservoir, &c. &c, a coach-house, two-stall stable, &c.; the domestic apartments are very capacious and numerous ...[30]

In 1831, Saull bought the property from a leather cutter and adapted it for his wine-importation business and fossil emporium.[31] It was only fifteen doors from the latest London landmark, the newly-completed General Post Office. This huge classical building was viewed by locals

---

the geologist Henry de la Beche. He had been first to describe the *Plesiosaurus*, and Saull shared his fascination with the new giant fossil reptiles. De la Beche was enamoured of all things French and was himself anti-clerical. He dismissed religious enthusiasm as "humbug" and, like Saull, saw salvation in science (McCartney 1977; J. A. Secord 1986b).

28 *Philosophical Magazine* 8 (Aug. 1830): 147.
29 *TS*, 12 Oct. 1832, 1; on the Geological Society Anglican consensus, Rudwick 1985, 31–32.
30 *MC*, 8 Feb. 1821, 4.
31 *Perry's Bankrupt and Insolvent Gazette* 6 (1 Jan. 1831).

and strangers alike with awe as befitted "the brain of the whole earth", channelling the empire's torrential volume of letters.[32] And that was the prestigious direction for museum visitors: 'close to the General Post Office'. It was this prominent position that made the venue so valuable. As the *Mining Journal* said, "So fine a collection as the present being thus rendered accessible, in the very centre of London", with its huge catchment, is what made it a must-visit site.[33] Here, Saull converted the lofts over the stables to house the collection.[34]

Relocating the whole business, presumably to house the new fossils, showed a huge commitment. The timing suggests that Saull moved to these larger premises precisely because he needed the space to accommodate his newest acquisition. He now bought one of the premier fossil collections in the country. It had belonged to the late James Sowerby, a talented engraver and collector, well known because he illustrated the publications of his rich patrons.[35]

Sowerby's museum, forty years in the making, included many unique 'type' specimens. The collection was an old-style cabinet with

> some thousands of minerals, many not known elsewhere, a great variety of fossils, most of the plants of English Botany about 500 preserved specimens or models of fungi, quadrupeds, birds, insects, &c. all the natural production of Great Britain.[36]

This was far more than Saull wanted. He was primarily after the fossils.

Sowerby had intended that his collection should illustrate the entire fossil life of England. Sowerby's sons had taken over after their father's death in 1822 and turned it into a paying museum in Mead Place, Lambeth. They had planned to re-locate the museum more centrally, making it a proper London money-making attraction,[37] but George Brettingham Sowerby I (his son) was in financial straits by March 1831:

---

32 Brady 1838, 37; Cruchley [1831], 43.
33 *Mining Journal and Commercial Gazette* 1 (7 Nov. 1835): 83.
34 *Notes and Queries*, 7th ser., 10 (6 Sept. 1890), 184.
35 On the Sowerbys mineral conchology and the larger questions their work raised about stratigraphic zoning, ancient environments, and the implications of comparisons of living and extinct forms for placing the poles in ancient times, see Elliott 1975.
36 Conklin, 1995; *St. James's Chronicle and General Evening Post*, 9 Apr. 1831, 1.
37 Cleevely 1974, 426–28.

hence the sale of his own as well as his late father's collection.[38] The private sales of James Sowerby's cabinet ran from 18 April into the first week in May 1831.[39] Saull snapped up most of the fossil and mineral portion.[40] He got the majority of fossils, including the good ones, the 'type' specimens of fossil invertebrates figured in Sowerby's *Mineral Conchology*, and he kept Sowerby's own identification labels on them.[41] What he paid is unknown, but, considering that Stevens's auction room in Covent Garden (a favourite for natural-history objects) shifted some of the leftovers for exorbitant sums,[42] it must have been substantial. With the liquor trade obviously flourishing, Saull was on a buying spree. His long pocket showed as he prepared to bid £40 (a labourer's yearly wage) for the fossil-seller Mary Anning's ichthyosaur from the Lyme Regis cliffs in May 1831—and even then he did not get it.[43]

In June 1831, Saull pooled Sowerby's "extensive" collection with his own and announced the new museum open. Founding such an institution did Saull's reputation no harm during the Whig ascendancy, with its 'steam-intellect' desire to promote 'useful' knowledge. The Whig evening paper, the *Star*, lauded him:

> Mr. W. D. SAULL, F.G.S. and F.R.A.S. of Aldersgate-street, the most liberal and public spirited friend of science in the City of London, having recently become the possessor of the extensive Geological Museum of the late Mr. Sowerby, Mead-Place, Lambeth, the whole of which has been stratigraphically arranged, with the addition of Mr. Saull's previous collection of fossils, and will be open for the free inspection of scientific gentlemen and friends, every Thursday morning, at his residence, as

---

38 Matheson 1964, 219. The auctioning of James Sowerby's specimens is not to be confused with his son George Brettingham Sowerby's own sales in 1828–33, advertised in *MNH* 1 (May 1828), 96. More of GBS's own collection was sold in 1831–33 by Thomas and Stevens's auction room on 22–26 Feb., and 14–16 Mar. 1831: *MC*, 12 Feb. 1831, 15 Mar. 1831; *Times*, 23 Feb. 1831, 8.

39 *St. James's Chronicle and General Evening Post*, 9 Apr. 1831, 1; 12 Apr. 1831, 1; 28 Apr. 1831, 1.

40 He evidently did not take the preserved birds, insects, fish, shells and left-over minerals because this "remaining portion" went under the hammer separately in June: *Times*, 9 June 1831, 8; Conklin, 1995.

41 They were still on when the British Museum acquired them: Anon. 1904, 322.

42 For instance, "among Mr. Sowerby's shells, *Mulleria*, £20, and *Voluta junonia*, £15" (Allingham 1924, 30, 84–85).

43 Knell 2000, 206. The surgeon Sir Astley Cooper bought an ichthyosaur from Anning late in 1831 (B. B. Cooper 1843, 2: 140), so Saull may have been pipped. On Anning's prices, Taylor and Torrens 1986, 143–46.

above. Geology, or Nature's own history of her own transitions and improvements, is now become one of the most popular, as well as most interesting, objects of general pursuit, and we consider public thanks to be due to Mr. Saull, for his liberality in thus promoting its study.[44]

Since Sowerby had been the de facto taxonomic expert on conchology and a describer and figurer for the works of many elite geologists, his collection would have been a draw for the "gentlemen". Hence "scientific gentlemen and friends" were Saull's invitees for the *Star*, which appealed to the liberal bourgeoisie in science and politics.

However, listen to Citizen Saull, liberty cap on, as a habitué of radical/blasphemy dens, summon a very different audience, the *sans culottes*. In Carlile's absence, the Rotunda, in a rotten state of repair, was run by his lover Eliza Sharples, assisted by Gale Jones, from February 1832. It was aflame with seditious and blasphemous harangues in these months. Here Saull, Hibbert, and Gale Jones would add inflammatory asides after Sharples's own lectures, "each in their usual strain of abuse of both Church and State", the spy added typically.[45] Here, too, the NUWC continued to demand universal suffrage and a free press; not, said Hetherington, that the powers would tolerate "such a proposition coming from 'the scum' (as they are called)".[46] The "scum" was Saull's target audience. Were working people to get the vote and take power, educational ventures would be needed to bring their schooling up to snuff. Indeed their "want of knowledge" made a proper rational scientific and economic education essential. Saull expanded on this after one Sharples lecture. Materialist reasoning was needed to counteract religious obscurantism, and science as a "force [was] fatal to that of tyranny and priestcraft. (Cheers.)". With the poor deprived of schooling, except by the local dame or Sabbath lessons, the people would find that it would add "more to their comfort and happiness to cultivate the sciences ... than to intrust [*sic*] themselves to the guidance of the priest, who deals only in mysteries". He

> concluded by volunteering his services to aid the cause of science and liberty, by public lectures, at any time or any where, and invited the audience to inspect his museum, which he very courteously and kindly

---

44 *Star*, 22 June 1831, 4.
45 HO 64/12, ff. 36–38, 47; *Isis* 1 (3 Mar. 1832): 59–60. Parolin 2010, ch. 8.
46 *PMG*, 5 Nov. 1831.

said should be open to them every Thursday, when he should be ready and willing to give them every information in his power.[47]

A familiar figure at the Rotunda, Saull had probably been offering to throw open his museum to working men from the start, but this report in February 1832 was the first evidence in print. Women, too, had probably been invited early on, but the first confirmation in print we get comes from 1833.[48] Nor is this surprising. The new historiography shows how active the women were in radical, blasphemy, and co-operative circles. They can no longer be written out as liberty-cap makers supporting their husbands but must be seen as more politically active shopkeepers, pamphlet sellers, theatrical demagogues, and jailed seditionists.[49] Saull was using the Rotunda and undoubtedly his other platforms to promote the new exhibition among the increasingly status-conscious working men *and* women.

Compared to Sowerby's original, the museum saw marked changes. Firstly, it was structured differently, for a different purpose. The whole lot, Saull confirmed, "is now stratigraphically arranged".[50] It implied that Sowerby's fossil animals and plants had been ordered another way, perhaps according to their relationships or some other criterion. So many collections, as Simon Knell says, were viewed simply as "an assemblage of unrelated objects, collected without direction and displayed without order or reason. Considerable curatorial input was required to turn collections into a resource for self-improvement".[51] The new stratigraphy was one such ordering principle, with its origin and direction indicators— to illustrate the "transitions" and "improvements" of life through time, while emphasizing (in radical hands) its perfectibility and material causation. The fossils were lined up in sequence, according to the strata they came from. Radicals elsewhere were equally emphatic that this was the correct approach. The British Museum would actually be censured by hostile radical witnesses during the Select Committee hearings in 1836

---

47 *Isis* 1 (3 Mar. 1832): 59–60.
48 *MM* 19 (25 May 1833): 117–18; *Lady's Magazine and Museum* 3 (Nov. 1833): 297.
49 Keane 2006; Frow and Frow 1989; Parolin 2010; B. Taylor 1983.
50 *Star*, 22 June 1831, 4; *Philosophical Magazine* n.s. 10 (Sept. 1831): 237; *Arcana of Science and Art* 5 (1832): 251.
51 Knell 2000, 92; M. Freeman 2004, 252. William Bean's fossils in Scarborough were displayed to show "taste", that is, for aesthetic effect: McMillan and Greenwood 1972, 152–53.

for not adopting this kind of chronological organization.[52] Of course, it was not only the radicals who adopted it: Gideon Mantell's museum, rivalling Saull's in size, had also been arranged in a temporal order, as Saull knew from his visits.[53] Mantell was now the "eloquent friend" whom Saull would quote in the *Mechanics' Magazine* about ridding the mind of prejudices as a prerequisite to studying geology. Although Saull was hinting with a Carlilean glint at more than his eloquent friend might have liked.[54]

More noticeable for working men was the entry price to Saull's museum. There was none—they could actually get in free, and without any formality. This was the second major difference from Sowerby's exhibition. Ticketed entry to the Sowerby museum was prohibitively expensive, at ten shillings for three months, or £2 yearly. This barred all but the wealthy elite.[55] Even Mantell's museum charged a shilling for admittance, and then only to entrants signed-in by a member of the Sussex Literary and Scientific Institution (of which it was part).[56] Saull's was a markedly different proprietorial attitude. His was not a money-spinning exercise but a democratization of transformative knowledge. Free entry was indicative of his socialist philanthropy and something all the newspapers would comment on. Accessibility was the watchword: no gentlemanly propriety was followed, no "introduction" required, which made entry so difficult in the Geological and Zoological Society museums. The "poor as well as rich" could turn up, "without any previous application", and all would be accompanied around.[57] But it was specifically working men whom Saull encouraged to visit—the power brokers of the expected socialist millennium, who needed to be educated for their new role. Or, as he put it on chairing a meeting of the Kingsland and Newington Co-operative Society to set up a Labour

---

52 *Report from the Select Committee on British Museum*, 1836, Parliamentary Papers, 14 July 1836: 21, 74, 78–79, 130–33; Desmond 1989, 148–49; McOuat 2001, 12ff.
53 *American Journal of Science and Arts* 28 (1835): 194–97.
54 *MM* 19 (25 May 1833): 117–18.
55 Conklin 1995; Cleevely 1974, 426–28.
56 Mantell 1836, 44. William Bean's private museum, in his Scarborough house, was only open six days in the season to the public, on "being properly introduced": McMillan and Wood 1972, 152–53.
57 *Preston Guardian*, 14 July 1855; Karkeek 1841a.

Exchange Bazaar in 1832, "in order to fit them for the great changes which are evidently coming upon the world".[58]

As a proselytizing socialist, Saull was in the vanguard of that wider movement in the 1830s to get free admission to 'public' buildings, not only museums, but also Westminster Abbey, St Paul's Cathedral, art galleries, natural history collections, and so on. Their exclusivity was becoming a national disgrace (in radical eyes), and he was shortly to start committee work under the radical MP Joseph Hume to petition Parliament to this effect.[59] Artisans found it inordinately difficult to gain access to institutions. Even visitors to the British Museum, which had long abolished the ticket system, had to be of "decent appearance", meaning a porter could have the final say.[60] Attire and demeanour were everything. One guide to museum planning twenty years later was still belabouring the point: "forbid the entrance of obnoxious and certain other persons; the rest of the public, if decently attired (hats, not caps, are generally required in France, except for soldiers and sailors), to be admitted either upon signature of name, address, and occupation, or in some cases without such formality".[61] But even a signature requirement was an impediment to the poorly educated and was known to be keeping them out.[62] The class restriction was often obvious, with the genteel preferred to the vulgar; and if the latter had any finer feelings (which the toffs doubted), they were certainly offended by the constant barriers, the need for countersigned letters, the payment, and the scrutiny.[63]

Then there was the price. Sixpence admission was enough to stop the 'lower orders' from coming to the Manchester Natural History Society museum.[64] The same was probably true of the tanner (6d) required at Norwich Museum and Liverpool Royal Institution Museum, and then they were only open one day a month. (This was revealed in the naming-and-shaming policy of Hume's Society for Obtaining Free Admission to Public Monuments and Works of Art—which, in 1843,

---

58 *TS*, 26 Sept. 1832, 2; *Atlas*, 30 Sept. 1832. 660.
59 Anon. 1837.
60 Hoock 2003, 259–60.
61 Quoted by Forgan 1994, 144–5.
62 *MM* 24 (5 Dec. 1835): 203.
63 Cash 2002.
64 Alberti 2009, 17–18; A. Secord 1994, 399.

still listed Saull's as the only truly free geology museum.[65]) Needless to say, children were barred,[66] and if the working classes could get in they were kept an eye on. That vandalism and theft were expected after their admission was suggested by the surprise that these things *had not* happened at the British Museum after working people were finally allowed in. This was to the dismay of some guards: "I am really sorry to say that not the slightest damage has been done to any one object in the whole Museum", reported one, "not a wing of a butterfly has been touched, not a leg of a spider has been broken, and there[fore] we have no plea to come forward with a recommendation to Parliament to abolish the new regulations."[67] Even the Principal Librarian was aghast at the "vulgar class" being let in and reasoned before a Parliamentary Select Committee that "people of a higher grade would not like *to come to the Museum* WITH *sailors from the dock-yards, who might bring their girls with them"*.[68]

To the next generation such reactionary attitudes seemed positively archaic:

> It was formerly said that educating the multitude would make every man a knave or a rebel; that introducing recreations among the populace would end in the tailors' and shoemakers' Saint Monday being extended to all classes of workmen and lasting till Saturday night. It was said that if Parks and Gardens were opened to the people, every tree would be cut up into walking-sticks, every flower-bed be trampled upon; that, if Museums were opened, the wings of every stuffed bird would be plucked, every glass-case broken, the geological specimens picked, and every curious picture in the books of the reading-room torn out.[69]

---

65 *CPG*, 15 Apr. 1843; Anon 1837. Saull was on the Committee of the Society. The Museum of Economic Geology in Charing Cross was, however, shortly to join it as freely accessible, when the newly-knighted Sir Henry de la Beche became its director. This museum was a government initiative to display the industrializing country's mineral resources, but it was targeted more at students, surveyors, and engineers than recreational visitors (Sopwith 1843, 8–9; J. A. Secord 1986b).
66 Bonney 1921, 2.
67 Anon. 1837, 6.
68 *MM* 24 (27 Feb. 1836): 430. J. A. Roebuck's petition from working men to Parliament in 1833 to have the British Museum and "all other exhibitions of science and art" open on the Sabbath was rejected (*Gauntlet* 1 [28 Apr. 1833]: 182–83).
69 *Reasoner* 26 (6 Jan. 1861): 1.

Nevertheless, at the time, prying open the museum doors required an almighty push. The overt class discrimination explains why Saull's precedent was applauded. Unimpeded access was rare, yet here was a warehouse museum free to absolutely everybody, and, astonishingly, "no personal introduction was required", which explains Thackeray's *National Standard*, in 1834, paying the ultimate compliment to Saull:

> All those, therefore, who contribute to elevate the moral character by the gratuitous diffusion of scientific knowledge are the benefactors of humanity, and as such, we hold that Mr. Saull deserves well of society, in doing as an individual what the French alone do as a nation—throw their museums and their lecture-rooms open to all the lovers of science, without distinction of either nation or rank in society; and it is hoped that such an example will soon be followed by other generous Englishmen, who love science for its own sake, and delight in smoothing its rugged approaches, and opening its temple to all.[70]

Two decades later, the *Civil Engineer* would *still* be holding France up as an exemplar and demanding wider access to English scientific institutions, citing Saull's "public spirit" for opening his museum.[71] But the "public" aspect of Saull's spirit was part of his socialist calling, and Thackeray's elevation of "moral character" part of his perfectibilist goal. The real target was artisan education for political ends. Still, the press now rated Saull's private facility as "essentially a public exhibition".[72]

Essentially public, but it *was* still private. Unlike state or civic museums, it was in private hands, and it was singled out for praise because many such museums were never opened at all to the public.[73] It contrasted, too, with the exclusivity of the professional bodies at the other end of the social spectrum. Access was coming to be seen as a right rather than a privilege by activists, so even the Geological Society came under pressure. Its museum was open to members (3 guineas dues, 6 guineas admission fees), who could escort guests, but, the liberal *Era*

---

70   *National Standard* 3 (18 Jan. 1834): 44–45.
71   *Civil Engineer* 17 (Feb. 1854): 41–43.
72   *Observer*, 27 Mar. 1842, 3. Fossil collectors are prone to being secretive, which also explains the praise for Saull. Even though his museum was only open one day a week, this was better than, say, William Bean's 15,000-specimen museum, which was open to the public for six days, at indefinite times, during the season (*Theakston's Guide to Scarborough* 1854, 131; McMillan and Greenwood 1972).
73   Such was the case of Hugh Miller's museum: Taylor and Anderson 2017.

newspaper carped, "As this society enjoys public apartments at the public expense, it should ... be thrown more freely open".⁷⁴ It was not to be. Such gentlemen's clubs for science specialists resisted the democratic trend, and the rising professionals later in the century only reinforced their exclusivity and left their museums ring-fenced and secure.

Saull would remain a geological outsider. The Geological Society's Star Chamber was a self-electing alliance of Anglican dons and London careerists. They came together as a professional unity government with its own agenda and social etiquette. We get some perspective on this by looking from the outside, from Saull's standpoint. Marginals came no more disparate than the three faces Mantell (himself a side-lined provincial) saw when he popped round to Saull's one day early in 1832. There he found a little group having tea. A more extraordinary sight could not be imagined: he was greeted by an eclectic mix of the millenarian and materialist. Saull was there: the Devil's Pulpit proselytizer, ecliptical swivveller and Owenite anti-capitalist. So too was the dirty little Jacobin himself, Sir Richard Phillips: the anti-gravitation demystifier whose astronomical algorithms could explain the seasons, shifting heat zones, and hemispheric quantities of water. Then came the strangest of them all: the tyro Thomas Hawkins, a twenty-one-year-old Somerset fossil collector extraordinaire (a youngster who had navvies shift a cliff to mine out an *Ichthyosaurus*), a wild, possessed millenarian, whose fossil "sea dragons" were interwoven into a visionary Mosaic past.⁷⁵ It is hard to imagine what the arch-materialists thought of Hawkins, whose tendencies towards hyper-Miltonic poetry made him semi-intelligible, and whose flailings about Pre-Adamites, the AntiChrist, and Voltairian infidels must have made them bite their tongues. All this was eclipsed by his revelatory visions of the ichthyosaur's world:

> Theirs was the pre-Adamite—the just emerged from chaos—planet, through periods known only to God-Almighty: theirs an eltrich-world uninhabitate, sunless and moonless, and seared in the angry light of supernal fire;—theirs a fierce anark thing scorched to a horrible shadow: and they were the horrible chimeras.⁷⁶

---

74 *The Era*, 16 Apr. 1843.
75 J. A. Cooper 2010, 50.
76 T. Hawkins 1834, 51; Carroll 2007, 2008 for a study of Hawkins as an "eccentric".

This apocalyptic hammerer denounced all who swore by "insensate Matter", those lost souls like Saull and Phillips in their "Paradise of Fools." If God was not speaking through Hawkins, He was speaking through the ichthyosaurs. The millenarian, awaiting the imminent Second Coming, was meanwhile damning "grotesque" notions of self-development. Although Hawkins accepted the obliquity explanation for formerly frozen ages, mankind had a more important "Orbit, the perihelion being with Adam, the aphelion with the Flood."[77] God knows what Saull and Phillips made of this. Being a fly on the wall at this meeting of millenarian and materialist minds would have given us an unprecedented insight into the fossil mediations and unruly exuberance of pre-Victorian palaeontological culture.

It shows why the professional gentlemen were trying to rein in geology. The urban gentry of the Geological Society effectively barred divisive talk of astronomy, mythology, Milton, Moses, and evolution. Their carefully policed science was uncontroversial and respectful. They described their work as a dutiful delineation of the strata. And by not ruffling social feathers they hoped to elevate and ordain their dubious new profession. The trio in front of Mantell stood for everything that was troubling. These *embrouilles* back-benchers, like Hetherington's 'scum', were never to be allowed near the star chamber, however much they envied the ruling coterie.

One thing materialist and visionary shared was the need for fossil museums, although not for the same reasons. This, too, had Hawkins practically babbling in tongues:

> Let us haste then to found sumptuous museums, which shall be as sanctuaries for the arts—the divine arts—until ignorance, driven to herd with bats and owls and every unclean thing, ceases to persecute them:— and let us raise noble galleries to receive the spoils of invincible science. Be temple and lower too devoted to their legitimate use, the Majesty on High should be worshipped of his creatures in the face of that spotless heaven which he made to be a figure of his incomprehensible glory and endless perfection.[78]

---

77 T. Hawkins 1834, 1–2; 1840, 1–4. The latter, so perplexing to commentators, could easily have been aimed at infidels, his freethinking tea companions, Saull and Sir Richard Phillips, while Hawkins's mention of a Golden Age—of permanent equatorial sun—hints at familiarity with Mackey's *Mythological Astronomy*.
78 T. Hawkins 1834, 30.

Rather than a hymn to His Majesty on High, Saull, with his Owenite faith in rational schooling, saw museums shape impressionable minds and ready them for a very different *socialist* millennium. Or, as "Brother Saull" told a trades' union meeting, training in his ideological facility would render a boy valuable "as a man".[79]

Saull had other forums for museum propaganda, most notably the moderate National Political Union (founded four months after the museum, in October 1831). This was an attempt to weld middle and working-class interest (to the benefit of the former), largely inspired by the Charing Cross tailor and radical co-ordinator Francis Place. He feared that NUWC extremism would derail the reform process. Even Hetherington, in November 1831, before the Reform Bill had passed, admitted that both unions were valuable and that the middle and working classes should co-operate to gain meaningful change.[80] Saull agreed, but many ultras were still trying to push the NPU to wider democratic ends. However, Place and the Whig moderates managed to keep universal suffrage and annual parliaments off the table and most working-class "Rotundanists" off the Council.[81] But the merchant Saull did make it on, and his house became a local enrolment centre for the NPU.[82] From the first the NPU pursued strategies close to his heart: by February 1832 the Council had opened a Reading Room, with Saull contributing to the costs.[83] And a weekly series of twopenny lectures on "Politics, Morality, and Physical Science" were projected—showing how necessary these were considered to be for an expanding electorate. Nothing can be "of more importance to the well-being of the community at large." Science was rigorous in its use of evidence, and "no subject shall be introduced unsupported by evidence, nor ungrounded on truth", and with listeners free to question and reason "the objects of the Union would thus be more materially promoted". Owen thought

---

79 *TS*, 22 Apr. 1835, 2.
80 *Radical Reformer*, 4 Nov. 1831.
81 Miles 1988, 186–90; Rowe 1970a, 39–40; Belcham 1985, 245–50.
82 Saull was elected on the NPU's foundation in Oct. 1831, and re-elected yearly; Rowe 1970b, document nos. 33, 34, 63, 66 (showing that Saull was proposed by Henry Revell, who was with him in the Southern Metropolitan Political Union); *MC*, 9 Feb. 1832, 3; *Carpenter's Monthly Political Magazine*, Mar. 1832, 299.
83 *Examiner*, 5 Feb. 1832.

twopence for the lectures too much. Saull went further and waived all admission charges for his science talks at the Union.[84]

More irons went into the fire in these frenetic months. Saull also agitated in the City for the Reform Bill. In September 1831, he was on the Guildhall Committee set up by the aldermen and liverymen to petition the Lords not to block reform.[85] With tensions rising and the third Reform Bill held up in April 1832, Lovett, Watson, and the ultras warned that arming was inevitable in the face of a feared military takeover. Enormous NPU meetings heard Saull call for passive resistance in the form of non-payment of taxes. Withholding tax to prevent "mutilation of the bill" (in Saull's words) became one of the NPU's policies of massive disobedience,[86] and, in this, they were backed by the NUWC.

Although a union designed to keep Earl Grey's eye fixed on reform, the NPU covered much more. Saull spoke frequently, and chaired meetings as often—on the Anatomy Bill (to enable medical schools to obtain legal cadavers and thwart the resurrectionists),[87] on removing the Irish tithes, on returning radical MP Joseph Hume to his Middlesex seat, and so on. Nor was sight lost of spies and the entrapment used by the Commissioners of Stamps' agents to catch news vendors.[88] Saull introduced Polish refugees fleeing after the failed rebellion against Tsarist rule and led three cheers for the eventual "restoration of Polish liberty". But always he would report back his NPU activities to colleagues in more radical venues, at the Optimist Chapel, acting as a sort of mole inside the moderate NPU.[89]

---

84 *MC*, 16 Feb. 1832, 3.
85 *Times*, 30 Sept. 1831, 3.
86 *Cobbett's Weekly Political Register* 76 (28 Apr. 1832): 247–52; *MC*, 26 Apr. 1832; *Examiner*, 29 Apr. 1832; Prothero 1979, 291–92.
87 *Bell's Life in London and Sporting Chronicle*, 29 Jan. 1832; *MC*, 26 Jan. 1832; *Examiner*, 29 Jan. 1832.
88 For his chairing of meetings to discuss these points, see *TS*, 17 May 1832, 1; 22 May 1832, 4; 26 July 1832, 2; 16 Aug. 1832, 2; 13 Sept. 1832, 1; 27 Sept. 1832, 1; *MC*, 23 Feb. 1832, 4; 17 May 1832; 16 Aug. 1832; 27 Sept. 1832; *Albion and The Star*, 18 Apr. 1833, 377, 403; *Examiner*, May 1832, 345.
89 HO 64/12, f. 96. *MC*, 22 May 1832; *TS*, 22 May 1832, 4. In 1836 he made a "manly and energetic speech" in celebrating the sixth anniversary of the Polish revolution (*TS*, 30 Nov. 30 1836, 2), and he was still contributing financially to the refugees in 1850 (*Reynolds's Weekly News*, 6 Oct. 1850).

6. Founding the Museum — June 1831

Even after the Reform Bill passed, he remained on the NPU council, sitting with Owen's sons, Robert and David Dale Owen.[90] The venue provides evidence of Saull's familiarity with the Irish nationalist, the "Liberator" Daniel O'Connell, a Kerry-born brilliant barrister and MP, democrat and former deist, now fighting for a repeal of the Union. Saull chaired O'Connell's NPU meetings on the government's Irish Disturbances Suppression Bill—a "Bill [that] resembled that monster at hell gates described by Milton"—which put the "poor famished peasantry", in Saull's words, effectively under martial law. While the Malthusian Whigs called for the Poor Law to be extended to Ireland (ministers were already contemplating stripping the indigent poor of outdoor relief—that is, welfare payouts—so forcing them to work for cheap rates or face the deliberately abominable new workhouses), Saull called for "justice, and not the cold hand of charity".[91] Saull, the anti-Malthusian, anti-Poor Law activist, condemned the suppression bill "in a wholesale way". Had the Secretary for Ireland been there, grumbled the Tory *Standard*, "he might well have exclaimed: 'Saul, Saul [sic] why persecutest thou the bill?'"[92]

Saull stayed at the NPU till the bitter end. The NPU's dissolution was already on the cards in 1833 (such unions had short life spans), with Saull chairing meetings to discuss its fate.[93] It limped on until 1834, long enough to see the Whig ministry itself dissolving, and the threat of the hated Wellington returning. Saull's last act here was to plead "for the people to convince the insane men" who supported the Tories "that they would ... not permit reform to be delayed".[94]

Not only was Saull's museum framed against this backdrop of heightened tensions and political lobbying, but his geology talks meshed with the reform hysteria at the radical chapels, the NPU, and

---

90 *Destructive* 1 (16 Feb. 1833): 23.
91 *Albion and The Star*, 18 Apr. 1833, 1; "Milton": *Morning Post*, 4 Mar. 1833. For Saull's chairing of O'Connell's meetings and speaking on the Irish situation, often at the Crown and Anchor, see *TS*, 4 Mar. 1833, 1; *MC*, 4 Mar. 1833; 25 Mar. 1833; *Standard*, 4 Mar. 1833; *PMG*, 9 Mar. 1833. He also condemned the Irish Poor Laws at the Cartwright Club and at the Guildhall (*TS*, 20 Mar. 1833, 3; 25 Mar. 1833, 1).
92 *Gauntlet*, 31 Mar. 1833, 128, quoting the *Standard*.
93 *TS*, 11 June 1833, 4; *Destructive*, 1 (15 June 1833): 159.
94 *MC*, 17 Nov. 1834; *TS*, 17 Nov. 1834, 2. On the protests at the thought of Wellington's return to power: *Times*, 20 Dec. 1834, 1; *TS*, 20 Dec. 1834, 2; *MC*, 20 Dec. 1834.

the countrywide Co-Operative Congresses. Often his blasphemous or co-operative harangue would devolve into a eulogy for science as he extolled his progressive museum and whipped up enthusiasm for listeners to visit. Having a finger in every radical and co-operative pie, Saull made sure his message about geological time, life's perfectibility, and mankind's destiny was broadcast widely through the radical world.

Geology Lecturing

Political and geological activism thus ran in tandem during these turbulent months. Saull's geological lecturing started shortly after the museum opened. The enabling climate for these set-piece talks was now complete: Phillips's astronomical explanations of planetary movements, Carlile's anti-Priestcraft naturalism and Cuvier's fossil ascendancy mated to Owen's perfectibilism. Saull would use the lectures to extract a higher moral meaning from the fossils, then invite listeners to confirm his deductions by studying the artefacts themselves. His first known geology lecture was at the Western Co-Operative Institute in Poland Street, and its date is significant. It was on New Year's Day 1832, the day his pseudonymous *Letter from a Student in the Sciences* was published, openly attacking religion as a *"despotism,* reigning tyrannically over the human mind"[95] (See Appendix 2). His last published attack on Christianity coincided to the day with his first known insurgent geological talk. From this point on, an ambiguously-infidel geology was to provide the anti-Christian's shield.

By this date, too, his shift to Owen's camp was complete. In 1831, Owen had taken out a lease on the grand hall in a spacious mansion at 277 Gray's Inn Road, near King's Cross—a former Horse and Carriage Repository—and made it the lecture hall of his "Institution of the Industrious Classes". In this "Great Room" Saull, himself on the Council of the Institution, delivered his first weekly lecture series on geology, with all profits going to the new "Missionary Society", which dispatched trained recruits to run Owenite branches in the provinces.[96] The "Great Room" was to be the headquarters of the tentacled Owenite

---

95 *PMG*, 31 Dec. 1831; [Saull] 1832a, 4.
96 *Crisis* 1 (1 Sept. 1832): 104; (29 Sept. 1832): 119; (15 Dec. 1832): 164; (29 Dec. 1832): 172.

empire throughout the country. Here Robert Owen himself lectured, and the Congresses of delegates from the Co-operative Associations of Britain and Ireland would meet. Saull was only one star in a co-operative constellation. Owen's son David Dale spoke on chemistry, as did their American fellow-traveller Henry Darwin Rogers. As a New York professor of chemistry, Rogers had just come over to England (in fact, he was the first American Fellow of the Geological Society[97]), and he would go on to pioneer surveys of Pennsylvania and Virginia. In 1832–33 he alternated with Saull in Gray's Inn Road talking on geology "much to the apparent satisfaction of their audiences", the two ploughing their profits into the social missions.[98]

The geology-fostering environment deeply affected Owen's own family. His boys were to take their love of emancipist science back to America, and David Dale Owen would become famous in his own right as the State Geologist of Indiana by 1836,[99] and go on to direct the geological surveys of Iowa, Wisconsin, and Illinois. Owen's followers who emigrated to New Harmony in Indiana in later years reported that David Dale's mineral museum—collected during his state surveys—was already three times the size of Saull's, now considered the standard.[100]

Saull's brandy depot museum had been crucial in this early Owenite career-building. His own lectures at the "Institution of the Industrious Classes" were "illustrated by many rare and beautiful specimens of fossil remains; among the rest of fossil palm, which is of very seldom occurrence," all taken from Aldersgate Street. And Robert Dale reported that Saull "kindly offered the use of specimens from his extensive Geological cabinet" to the others, so they all provided hands-on, illustrated talks in 1832–33, thanks to the museum.[101]

Saull was deeply embedded in the Gray's Inn Road institution, bureaucratically and financially. The backers thought of buying the leased premises to put it on a more permanent basis, but apparently

---

97 Henry Darwin Rogers was initially elected FGS on 1 May 1833. I thank Wendy Cawthorne at the Geological Society Library for information on Rogers and his backers. Gerstner 1994, 22.
98 *Crisis* 1 (8 Dec. 1832): 159; (15 Dec. 1832): 164; (29 Dec. 1832): 172; (5 Jan. 1833): 174; *PMG*, 22 Dec. 1832.
99 *NMW* 3 (29 Oct. 1836): 4; Horowitz 1986.
100 *NMW* 12 (8 July 1843): 10.
101 *Crisis* 1 (15 Dec. 1832): 164; (5 Jan. 1833): 174.

it was too exorbitant.¹⁰² It was a large venue, which hosted the Third Co-Operative Congress in April 1832, a week's jamboree of the country's co-operative groups, with Saull chairing meetings.¹⁰³ Optimism was running high; they started a new paper that month, the penny *Crisis*. It provided a vehicle for Owen's lectures (and soon enough, Saull's), as well as weekly intelligence. The illegal market place was already crowded, and barely had it started when *The Thief* (itself a startup, but run by the more *"light-fingered* gentry") was hooting at it:

> The *Crisis*–Rhymes with the *Isis* [another inflammatory penny print, founded in February 1832 by Eliza Sharples], and seems of the same kidney, edited too by Mr. Owen, (Oh! name unmusical to tradesmen's ears!) who talks of Co-operative Congress, explains his principles by a *ball* and *lecture*!!! and professes to sell "Truth and Happiness" price one penny!¹⁰⁴

Nonetheless the less light-fingered *Crisis* fared well. Early circulation was boosted by philanthropists buying batches, a hundred copies a time, to distribute freely. Twenty thousand copies of early issues were said to have sold.¹⁰⁵ Robert Owen was editor, and, from November 1832 until April 1833, his son Robert Dale joined him. But a revenue drop caused Robert Dale to give the venture to the printer before leaving for the American New Harmony community in Indiana that April. And, while the new proprietor made efforts to improve the type and content, sales began to flag.¹⁰⁶

Saull did more than emphasize the geological proofs of perfectibility in print. He helped perfect the co-operative system in real time, sitting on the Council of the new "Equitable Labour Exchange" in 1832, run from Gray's Inn.¹⁰⁷ This bartering bazaar was designed to cut out the middle man and ease the unemployment among London's artisans and shopmen. Labour notes were issued for items (valued by their material plus labour costs), and these notes could be exchanged for equivalent

---

102 *The Satirist; or, the Censor of the Times*, 5 Feb. 1832.
103 Carpenter 1832, 78; Claeys 2005, 4: 77. He similarly officiated at the Fourth Congress in October 1832, *Crisis* 3 (19 Oct. 1833), 64; *Lancashire and Yorkshire Co-Operator* n.s. no. 10 (n.d. [Oct.1832]): 23.
104 *Thief*, 21 Apr. 1832, 1.
105 *Cosmopolite*, 28 Apr. 1832, in HO 64/18, f. 606.
106 *Crisis* 1 (27 Oct. 1832): 138; 2 (27 Apr. 1833): 125.
107 Claeys 1987, 54–55; J. F. C. Harrison 1969, 72; Holyoake 1906, 1: 105.

goods held at the Bazaar, thus avoiding the use of capital altogether.[108] The notes even made their way into the old immoral world: they were so prevalent for a while that theatres accepted them. Carlile too, though he deplored Owenism, took them as half payment on books.[109] Although items at first flooded in (and by September 1832 Owen reported they had "large stocks of goods already in the Bazaar"[110]), it was not a wholehearted success. Nor was the attention it attracted always positive: a utopian absurdity, one critic thought, "which is to pave the streets with penny loaves and roof the houses with pancakes, not to mention the licence it affords with respect to one's neighbour's wife".[111] Some said it was already in a "dying state" by New Year 1833, although in fact they were trading articles to the value of "37,000 hours per week" at this time, but they still decided to merge it with the Co-Operative Society.[112] Saull helped set up local Bazaars besides, for example the Kingsland and Newington Co-op Labour Exchange,[113] so he was well placed to assess their shortcomings. Partly, it was that the labour notes were devaluing as trade goods or the choices dwindled. Then there was a ludicrous inequity in the swaps, as he later recalled, with "some articles of food being wanted much more frequently than others. The baker would be overpowered with articles which he did not want".[114]

Things were in a bad way by the end of 1832. As the spy reported, "a Rich Organ which they had erected on the premises for Balls, Vocal Performances &c together with Chandeliers, Ornaments &c has been seized by the Commissioners of Pavements for the Rates."[115] The local bazaars went the same way. Take William Benbow's at 8 Theobald's Road, a huge, ramshackle place that could hold 2,000, and sometimes did.[116] Here the co-operators—Lovett, Cleave, and Watson—half-splintering

---

108 *Examiner*, Aug. 1832, 551; Hayes 2001; Abrahams 1908; Oliver 1958.
109 *PMG*, 8 Dec. 1832. Theatres: McCabe 1920, 80.
110 *MC*, 25 Sept. 1832. Others venues followed suit. The Gothic Hall Bazaar in the New Road was reportedly turning over £350 worth of stock a week in August 1832: *Cosmopolite*, 11 Aug. 1832 in HO 64/18, f. 663.
111 Abrahams 1908.
112 *TS*, 1 Jan. 1833, 2; *Cosmopolite*, 5 Jan. 1833, in HO 64/18, f. 733.
113 *TS*, 26 Sept. 1832, 2; *Atlas*, 30 Sept. 1832, 660.
114 *The Star of Freedom*, 5 June 1852; *The Journal of Association, Conducted by Several of the Promoters of the London Working Men's Associations* 1852, 182.
115 HO 64/12, f. 179 (23 Nov. 1832).
116 HO 64/18, f. 108.

from their NUWC colleagues, set up, or rather climbed up, for they met in the room above the NUWC. Not only was there physical proximity here, but a mongrel mix of radical and Owenite objectives. The close contacts were revealed as the NUWC gave the co-op schismatists money to remedy their dilapidated rooms above, accessible only by ladder. Of course, this co-operative faction remained to the left of the patrician Owen, and, in particular, they hated his pandering to aristocrats. It did not help to see society ladies drop in to the Labour Exchange,[117] or the King be invited as a patron of some new job scheme. But their bazaar, too, faltered and collapsed just the same, late in 1832. And between them, these defunct bazaars left a lot of worthless circulating labour notes.[118]

On the NUWC side, Hetherington in the early thirties stood firm and insisted that political power must precede social perfection. He argued for the same rights as Saull did—suffrage, short Parliaments, ballot, and no property qualification—but insisted that these political gains must come before co-operation could be contemplated. He split opinion and sparked public debates, and made the *Poor Man's Guardian* essential for the more pro-active political wing. For a moment (it was short-lived), he became hyper-critical of Owen, who "exhibits a strange perversity of mind in expecting to realize his political millenium [sic] before working men are placed on an equal footing with the other classes". Losing faith in Owen's idealism, he found Owen's tolerance towards the oppressors, the aristocrats, and capitalists "preposterous", which militated against any immediate political rapprochement.[119]

Saull and Hetherington remained the best of friends, even as Saull spoke up for Owen's schemes and Hetherington demanded a prior political emancipation. It infuriated Hetherington to see the "the benevolent Owenites ... 'dancing jigs at two-shilling hops', while thousands and tens of thousands of their poorer fellow-countrymen are pining in want and destitution".[120] Owenism put the cart before the horse. The aristocratic masters and middle-men would never let co-operation work, and only when "the working classes succeed in obtaining political power" could Owen's exchanges be implemented. Worse, the house

---

117 *Lady's Magazine* 3 (Nov. 1833): 297.
118 HO 64/12, ff. 59, 76, 79, 83, 145, 179.
119 *PMG*, 14, 21, 28 Jan. 1832; 6, 9 Feb. 1832; 22 Sept. 1832.
120 Royle 1998, 52; Claeys 2002, 175–82; *PMG*, 25 Dec. 1831, 14 Jan. 1832.

was dividing against itself: Owenites had split the cause and sapped its strength, which served only to "paralyze the nobler efforts of others". While many wanted to see radicalism and co-operation "go hand in hand",[121] for Hetherington, in the early thirties, perfecting man came second. That said, Hetherington never actually let go of the Owenite doctrines that circumstance creates character, and, therefore, that a better society would produce better people;[122] and he took from Owen his moral-force beliefs. These guaranteed his later return to the fold.

Meanwhile Owen's own house was in trouble. After a fracas with the Gray's Inn building's owner, who used an axe-wielding mob to regain entry, the co-operators were evicted. At the time, Saull was actually in the process of valuing the fixtures ready to move.[123] In February 1833, they rented new premises at 14 Charlotte Street, Fitzroy Square. With its inner court 16 by 120 feet and corridors all round, it could accommodate the Exchange stalls and, in April, artisans started filling them up.[124] At the same time, the managers tried to cut out the banks by starting a "United Trades' Loan Fund", where tradesmen could obtain credit to purchase raw material. Saull was (inevitably) its treasurer.[125]

Barely had they finished setting up the new Charlotte Street auditorium before the sparks started to fly over Saull's deeper scientific views. The *Crisis* now gave the first full-blown account of his evolutionary Owenism, with its monkey ancestry for mankind.

---

121 *PMG*, 14 Jan. 1832, 22 Sept. 1832.
122 *Reasoner* 7 (5 Sept. 1849): 152.
123 *Crisis* 2 (2 Feb. 1833): 26–27.
124 *Crisis* 2 (16 Feb. 1833): 42; 2 (30 March 1833): 95.
125 *Crisis* 2 (18 May 1833): 149; (8 June 1833): 174, 175; (10 Aug. 1833): 248.

7. Monkey-Man —The Bristol Lecture 1833

... one of the most impressive and interesting lectures that has ever been delivered in Bristol.

Carlile on Saull's outrageous evolutionary speculation.[1]

Having offered to lecture to provincial co-operative branches, Saull found himself talking geology in Bristol in August 1833. He was either invited—probably by the First Bristol Co-operative Society, in Old Market Street[2]—or he was in Bristol on business and opportunistically speaking at the co-op. The Atlantic port, with its shipping ties to Bordeaux, Spain, Portugal, and Madeira, was at the centre of the wine trade. Here Saull would have come to pick up hogsheads of claret, sherry, and port, or rum from the West Indies. Perhaps that was what brought him here that summer; certainly "it was his custom to take every opportunity of delivering lectures on geology" while travelling on business.[3] Now, finally, we have a transcript of one of his lectures, reports having been inserted into Carlile's *Gauntlet* and another published in the Owenite *Crisis* by its new editor.[4]

Most Saull lectures we know only by title. His talks in back-street London[5] would be simply listed as "Geology", or "Evidences of

---

1 *Gauntlet*, 29 Sept. 1833.
2 *Crisis* v1 (14 July 1832): 71–72.
3 *Monthly Notices of the Royal Astronomical Society*, 16 (1856): 90.
4 *Crisis* 3 (5 Oct. 1833): 36–39; *Gauntlet*, 29 Sept. 1833, 529–33.
5 Throughout the thirties, he would speak at the Western Co-Operative Institute off Oxford Street, at Owen's Institution 14 Charlotte Street, the Society for the Acquisition of Useful Knowledge in Bedford Square, the Great Tower Street Mutual Instruction Society, the Rational Institution in Curtain Road, the East London Branch of the Association of Rational Socialists in the Mechanics' Hall of

Geology", or, in one case, "Geology in reference to Human Nature". In another we have a bald strap line: his threepenny talk at Charlotte Street on 2 December 1834: "Lecture on the FORMATION of the EARTH, and the Real as well as Probable Revolutions it has undergone to produce the present state of Organic Nature".[6] Only the Bristol lecture was ever apparently reported verbatim. So the sort of geology he was promoting in the co-operatives and in his museum has to be gleaned from this.

Standing back for the moment, we might consider how geology was used and received at the time in order to understand Saull's uniqueness. The socialists' geological punch in the 1830s largely came from their emphasis on the antiquity of the earth and the evidence of death before Adam. Both were seen, if not to de-legitimize scriptural literalists, then at least to make them uncomfortable. Hence the *Church of England Magazine*'s innuendo about socialists outraging Revelation by starting their Sabbath sermonizing with a geology lecture.[7] It was part of the reason that geology was considered necessary for a socialist education.[8] The other part was their Enlightenment faith in the redeeming value of science: following nature's law would lead to moral elevation. Not merely moral, but social: materialist science was to be the immovable bedrock underpinning the new co-operative system of harmonious, class-less social relations. As so often, the "is" of nature was being made a justification for the "ought" of politics.[9] The London Owenites would advise branch lecturers to teach geology, chemistry, and astronomy to demolish the props of the old immoral world, and to make geology one of the axioms of a rational school education.[10] Undermining Genesis would undercut rival pulpit-power, which had a tenacious hold in a city with 400 churches.[11] By proving a far-distant "Age of Reptiles" (as Saull's friend Mantell was calling it), so contested in literalist Christian circles, and the primeval prevalence of death (so graphically displayed in fossils), socialists could argue that death was not due to the wages

---

Science in City Road, Finsbury, at the Social Institution, 23 John Street, Tottenham Court Road, as well as lecturing the Bristol and Leeds socialists.

6 *NMW* 1 (29 Nov. 1834): 40; 3 (12 Nov. 1836): 20.
7 *Church of England Magazine* 9 (1840): 120.
8 *NMW* 9 (13 Feb. 1841): 91.
9 Bloor 1983.
10 *NMW* 6 (24 Aug. 1839): 704; (5 Oct. 1839): 789–91.
11 *Cosmopolite*, 29 Sept. 1832, in HO 64/18, f. 652.

of sin. Animals died before Adam transgressed, the first appearance of death had nothing to do with a mythical fall of man. Such subversive claims made fossils particularly ticklish for many evangelicals and biblical literalists.[12]

Owenite organs featured a plethora of reports on the antiquity of the globe to expose the "errors" of the priests, on Mosaic "Days" twisted into thousands of years, on death entering the world with fossil carnivores, ultimately on the "false" notion of mankind's Fall and the "true" geological base of socialism's ever-perfecting world.[13] This was becoming the stock stuff of socialist demagogues to counteract the placard-carrying Christian missionaries outside their halls. Geology was necessary for socialism to prove that perfectibility, not depravity, maketh man.

Geology was practically placed in opposition to the pulpit. Sometimes this was literally so, as for an Edinburgh social missionary who would later report:

> We have just established a geological class, from which we anticipate good results, especially when we consider that we are surrounded by one of the most favourable geological localities in the kingdom. Will it not be delightful to march, en masse, with hammer in hand, to some favourable spot, and there read "sermons in stones," &c., on some fine Sunday morning? what a difference to being pent up in one of your cold dissenting chapels.[14]

Geology was pushed as a propagandist tool. Could it not convert the faithful unaided?, it was asked, after an Irish Catholic-turned-socialist started sending museum specimens to Owen's Institution.[15] As a result of this multivalent justification, geology was invariably lashed to atheism in anti-blasphemy rags: here mosaical was pitted against mineralogical, and the "low-minded" were condemned for wresting false science to

---

12 Biddulph 1825, 123–39; *Christian Observer*, Feb. 1829, 91–96; June 1839, 345; *NMW* 6 (27 July 1839): 646. O'Connor 2008, 210, for the *Christian Observer*'s editorial angle on geology.
13 *NMW* 1 (12 Sept. 1835): 364; 4 (20 Jan. 1838): 101; (16 June 1838): 268; (22 Sept. 1838): 389; 6 (27 July 1839): 646; (5 Oct. 1839): 788; 7 (16 May 1840): 1205; (6 June 1840): 1280; 9 (27 Feb. 1841): 128; (17 Apr. 1841): 247; 10 (14 Aug. 1841): 55; 11 (21 Jan. 1843) 243; 11 (18 Mar. 1843): 294.
14 *NMW* 11 (24 June 1843): 434.
15 *NMW* 10 (25 Sept. 1841): 95.

blasphemous ends.[16] Infidels were known to be twisting geology to undermine scripture[17] and to weaken the moral arm of the state. In the end, "Geology was suspect because it provided the easiest infidel ammunition for factory activists, as the shop floor testified."[18] With the science so much the rage, and so "perverted to infidelity", it became essential for ordinands to master it in order to reassure their flocks.[19] With radicals and co-operators trying to demolish the tithe system and disestablish the Church, breaking its "Adulterous Intercourse" with the state,[20] socialist lectures were expected to be subversive, irreligious, and concentrate on antiquity and death. Saull did not disappoint; he went much further.

Even if geology was a legitimate science, the fear for the faithful was always that some clever Voltaire would make great play of the successively recreated worlds, as an Oxford don told the *Times*.[21] And focussing our microscope on Saull's stratum of society, we find plenty of noisy Voltaires. At one end was the extremist Julian Harney, the future Marat of Chartism, with his red cap of liberty, waving his dagger during public meetings to make a point, who started from first geological principles to claim his red republic.[22] At the other was the mild moral-force Saull, for whom a self-progressing fossil life pointed to Owen's future perfect society. These activists mined out all the political ore they needed.

One other premise marked many socialist lectures: that some inherent power in nature underlay its self-propulsion. Primitive man was himself "generated by Nature". It was said time and again, and it made opponents shudder. Protagonists asked what it meant. To Benjamin Godwin, combatting Bradford's infidels in his *Lectures on the Atheistic Controversy* (1834), the assumption that "unintelligent matter" could rearrange itself to make a human was an "absurdity". Nor was the metaphysics any less gobbledygook. If, talking of "the necessary laws of

---

16 *The Age*, 8 Jan. 1837, 5.
17 *Atlas*, 2 July 1837, 424; *Albion and The Star*, 16 Sept, 1834, 3.
18 *Nonconformist*, 22 Nov. 1848, 888.
19 *Atlas*, 12 Nov. 1842, 730.
20 HO 64/12, f. 152.
21 *Times*, 26 June 1845, 5.
22 *Democratic Review* 1 (June 1849): 9–10. On Harney: Epstein 1994, 19–20; Lovett 1920, 1: 207; Holyoake 1905, 111.

nature", "Necessity" is defined as the "infallible connection of causes with their effects", then, reasoned Godwin, the effect, intelligent man, must require an Intelligent cause. Of course, many street atheists now reasoned that matter was not "unintelligent", and that thinking matter gave rise to thinking man. But, for Godwin, an "innate original tendency to a higher state of being" was an absurdity.[23] Others repeated it in the Literary and Scientific Institutes: there could be "no self-origination". "Nature herself cannot accomplish such a result".[24]

The trouble was, neither Carlileans nor Owenites got nearer to an explanation of this self-development. The *Crisis* could only criticize religions

> where man and woman are spoken of, not as what they really are, organized substances generated by nature, like all other substances, but as having been created as if apart from nature for the purpose of being condemned from birth to toil in want and misery, as responsible beings, mystically said to be composed of a body and of a soul...[25]

But exactly *how* were they "generated by nature"? It was not good enough to moot an Owenite environmental cauldron, as in Logan Mitchell's *Christian Mythology Unveiled*, which posited that "Nature *always* produced the animals and plants that were proper for the climate and soil".[26] The demagogues invariably ended up where they started, looking at the extinctions and productions and invoking an "energy in nature, by which new species are brought into being".[27] What on earth was an "effort of Nature" anyway? asked the unconventional Universalist Rev. James Elishama Smith, the wild millenarian (and incoming *Crisis* editor), who now saw Christ's coming heralded by an earthly Owenite millennium of equality and justice. Here was a socialist preacher who hated Carlile's "effusions" and considered an "omni-active" God explanation enough of nature's geological "revolutions".[28]

> Even Mr. Saull [said Smith], who denies the existence of spirits, though he acknowledges the reality of liqueurs, uses this language, and tells

---

23   B. Godwin 1834, 175.
24   F. J. Francis 1839, 162–63.
25   *Crisis* 4 (14 June 1834): 78–90.
26   Mitchell n.d. (ca. 1842), 234–35.
27   *FTI* 1 (1842): 7.
28   *Shepherd* 1 (27 Sept. 1834): 40; 2 (15 Feb. 1837): 33–35; 4 (13 Jan. 1838): 94–95.

you most gravely, like an old man telling about a ghost, or a piece of witchcraft, that Nature made man by means of laws! Ask him if ever he saw Nature or her laws, and his eyes will look as round and startling as dozes [sic] of castor oil swimming in gin.[29]

The trouble was, when Saull did proffer something more of an explanation, Smith loathed it. Saull's solution was simply too shocking for the more religious millenarian Owenites.

The Origins of Mankind

A major debating point for the clergy remained the origin of man. It figured in those set-piece debates between rationalist Owenites and Christians in public theatres, which were a feature of the 1830s. Preachers raised human origins as a stumbling block for the "atheists". Protagonists goaded Owen: if humans are "a production peculiar to our globe", so that when the planetary orbit shifts, "the human species will change, or will be obliged to disappear", how to explain it? Owen was faced with an "absurdity", that somehow infant humans had been thrown up by the earth without parents.[30]

In Saull's circle, the idea of man generated from nature was a given. But geology had overridden the old Enlightenment ideas of Holbach: no longer were the motion of atoms responsible for emergent life with its new qualities, as he had thought. Infidels now gave these atoms themselves immanent qualities—they were self-organizing and thinking. And Holbach's nature as a "dynamic chaos"[31]—balanced, eternal, and possibly coeval with mankind's origin—had become progressive, directional, and finite: a rising series of fossil species, each originating at a certain point in the rock strata.

Mankind's appearance was the burning issue. Saull, in his geology lectures, was to suggest that humans had not sprung from the ground readymade, like Milton's lion "pawing to get free" of its earthly cradle. This despite Volney's *Ruins*—that ubiquitous bible used as a warm-up act before Sabbath blasphemy lectures right into 1833[32]—which had

---
29 *PS*, 1 Oct. 1842.
30 Owen and Campbell 1839 [1829], 77–78.
31 Bowler 1974, 164.
32 HO 64/15, f. 171, in this case at the Borough Chapel.

man "formed equally naked both as to body and mind," and thrust "by chance upon a land confused and savage. An orphan, deserted by the unknown power that had produced him".[33] Carlile had modified this a tad, suggesting that the first race of men must have been mute and uneducated, and possibly their bones might still be found on the unexplored ocean floor.[34] Both scenarios were lame and old-fashioned, and neither took account of a rising fossil life, producing a gradation of species, nor of its esoteric contemporary corollary, an 'evolving' lineage.[35]

When Saull turned up in Bristol, on either 22 or 23 August 1833,[36] he found a riot-torn city, with the incendiary aftermath of the Reform riots visible in the burnt-out bishop's palace. Workers and trades people were his usual audience, and not only men. Like the Rotundanists, he welcomed women and children. If his venue was the Bristol Co-operative Society, this alone would have made him obnoxious to the city fathers. The bruised Tory merchants deplored the workers' halls springing up where "the operative classes are encouraged to turn literati", when it was clear that the "covert design" of their demagogue leaders was to stir up "political agitation". The anti-radical *Bristol Job Nott*, founded to emasculate the burgeoning pauper press, even viewed libraries with suspicion, and advised operatives to sit at home quietly studying their Bible.[37] With street vendors surreptitiously shifting unstamped papers, and the Bristol NUWC branch gaining traction,[38] the patrician press hysterically warned that "Hetherington and Carlile in their 'poison shops' offered that 'black draught' which brought down on its victim 'discontent, sulkiness, sabbath-breaking, scoffing, hatred of the law, of kings, magistrates, and all superiors'."[39] While the city fathers at the elite Bristol Institution screened their mechanic audiences and fed them a safe science based on miracles, wise design, and the creation of everything in its proper place—presenting this knowledge in patronizing form as a

---

33 Volney 1819, 17.
34 *Prompter* 1 (1 Oct. 1831): 820–22.
35 R. Phillips 1832a, 51.
36 The *Gauntlet*, 29 Sept. 1833, 529, placed his lecture on "Thursday, the 23d ult.", but Thursday was 22 August and Friday the 23rd.
37 *Bristol Job Nott*, 24 Jan. 1833, 233; Klancher 1987, 184 n. 25; Murphy 1994, 146.
38 Rowe 1970b, document no. 27.
39 Hollis 1970, 143.

"boon, emanating from the superior to the inferior"[40]—Saull's intentions at the Co-operative was completely the reverse.

Addressing what Tories considered the violent underbelly, Saull provided a mentally liberating science for the marginal men. It was exactly as the *Job Nott* feared. He even kicked off his talk by assuming that "all present are decided reformers". They would therefore appreciate his "enlarged views" of science, which were designed to "annihilate from amongst men the present extreme amount of ignorance, poverty, and consequent crime and misery".

Saull started his two-hour lecture by attacking the religious myths legitimating aristocratic domination—those biblical fantasies taught (as Owen complained) from the cradle. These "blind guides" were "trammels" and "worse than useless": they were "absolutely mischievous" socially-controlling devices. Not only did they not "produce sound morality, social happiness, or political elevation", to the contrary, they have "invariably tended to uphold the powers of the ruling few, at the expense of the welfare and happiness of the oppressed and deeply-injured many".[41] In a year which saw radical clamouring for the removal of tithes, the de-funding of the "Government Clergy",[42] the disestablishment of the Church (which Saull would campaign for[43]), and the removal of Anglican monopolies on the services surrounding births, marriages, and death, this would have chimed. But go further, Saull said: cast off all "systems of religion, which are nothing but phantasies of the fever-excited brains of the various religious enthusiasts" designed to intimidate the poor, dispossessed, and disenfranchised.

Having established his credentials, Saull now proffered a solution. "What, then, is the course we should pursue, to counteract these direful effects?" The structures he put in their place were the "immutable truths" of nature, now seen as the progressing and perfecting march of life and mind. Here, Saull's geological infidelity manifested on multiple counts. Not only in his belief in the "immense distances of time", the aeons to lay down the thousands of feet of strata from ancient sediments—all gentlemen geologists now accepted such a 'deep' time. Nor only the idea

---

40 Neve 1983, 188.
41 Saull 1833a, b.
42 HO 64/19, f. 158; *The "Destructive," and Poor Man's Conservative*, 2 Feb. 1833, 2.
43 *TS*, 28 Apr. 1835, 2.

of a "regular advancement in the production of animal and vegetable life" shown in the rocks—and he illustrated it by pointing to the ancient ferns comprising Britain's "great coal-fields", and the rise of huge saurians, some reaching "the enormous length of eighty feet". Rather, the "enlarged" views from which to draw new social conclusions, and the real 'infidelity' to make even gentlemen geologists blanch, came next.

The fossil ferns from the coal districts were the prize exhibits in Saull's cabinet, so he might have brought some with him to Bristol. Since these ferns only flourished "in the very hottest climates", Britain must then have been a tropical country. The same was true of the corals in Jurassic rocks; because corals now only grow in warm atolls, these conditions must have prevailed in Jurassic times in Britain, "namely, shallow salt water and a very hot climate".[44] He was suggesting that changing local conditions had "produced" these creatures as needed—in line with Owenite environmental thinking. And what caused these changing conditions? Given that Saull had just (June 1832) republished his mentor Sir Richard Phillips's *Essay on the Physico-Astronomical Causes of the Geological Changes on the Earth's Surface*, we know that he had in mind a shift in planetary axes.

When, in this Saull-edited work, Phillips said "the strata prove the gradual evolution of all things", we must beware of treating him as a modern. His mooting of the "origin of Species" and "progressive evolutions" of life reflected an older Enlightenment mindset. And he had a distinct mechanism to explain them. Phillips's theory predicted that, with the long-term swinging of the earth's axis, the sea would rush in to submerge the northern lands as it drained from the south, or vice versa—hence the strata alternated every 10,465 years between sediments of marine and continental origin. This immersion, in his view, was the means "of restoring an exhausted world and improving it". Each immersion flushed in vivifying erosional material. And the more immersions, the more finely ground the erosional material became, releasing ever more nutrients. The result was that this ever-refining "pabulum" "may have resulted [in] the improvements and refined complications of animals". He seems to have imagined the whole

---

44 Saull 1833a, b.

hemisphere's fauna and flora being regenerated at a stroke, each time in more "improved" form. There is no sense of ancestry here, no lineage; like so many Enlightenment thinkers he accepted a sort of spontaneous re-emergence, but now of the entire ecosystem. These revivifying fluids "must evolve, *at once*, every thing that is possible" (my emphasis), producing at each turn a "wonderful gradation of being". And because there was no necessary lineal progression through the strata—just improvements and changes—he believed that human fossils might still be unearthed in older rocks among the tropical animals.[45]

Phillips's impact on his protégé was profound. One facilitating factor in Saull's solution to human origins might have been Phillips's compassion, which extended from the poor to all suffering life. Pain, disease, distress, being "common to entire animal nature", left Phillips the vegetarian seeking the freedom of all animals from subjugation. He pushed humans and animals into the same category and integrated the liberation of brutes into a search for mutual justice on earth,[46] which could have been a valuable heuristic to Saull's search for life's relationship. But for all that, Saull had outgrown Phillips's Enlightenment spontaneity, the 'all-at-once' productions. We only appreciate Saull's difference as he moved in his 1833 talk to the emergence of "hot-blooded animals", which appeared after the colossal saurians and culminated in the appearance of cave bears and hyaenas in Britain. And here he envisaged another giant, "the fossil elephant, which, by the teeth or grinders, and bones, we know was at least twice the size of the largest elephant of the present day".[47] (Gigantism introduced a sense of awe in these talks, as if to reinforce the immensity and difference of the past. It served the same function in his museum, where the grinders were on display, as well as the remains of "immense rhinoceroses". He was to exhibit some fossil hippo grinders at the British Association for the Advancement of Science the following year.[48]) With the great beasts' passing came the last

---

45 R. Phillips 1832a, 47–48, 51–53, 70; Rupke 2005 on this 'autochthonous' emergence. Rarely was any attempt made to plumb this 'spontaneous' origination of species. One exception, discussed by Topham 2022, 364–65, was a Manchester Owenite, Robert Whalley, who invoked a crystallization-based explanation in 1835.
46 Duthie 2019, 86–91.
47 Saull 1833a, b.
48 *Literary Gazette* 922 (Sept. 1834): 637; *Athenaeum*, 27 Sept. 1834, 715.

and "most singular of animals, 'man'."⁴⁹ That "man" was an "animal" should have set orthodox tongues wagging. How he arrived on the earth certainly did.

Saull envisaged man "emerging or advancing, perhaps, from some of the simian, the ape or monkey tribe, educed by circumstances over which neither they nor he could have the least control".⁵⁰ That was his clearest materialist statement on the external agencies generating man from a monkey. No longer were humans generated directly from the earth. He seemed to be invoking some sort of bloodline, as the monkeys moved 'up' to become men.⁵¹ Mankind was the monkey perfected, in line with Owen's perfectibility doctrine. Saull had said as much in his 1832 *Letter from a Student in the Sciences to a Student of Theology*, that the "strange animal called man made his appearance, emerging many steps in advance of the race of the Simians, who had inconsciously been his precursors."⁵² But that letter was anonymous. Out in the open now, this belief was to land him continually in hot water for years to come.

There was no Creation here. Nor was "man" *sui generis*, conjured up by animating fluids with the rest of nature. He issued naturally from monkeys or apes—he had a primate pedigree. But like Phillips and the ideologues before him, Saull was adamant in denying any miraculous intervention in the production of man, the sort taken for granted in gentlemanly society—an intervention which reinforced the authority of the Church over the souls so created. Notice, too, the congruence of his wording about man's evolution by natural agencies outside his control, with the Owenite dictum (run on the masthead of the *Crisis*): "IT IS OF ALL TRUTHS MOST IMPORTANT, THAT THE CHARACTER OF MAN IS FORMED FOR–NOT BY HIMSELF". Environmental and cultural forces shaped 'evolution' as much as human character. This showed the need for correct co-operative

---

49   Saull 1833a, b.
50   Saull 1833a, b.
51   This might, as Hodge 1972, and 2005, 112, portrayed it, be envisaged as multiple parallel lineages. As the monkey line advances to become human, so an independent lineage with lower primates at the top moves up, and these become monkeys. The lineages all had separate starting and end points. There was no common ancestry, that is, no branching tree. Bowler 1984, 80, for a diagrammatic representation.
52   [Saull] 1832a, 6; Saull 1833a, b. Saull's ape hypothesis was long recalled in socialist histories (M. Beer 1921, 330).

conditioning to continue the process for the moral regeneration of mankind.

The frisson between radicals and co-operators played out in Saull's evolving nature. Here sovereign, self-reliant, and self-transforming life-forms (the radical component) still needed the external progressive push from changing ecosystems (the Owenite environmental component), itself driven by large-scale planetary wobbles.

Mankind's lowly ancestry and rise from savagery was seen to justify the drive to the social millennium. "From the gradual progress in nature towards perfection", it followed that the animal man, "as a part and parcel of that nature", should be advancing too. But people were being held back by the old "barbarous" doctrines which "crafty" religious teachers "force ... into the mind of youth": of humans as a sinful species sunk in depravity, requiring salvation, the degenerate relicts of once blessed beings. Men of the cloth remained blind to the real material beauty around them. "While these practices continue, and while we pay those the most money who continue to propagate them"—the obligatory hit at the lord bishops who blocked reform and the tithe payments angering non-Anglicans—"we may look in vain for any great advance in moral or social improvement." The answer lay in the rational school initiatives being launched by the London Owenites. In these, nature and practical science were cherished, religions were seen as culturally relative, and mores were contingent developments. Had the populace been educated rationally, and properly empowered, the social transformation would already have been complete. People would be responsible and ready to vote. Thus Saull planted his infidel, co-operative, and radical colours on the sedimentary strata. As a rich merchant, he was no real communitarian but a great respecter of property, with faith in continuing parliamentary reform and eventual working-class suffrage. And here he pointed his finger at the Bristolians, still smarting after the Reform Bill conflagration. The arsonists who fired their Town Hall and city centre during the riots were

> so ill-informed, that they know not how to enjoy freedom and liberty, and the proof of it is as clear as the sun at noon-day; for had the minds of the people been prepared [by a secular socialist schooling] ... you would

not have had the afflicting prospect you now behold, namely, that of seeing part of your beautiful city in ashes!⁵³

The Bristol riots had stunned the nation and were a talking point for years. They generated the same sort of shudder that his monkey parentage did: both were assaults on a civilized Christian nation. On the other hand, Saull's thorough-going materialism was effusively received by an audience bombarded by such infidelism, who saw it as one of the best lectures ever delivered in the city.⁵⁴

This talk was probably his stock lecture, one he had been delivering for a couple of years. That might explain why there is no mention of the French transformist Jean-Baptiste Lamarck: the lecture was possibly being read unmodified, despite recent developments. Saull knew of the wealthy barrister-turned-geologist Charles Lyell's influential *Principles of Geology*. This had exhumed the poor Frenchman from his pauper's grave the previous year (1832) in order to unceremoniously re-inter him. The British were made painfully aware of Lamarck's shameful science from reading the second volume of Lyell's *Principles*. Nothing really prepared a genteel nation for Lyell's evocation of Lamarck's *"force of external circumstances"* causing chimpanzees to stand erect to be counted savage men, and then his *"tendency to perfection"* to ensure their intelligence grew to civilized standards.⁵⁵ Lyell set teeth chattering. His long refutation acted like a red rag to the religious bull—hysterical reactions in sober papers became the norm. England might have been backwards in its natural theology (and rather laughed at in Europe), but it was not only traditionalist passions that were inflamed. Even the *Atlas*, that huge (and hugely expensive at 1s) Sunday paper aimed at well-to-do liberals, was beside itself. The popular rag had its rough edges, and occasionally transgressed "the rules of courtesy" in its reviews,⁵⁶ but this time it went over the top. It railed against the "absurd creed of those grovelling idolaters who desecrate the temple of Heaven, and tear thence the Deity; who profane the altar of earth, and banish then the pure incense of creation...". "Ridicule is wasted on such egregious absurdity...", it said of the French *enfant terrible*.

---

53 Saull 1833a, b.
54 *Gauntlet*, 29 Sept. 1833, 529, 532–33.
55 Lyell 1830–33, 2: 14–16.
56 [James Grant] 1837, 2: 128; Bourne 1887, 2: 45.

> "We came from nothing," is the doctrine of LAMARCK ... Our journey is long, our progress uncertain; but, guided by the *'tendency to perfection,'* struggling with our other deity, *'the force of external circumstances,'* we shall become gods from men, as we became men out of monkeys, and monkeys out of the monad...

"How long", it asked, like a terrier thrashing a rat, "would the priests of transmutation require to convert the blue-faced ape of the zoological garden into a BROUGHAM"? Ultimately Lyell's dishing of Lamarck was applauded, by using the squash-and-bash slang of the day: the "absurd doctrines of the LAMARCKIANS are absolutely squabashed".[57] Because of Lyell's over-exposure, "Lamarck" became a pejorative word. The critics got carried away. The *Belfast News-Letter*'s use of screamers said it all: "Pro-di-gi-ous!!" and "rank nonsense!". None doubted that such transformist views were "Atheistical" (in the *Patriot*'s headline),[58] because so much contemporary natural theology rested on the wisdom of Creative action in designing and introducing consecutive species. They were introduced, not by "a transmutation of species", which was a "phrenzied dream", but "by a provident contriving power".[59] Critics insisted there could be no innate, "original tendency to a higher state of being". "Nature herself cannot accomplish such a result". There could be "no self-origination", no self-empowering emergent life, no drive from below. Any claim to the contrary was subversive to conservative society and, given the state of natural religion, sacrilegious as well.[60]

Theological critics stood indebted to Lyell (perhaps more than he wanted or expected). The Regius Professor of Modern History at the University of Oxford, the Rev. Edward Nares, in his *Man, as known to us Theologically and Geologically* (1834), shuddered at the thought of "man as a mere development"

> and therefore feel myself, as others should do, greatly indebted to Mr. Lyell, for taking our part against Lamark [*sic*], who would have made ... nothing but apes, and monkeys, and ourang-outangs of us; or even worse, a mere expansion of organic particles. Dr. Macculloch, speaking

---

57 *Atlas*, 12 Feb. 1832, 107–08.
58 *Patriot*, 18 Apr. 1832, 8; *Belfast News-Letter*, 10 Apr. 1832.
59 *DPMC*, 18 Jan. 1834, 402, reporting the Rev. Adam Sedgwick's discourse at Cambridge. Hilton 2000 for an analysis of Anglican Providentialism.
60 F. J. Francis 1839, 162–63; B. Godwin 1834, 175ff.

of Lamark's system, is puzzled to say whether it were the effect of Epicurism, disease, or imbecility.[61]

The critical adulation of Lyell—and the fact that he had flatly denied a systematic and continual progression in the fossil record in order to undercut Lamarck[62]—might have made Saull reluctant to modify his talk. In his museum lectures a few months later, Saull countered that

> whatever Mr. Lyell may say or write to the contrary, there appears to be an uniform law, proceeding from the more simple to the more complex, from lower to higher gradations of intellect, from the zoophite to
>
> "The diapason closing full in man."
>
> It is a singular fact, that the order of creation, as we find it in the different strata, from the transition to the tertiary formation, should be the same as in the development of the human foetus.[63]

Saull clearly knew Lyell's book. He even knew Lyell himself, or, at least, Saull was attending Lyell's *soirées* by 1835 with the Geological Society elite.[64] But even if the Bristol talk (which was simply the first geological lecture reported in full) came twenty months after Lyell's book, all of this suggests that it was not about to be modified to take in Lyell. The thesis would remain 'as is'. Recall the *Crisis* editor's claim that "Our friend [Saull] has an unconquerable tendency ... to trace the genealogy of his own species up to the monkey tribe", which sounds as if these heresies had been expressed for some time prior to Lyell's fingering of Lamarck.[65]

Even though Saull's perfecting fossil series and ape ancestry were not based on Lamarck, and Saull was evidently reluctant to namecheck Lyell's work, the Gallic reprobate Lamarck became better known even in the underworld because of Lyell. The ultra-radical *True Sun* set off squibs shortly after Lyell's book was published in 1832:

> Oh! marvellous Sage! oh! wonderful Lamarck!
> Prover of every thing—and more besides;

---

61 Nares 1834, 165–66.
62 Bartholomew 1973, on Lyell's aesthetic revulsion of a bestializing ape ancestry; J. A. Secord 1997, on the meaning of Lyell's *Principles of Geology*.
63 *National Standard* 3 (18 Jan. 1834): 44–45.
64 Morrell 2005, 137.
65 *Crisis* 3 (5 Oct. 1833): 36.

> Describer of each creature in the ark,
>   Discloser of all truths that nature hides!—
> ...
> That prodigy of beasts, the camelopard,
>   was born, at first, in nature's *medias res*;
> Its neck grew long—because the brute strained hard
>   To crop the foliage of the lofty trees!
> There's not a thing but habit alters quite;
>   None wear the shapes in which they first began;
> Then say, Lamarck, thou star in Nature's night—
>   May not your theory extend to Man?
> ...
> For what were Lords invented? Do you think
>   That Nature made them for no other uses
> Than just to talk about "destruction's brink,"
>   To plead for tithes, and to resist abuses?
> Were Bishops sent us for this simple reason—
>   To eat of turbot, and to drink their fill?
> Or to commit, against the people, treason—
>   To scoff at millions, and to "trip" the Bill?[66]

From giraffes to bishops, it was grist to the radical mill. The political appropriation continued, with "Monsieur Lamarck" made to serve up some fine farces on pernicious habits perverting the lordly species.[67] Did Saull read the *True Sun*? Of course he did—it was the ultras' own evening paper.[68] It had a talented and fearless staff, including the editor John Bell, a hack whose "whole heart and soul was anti Church and State",[69] and John Thelwall, whose trial for high treason in 1794 had given him his radical cachet. But it is best remembered for its Parliamentary reporter in 1832: Charles Dickens started his newspaper career here. It had a good circulation during the Reform Bill furore, but it fell off sharply after the bill's passing.[70] In difficulty by October 1832, the paper was kept afloat by a massive subscription drive, organized by Saull, Watson,

---

66   *TS*, 9 Apr. 1832, 3.
67   *TS*, 11 Apr. 1832, 4.
68   HO 64/12, f. 163 (2 Nov. 1832). The spy was reporting on it as the only stamped paper supporting the ultra's ideals.
69   *PS*, 23 July 1842.
70   The paper never ran adverts, which were pointless when its clientele could not afford the goods, so advertising revenues provided no fall back ([James Grant] 1837, 2: 110–11).

Cleave, Lovett, and Hetherington, that is, the usual suspects.[71] So, yes, the *True Sun* was Saull's paper, with its Lamarckian squibs.

Robert Owen himself knew devotees of the heretical Frenchman. In America, he had set up the New Harmony colony with William Maclure,[72] a proficient geologist who had travelled widely in Europe. Although he was later dubbed "The Father of American Geology", Maclure actually published little on the subject. What he did publish showed his other side. He was a communitarian whose pamphlets on social matters were collected up into three volumes. One contained a speculative piece on the meaning of geology, whose import was to show that the creature-entombing rocks attested to a universal progressive flux and that the rise of life from its spontaneous origins supported "the anti-christian suppositions of the naturalist Lamark [sic]".[73]

In blasphemy circles, the idea of monkey origins was suitably cynical and nihilistic but far from shocking. Saull could have been primed by any number of notions floating around in disreputable places. For example, the pirated Lawrence, having shown in *Lectures on Man* that a peculiar larynx prevents the ape from talking, left it open for Carlile to suggest therefore that "*Man* was a mere *ouran-outang* before he began to speak".[74] Since, in polite society, apes were met with expressions of disgust, as a grotesque mockery of the divine countenance, what better than to squash such arrogance than with this hideous ancestry? It was the anthropological equivalent of Carlile's cynical line, that the human body rots to a dunghill.[75] And Carlile's imprisoned shop assistants saw apes and humans present "every progressive step" towards the intelligent summit and outrageously asked why souls, having been granted to

---

71 TS, 25 Oct. 1832, 1; PMG, 27 Oct. 1832; *Examiner*, 28 Oct. 1832. The paper rose and fell with radicalism itself, so that, by the late thirties, it was among largest circulating London evening papers, selling over a quarter of a million copies (*American Almanac and Repository of Useful Knowledge for the Year 1839* [Bowen, Boston], 71).
72 Armytage 1951; J. P. Moore 1947. On the more recent revisionism arising from the publication of Maclure's travel journals: D. R. Dean 1989; Torrens 2000. Maclure knew Volney; they had independently explored the same regions of North America and Maclure's geological map was actually superimposed on Volney's near-identical one: G. W. White 1977.
73 Maclure 1838, 3: 175–78.
74 *Republican* 7 (27 June 1823): 829–30.
75 Carlile 1821, 98, 132.

black peoples, were denied to apes.[76] In this, they should have known better.[77] But the orang's closeness to man was their polemical point.

Respectable commentators with their ears to the ground knew of Dr Darwin's and Lamarck's "numerous living followers", and how even some sound naturalists were flirting with this "atheistical school".[78] But the real "atheistical school" was several strata lower in the social column, where Saull's infidel sympathies made acceptance of a self-developing nature easier. And despite the possible sources of nature's transformism applauded by the pauper press and panned by anti-infidels—the lampooned Lamarck, "monstrous" *Telliamed*, or "mass of trash" uttered by Lord Monboddo about men once having had tails[79]—we do not really know where Saull's monkey came from. But for subsequent events, mankind's rise from ape-like savagery, Saull's immediate inspiration, it seems, came from a dream.

Sir Humphry Davy's dream, to be exact—for it was his posthumous *Consolations in Travel, or, The Last days of A Philosopher* (1830) that most struck Saull. The fashionable chemical philosopher had been fêted by high society, showing how far he had come from his humble Cornish origins. Davy had long abjured all sceptical notions. Indeed, his was the view from the other side of the prison bars. Where Davy, the rising chemist, the discoverer of potassium and sodium, had been called in to disinfect Newgate gaol, Saull would be visiting the recidivists within. Davy's brother and biographer recorded how Davy the swell grew hostile to a "repulsive" materialism, which would make "mind ... the result of organisation". Sir Humphry, the President of the prestigious Royal Society and immortalized even in his own age by his Davy lamp to protect coal miners, finished the *Consolations* on his deathbed in Rome in 1829. He was only fifty. The dying man, bequeathing the manuscript to his wife, trusted that it would "give encouragement to timid minds not to yield to the irony and scoffs of the gross materialist and atheist".

---

76   *Newgate Monthly Magazine*, 1 (1 Nov. 1824): 99.
77   Their racism is the more surprising as another incarcerated in Dorchester gaol with Carlile was his friend, the black revolutionary and anti-slavery campaigner Robert Wedderburn. He was the son of a black slave and Scottish slave owner. He was also another blasphemy-chapel owner and a powerful speaker who fired up sympathy for enslaved people: Wedderburn ODNB; Prothero 1979, 110; *Republican* 4 (8 Sept. 1820): 40.
78   Rennie 1834, 73, 84–86.
79   Good 1826, 2: 89–90; Murray 1831, 23.

How ironic that Saull should extract such a contrary message. But even sympathizers worried about some of Davy's dreams about the ancient world. In Davy's dreamscape, the whole of human history was telescoped into a series of dioramas. His images were like a "display of fire-works, which dazzle and confound without enlightening the senses", said another, "and leave the spectator in still more profound darkness."[80] Not so Saull. The book's impact was enlightening in quite another way, and the emotional appeal of Davy's vision of mankind's rise from savagery might have influenced Saull's subsequent career swing to ancient archaeology in the 1840s.

Davy's reverie saw him imagining the successive creations of life. For Saull to imagine *actual* transformations, a bloodline, was far bolder. And, needless to say, there was no monkey ancestry in *Consolations*. That was Saull's blasphemous interpolation in the gaps. Sir Humphry was being co-opted and desecrated. In his 1833 talk, Saull paraphrased Davy's dreamscape. After the rise of life through untold geological aeons, finally

> Man appeared as a naked savage, feeding upon wild fruits, or devouring shell-fish, or fighting with clubs for the remains of a whale which had been thrown upon the shore ... In the next epoch, imagine a country, partly wild and partly cultivated, when men were covered with the skins of animals, and secured their cattle in enclosed pastures. Some were employed in tilling and reaping corn, and others were making it into bread. Huts and cottages were erected for shelter, and some were furnished with the mere rude and humble conveniences of life ... they owed their career of improvement to the influence of a few superior minds amongst them: one taught them to build cottages, another to domesticate cattle, and another to collect and sow corn and seeds of fruits.[81]

Such a notion, startling to a genteel readership, was almost commonplace in the Rotunda. As the cynical Eliza Sharples said here in 1832:

> To suppose man made perfect, and endowed with superior intelligence in the garden of Eden, is to suppose that which is contrary to the principles

---

80 Paris 1831, 2: 371; Davy 1836, 1: 149–150, 385; 2: 88, 92, 385, 400. Even Lyell was influenced by the *Consolations*. Secord speculates that Davy's blast at an "absurd" transformism could have caused Lyell to add the anti-Lamarckian volume to his *Principles* at the eleventh hour: J. A. Secord 1997, xxxi; 2014, ch. 1.

81 *Crisis* 3 (5 Oct. 1833): 36–39; *Gauntlet*, 29 Sept. 1833, 529–33, paraphrasing Davy 1830, 18–30, 143–51.

of natural history. All infancy is in its nature alike animal, and devoid of idea or mind. The human infant of a year old has no more, nor perhaps as much mind, as any other domestic animal of that age. The human infant, nursed by any other animal, and deprived of human society, would be near akin to the animals from which it associated; so that an idea of perfection in the first man, is as unreasonable as the idea of perfection in an infant, not nursed and educated in human society. If there ever were a first man, we may seek the nearest picture of him among the wildest and least associated people; and he must have been much more like a monkey, than like the shaven, and shorn, and washed, and educated man of the present day, with all his defects.[82]

Thus, for a dedicated Rotundanist such as Saull, jumping up to slate Church and state after many a Sharples talk,[83] pre-fixing a monkey parentage to Davy's dream might have seemed obvious. Perhaps, too, he recalled the much-maligned Scots jurist Lord Monboddo, who imagined humans in different states of language acquisition across the globe today. Judge Monboddo's contemporary judicial methods of weighing evidence about tailed humans are only now being reappraised seriously, so he can be fairly relocated back into his age, rather than being dismissed as a crank.[84] Many dissidents certainly took him seriously at the time, and Saull might even have heard George Birkbeck talk on Monboddo in his lectures at the London Mechanics' Institution.[85] So little was known of the world's monkeys or apes that it was easy to joke, as the essayist and poet Leigh Hunt did, that men with tails would differ no more from us than a savage would from a philosopher.[86] Whatever, Saull's similar bit of bestiality would have had Davy turning in his grave.

The *Crisis* published Saull's talk. But its new millenarian editor, the Rev. J. E. Smith, was clearly troubled by it. He could agree on mankind's original barbarity.

> When man came from the hands of God he was a mere savage, a naked ignorant savage; and so ignorant that he did not know that he was naked. He neither knew good nor evil; saw no difference between virtue and vice, the Lord had made him so ignorant.

---

82   *Isis* 1 (14 Apr. 1832): 145–46.
83   HO 64/12, f. 47.
84   Sebastiani 2022.
85   *MM* 7 (16 June 1827): 374.
86   *Leigh Hunt's London Journal* 2 (8 Apr. 1835): 106; Good 1826, 2: 89–90.

The two co-operators, Saull and Smith, materialist and millenarian, worked together in the shared belief that mankind's progress was hindered by those "advocates of oppression and ignorance, who find it to their advantage to keep the poor in subjection".[87] Social redemption would be achieved once the unholy grip of tyrants and priests had been broken. But the fellow-travellers would not travel very far together. The real problem for Smith was that monkey.

---

87   J. E. Smith 1833, 100, 102.

8. The Antichrist and the Shaven Monkey

> What does a little insignificant imp like man—sprung from a baboon, as some imagine—what does such a shaven monkey know of the secrets of eternal and infinite nature?
>
> Saull's universalist co-worker at the labour exchange, the Rev. J. E. Smith.[1]

Character assassination was the stock-in-trade of anti-infidel tirades in an age when 'refinement' was all. For a generation, Christian youths would be taught that the "atheist", by choosing an orang for his "grandsire" and denigrating his ancestry, showed appalling taste, besmirched his character, and revealed his self-loathing. The "awful wickedness" of such beliefs risked ending in immorality, for "The son of an ape cannot surely be far removed from his adopted father." No such reprobate, therefore, is "fit for the society of men".[2]

But the righteous were not alone in their fear. Some Owenites themselves were flummoxed by Saull's step. A few simply skirted his obnoxious monkey by visiting a plague on both houses—the "defective" Mosaical theory of man's origins, *and* that of the French-inspired ideologues who looked into "the wide bosom of nature for those occult causes which have brought him into existence".[3] Most infidels, though, favoured some sort of naturalistic explanation. Lawrence's *Lectures* would be read as the lesson before a Sabbath blasphemy sermon, to prove that humans were produced "totally without the assistance of

---

1 J. E. Smith 1833, 187.
2 *The Juvenile Instructor and Companion* 3 (1852): 29–33.
3 NMW 6 (14 Dec. 1839), 950–51.

a first cause as the Superstitious Nonsense of the Clergy dictate".[4] But Saull's monkey moonshine was too provocative even for some diehards, most notably the outspoken Universalist, the Rev. James Elishama Smith, newly arrived in London. Smith was to become one of the Owenites' main lecturers, often talking alongside Saull, yet he poked so much fun at the monkey man, and with such good-natured vehemence, in every organ he went on to edit, from the Crisis to the cosy Family Herald, that, ironically, he gave Saull his widest exposure. In effect, Smith's whimsy, in the Penny Satirist and his high-circulation journals, brought Saull's peccadillo to middle-class attention. Nowhere does Saull himself actually mention 'monkey man'; it was rather the constant imputation in Smith's spoofs that caused the concept to stick. Clearly the monkeys got under Smith's skin, leading him to become Saull's friendly nemesis for life.

Smith and Saull would share lecturing venues and political goals, and they came to work side-by-side in Owen's "Institution of the Industrious Classes" in Gray's Inn Road (in 1832–33) and Charlotte Street (1833–34). But Smith was the unlikeliest bedfellow. Originally a Presbyterian, he had gained a theology degree at Glasgow. But he was a rebellious soul who came to reject Calvinism. Even in Glasgow, he had become a millenarian, expecting Christ's thousand-year reign momentarily. There followed two years with a Southcottian sect, the "Christian Israelites", "one of the most disreputable and outcast of all the sects in England".[5] The sect was led by John Wroe—the prophetess Joanna Southcott's successor—a visionary of "savage look and humpback".[6] The Southcottians used trances and visitations to unveil the truth, and Southcott had revealed herself as the Bride of the Lamb (that is, Christ's bride), described in Revelation. Even in London, where he arrived in August 1832, Smith would continue to endorse prophecies.

How the originally apolitical Southcottians came to align themselves with freethinking radicals is explained in Philip Lockley's reappraisal of the sect's fragmentation. Smith imbibed the idea that the millennium might not be a sudden irruption but be preceded by an improving

---

4   A reading of Lawrence's Lectures held before Smith's lecture on the Trinity at the Rotunda: HO 64/12, f. 180.
5   J. F. C. Harrison 1979, 143.
6   W. A. Smith 1892, 51.

society, and this demanded reformist work in the community. Doctrinal changes reinforced this belief: the divine agency was now seen to spur human behaviour, to remove the unjust as a prelude to the millennium, which again pushed Smith towards the radicals. Even inside the sect, Smith had made contact with local freethinkers, whom he saw doing God's work, and thus began his contingent pact with deists and atheists to replace the corrupted church.[7] Yet, through it all, Smith continued to view his supposed radical soul-mates as the unconscious agents of Divine action.

To the thirty-year-old Smith, the great Babylon of London was a "monstrous" Hades, so huge, with its million and a half population, "that I cannot go into town without spending several hours in walking".[8] Best of all, it had irreligious chapels-a-plenty, so he was able to make a living by preaching his blasphemies, only now he blended ultra-radicalism into the heady mix. The street audience was receptive to heresy, but then "The hostility against the Church is dreadful [in London]. I never see a clergyman in his canonicals on the street. They dress like other men, and pass unnoticed." In the smoky city, Smith was sure that infidelity "will turn the Church upside down."[9] In the irreligious chapels, he was still introduced as "Reverend". The title evoked power and authority in a sermon-controlled society, and it was arguably all the more necessary to lend credence to his biblical blasphemies.

Smith was about as far from respectable as one could get and still call himself a "Christian". Not that the spy considered him a Christian. The undercover agent had been alerted to Smith's appearance on the London scene. His report for 8 October 1832 warned the police:

> I have been aware for some time that a young man who calls himself the Revd J E Smith A.M. has been associating with Carlile, Taylor, Saul [sic] and the Lady of the Rotunda and that he intended to deliver Lectures on the "fallacy of the Christian Religion"! He has commenced them at the Chapel in Chapel Court Boro which Sometime ago stated was taken ... to hold Meetings of the Union [NUWC] ... Smith is a young Man, who has been bred in the Scotch Church as a Presbyterian Minister and is now an Infidel ... from his language and having heard Taylor, Hibbert, Gale Jones

---

7 Lockley 2009, chs. 6–9; 2013, chs. 7–9.
8 W. A. Smith 1892, 89; 1833.
9 W. A. Smith 1892, 94.

and others on Theological subjects, I never heard a More determined abuse of Christianity or its Principles than Smith delivers ... [10]

Smith being the worst of the lot was saying something. Here was a millenarian who saw the Devil and God as one. (Smith deduced this, somewhat prosaically, from the misdemeanours of Southcottian sect leaders: one, accepted to be doing divine work, was caught in a sex scandal, and lied, which implied that God and Satan must be the same.[11]) These two spiritual extremes combined to "make nature, the true goddess, and the only supreme Deity", with its pleasure and pain, love and hate, good and evil. And it was from nature, not priestly injunctions, that our morals derive. But then he had no time for priestly doctrines or any divisive faith. He cared not if the Bible was divine or man-made, and liked revelation no more than "an old ballad". Hell, for him, was a "monstrous overgrown delusion".[12]

> I set my face against the God of lies, and his lying gospel, which the clergy preach; and I wait for the birth of the God of truth, which is promised in the scriptures under the name of a child—"Unto us a child is born..." &c., The true God, the God of peace, is not yet born; the Devil his father is still on the throne, and the clergy are his instruments of deception; but as soon as the child is born we shall have no more priests, no more temples. These two Gods are both of them in the Bible; the clergy worship the black one, and I am giving birth to the white one.[13]

Smith was not a man to flinch at heresy, scientific or theological. He saw Genesis as purely allegorical. He thought it ludicrous to consider the planet six millennia old: "it may be a million", even if "human society or man is not above 6,000 years old". His faith rested in geology, which had forced its way through a "forest of opposition", all because it "proves from undeniable facts, that this world is much more ancient than it is represented in Scripture, and that it had been peopled by myriads of living creatures thousands of years before man was made upon it."[14]

So said Smith in his packed lectures in that radical-blasphemous hotbed, the Borough Chapel. His fiery talks here were packaged into

---

10 HO 64/12, f. 150 (8 Oct. 1832); Prothero 1979, 262.
11 Lockley 2013, 162–63, 174.
12 J. E. Smith 1833, 9, 34–35, 45, 223.
13 J. E. Smith 1833, 128.
14 J. E. Smith 1833, 42, 230.

book form by the Owenite publisher B. D. Cousins and sold under the provocative title of *The Antichrist* (1833). The Borough venue itself was commodious, notorious, and crammed with enthusiasts. The spy on one visit reported it full to the brim (it held 800), with "100 females, same number of youths of both sexes".[15] We know that it was radical audiences specifically that made the auditorium profitable because the landlords advertised the lease in the *Poor Man's Guardian* and *Crisis*.[16] And being taken over by deists, millenarians, and NUWC activists made it the regular haunt of spies. Here, in 1832–33, Smith's Sunday "Antichrist" lectures were interspersed with Saull's Friday ones "on Astronomy, Geometry, Gasometry, Chemistry &",[17] and the NUWC's Tuesday meetings. So, in a typical week at the Borough, you could hear Smith on Jesus as the Antichrist, the venerable firebrand George Petrie on "the enslaved and wretched condition of the working classes",[18] and Saull on the astronomy underpinning his evolutionary palaeontology. Saull would be present, too, on a Sunday, sometimes commenting on Smith's "Antichrist" sermons, and "promising to assist the lecturer on all other occasions".[19]

One November 1832 night, Smith sermonized on "Nature". All animals as well as the earth existed by chance, he said. Although the spy could not understand Smith, Saull did, and stood to say at the end that he "did not go far enough," possibly meaning not as far as monkeys, and he was "much applauded".[20]

So infamous was the Borough Chapel that Saull's friend George Petrie was currently immortalizing it in a sort of addendum to his poem *Equality*. Petrie was one of the more revolutionary elements inside the NUWC, an old soldier who thought the people should be drilled, and he drew up plans to move on the Bank and the Tower of London in 1833,

---

15 HO 64/12, f. 170.
16 *PMG*, 4 May 1833; *Crisis*, 1 (8 Sept. 1832): 108.
17 *PMG*, 24 Nov. 1832; *Prompter* 1 (27 Aug. 1831): 752. NUWC weekly meetings were held here from 1832 to 1834.
18 For example, *PMG*, 17 Jan. 1833.
19 HO 64/12, f. 188 (17 Dec. 1832).
20 HO 64/12, f. 177. Smith took Holbach as his text. The spy reported more of Saull's speech, but garbled it, suggesting that the talk was as unfathomable to him as Smith's. This 18 Nov. 1832 "Nature" sermon of Smith's is not in the *Antichrist*, which otherwise contains this 1832–33 Borough series. It is, however, abstracted in the *Cosmopolite*, 1 Dec. 1832, in HO 64/18, f. 728.

shortly after penning the poem.²¹ As such, he remained a target for the intelligence services, and we find the spy at one point pumping him for information.²² Petrie joined Saull in the Owenite movement. They worked together in the BAPCK²³ and on the news vendors' fund. The poem, *Equality*, was actually proudly pinned up on the walls of Saull's museum. Nor did the veneration stop there. Shortly, in a strange twist, Petrie himself (skeletonized, anyway) would end up as one of the more macabre exhibits in Aldersgate Street.

Blasphemy poetry was always subordinated to politics, and *Equality* was heralded as the liberationist verse of the age:

> Though slow, yet firmly we proceed,
> The Borough Chapel takes the lead,
> And Reason's sons assemble there
> To feast on knowledge, not on pray'r;
> To scan the rights and wrongs of man
> On Nature's, not vile Custom's plan;
> That is to say, the sacred cause
> Of Equal Rights and Equal unveil.
> Not charter'd rights, nor rights divine
> Of kings, or lords, or holy swine;
> But rights of all who dare be free,
> Rights founded on Equality!²⁴

The Borough Chapel materialists, at least, were hostile to Smith's religious millenarianism, deriding Southcottism as so much "Fanaticism",²⁵ and presumably Saull shared this aspect of their distaste. Unlikely bedfellows or no, the millenarian theologian Smith and atheist geologian Saull did share an interest in the "Devil's Chaplain". With Saull visiting Taylor in Horsemonger Lane gaol,²⁶ Smith took over Taylor's Rotunda slot in 1832–33. Smith told Carlile that he was as one with Taylor on the astronomical roots of Christian myth and that he saw the Bible as *"a divine piece of*

---

21 Prothero 1979, 293–95; Rowe 1970b, document no. 31.
22 HO 64/15, ff. 105–06 (11 Feb. 1834).
23 J. F. C. Harrison 1969, 199; Hollis 1970, 100–01.
24 *PMG*, 23 Mar. 1833.
25 HO 64/12, f. 170.
26 Saull supported Taylor to the bitter end. Taylor, discharged from prison, re-started his astronomical talks at 8 Theobald's Road, with Saull and the fishmonger John Pummell as his support group, but his moment had passed: HO 64/15, f. 148 (18 Feb. 1834).

waggery".²⁷ He now sank to the blasphemous depths. While Taylor had made the Bible an astrologer-priest's invention, with the heavenly drama cast as human parable, Smith saw the drama communicated in visions to the ignorant scribes, who never understood the astronomical key they held. Smith portrayed himself as bringing about "a completion to [Taylor's] system". He would show how much more the allegorical nature of scripture could reveal about science and belief, while damning the ignorant priestly "pretender to sanctity [as] a hypocrite and a curse to society".²⁸

The Home Office was now keeping close tabs on Smith. The spy was shocked to hear, as McCalman quotes, "that God was to be found in excrement as much as anywhere". No wonder the informer considered Smith's language "the most vulgar and blasphemous I ever heard."²⁹ But Smith was evidently spellbinding as he unlocked biblical secrets and rubbished the taboos surrounding Hell and Heaven. His London lectures, having muted the extreme messianic message and mixed in a new inflammatory radicalism, were wildly applauded. And Smith himself was dashing. He rivalled Taylor in charisma, with his piercing eyes and charming manner. As a result, his Borough and Rotunda audiences again attracted many ladies—not only radical firebrands like Anna Wheeler, but middle-class matrons, some of whom went on to finance his lectures.³⁰

Saull and Smith stood shoulder to shoulder on politics. The *Antichrist* dramas were as much social subversion as theological deconstruction. No headier mix would appeal to Smith's mechanics in the Borough, who were, he said, mostly "Infidels". Now the only 'fall of man' he wanted to redeem was the people's collapse into penury. With no Fall, there was no Saviour of the elect, or damnation for the rest, and so no reason for priestly repression. If man had fallen, it was "into the hands of tyrants and priests, tax-gatherers and tithe-gatherers". The stamp duty, that attempt to stamp out the pauper press, was equally traced to its Edenic

---

27 Quoted by McCalman 1992, 63; J. E. Smith 1833, 36.
28 J. E. Smith 1833, 36, 210–11.
29 McCalman 1992, 63–64; Saville 1971, 120–21; W. A. Smith 1892, 86–87, 205.
30 Saville 1971, 120; W. A. Smith 1892, 90, 204–05; Pankhurst 1954.

root. Who forbade Eve to eat from the tree of knowledge? The Lord, an aristocratic "tyrant", prohibiting or levying his own tax on knowledge.[31]

Smith's theological attack on the stamp duty mirrored Saull's frontal assault. The profits from Saull's concurrent lectures on astronomy in the Borough were going to support the jailed news vendors. Here, too, he echoed the demand for an end to the newspaper tax: English law does not recognize a plea of ignorance, yet it taxes

> knowledge so heavily, as to make it utterly inaccessible to the great body of the people.—(Loud cheers.) The government would neither instruct them, nor allow them the opportunity of securing knowledge themselves. Yet they saw large sums of money taken from their pockets for the purpose of public education—and ... he asked where these funds were? Divided among the aristocracy! The people should combine together, and demand as their right, to have schools and seminaries of instruction of their own; they should support the unstamped Press, which stood forward in their cause.[32]

So there seemed a lot of common political ground. This increased as Smith moved to Owen's Institution in Charlotte Street.[33] Smith had witnessed the opening of the Labour Exchange in autumn 1832, and wrote home about the labour notes and bank for swapping goods. And "by this system they contemplate the total abolition of all gold and silver currency and accumulated wealth—the root of all evil. And this they call the millennium." Without drawing breath, he continued:

> A great many of them are Atheists. Atheism is quite common in London—pure Atheism. A gentleman, a clever man and a man of learning, lately told me that Atheism, in his opinion, was the most rational system he met with.[34]

That a freshly-arrived Glaswegian millenarian should be so struck is not surprising, given the number of Owenites and infidels in the city. In Smith's estimation, extreme views like Saull's, that religion was a *"despotism"* that encouraged nothing but "insanity", were quite prevalent.[35]

---

31 J. E. Smith 1833, preface, 95, 103, 111.
32 *TS*, 6 Aug. 1834, 2.
33 HO 64/15, f. 171 (22 Apr. 1833).
34 W. A. Smith 1892, 81–84.
35 [Saull] 1832a, 4.

By the time he moved to the heart of Owen's institution, in the summer of 1833, the Labour Bank was waning. But Saull and Smith, atheist and universalist, could mediate their religious difference through their support. While Saull was helping to run the Exchange, Smith was devising a more radical route for the failing enterprise, one that gave bread to all and made the breadline a thing of the past. Smith had not only become a land nationalizer, but he thought that "all productions of public utility or of general demand" ought to be in common ownership. Thus, agricultural labourers were due their share of manufactures, and the industrial workforce their share of farm produce. For Smith, it was those unique works that were made outside of day labour—paintings, sculptures, mechanical inventions—that could be bartered. With everybody having the necessities, this "exchange of luxuries" would give life its spice, and the surplus would go to freely-accessible museums, like Saull's.[36]

A common commitment to perfectibility and the social millennium kept the two men close for a year or two. The future protagonists shared a belief that man had not fallen, but risen from savagery, "toward civilization and refinement".[37] In Smith's view, the struggle of Nature's extremes, with man jostling between good and evil, helped to perfect him, where for Saull that was achieved by a clement nurturing environment. The new recruit's Universalism thus chimed with the old Owenites' perfectibility. Their non-responsibility doctrine, in which culture determined character, was matched by Smith's take on St Paul: "it is not he that sinned, but *sin* that dwelleth in him", so the wickedness, as Owenites agreed, was not man's fault.[38]

Like Saull, Smith saw mankind's rise as a successive stripping of priest-induced ignorance, a shedding of those "fables and fairy tales, superstitious rites and creeds ... and gods of every shape and every size." The two agreed on the material ascent of life; or, as Smith had it, on life's goal, where you have:

> no law, perfect liberty, the real light of science, the sun of righteousness, real radical reform, the true and the only Millennium, in which there are neither priests nor law givers, tithes nor taxes. All this is the true

---

36 *Crisis* 3 (15 Feb. 1834): 201–02.
37 J. E. Smith 1833, 102.
38 Saville 1971, 123–24; J. E. Smith 1833, 72; Prothero 1979, 262.

progress of nature upward to perfection, emerging from the lowest grade of animal and intellectual nature...[39]

This directionalist tendency saw the two men closest in 1832–33, but it remained fundamentally a working amalgamation forged in the white heat of the moment. Smith's sermons had man coming "from the hands of God ... a mere savage". However, Smith understood "God", and it could even have been a personification of nature, for Saull, the materialist, there was no Maker, and no personification was necessary. The two might be casting off the unholy union of priests and "tyrants", fighting against a state that would tax the population into ignorance. But Smith could never go so far as to cast off God and Demon, however analogized and de-personified. The reverend might even have had sympathy for infidelity as a form of ultra-Protestantism[40]—in fact he had more sympathy for infidelity than Protestantism—but removing the miraculous roots of the millennial march remained a step too far for him.

The miraculous process led to the "birth of the God of truth", the thousand-year Millennium. The literalness of Smith's Millennium, when "the Messiah, 'shall make an end of sin, and finish transgression, and bring in everlasting righteousness'", would have been anathema to Saull. His faith was in mankind's secular perfection, where the democratization of rational knowledge swept away such religious phantasms. Nor would Saull ever credit the miraculous. Yet, for Smith, "everything is a miracle". And those of the pagan "magicians and sorcerers" were as valid as the apostolic healing miracles. They were tapping into arcane sources of power, just as modern mesmerists did. Miracles were part of Nature, indeed they *were* Nature. It was "preposterous" for infidels like Saull to doubt them, for what "does such a shaven monkey know of the secrets of eternal and infinite nature?"[41] While co-operators might be unconscious agents fulfilling Divine Prophecy, Saull had clearly tipped over to the dark side, the side of the apes.

---

39   J. E. Smith 1833, 6, 199.
40   J. E. Smith 1833, 100, 185.
41   J. E. Smith 1833, 180, 183–84, 187, 227, 234. The "shaven" aspect was perhaps significant to Smith, who, as a Southcottian, would have been bearded at a time when it was unfashionable (W. A. Smith 1892, 53).

We can start to understand how different the "Christian" millenarians were inside the Labour Exchange, and thus why Smith should ultimately target Saull's monkey. Infidels had not conceived nature's process correctly. Smith believed "all nature to be conducted upon a systematic plan", in which Evil and Good, Devil and God, work in mystical sway to hone human knowledge in order to achieve millennial perfection, when "man can become as God". Saull mocked this with his mechanically derived monkey-man. He was misreading Nature as merely autonomous. His Nature "is a dead God; he acts, but knows not what he is about; he is a sort of somnambulist." How was the Millennium to be reached? By a mix of true science and true religion. Smith swore by science. He believed that science's progress "within these few past years has withdrawn the veil of futurity", increasing "the speed with which the Sabbath is approaching". Science was to be a major contributor to the millennial government, but it had to be the right sort. Saull's was putting this blissful state in jeopardy. If "It is nature only that we ought to acknowledge as perfect God, and she is to be found within each of us",[42] then the bestiality of Saull's monkey ancestry damned itself.

Smith was dynamic, engaging, and co-opting. When he entered Charlotte Street in 1833, Owen's house organ, *The Crisis*, was flagging, "a lean and haggard-looking starveling".[43] So Smith took over: the printer gave him the editorship, starting with the third volume (7 September 1833), just after Saull's Bristol lecture. The paper was redesigned, with a new moralizing masthead, graphically depicting the ramshackle old world of higgledy-piggledy high street housing in opposition to Owen's rationally-ordered factory-like 'scientific' building. Smith inherited a *Crisis* selling some 1250 copies a week, and claimed to have "reared it to manhood and respectability", implying a ramped-up circulation.[44]

There was no stopping Smith. He began a Sunday morning lectures series in the Charlotte Street Institution from June 1833, alternating with Owen, who talked in the evening. With Owen often being away, Smith started taking over the evening slot as well. He was now introduced each week by the levelling title "Mr. *Smith*". Gone was

---

42 J. E. Smith 1833, Preface, 45, 124–5, 133, 171; Saville 1971, 124–25.
43 *Crisis* 4 (23 Aug. 1834): 154–55.
44 *Crisis* 4 (23 Aug. 1834): 154–55; W. A. Smith 1892, 101. According to Saville (1971, 126–28), the *Crisis* became "a much livelier paper", but I am not sure this is true.

the pompously authoritative "Rev." Even if the lectures now veered more towards secular moralizing, from the 'Infidel' perspective things looked ominous. Smith's sermons took pride of place on the title page of the *Crisis*. Now, it seemed, justification of the Owenite community rested as much on a Universalist Bible reading. The materialists saw the paper's cover given over to mystical musings and the interplay of devils and gods in Nature. Under the Owens, the paper had focussed on social regeneration and re-orientation through Labour Exchanges. The paper had eschewed theology, simply taking side swipes at the Church temporal: its missionary failures, idolatry, public funding for new churches, clerical intolerance, and the interminable "water, weeds, mud, mire, and reptiles" of sectarianism. Always critical, its motto was "truth without mystery". They noted the untrustworthiness of the Bible, and unpicked morality from religion in order to undercut the belief that it was "the fear of the eternal fire alone that saves us from being brutes".[45] It had all gone to reinforce the public perception of Owenites as thorough-going sceptics.

Under Smith it seemed "Truth without mystery" had become 'truth within mystery'. Nature had been turned upside down to become "a genuine mystic".[46] In the critical weeks, late September to early October 1833, when Smith was planning his critique of Saull's monkey theory, his sermons embraced Christians and atheists, Public Worship, Primitive Christianity, and Revelation. This millenarian mysticism and assault on atheism was a radical departure for the *Crisis*, and the materialists were furious, those men who "had put off religious belief, torn the garment, cast it away, followed after it, trampled on it, [and] gloried in their nakedness".[47] There was no ghost in Saull's machine, earth was his location for universal salvation, not heaven. But the Antichrist was now among them, claiming that minds do not die, any more than

---

45 *Crisis* 2 (16 Mar. 1833): 77; 1 (1 Sept. 1832): 104. Even Owen's offer to the millenarian Edward Irving's "harmless religious enthusiasts"—persecuted and homeless—to allow them the use of his Institution had to be *apologized* for in the *Crisis*. It was partly justified on the grounds that some of Irving's intelligent laity would come round to Owenism (*Crisis* 1 [12 May 1832]: 26). According to Saville, Smith heard Irving preach in the Labour Exchange, and it was through Irving that Smith came into contact with the Owenite community in the first place (Saville 1971, 120).
46 *Crisis* 3 (14 Sept. 1833): 9–10.
47 [Somerville] 1848, 413–14.

matter. They simply rejoin "the great ocean of power or consciousness, or will, which pervades the universe, and which is indestructible."[48] Worse was to hear Smith claim that infidels had no system, nothing positive to replace Nature's revelation. The infidel "deprives nature of intelligence as a whole ...With him it is chance; he has made Nature as God is represented to have first made man, perfect outwardly—all the material organization complete; but the breath of life, or intelligence, is wanting." One had only to analyze religions to demonstrate "that they are a grand revelation of nature in a mystery; and by thus systematizing religion ... I bring to life and conscious intelligence the dead god of the atheists."[49] Worst of all for the Labour Exchange activists was to hear the spiritual interloper pompously preach that

> Infidelity can never become a source of action; there is no impulse, no enthusiasm, no life in it. Infidel is a most repulsive name; a faithless man is a selfish, solitary, unsocial lump of inanimate matter; put a spark of faith in that lump, and he lives and acts; for then he has an end in view, he looks forward.[50]

This must have been horrifying to 'infidels' like Saull. No unsocial, inactive lump himself, he was putting money and energy into the Labour Exchange to make it a success in supplanting the capitalist middle-man. His infidels were actively driving towards their own earthly Owenite millennium of social harmony.

On the eve of Saull's monkey lecture, Smith was marking the atheist's card. So far had the *Crisis* departed that his editorials were calling them out, with their dead matter, their "defective system of nature".[51] In line with this, one of Smith's first acts, on 28 September 1833, two weeks after taking over the *Crisis*, was to publish the transcript of "Our friend" Saull's geology lecture. This gave Smith the chance to editorialize, and rather underhandedly start his scoffing attacks on Saull's monkey forebear.

Smith totally agreed with Saull on the "graduated scale of creation from the lowest to the highest." Geology proved the point, and this advance "went on in man himself, and will continue to do so for ever."

---

48   *Crisis* 3 (21 Sept. 1833) 12–13.
49   *Crisis* 3 (5 Oct. 1833): 40.
50   *Crisis* 3 (4 Jan. 1834): 145–46.
51   *Crisis* 3 (5 Oct. 1833): 40.

The material and moral were a seamless stream, with Owen accelerating the process.[52] The monkey was the insuperable problem, causing Smith to ask facetiously whether Saull's "unconquerable tendency" to make him a great grand-parent stemmed from "philanthropy or misanthropy, gravity or jocularity". Smith was imputing motives for his own ends. Was it disguised misanthropy, an attempt to degrade our sublime dignity? even by one whose philanthropy was spent in raising human dignity. Was it jocularity, jesting with funny monkeys to prick Christian pomposities?[53] Given the prevalence of freak shows featuring hairy men, and theatrical "monkey-man productions, often featuring a character named Jocko in the lead",[54] was it a rival hoodwinking show to pull in the punters? But while such freak shows were invariably fraudulent, Saull was serious. Yet he seemed to be teasing out deep-seated fears, of the threat of human bestialization, even in his most heterodox ally.

Being among infidels, Smith could disingenuously claim that it really mattered not "whether we were originally baboons, or savages of human shape divine; and we have no objection to concede to the simian tribes the cheering prospects of one day rising to the enjoyment of intellectual and scientific powers". Disarming this might have been, but it belied the fact that Smith would actually mock Saull's monkey man for the next twenty years. For an unconventional soul, Smith now took some pretty conventional pot-shots at Saull. Species were discrete, there were no signs of one blending into another. Indeed mankind had only recently appeared, as if unexpectedly—"and this suddenness ... is rather a formidable argument against the supposition that Nature gradually converts one genus of animal into another—an oyster into a lobster; a lobster into a seal; a seal into a dog; and a dog into a monkey, &c." Each animal was, therefore, "an original formation of Nature", with man the last formed.[55] However animals arose, it was not one out of the other.

The crux for the millenarian was that the atheists were blind to the *anticipating* nature of the environmental drive.

---

52 *Shepherd* 1 (18 Oct. 1834): 61.
53 *Crisis* 3 (5 Oct. 1833): 36.
54 Qureshi 2014, 266.
55 *Crisis* 3 (5 Oct. 1833): 36–37.

> Certainly, as Mr. Saull very justly observes, [humans] originated in circumstances over which they had no control; but these circumstances have been uncommonly wise and intelligent circumstances; they have bestowed the greatest perfection of body where they have bestowed the greatest perfection of mind. The beauty of the human frame is not more pre-eminent than the versatility of its corporeal faculties—the hands, the feet, the organs of speech, the erect position, are all in harmony with the intellectual supremacy of man. Circumstances have never yet bestowed an intellectual brain on an animal with limbs like a horse, or trotters like a sheep. But why should they not have done so?[56]

Not quite so unconventional now. The "perfecting" agent pre-planned everything—so nature "produced the proper food in abundance before she produced the animal which was to subsist upon it", said Smith in one of his front-page sermons. He seems not to have understood the fossil record too well, for he imagined that after plants came herbivores, which "lived long time in joyous tranquillity, and attained an immense size," because there were "no carnivorous animals to destroy them".[57] Even if his fossil zoology was skew, the point was nature's anticipation. He saw culinary conditions being tweaked to meet the needs of incoming species, not the periodic updating of species caused by naturally-changed environments. And there is no telling how Smith envisaged the process: that was not his problem. His conventional concern was to show that nature was continually re-arranged for "some wise end."[58]

For Smith, the "beauty" of the body spoke of wisdom. How different the materialist's language. The hard-bitten Carlile thought humans no more wonderful than vegetables. Man was so many organized atoms, and nature was indifferent to his pleasure or pain. Each atom might be immortal, Carlile wrote from his jail cell, but after his body has rotted "like a dunghill" those atoms "can retain no sense of a former existence."[59] Such nihilistic thoughts drew cynical satisfaction. What better levelling sentiments for the hordes of hovelled poor, than to know that aristocratic opulence was to end in the same dunghill? Not for Smith this misanthropic metaphor, any more than a jocular monkey one. His intelligent conditions were guiding man so he "can become

---

56 *Crisis* 3 (5 Oct. 1833): 36–37.
57 *Crisis* 3 (21 Dec. 1833): 129–30.
58 *Crisis* 3 (21 Sept. 1833): 12–13.
59 Carlile, 1821, 43–44.

as God". But while Smith talked superficially like the atheists of *nature* "ordaining" the progress of both animals and society—which itself unfolds like a butterfly from a cocoon, suggesting that the development of life and society were all of a piece, part of "one grand unity", as they were for 'atheists'—underneath lay a "beauty and wisdom". It spoke of a pro-active intelligence, a "breath of life", and that Saull denied.[60]

A monkey was a "caricature of humanity" in Smith's eyes,[61] just as Saull's soulless transmutation was a caricature of creation. The "hands of god"[62] fashioning mankind were intelligent conditions, and monkeying with these with misanthropic or jocular motives was the real blasphemy. A nonplussed Saull now found himself castigated by the millenarians of his own party.

Saull's talk was much more to materialist tastes. The irreligious Owenite core had little truck with Smith's wishy-washy universalism, but the man who really loved Saull's lecture was Carlile. Having called for men of science to proudly proclaim their materialism, how could he not? And coming from his own acolyte only made it sweeter. Whatever his political divergence, Carlile was never one to bite the hand that fed him. Fresh out of Giltspur Street prison in September 1833,[63] he ran Saull's lecture in his *Gauntlet*, a threepenny weekly started from his cell. To him it was "one of the most impressive and interesting lectures that has ever been delivered in Bristol."

> It is well known that Mr. Saull is one of the few that has honestly and fearlessly stood foreward in the exposing of error, folly, and ignorance, and in support of the great and glorious truths which we doubt not, as man progresses, will become more universally accepted ... Persecution and hypocrisy, we believe, may yet for a while retard the progress of human improvement and human happiness. But if such men as Mr Saull will only persevere and co-operate with each other, we feel assured, using the words of that gentleman, neither kings, priests nor lords, can withstand the intelligence of the people; and that, ultimately, truth, justice and humanity must be erected upon the ruins of kingcraft, priestcraft, and all those other evils which, at this moment, unhappily afflict the world.[64]

---

60 *Crisis* 3 (21 Sept. 1833): 12–13.
61 *Crisis* 4 (19 Apr. 1834): 9–11.
62 J. E. Smith 1833, 100.
63 Wiener 1983, 176–99.
64 *Gauntlet* 1 (29 Sept. 1833): 529.

9. Damned Monkeys

> His wink impertinent, his saucy stare,
> His grin ridiculous, his careless air,
> His more than idiot vacancy of face,
> His monkey arts, and baboon-like grimace.
>
> William Benbow in *Crimes of the Clergy*, adding insult
> to injury, satirizing a debauched bishop.[1]

For many, it was the monkey that ratcheted up the real horror. The concept of 'monkey' came with a lot of unsavoury baggage, particularly in the years preceding Saull's 1833 shock. No creature, bar the serpent, came with more evil biblical report, and that, mixed with rampant anthropocentrism, coloured the popular perception. It was perfect for spoofing, and often the pompous were the satirical butt. One archetypal joke, long circulating in middle-class circles about London Zoo's wanderoo monkey, exquisitely captured this: the black monkey whipped off the wig of a passing bishop, which it then "profanely transferred from the sacred poll to his own".[2] To make a monkey of a bishop touched on so many uncomfortable themes. But this Gillrayish image ultimately underscored the more than metaphorical irreverence of the monkey's grimacing, human-aping world. Of course, the gibbering, pilfering, comical image made monkeys perfect subjects in political satire.

Even *John Bull* had undergone a face change from stout-hearted yeoman to a broad grimacing baboon.[3] But more usually the radical point was a lack of morality displayed by ministers and macaques alike. As the sharp-tongued Eliza Sharples said at the Rotunda, Court and Church were like Bartholomew Fair, a "raree-show ... with a

---

1 Benbow 1823, 82.
2 [Broderip] 1838, 92; Broderip 1847, 242.
3 Parolin 2010, 129.

large number of monkeys, gorgeously dressed menials, clamour and clangour, confusion and cheat, and a general waste of time."[4] One seriocomedy running among co-operators had "monkey, king, or bishop" as glutinous consumers, producing nothing, but stealing their fellows blind.[5] The blunt moral of filching monkeys as a metaphor for capitalist thieves or debauched bishops was recycled endlessly.[6] So if monkeys *had* evolved into men, the progress towards morality had been palpable, and it was destined to be extended into an Owenite future.

The city itself provided a distorting lens through which urban monkeys were judged. They were the slum dweller's accomplice, the degenerate drunk, the pomposity-pricking mimic. They were ceasing even to seem exotic.[7] Jerry the Satyr (showman's slang for a gaudy-faced mandrill) was a favourite at Cross's Menagerie. Here he sat in a chair with his "glass of sling", puffing on a pipe, an "odious ... looking monster".[8] Complaints were common that the streets were "infested" with vagabond Italian boys, picking up pennies grinding organs with their monkeys.[9] Decent women feared to walk alone in the Strand or Pall Mall, because of the "blackguardism that ... crowded round the barrel-organ and the monkey".[10] Monkeys thus became associated with 'street arabs', as they would soon be slated, and the menacing poor.

Even worse were the fighting dens, where baboons, with their ferocious canines, were pitted against bull-terriers. The most famous in the 1820s, Jacco Maccacco, tore apart a succession of prize-fighting dogs at the Westminster pit. Here the classes were forced to mix promiscuously—the mingling scene was even painted by Landseer, and a Cruikshank print shows swells and rabble crowding round the scene of

---

4   *Isis* 1 (8 Sept. 1832): 474.
5   W. Thompson 1824, 199–200.
6   *Union* 1 (1 Apr. 1842): 4–5; *NMW* 8 (8 Aug. 1840): 90; *Lancashire and Yorkshire Co-Operator* (24 Dec. 1831): 4–5; *Crisis* 2 (24 Aug. 1833): 267–68.
7   On the influx of exotic animals into Victorian England, largely as a result of empire, most ending up in aristocratic hands, see Simons 2012; Grigson 2016.
8   Anon. 1830, 216–17; Broderip 1847, 94. In 1831, when the menagerie moved from its "murky dens" to the leafy Surrey Zoological Gardens in Walworth, the "hard drinking" Jerry died within months, a symbol of old debauchery paving the way for Victorian sobriety (*Mirror of Literature* 19 [5 May 1832]).
9   J. T. Smith 1839, 135; C. Knight 1841, 1: 422–23; Mcallister 2013.
10 C. Knight 1864, 2: 26–27. Visitors were astonished at such "sights of daily occurrence" (An American 1839, 50).

carnage.¹¹ The shrieking spectators comprised the gamut from dustmen and lamp-lighters to "honourables, sprigs of nobility, M.P.'s, ... all in one rude contact, jostling and pushing against each other."¹² Monkeys were forcing class 'miscegenation' at these bloody contests, tearing aside barriers as they tore apart dogs, which made them even more socially suspect. Judgmental attitudes and a civilizational yardstick meant that monkeys came off badly, particularly the "savage Baboon, whose gross brutality is scarcely relieved by a single spark of intelligence".¹³ They were culture-debasing, class-mongrelizing creatures, perfect evolutionary grist perhaps for co-operative "scum", but disdained by polite society. When the Drury Lane and Covent Garden theatres fell on hard times, the sawdust arenas returned, and, to the disgust of cultured patrons, melodramas and Shakespeare gave way to mandrills and acrobats.¹⁴ What Saull saw as monkey stealth promising a brighter human future, the literati saw as the ever-present threat of social degeneration.

At the new Zoological Gardens in Regent's Park, monkeys were symbolic of the safely caged 'lower' orders. Gentility was the hallmark of the promenading Gardens, not surprisingly because aristocrats were partly responsible for the zoo's founding in 1826.¹⁵ The riff-raff were excluded, the gate fee of 1s ensured that.¹⁶ Even then, visitors had to have a Fellow's recommendation. Thus in leafy surroundings, the well-heeled could stroll in peace, "without that nasal offence whereby one is always afflicted in confined collections." And behind bars were the monkeys, different species crammed together pell-mell "like slum-dwellers", and they behaved accordingly, frequently attacking one another.¹⁷

If disgusting monkeys were becoming better known, apes were another matter. In 1833, when Saull was lecturing, few Londoners had ever seen one.¹⁸ Even then only two types were known, the chimpanzee

---

11   George 1952, 478.
12   Egan 1821, 258–62; J. Brown 1858, 315–17.
13   E. T. Bennett 1829, 141–42, 144; Ritvo 1987, 34.
14   Tristan 1980, 180.
15   Desmond 1985a, 153ff; Åkerberg 2001, 64–68.
16   Minutes of Council, Zoological Society, London, MS, 1 (5 May 1826 to 4 Aug. 1830), ff. 27, 39; Desmond 1985a, 228–29; Åkerberg 2001, 77; Cruchley [1831], 101–03. On the opening up of the gardens from the 1840s, see R. Jones 1997.
17   Charman 2016, 102; Mudie 1836 2: 310.
18   A baby orang-utan had been exhibited in Piccadilly in 1831. Its demeaning caricature of humanity was so off-putting to one "lady of quality" that she turned

and orang-utan, and no adult of either had been seen alive in town. The few tiny tots that had been brought by sea captains and exhibited before the later 1830s had quickly perished in the cold.[19] Those that did survive for a while, like the zoo's chimpanzee Tommy, in 1836, were dressed up and forced into human ways, in his case in a Guernsey shirt (necessary as much for contemporary modesty as warmth), but unseemly clothes merely enhanced his lowly human-mimicking status.[20] Only privileged visitors—aristocrats, savants, and reporters—were allowed behind the scenes to see him, and the eighteen-month-old caused a sensation in the penny dreadfuls. He engendered queasy feelings because of his wrinkly, hairy, parody of a human face. Even the zoo's sympathetic vet confessed he had to overcome his feeling of "dislike, and almost of loathing, when he paid him his usual morning visit".[21] It was the same with the zoo's succession of baby orangs, the first of which went on display in 1837.[22] Away from the public gaze, these babies were 'presented' to aristocrats as one would 'present' a dolled-up commoner at court. One was even 'presented' to Queen Victoria, although, as if to show that class connotations extended to the simian orders, the keeper did not put on her cap, "as he was afraid it might be thought vulgar".[23]

A perceived coarseness was ever-present in these accounts of apes, something enhanced by their working-class clothing, artisan's cap, and sailor's shirt.[24] It was exacerbated by accounts of the ape's amoral behaviour, which called up derogatory images of the 'visceral' working classes. Stupidity also marked them out for some. The aping came without intelligence. London Zoo's vet—himself an expert on domestic

---

    her face: *Cosmopolite*, 19 Jan. 1833, in HO 64/18, f. 734.
19 [Rennie] 1838, 1:63; W. C. L. Martin 1841, 403, 408; Youatt 1836.
20 On dressing apes and forcing human behaviours on them, Ritvo 1987, 31.
21 Youatt 1836, 273.
22 The first was a three-year-old, bought on 25 November 1837 for £105, only to die on 28 May 1839: "Occurrences at the Gardens", Zoological Society, London, MS, 28 May 1839. The ZS had had a baby orang before, but it died before it could be shown: *Proceedings of the Committee of Science and Correspondence of the Zoological Society of London*, Part 1 (1830–1831), 4, 9, 28, 67.
23 Rev. R. S. Owen, 1894 1:193–4; Scherren 1905, 85. On the divergent perception of orang-utans, caged and wild: van Wyhe and Kjærgaard 2015.
24 In the *True Sun* (16 May 1834, 3), an image of a Guernsey shirt under a "threadbare frock" coat was used to spoof the "Gentlemen's Fashions for May", which shows how unseemly such apparel was. *NMW* 9 (6 Feb. 1841): 78, pictured a smuggler in one; Mayhew 1861–62, 1:66.

animals—thought Tommy's mental capacity no higher than a farm animal's.[25] Charles Lyell would not even give apes a dog's sagacity, but then he had probably never seen an orang and he was grinding an anti-Lamarckian axe.[26] The dog was the yardstick of intelligence: obedient and devoted, even 'spiritual'—something that could never be said of apes. Some were even comforted to think that their dogs would join them in heaven.[27] There *were* naturalists who spoke up for the tiny ape's "prudence and forethought".[28] But even this came with a caveat. It only applied to the mentally agile young—with age, and a growing bestial physiognomy, the adult chimpanzee becomes "nothing else but an animal, gross, brutal, and untractable".[29]

Probably it made sense to Saull, depicting dim, servile, thieving, hovel-apes perfecting into smart, moral Owenite autodidacts. But high society, with its heraldic pomp and respectable ancestry, looked for something more regal in its blood line and was hardly going to be receptive. In 1830, as Saull started his evolutionary talks, Satyrs and Troglodytes were often little more than freak-show exhibits; the growing trade in freaks—side-show abnormals pushed aside as grotesque and exploited as "not 'us'"[30]—only confirmed their status. Cartoonists would make great play of apes in their lampoons of national hate figures: negroes, Irishmen, revolutionists, and so on.[31] Apes themselves were something to be despised. These mimics were playing tricks with accepted norms. To say, as Saull did, that they were actually our grand-parents disrupted the fixed social boundaries. He was trying irrationally and suspiciously to make an ostracized hate-figure into a family relative, which, if nothing else, was unabashed social effrontery.

For Saull's infidel cadre a sub-artisan ancestor was acceptable: it showed progress as the ape was pulled up by its bootstraps.[32] But to the

---

25 Youatt 1836, 274; Ritvo 1987, 35–39 on the competing claims of dogs and apes.
26 C. Lyell 1830–33, 2: 61.
27 Epps [1875], 558, 560–61.
28 Rennie 1838, 70; Broderip 1835, 164.
29 *Edinburgh New Philosophical Journal* 30 (Jan. 184): 7; Richard Owen 1835, 354–55.
30 Tromp 2008, x.
31 Curtis 1997, 102. For an attack by the Chartist Henry Vincent on a hated dignitary, the Mayor of Newport, for his simian looks and moral defects (accompanied by a woodcut picturing him as a chimpanzee), see Scriven 2012, 178.
32 Saull's friend, Pierre Baume, was to buy a monkey himself years later: Cooter (in press) *The Man Who Ate his Cats*, ch. 19; Holyoake 1906, 1: 219.

upper crust, the notion that they had climbed from ragged-trousered satyr to Hetherington's "scum" to get where they were was repellent. Contemporary images of the immoral and irrational ape, mocking the divine countenance, made casting it as a blood relation an uphill struggle. Apes were grotesqueries with a dimmed intelligence, a thieving nature, and a farcical face, creatures sculpted in jest, as Smith seemed to imply, or in blasphemous derision.

If anything, the monkey image was even worse for men of the cloth at this moment, for biblical exegetics suggested that apes might even have a Satanic strain to them.

From Suspect to Satanic—A *Monkey* Bible

The monkey's religious reputation was plummeting. Satyrs had always been the hairy demons of the Bible, a sickening amalgam of man and beast, and monkey satyrs had long been in bad theological odour. In medieval manuscripts and bestiaries, they had a malevolent aura, but the ape's meaning had slowly transformed through the centuries. Portrayed initially as Lucifer, even as the tempter Christ in the desert, the ape ended up representing a lustful sinner, the fallen man, himself hunted by Satan. As boundary creatures, they stood not only between man and a mocking, soul-less world, but between sin and redemption. However, this diabolic image had all but faded by the Renaissance, as the ape was secularized as the fool, now to be "regarded with less horror and more bemused tolerance".[33] After this, monkeys settled down to caricature the stupid side of human creation.

Then in the early nineteenth century, the monkey suddenly regained its old infernal garb. It came with the publication of perhaps the most famous biblical commentary of the century, the polyglot Wesleyan Dr Adam Clarke's eight-volume text (1810–1826). Dr Clarke was a scholar of prodigious learning, whose familiarity with ancient scripts meant that his linguistic studies took him to some startling areas. The beast that tempted Eve, the *Nachash* in Hebrew, was not a serpent, in his view. It was an ape.[34] And after tempting Eve, the orang was deprived of its

---

33 Lach 1970, 2: 177; Vadillo 2013.
34 A. Clarke 1837, 1: 46–47.

voice in punishment and cursed to drop onto all fours. Clarke was no ordinary Methodist. A lover of science, he pushed it rather flamboyantly into his criticism, and, quite atypically, he reached out to the gentlemen savants. Like Saull after him, he became a fellow of the Geological and Antiquaries Societies, and this engagement meant his views broke out of the narrow Wesleyan confines. His analysis was notorious among ministers and engendered volumes of heated commentary. Orientalists rejected a baboon in Eden, while among naturalists Dr Clarke's Eve-seducing ape was equally scorned.[35]

It seemed that devilry was to be added to debauchery. The Carlileans milked his cursed ape for all it was worth, and the *"monkey* Bible" in the mid-1820s had become a laughing-stock.[36] Carlile plundered Clarke mercilessly, and hardly a *Republican* went by without a citation. But it was one of Carlile's assistants who had the last laugh. The "humble mechanic" John Clarke (the Dr's namesake) was himself an ex-Methodist, and now a scoffing infidel. He too knew chapter-and-verse: "our walking Bible", they called him,[37] although, on being sentenced to Newgate prison, the "walking Bible" found himself "chained to a certain place, as Bibles of old were".[38] The 'chained Bible' made good use of his prison term. Despite the terrible conditions, he wrote a series of sixteen letters to Dr Adam Clarke. These were published piecemeal, and then as a 316-page book, known to the faithful as *Letters to Dr. Adam Clarke* (1825). The leading radical publishers brought out their own editions and the book was reprinted constantly, with a result that *Letters* became standard infidel fare and could be seen in every radical catalogue.[39] The atheists now had their own elaborately deconstructed commentary, which Carlile considered "one of the best examinations of the Bible extant".[40]

For partisans, the 'chained Bible' had bested the good Doctor. Irreverence laced with erudition marked his onslaught on the ape tempting Eve, or eating dust, or going around on its belly. His *Letters*

---

35 *MNH* 2 (Mar. 1829): 118; Richard Owen 1850a, 240–41; Richard Owen 1849–84, 1: 151–52. For the original criticism: Wait 1811; Bellamy 1811.
36 J. Clarke 1825, 75.
37 *Newgate Monthly Magazine* 1 (1 Oct. 1823): 61; *Republican* 10 (30 July 1824): 124. There is little on John Clarke in print; the best account is McCalman 1975, 76–78.
38 *Republican* 11 (11 Mar. 1825): 305.
39 Saull's friend James Watson printed his own edition, as did Joshua Hobson—the Owenites' printer—and James Guest in Birmingham.
40 *Republican* 14 (4 Aug. 1826): 128.

became a freethought classic, and Dr Clarke's commentary was the excuse for so many other dissidents to make a monkey of biblical exegetics.[41] Had *they* suggested that Eve was "seduced by a baboon", one atheist ventured, they would have been incarcerated for blasphemy.[42] Such was the feeling among Saull's infidel friends.

Incessant squibs meant that the sacred seemed to be being profaned by the mere mention of the word "monkey". At this moment, an infidel invoking an ancestral monkey might sound like he was jesting, as Smith hinted, or more seriously that he was invoking our satanic origins. Dr Clarke was killed by the cholera outbreak in 1832, just as Saull started promoting his monkey-man, but the revered Methodist had enabled the pauper presses to add more ignominy to the ape's sordid aura. It left mankind's monkey heraldry multiply suspect in many eyes, just as Saull was advocating it. Saull gave an unwholesome new meaning to man's fallen estate, or to being born in sin. A Satanic origin was the most damnable of all ancestries. There was something sordid, distasteful and, for the devout, sacrilegious about the suggestion, especially as it fell from the mouth of an indicted blasphemer. By racking up the profanation, Saull was making it difficult for any but ultra-materialists to join him at this moment.

---

41 Benbow 1823, 159; *Isis* 1 (12 May 1832): 211; *NMW* 8 (18 July 1840): 34; R. Cooper 1846, 45–46; *Reasoner* 8 (30 Jan. 1850): 25.

42 *Investigator* 1843, 10; *Reasoner* 16 (5 Feb. 1854) Supplement, 104–05.

10. An Appeal to the Revolutionary Enemy

Saull confessed that he was better appreciated on the Continent. Unlike most of his confreres, he had a French connection, being known to the Parisian savants. This gave him a unique dimension—indeed, at times he deliberately talked over the heads of the English and made a direct appeal to what some still considered the national enemy.[1] His European connection was two-fold: as a wine and brandy importer he was well travelled and often in Paris, and it was because he was so successful in this trade that he could afford to travel to France frequently. Further, he was Robert Owen's right-hand man by the early thirties; and Owen, venerated in French republican circles, but not speaking a word of the language, would be accompanied on political trips by Saull as his interpreter.

Since Owen had long known the Parisian savants, he might actually have introduced Saull, either in France or in London. And, in Paris, Saull—as a new museum proprietor who had his own spectacular coal-age ferns—would undoubtedly have visited the collections in the Muséum d'Histoire Naturelle (an institution with an allowance of 425,000 francs, as the *True Sun* reported to shame the niggardly British government for its neglect of institutions at home).[2] It is possible that Saull was in Paris in 1834,[3] when a behind-the-scenes visit could have

---

1. For Francophobia among the career zoologists at this time, see Desmond 1985a, 174–76.
2. *TS*, 7 Dec. 1835, 1. It also reported that "a most complete collection of fossil plants has been placed in the geology department".
3. Talking to the Trades' Unions in April 1835—and inviting its delates to his museum—"Brother Saull" said that he had been in France "lately" (*TS*, 22 Apr. 1835, 2).

been arranged by the great fossil botanist, Adolphe Brongniart, the new professor of botany at the Muséum.

Certainly, Saull was elected a member of the Société Géologique de France (founded 1830) on 19 May 1834. He had been put up by the one of the society's founders, Ami Boué.[4] Boué was an Edinburgh-educated geologist, which undoubtedly made him familiar with Owen's New Lanark enterprise (visiting it was *de rigueur*). And having discovered human bones near those of extinct fossil mammals in 1823,[5] Boué was one of the first to accept a deep human antiquity, putting him in sympathy with Saull. Boué was also censored for his anti-Christianity, and was "a transformist of some sort",[6] which must have strengthened their rapport. In common with many geologists and fossil zoologists coming out of Edinburgh at this time,[7] he was a 'gradualist', sceptical of geological catastrophes and loathe to allow Mosaical intrusions into geology.

As in London, so in Paris: Saull's museum spoke for him.[8] On being elected, Saull invited the Société savants to his exhibition in London, and many of them were to come over the years.[9] He also told them that he was working up an essay on astronomical causes of geological events and that he would submit it to them direct. He fully intended to pass over the British geologists, obviously expecting a better reception in Paris. The Société was more egalitarian than its British counterpart and receptive to planetary explanations. As a consequence, Sir Richard Phillips's astronomical views were better received.[10] By contrast, Saull complained that the London Society, with its empirical emphasis, shunning anything that smacked of cosmic and creational explanations, showed a "great reluctance" to investigate these larger "laws".[11] He found himself in effect excluded. Just as Whigs and Tories came together in the country to resist the wilder radical demands, so it was inside the gentlemanly Geological Society. Here a ruling elite of Anglican

---

4   *Bulletin de la Société Géologique de France* 14 (1833–4): 586.
5   Grayson 1983, 117.
6   Corsi 2021, 352; Fisher 1866, 2: 143–44.
7   Corsi 2021, which updates J. A. Secord 1991, and Jenkins 2019, 88–150, on Edinburgh's "Lamarckians", a catch-all category in need of being broken down.
8   It was mentioned by Boué 1836, 2: 555.
9   *Bulletin de la Société Géologique de France* 15 (1835): 67.
10 R. Phillips 1832a, 36; Marcel de Serres in 1836 was taking a similar astronomical approach to large scale geological events (D'Archaic 1847, 18). Rudwick 1985, 28.
11 Saull 1853, viii.

dons and London careerists closed ranks.[12] This ministerialist cabinet of geology steered clear of divisive issues and insisted on empirical papers dealing mostly with stratigraphy. In so doing they could present a united front as purveyors of incontrovertible facts.[13] All talk of creation, causation, and cosmology was tactfully ruled out to leave the science circumscribed and safe. To them, stratigraphic facts were uncontroversial; they were the rocks on which the dons and divines justified holding their mace of power. But obstreperous backbench radicals had a larger purview, which embraced astronomical causes and fossil filiation. For Saull, Sir Richard's orbital explanations of stratal regularity and alternate marine inundations fired him precisely because it opened the great question of origins.

It all came out in his book—the one promised to the French. This would appear in 1836 as *An Essay on the Coincidence of Astronomical & Geological Phenomena, Addressed to the Geological Society of France*. As the title suggests, it was explicitly presented to the French *savans*, in an attempt to gain a hearing.[14] Such a monograph would never have been sanctioned by the Geological Society, so he had to print it privately. In the *Essay*, once again, Saull invited "my brother members" of the Geological Society of France to his museum, which he suggested was set up stratigraphically to prove Phillips's theory of planetary oscillations driving global changes and faunal progress. Everything about the *Essay* would have been hated by the dons and divines at home. Not merely the spontaneous chemical origins of minute corals on the earth's original pulverized granite substrate; worse for the cosmically-averse geological gentry was the alternating hot and cold climates as a result of the planetary wobbles, with the poles migrating, and the oceans moving from one hemisphere to the other, leaving rock strata composed of sea sediments here and dry land there.[15]

Although Saull pitched his *Essay* at the French savans, it actually attracted attention on the fringe at home. Because he had directed the gaze away from the "bosom" of the earth towards the heavens, the best-known astrologer of the age, "Zadkiel" (Richard James Morrison),

---

12 Morrell and Thackray 1981, 2.
13 Rudwick 1985, 25; J. A. Secord 1986a, 22; O'Connor 2008, 18.
14 *Bulletin de la Société Géologique de France* 17 (1835–6): 151.
15 Saull 1836, 3, 5, 12–19, 30.

hailed it as "the only possible theory of Geology which (*as far as it goes*) can be true". That "*as far as it goes*" was the rub. Zadkiel's *Horoscope* specialized in self-help science—astrology, phrenology, physiognomy, and predictive meteorology—but his underlying goal was to blend geology, astronomy, and astrology into a unified science, and Saull provided the key. Zadkiel was looking for respectability for his dubious science, trashed as charlatanry almost universally. He was an anti-Owenite, and bent astrology towards Christianity by blending in an element of free-will. For him "*one* only" geo-astronomer among the puffed-up Geological Society fellows had set his "course towards the pole star of universality in nature's causes" and that was Saull. His swinging poles, producing successive tropical and ice ages—and for the astronomical part Zadakiel gave Mackey's *Mythological Astronomy* the credit—set the foundation for this unified science. Zadkiel was to stretch the causation from Saull's "geological phenomena up, through the operation of meteorological causes, to those astronomical affections", and onward to astral influences.[16] He was piggy-backing to further his scientific respectability. Both Zadkiel and Saull, in their own ways, were trying to claw back some dignity and claim authority by standardizing around the scientific norms of uniformity, causality, and prediction. But Zadkiel's endorsement was hardly the sort that Saull wanted.

Ten days after his induction into the Société in 1834, Saull forwarded a parcel of fossils from Brongniart to Gideon Mantell. This suggests that Saull had carried the package back with him from France. Whether he had or not, he was hereafter a Continental conduit to English provincials.[17] More to the point, it shows his familiarity with Brongniart personally. But the real payoff for Saull in Paris was Brongniart's ultimate accolade. The Parisian expert made one of Saull's prize tree-fern trunks from the Oldham coal mines a new species, christening it after him, *Sigillaria*

---

16   Morrison [Zadkiel] 1841. On Morrison: Latham 1999, 176–77; K. Anderson 2004, 101–05.

17   J. A. Cooper 2010, 57. Since Mantell sent packages with people going to Paris, it is possible Saull carried this one back. Saull was a go-between for other Continental collectors: for example, in 1838 he presented a thousand Tertiary fossil shells to the Zoological Society of London on behalf of the Turin invertebrate palaeontologist Giovanni Michelotti: *MC*, 13 Dec. 1838, 2; *Proceedings of the Zoological Society of London* 6 (1838): 167. For his part, Saull sent books to Michelotti (1841, 10).

saullii.[18] Such immortalization would raise Saull's intellectual stock at home and make his museum more of a personal shrine. I suspect Saull sent or brought the best of his coal shale fossils to Paris for description, because Brongniart was publishing his monumental *Histoire des Végétaux Fossiles* in parts at the time (1828-37). Brongniart also "received" from Saull fossil fronds of a related fern found in the same slates, *Pecopteris*.[19] That said, these were probably all specimens bought in with Sowerby's collection, but Saull had gone, tellingly, to the French rather than the English for identification.

With a fossil named for him, Saull saw his standing improve. Nothing could better enhance the museum's reputation. This French appreciation was always the most welcome. Saull's biography would be published on the Continent and, eventually, his eulogy was to appear in Michaud's *Biographie Universelle*,[20] to contrast with the paltry obituaries at home. He was no prophet in his own land. Of course, the French did not really know him, making their biographies factually dubious: thus, Michaud glamourized him as the "Scion of a family of squires or knights, native of Devonshire"! But still it showed his greater impact on the Continent. As a result, his museum was listed in French *dictionnaires* and tourist guides, hinting at his European reputation.[21] It led to the ironic fiction of his 'aristocratic' pedigree, on account of the 'Devonshire' middle name (simply his mother's maiden name). But the polite fiction probably eased his slide through French scientific society.

As a result, when the antiquarian and flint-tool finder Boucher de Perthes visited Aldersgate Street he "expected to meet a duke", only to be greeted by "an honest merchant of the City, whom I found in his store putting on his shoes". Still the visitor was awed by the exhibition: none of the medals, Roman pots and Celtic axes on the ground floor prepared him for the fossil riches above: here was a lavish collection of

---

18 Brongniart 1828[–1837], 1: 456–57, pl. cli. The dating is problematic: the *Histoire* carried a date on the title page of 1828, but parts were added continuously through to at least 1837. Saull's *Sigillaria* was on page 456–57, and this part was published in 1836. Other illustrations of *S. saullii* occur in Mantell 1844, 1: 135; 1851, 32–33.
19 Brongniart 1828[–1837], 1:348–9, and pl. 121, Fig. 1, for Saull's *Pecopteris plumosa* fronds.
20 Michaud n.d., 38: 47.
21 Duckett 1853–60, 12: 412.

plants and fruits coming from the collieries of England, plants and fruits whose analogues do not grow any more but in the torrid zone. He pointed out to me a fossil tooth of shark of approximately ten centimetres length, having belonged to an animal from sixty to eighty feet. A femur of fossil elephant (cast) being close to one metre length, representing an animal of more than fifteen feet high.[22]

The Radical French Connection

The ennoblement belied "Citizen" Saull's subversive salutes at home to the past French revolutions. Red cap on, he would contribute to the fund for the widows and orphans of those who fell during the July Revolution of 1830.[23] Thereafter he would celebrate its victory over despotism at annual dinners, at which he would sing the *Marseilles*.[24] Anniversary commemorations were a way for the disparate radical and Owenite groups to consolidate around a mutually-beneficial event. Whether it was Tom Paine's or Robert Owen's birthday, the July Revolution, the celebration of trial by jury, or to erect monuments to old jacobins,[25] Saull was present every year, often in the Chair.

There remained immense interest in revolutionary sources in the pauper press, as, indeed, in the French Revolution's leaders. When Robespierre's sister died in 1834, it was expected she would leave evidence to dispel the middle-class calumnies about his supposed Terror. His "memory blasted by all the assassin pens" would be retrieved and returned to show a virtuous man, the *Poor Man's Guardian* prophesied.[26]

Radical funerals, too, took on a French aspect, making them suspect to a genteel nation. The unions and Friendly Societies had taken the last rite into their own hands, saying a solemn farewell to their members, with processions, bands, and choirs. Now the socialists were to take over the rite as they buried their own dead, often in the face of fierce opposition from the clergy. Not least, the pomp was valuable to attract the poor to the cause, and a "lavish ritual available even to the simple

---

22 Perthes 1864, 416–18.
23 *The Star*, 18 Aug. 1830, 3; 23 Aug. 1830, 1; 24 Aug. 1830, 1; 25 Aug. 1830, 1.
24 *Republican* (Hetherington), 25 June 1831, 8; *PMG*, 6 Aug. 1831.
25 For instance, the monument to commemorate Thomas Muir the Scottish Jacobin transported for sedition in 1793–94 (*MC*, 16 Mar. 1837, 2).
26 *PMG*, 9 Aug. 1834; 13 June 1835.

working man was considered a trump recruiting card."[27] As the radicals perfected their own final rite of passage, the tricolour-waving London citizenry further modified it by introducing French-style funeral orations. They celebrated radical life in death Continental style. When the old shoemaker and London Corresponding Society founder Thomas Hardy died in 1832, Saull urged the National Political Union to attend his funeral procession in force. It was to process through London from Pimlico to Bunhill-Row, where "his obsequies should be celebrated as the French were accustomed to celebrate those of their great men."[28] Among the entourage, with many mourners sporting tricolour ribbons, walked the spy alongside Hetherington. In front of them was

> the Barouche of Hunt in which he sat while Saul [sic], Thelwall, Gale Jones and others Rode in Carriages or Black Coaches ... We went with a very great crowd to Bunhill Fields Burial Ground were [sic] Thelwall delivered a long lecture over Hardys Grave.[29]

This set the pattern for Owenite funerals. These were important events, for with celebration came cohesion, while the whole shebang acted to raise the Owenites' profile.[30] No lonely burial here; the fallen were treated to a full send-off. The religious aspect was shunted aside—dwelling not on a future state but a past one well lived in aid of the community, with the emphasis on the value of comradely morality and shared values. The radicals and Owenites developed this set of customs in a hostile culture by working up these Gallic-style obsequies. Saull would make the point at the graveside of numerous old firebrands over the years. He would follow "the excellent example of our enlightened brethren in Germany, France, and other countries on the continent of Europe, who ... speak to the assembled friends and relatives of the virtues, the patriotism, or the philanthropy of the deceased", thereby firing up the living to continue the good fight.[31]

Even French revolutionary failings could provide Saull his moral. At one debate in Theobald's Road, on abolishing the Established Church,

---

27 Yeo 1971, 102.
28 *MC*, 18 Oct. 1832; *Examiner*, 21 Oct. 1832; *Bell's New Weekly Messenger*, 21 Oct. 1832, 439. On Thelwall's own funeral, see Thelwall 1837, 430.
29 HO 64/12, f. 157; *Cosmopolite*, 20 Oct. 1832.
30 Nash 1995a, 158–62; A. Taylor 2003.
31 Saull 1838a.

he used France as a cautionary tale: "we must be careful not to pull down one house before we [have] built another." He knew from his trips that "France was evidently not prepared to realize the advantages of the revolutions which had taken place there". Rational institutions must be in place before the irrational ones are pulled down, lest reactionary elements fill the vacuum. Disestablishing the church, which he supported, would be successful once a proper secular schooling was accepted, which required a cultural shift.[32]

Even more did he use these trips to point up British backwardness, especially in its religious toleration. He reverted to this time and again, scoring political points on a variety of fronts. As a Common Councillor for his Aldersgate ward, he complained that potential delegates had to swear an "odious" oath, to abjure the "damnable and heretical 'Catholic' faith". Such an "abominable" sectarian ban showed that Continental countries were "far before us in religious toleration; he was a member of a learned society in France [Société Géologique], but on his election he was not asked his creed", he said.[33]

Comparisons with the French peasantry were equally unfavourable to their British counterparts. The "degraded condition [of] our unhappy countrymen" stood in contrast to that of their French cousins.[34] After one trip with Owen to France,[35] Saull's home-coming lecture on "Geology, in connection with the social and moral improvement of the people" unravelled into a panegyric on French institutions and peasantry. The Francophile reported that there were "no turnpike gates, no hedges shutting out the weary and footsore passenger from a view of the green fields". This gave him a view of a land cultivated more productively in small patches, which reduced poverty (because there was no law of primogeniture, so all sons received a patch to grow their staples). And the land's "produce is not as with us, carried away by drones who eat, but labour not".[36] Nothing would spoil Saull's idyll. No mention here

---

32   *TS*, 28 Apr. 1835, 2.
33   *TS*, 26 Dec. 1835, 4; Saull 1837.
34   *NMW* 8 (22 Aug. 1840): 124.
35   In 1837: Claeys, 2005, 9: 274–76. Saull tried and failed to get Baume discharged from Fleet Prison to go with him on this trip: P. Baume to W. D. Saull, 3 and 9 June 1837, Manx Museum MM 9950 uncatalogued: I am indebted to Roger Cooter for this. On Owen in Paris: *NMW* 4 (28 Oct. 1837): 3–4.
36   *NMW* 3 (30 Sept. 1837): 397–98.

of the new French "profitocracy", the "crafty middlemen" who had subsequently stolen the revolution from the artisans who had manned the barricades.[37] Nor was there mention of the poor being robbed, despite NUWC complaints about France's bourgeois plunderers. Perhaps a rosy view was best for a wine wholesaler, a middle-man himself, with shopocrats now being placed alongside aristocrats as hate figures, and profits becoming the new radical target. Rose-coloured spectacles left his view of *la grande nation* undimmed. A rejected prophet in his own land, and garlanded in Paris, Saull needed France.

---

37   Weisser 1975, 39–41; Dinwiddy 1992, 213, 217; *PMG*, 3 Aug. 1833, 248.

11. Creation on the Cheap

The blasphemy chapels and Owenite halls attracted a particular sort of aggrieved, politically-astute, and inquisitive artisan. Here politics, economics, science, and irreligion were bartered, with audiences participating in the free-for-all. What snippets of irreligious geology attendees picked up, often incidentally, was fine-tuned to this reformist platform. And for the real enthusiast, there was always the extensive LMI library, or the cut-price pirate editions of Toulmin's *Antiquity and Duration of the World* and Lawrence's *Lectures on Man*. Leather-bound gentlemanly tomes had no shelf space here, but urban agitators could access a plethora of cheap pamphlets and books falling from the presses operated by Carlile, Hetherington, Watson, Cleave, Brooks, and others. Even so, few could afford the shillings, in which case the aficionados would club together. Anne Secord has discussed how the rural artisan-botanists chipped in to buy texts and used the pub as their club house. Urban operatives were better paid, and could fork out the shilling fee and "trifling" ha'pence a week to join the Operatives' Literary Association (founded 1834), which put 'Mirabaud' (Holbach) and every other destabilizing work within reach.[1]

If that was still too expensive, mechanics had other leisure choices. Workmen in town mostly got their meals in pubs,[2] but, by the 1820s, the proliferating coffee houses catering to operatives were the preferred haunts of the radical intelligentsia, many of whom were temperance advocates. Away from the prying eyes of the licensing justices, the coffee houses could stock the latest penny trash and provide a talking shop. For Christina Parolin, these 'free-and-easies' were the archetypal "public

---

[1] *PMG*, 16 Aug. 1834. A. Secord 1994. A skilled artisan, say a compositor on London's dailies, could earn £2 8s a week, six times a farm labourer's wage (*LMR* 1 [13 Nov. 1824]: 28).

[2] Lovett 1920, 1: 32.

sphere", now to become hotbeds of sedition and materialism as the political temperature rose.³ The Operatives' Literary Association met in Presley's Coffee House on John Street. An NUWC branch held meetings in the Hope Coffee House in Snow-hill. Police spies crawled from one to another, keeping their ears to the ground. William Lovett, having started off as a cabinet maker attending a book-sharing "literary association" in Gerrard Street—which first fired his intellectual curiosity—had gravitated to Tom's Coffee House in Holborn, where he heard Carlile, Taylor, and Gale Jones, and to Lunt's on Clerkenwell Green. By the 1830s, he had his own coffee house in Hatton Garden. Here, coffee by the pint came with the unstamped press free to peruse. His Reading Room with several hundred volumes led to a "commodious" Conversation Room where issues could be thrashed out from five till eleven nightly.⁴

But the more immediate and exciting entertainment was to be had in talks. And much of the dirt cheap or even free science lecturing was now dominated by Saull's ubiquitous appearances. His venue of choice switched to "Watson's", as the spy called it. Saull's friend James Watson had cut his teeth as a Carlile shopworker. He had already spent a year in Coldbath-fields jail for selling Palmer's *Principles of Nature* to an undercover informer. According to Saull, Watson had worked at Owen's Bazaar after Carlile's shop.⁵ Then in 1831, when Pierre Baume moved out of his Optimist Chapel in Windmill Street, Watson moved in, and he renamed it the 'Philadelphian Chapel'.

Watson had already bought out Baume's printing and book-selling premises—the sale sponsored by the renegade old Etonian Julian Hibbert, still using his family's wealth to finance freethought. Like Hibbert, Watson was abstemious, but grave, showing none of Hibbert's *joie de vivre*: a bit too cold for the coming generation of hotheads. "Watson was of the old Puritan type of our great Cromwellian time, such a man as Ireton [the commonwealth commander and Parliamentarian who signed Charles I's death warrant], simply wise, serious, and most earnest".⁶ But despite his fearsome reputation, one broad-minded Congregationalist, stepping into his shop one day, and fearing that

---

3 Parolin 2010, 225.
4 *PMG*, 25 Jan. 1834; 28 Feb. 1835; Lovett 1920, 1: 35, 37; NUWC: *PMG*, 11 Oct. 1834.
5 *Reasoner* 16 (5 Feb. 1854) Supplement, 97–112.
6 Linton 1894, 38.

"righteous retribution should bring the bells down upon his head", was surprised to find Watson "mild and temperate, transparent and honest".[7] Watson was another who split his time between the NUWC and Owenites, hence at the Philadelphian, the weekly topics covered the gamut: Sunday: philosophical discussions; Monday: NUWC meetings; Tuesday: co-operative subjects; Thursday: theological debate. And lest the last should be misunderstood, adverts pointed out that it was to examine "the claims of a certain Book to infallibility", to which end the faithful appealed "to the inquisition and ... torture for its support."[8]

Advertisements for Philadelphian meetings in the *Poor Man's Guardian* gave nothing away bar lecture titles. But police reports reveal that Saull was lecturing here, and at least two of his more vituperative talks the nark thought worth reporting. The first was in March 1832, amid the Reform turmoil, as Benbow and Hetherington were taken into custody. (And as NUWC members practised their "sword exercise" in Goodman's Yard, behind the Tower, fearing the worst,[9] armed with Colonel Macerone's hot-off-the-press *Defensive Instructions for the People* with its lance-making help for street warfare.[10]) At such a moment, Saull, undeflected, spoke on "Creation" at Watson's. By this, he meant the production of species without any supernatural aspect. His vehement tone matched the political mood. The talk had "a Strange harshness", reported the spy, "I never saw or heard so much abuse."[11]

Again, in September 1832, Saull lectured on the "The influence of Science upon the future condition of Mankind". Only one part did the spy think worth passing on to the police, and that was Saull ploughing his old furrow. He advocated

> Principles of Materialism against any Religious or Superstitious systems got up by the tyrants of Mankind to rob them and keep them in ignorance and went to prove that the best way for Men to act was to leave all such

---

7   *Reasoner* 16 (15 Jan. 1854): 37.
8   *PMG*, 3 Sept. 1831; Prothero 1979, 261.
9   HO 64/12. f. 67 (26 Mar. 1832), "sword exercise" underlined by the police, to raise Home Office awareness of it.
10 Macerone [1832]; *PMG*, 10 Mar. 1832. Macerone's pamphlet mooted the fall of the Whigs, necessitating armed resistance to the returning anti-reform Tories. As Hollis (1970, 41–42) says, the pamphlet was "alarming in working-class hands". Watson displayed his Macerone lance at the Philadelphian chapel (Barker [1938], 22–23).
11 HO 64/12, f. 67 (26 Mar. 1832); *PMG*, 24 Mar. 1832.

foolish Dogmas to the wind and follow their own wants as Nature dictated to them.[12]

This, given the times, was hardly incendiary, but it fitted the pattern of resistance to state authority. It was the month when the fearsome Bronterre O'Brien took over editing the *Poor Man's Guardian* to ratchet up attacks on the profiteering middlemen, from shopkeepers to merchants. At the same time, Lovett became a national figure by orchestrating resistance to the draft—the "Militia Laws", which saw the poor forcibly recruited into local militias—and, having resisted himself, he found all his property seized, in what was damned as "undisguised, downright, absolute robbery".[13] Saull's remained a more passive, scientific undermining of authority. Through all, he stood resolute, undeviating, with his eyes on the "Creationist" ball—and un-arrested.

It seems that Saull often spoke at the Philadelphian after a lecture, denouncing Christianity.[14] Watson was another who mixed co-operation, Carlilean freethought and NUWC class-consciousness, but his activism was quite different from Saull's. Imprisonment was an occupational hazard for seditious printers and vendors, and borne with fortitude as a badge of honour. There was the odd court triumph. In 1832, Watson (with Benbow and Lovett) was acquitted of unlawful assemblage and riotous behaviour on the 'Farce Day'. (The government, to assuage the Almighty in the face of the coming cholera, had ordered a national 'Fast Day', inflicting still more punishment on the poor, and a loss of a day's wages. A radical procession to protest it had been met by police truncheons, and subsequent prosecutions.[15]) In 1833, Watson was fresh out of jail, having served six months in Clerkenwell for selling the *Poor Man's Guardian*. Freedom did not last long, for he was back in prison in 1834, two months after his marriage. As Saull reminisced later, "He [Saull] had not suffered personally as his friend Watson had done", at least in terms of incarceration. Saull was very careful to keep out of prison, but with prosecutions piling up he was to expend ever increasing sums on court costs for comrades-in-arms.[16]

---

12 HO 64/12, f. 142 (18 Sept. 1832); *PMG*, 15 Sept. 1832.
13 Wiener 1989, 27; Lovett 1920, 1: 66–68.
14 HO 64/12, ff. 96, 105.
15 Lovett 1920, 1: 80; Benbow 1832; Hollis 1970, 53.
16 *Reasoner* 16 (5 Feb. 1854) Supplement, 97–112; Linton 1879.

With Hibbert's backing in 1833, Watson's Windmill Street presses started churning out usable Enlightenment philosophy in twopenny parts: Volney's *Ruins of Empires* sold at 2*d* a sheet weekly from August, or three numbers stitched for 6*d*—and it was still a "beautiful little edition". Then came a uniform edition of Mirabaud's (that is, Holbach's) *System of Nature* in twopenny numbers.[17] Others later recalled that "before there was so much talk of cheap printing, our friend was printing and publishing cheap".[18] But Watson could only do it because he was so heavily subsidized. Otherwise, it meant going into debt, as he knew well from Carlile's case. It was only a bequest on Hibbert's death in January 1834, which Saull oversaw as an Executor, that wiped out Carlile's £492 losses.[19] Watson no less benefited from Hibbert's munificence. Watson and Hetherington were left 450 guineas apiece in Hibbert's will. Hetherington invested in his first steam press, a machine that could speed up production ten-fold, turning out 2,500 sheets an hour. Hetherington and Watson were fellow publishers who never fell out with one another, indeed they worked in conjunction. Harassed by the authorities for refusing to pay fines, Hetherington made over his *Poor Man's Guardian* press to Lovett, who installed it on Watson's premises, then Hetherington sold his book stock to Watson. But it proved an unsuccessful manoeuvre, for the police raided Watson's and confiscated £1500-worth of print, type and stock in 1835.[20]

Nonetheless, Watson's wholesale depot was probably the most important radical warehouse in the country in the 1830s–1840s. He shipped the subversive works of fellow unstamped publishers and his own stock-in-trade: Volney, Palmer, Holbach, and Paine.[21] With Holbach never out of print, or out of financial reach, the fundamental question of origins, taboo in gentlemanly geological circles, remained alive on

---

17   *PMG*, 10 Aug. 1833; 23 Nov. 1833; also 16 Feb. 1833; Linton 1879.
18   *Reasoner* 16 (5 Feb. 1854) Supplement, 97–112.
19   Julian Hibbert, Will, Public Record Office, National Archives, PROB 11/1827, 14 Feb. 1834. *Satirist*, 30 Mar. 1834; *Patriot* 2 Apr. 1834, 113. Carlile used his windfall to put out a new "pocket edition" of *System of Nature* in 1834, but it still cost seven shillings—a Tolpuddle Martyr's weekly wage: *PMG*, 20 Sept. 1834; Watson also published a twopenny *Life* of Holbach in 1834: *PMG*, 12 Apr. 1834.
20   See Hollis's (1970, 130–31, 162, also 126) unravelling of events. *PMG*, 8 Aug. 1835. *Reasoner* 16 (5 Feb. 1854) Supplement, 97–112.
21   Then came titles on atheism, communism, Owenism, anti-Church tracts, and later the *Reasoner* as well as the *Free-Thinker's Information for the People*.

the street. Political titles were snugly shelved alongside pamphlets and digests on subversive science. To take some examples: his *Facts versus Fiction: An Essay on the Functions of the Brain* was a twopenny pamphlet culled from Lawrence's *Lectures on Man*.[22] This was ostensibly to test whether "the mind of man is a separate and distinct principle residing in his corporeal fabric, or simply the function of an organ—the result of the action of the brain". But Watson's introduction makes plain that the result would counteract the "wild and imaginary notion" of a soul enjoying a paradisaical afterlife to compensate for the brutality of this one, a "fantastical" idea necessary to the apparatus of social control used by repressive governments.[23] Other offerings included the American Dr Thomas Cooper's anti-Mosaic *Connection between Geology and the Pentateuch*[24] and *Galileo and the Inquisition*.[25]

Watson later published an address that caused a kerfuffle in the medical press, the young physician William Engledue's *Cerebral Physiology and Materialism*. This was another clever piece of piracy. Seditious publishers kept an eye on the medical press, as they had done in Lawrence's day. Many GP-supporting medical papers were themselves radical, as they fought the nepotistic hospital consultants and unreformed medical corporations. So, when Engledue's iconoclastic address to the Phrenological Association, published in the *Medical Times* raised hackles[26]—for rejecting souls, spirits, and mind as too mystical, and for making the brain 'secrete' thoughts as the mouth does saliva[27]— the pirates took notice. At a stroke, Engledue had undermined, in Roger Cooter's words, "two decades of carefully rehearsed, endlessly reiterated, and successfully sown rhetoric on phrenology's harmony with socioreligious views."[28] By invoking the non-responsibility of man for his actions, Engledue would let the critics tar phrenologists as atheists who denied all "religious obligation". His talk led to mass resignations

---

22 *PMG*, 25 Aug. 1832.
23 Lawrence 1840; first edition: *PMG*, 25 Aug. 1832.
24 Thomas Cooper M.D. 1837 was challenging Yale Professor Benjamin Silliman's "Mosaic" bowdlerising in his American edition of Robert Bakewell's *Introduction to Geology*. Cooper's riposte was re-published by Watson: *Reasoner*, 3 June 1846, 16.
25 *NMW* 11 (28 Jan. 1843): 252.
26 *Medical Times* 6 (2 July 1842): 209–14. Desmond 1989 and Underhill 1993 on the medical politics.
27 Engledue 1843, 4–9.
28 Cooter 1984, 94.

from the Phrenological Association and demands that the phrenological society in future exclude all reference to "materialism".[29] Man not being responsible was, of course, the Owenites' doctrine. And Owenites were the first off the blocks, reprinting his address in their *New Moral World*.[30] It was too good for Watson to pass up. He got out a 32-page, fourpenny pamphlet of the Address in November 1842.[31] And he capped it by attaching a provocative letter from the London University physiologist John Elliotson, arch-materialist and mesmerist, who viewed the brain as so much thinking-matter. He agreed in the bluntest terms that cerebral stuff needed no 'soul' to help it spew out thoughts.[32]

All these examples show that the street tracts had to be terse, non-technical, cheap and pointed. And opportunism played a huge part: as political issues flared, or addresses fired up the orthodox, the pauper presses were prepared to capitalize with immediate effect.

The Mechanics' Hall of Science

The lease of the Philadelphian Chapel expired in October 1832, and, in another typical reversal, it was snapped up by Finsbury's Baptists.[33] As a result, Watson used Hibbert's bequest to move his "Cheap Publications Warehouse" to "more eligible premises" in Commercial Place, off the City Road, Finsbury, in March 1834. Here he set up a circulating library of rationalist books, charging 1*d* a week.[34] The spy reported that he had revamped his shop front "to more expose his business as a Pamphlet seller of which he boasts a very great increase lately".[35]

Attached to the shop was a huge, barn-like hall, and, through the winter of 1833–34, Saull and his lifelong friend, the radical apothecary

---

29 *Medical Times* 6 (2 July 1842): 266–67, 295; *Lancet*, 2 (13 Aug. 1842): 702.
30 *NMW* 11 (17 Sept. 1842): 94–95, 102–03, 118–19.
31 He undercut and beat by weeks the medical publisher Bailliere's one shilling version of the Address: *Medical Times*, 7 (17 Dec. 1842): 194.
32 *NMW* 11 (12 Nov. 1842): 164. Watson then advertised his fourpenny pamphlet in the medical press: *Medical Times* 7 (3 Dec. 1842): 162. Henry Atkinson, who was to introduce atheism to the middle classes with his and Harriet Martineau's *Letters on the Laws of Man's Nature* (1851), was working with Engledue at this point and praised his address as "truly philosophical"—meaning seriously disturbing to conservative factions (*Medical Times* 8 [5 Aug. 1843]: 294).
33 HO 64/12, ff. 149, 161.
34 *PMG*, 22 Mar. 1834.
35 HO 64/12, f. 170.

Thomas Prout, along with others, mucked in to help Watson fit it out as a new venue. This was one that would prove, finally, to be long-lasting. They branded it the "Mechanics' Hall of Science".[36] The title "Hall of Science" is telling—'Science' was now viewed in these infidel circles as a rationalist saviour, an explanation of the true meaning of life and society. This was not the first use of the name, which would come to encompass all later socialist venues, but it was the first and most prominent London "Hall of Science".[37] With Watson and Saull behind it, the "Science" in the title had an overt anti-Christian meaning. Despite this, the original idea was to make it an auditorium for a sparkling lecturer, the former fustian-cutter and factory worker, Rowland Detrosier. He was a brilliant orator, a deist who believed in design, and another interested in working-class education. His star was rising, and although young and self-taught in the sciences, he was already sitting in the Secretary's seat as Saull chaired mass NPU meetings.[38]

Julian Hibbert's bequest made the hall, but it was tainted in orthodox eyes. This was the Hibbert who had caused outrage by refusing to swear on the Bible in court. Was not this Hall of Science, cried the failing Tory *Albion and the Star,*

> established with the aid of funds bequeathed for the purpose, by a professed Atheist who, but a few months before his death was, in consequence of a public declaration of his infidelity, dismissed from a Court of Justice as unqualified to give evidence upon a trial?

It was enough to damn the hall. The evening *Albion*, as one of its last gasps, decried the "pestiferous harangues of revolutionary demagogues" in Commercial Place, with all their "absurd doctrine of the 'majesty of the people'". Under "the insidious pretence of diffusing knowledge", these defilers spread "the destructive principles of blasphemous infidelity" in the hope of scouring from the "breasts every virtuous and social feeling". True science reinforced religion, these were "babblers of a jargon they pretend to call philosophy". And, with Watson's "beggarly shop"

---

36 Linton 1879, 19; Royle 1976, 104.
37 Frances Wright ran a "Hall Of Science" in Broome Street, New York, the old Ebenezer Church, from March 1829 (*Free Enquirer* 2nd ser. 1 (25 Mar. 1829): 174. In Britain, *The Poor Man's Advocate* (25 Feb. 1832) was discussing a "Mechanics' Hall of Science" in Manchester in 1832.
38 G. A. Williams 1965.

festooned "with obscene and indecent prints, profane and blasphemous placards", shock turned to outrage. For the *Albion*, this disdainful trash was enough "to demoralise and deprave the rising generation."[39]

At the gala opening on 7 April 1834, the spacious 2000-seat lecture theatre was packed to the gills. Owen in the chair lauded the institution's aim of placing "rational education within the reach of the working classes". Detrosier then made an eloquent speech, denouncing caste, war, nationalism, and the ignorance in which men were kept by Church and state. Saull seconded his resolution on the need for "really useful knowledge"—that is, class-useful knowledge—to arm the working people.[40] But Detrosier did not enjoy his platform for long. He died of a "chill" (presumably a lung infection) a few months later, aged only 34.

True to its aims, the Hall hosted talks turning on the origins of Heavens and Earth, equality, the education of women, meteorology, astronomy, prison reform, the air pump, colonization and capital punishment, anything on a Sunday, screamed a disgusted Tory press, "except religion". As such it is "a place which the local authorities ought long since to have indicted as a public nuisance and a scandalous disgrace to the neighbourhood".[41] It was large enough for fortnightly social festivals, with the Owenites enjoying dancing, music and dining. And Owen's right-hand-man, Saull, as usual would talk on the rise of fossil life and its astronomical driver, subjects always twinned in his eyes.[42] The authorities' antipathy showed again in the *Morning Herald*'s account of the Irish Nationalist Daniel O'Connell's opening salvo here in favour of Irish municipal elections (in which, needless to say, he was backed by Saull):

> A public meeting of the idle and dissolute of the metropolis, but purporting to be one of the electors and inhabitants of the borough of Finsbury, was held last night at a barn-looking place, called the "Hall of

---

39 *Albion and The Star*, 5 June 1835, 2. On the faltering Albion, [James Grant] 1837, 2 ed. 2: 112
40 *TS*, 8 Apr. 1834, 2, 6; *Republican* (Hetherington), 13 Apr. 1834; HO 64/19, f. 395; Johnson 1979 on "Really Useful Knowledge".
41 *Albion and The Star*, 5 June 1835, 2.
42 *NMW* 4 (6 Jan. 1838): 85; 5 (26 Jan. 1839): 224.

Science," in the City-road, at which O'Connell re-opened his campaign against the Lords, with a ferocity surpassing all his previous attacks.[43]

Such attacks graphically demonstrate the sneering attitudes of the authorities to this non-cap-doffing and nose-thumbing "Mechanics' Hall of Science". And Saull's talks here, on corporate reform, anti-creationist science, and the production of destabilizing knowledge,[44] only reinforced their fears. From now on, this, with Owen's "Institution of the Industrious Classes", was to be Saull's regular haunt.

---

43 *Standard*, 28 Apr. 1837; *Courier*, 7 July 1836, 4. Saull too demanded the reform of the Lords (*TS*, 2 Dec. 1835, 2).
44 *TS*, 28 Aug. 1835, 2; *NMW* 6 (26 Oct. 1839): 848.

12. Making Sense of the Museum

his museum would be a sealed book to the many, were it not for his lectures

The *National Standard*'s reporter, commenting on Saull's talks as he accompanied his museum guests.[1]

Since Bristol's was a stock Saull lecture, listeners at many radical venues in London in the thirties probably heard something similar. It played well in Owen's headquarters, where it was delivered to such "thronged" audiences that it was expanded into whole courses.[2] It, or variants of it, monkey and all, ran at the "Institution of the Industrious Classes" in Gray's Inn Road, and at its successor in Charlotte Street, then, later in the thirties, at the Social Institution in John Street, off Tottenham Court Road. And Saull's other platforms read like a 'where's where' of activist haunts: the Western Co-Operative Institute in Poland Street; Chapel in High Street, Borough; the "Society for the Acquisition of Useful Knowledge" in Store Street; the Great Tower Street Mutual Instruction Society; the Rational Institution, Curtain Road; and his favourite, the Hall of Science in Commercial Place, Finsbury.[3]

Even the clergy attended. 'Know thy enemy' was presumably the explanation for some vicars sitting in on Saull's addresses. Many already thought science suspect, and Saull's Sunday onslaught, mixing socialism, monkeys, and Heaven on Earth, simply ratcheted up their fears. In one case, the result was a strengthened primer for Church of England ordinands on how to combat an infidel geology. The Rev.

---

1   *National Standard* 3 (18 Jan. 1834): 44–45.
2   *NMW* 3 (26 Nov. 1836): 37; (10 Dec. 1836): 53.
3   *PMG*, 31 Dec. 1831; 24 Nov. 1832; *Crisis* 1 (15 Dec. 1832): 164; (29 Dec. 1832): 172; *NMW* 1 (29 Nov. 1834): 40; 3 (12 Nov. 1836): 20; (19 Nov. 1836): 29; 4 (28 Oct. 1837): 5; (6 Jan. 1838): 85; 7 (6 June 1840): 1290; *Shepherd* 1 (1 Aug. 1835): 392; Roebuck 1835b, 16; *PM* 1 (6 May 1837): 224.

Johnson Grant, minister of Kentish Town Chapel, broadened his advice for those taking holy orders in such a hostile climate:

> ...let it not be forgotten, that all proceedings with which all the Socialists desecrate the Sabbath, and outrage revelation, invariably open with a lecture on geology, in which the concessions of philosophers are triumphantly re-echoed, as if the foundations of revealed truth were shaken, and the authenticity of Scripture given up. Now, what are ingenious, and, to the geologists themselves, satisfactory explanations, are infidel sneers to the Socialists, and disheartening alarms to the great body of believers. Hence we are tempted to tremble for an edifice where the supporting pillars are thus shaken, and to exclaim with the Psalmist (Ps. xi. 3), "If the FOUNDATIONS (of our holy faith) be destroyed, what can the righteous do?"[4]

Saull's talks were invariably illustrated by fossils from his collection. At Owen's social festivals, Saull would talk geology using his more visually stunning exhibits as a prelude to the dancing.[5] So the museum was regularly signposted. But it was the lecture *in situ*, in Aldersgate Street, in the spacious wine depot, surrounded by crinoids and palms and gigantic saurians, that had the greatest impact. Here the wine merchant's explanation gave the fossil sequence its meaning.

A reporter from William Makepeace Thackeray's *National Standard* visited it a few months after the Bristol talk and was astonished:

> This gentleman has fitted-up the upper part of his house, in Aldersgate street, as a geological museum. As a private collection it is immense, and does great credit to the taste, learning, and liberality, of its possessor. It contains great varieties of all the known rocks, and specimens of nearly all the known fossils. In those of the secondary formation—in which is included the coal measures—the museum is peculiarly rich; and, fortunately for science, Mr. Saull does not place his "candle under a bushel," nor, like a miser, lock up his stores. With a liberality which does him honour he opens his museum to the public every Thursday, at eleven, when all are admitted gratis; no personal introduction is required:

---

4   Johnson Grant 1840, xiii-iv; reproduced in the *Church of England Magazine*, 9 (15 Aug. 1840): 120; and in *NMW* 8 (5 Sept. 1840): 159, where the editor noted that this "seems to have emanated from the attendance of the Editor of the *Church Magazine* at Mr. Saull's lecture in the Institution, John-street".

5   *NMW* 3 (20 May 1837): 235.

it suffices to say, that the love of science, and the desire of improvement,
is the motive, to ensure a hearty welcome.[6]

For Saull, a City liquor trader, an indicted blasphemer, lacking education, 'character', or noble bona fides, a museum of rich merchandise, like an FGS badge, provided his credentials of "taste, learning, and liberality". But the guests, left to themselves, could only stare at indecipherable bones and unfathomable fossils. A wild scatter of perspectives brought from innumerable social angles could end without any value being drawn from the bare bones. As the *National Standard* said, his assemblage would be a closed book, or so much fossil gibberish, "were it not for his lectures". The rocky imprints of palm fronds or the giant saurian limbs were mute witnesses whose story had to be coaxed out. Like a good barrister, Saull interrogated them to get the narrative he wanted. The lecture brought the exhibits alive, made them relate a radical story with a moral. He scrupulously laced the fossil sequence into an imagined 'evolutionary' parade, whose "progress to perfection" justified the Owenite call to social arms.

"When the company are assembled", continued the *National Standard*, "Mr. Saull commences his lecture". He runs through the rocks in time sequence, using modern seascapes to help the mind's-eye, starting with ancient microscopic "zoophytes", like the corals of a modern atoll, building their rocky structures. He then walked, physically and metaphorically, along the shelf, through time, following the fossil stream, "showing, step by step, the geological changes, and the consequent changes in organic nature", working his inexorable way to mankind. 'Lower' animals merged into 'higher' ones, just as the human foetus at first resembles "a kind of worm" but then naturally acquires gills, its skin is soft and naked, "like the mollusca", "after which we successively become fishes, reptiles, birds, and mammalia."[7] Having "passed through all these gradations"—just as the fossil species did—the foetus emerges at birth in its highest form, the human being. Some

---
6   *National Standard* 3 (18 Jan. 1834): 44–45.
7   *National Standard* 3 (18 Jan. 1834): 44–45. Gould's 1977 history of the concept of human foetal stages recapitulating 'ancestral' developments concentrates on the social implications being drawn from them later in the century. Much earlier, however, the levelling implications served London's republicans, who were probably only too happy to portray the embryonic Gulielmus Rex emerging from a reptile.

radical comparative anatomists believed this was literally, or, at least, analogically so: that we were fishes, then reptiles, then lowly mammals in the womb. We were then born as the 'highest' form, humans—the embryo literally encapsulated our ancestry, and recapitulated our fossil ascent. And although the great Tory comparative anatomist Richard Owen at the College of Surgeons was to quash the very idea as an absurdity,[8] Saull seems to have subscribed to it.

The proprietor, said the reporter, has arranged his "temple" so that all these natural "facts are demonstrated". But being a socialist he did not end there. The report cryptically moots mankind not necessarily being the end point, however the paper conspicuously failed to follow Saull on the final Owenian process, perfecting the social man.

Geological Education

> ... the revelations of GENESIS and GOSPELS are, to stationary blind faith, precisely what the revelation of GEOLOGY and PHYSIOLOGY, are to progressive understanding, and ... these two sciences become, therefore, indispensably necessary to all rational schools from which the Genesis and Gospels are to be excluded in order that pupils may be taught rationally, instead of mystically, WHERE THEY REALLY ARE, AND WHAT THEY REALLY ARE.[9]

Robert Owen's plans for rationalist primary education in the city were now coming to fruition.[10] Education of any description was a desideratum in England at the time. It was estimated in 1835 that of the four million children in the country under 15, half received no education whatever.[11] Saull chaired the meetings in Owen's Institution in Charlotte Street to design this schooling for the co-operators' infants. He "denounced the degrading effects of the miscalled education given at our charitable institutions and public seminaries, and avowed his intention of devoting his attention and aid to the proposed school for

---

8 Sloan's introduction to Richard Owen 1992, 62–63; Desmond 1989, 52–53, 58, 337ff.; Evelleen Richards 1994, 392–404; 2020, ch. 4, has exposed the ramifications of Owen's assault on the concept.
9 *NMW* 6 (5 Oct. 1839): 789–91.
10 J. F. C. Harrison 1967. On Owen's New Lanark school curriculum: Hutchison 1835, 511.
11 *MM* 23 (4 July 1835): 271.

the children of the Socialists." And, as always now, he "announced his intention of bequeathing his valuable museum for the benefit of the rising generation" as the visual learning element in this 'rational' process.[12] Such learning could be really useful, not least for breaking the gentry's grip and stopping them from training stooges for livery or service, or "to wait on them behind their carriages".[13] But more, by giving "the poor man's child that rational education" he or she "could be rendered, moral, intelligent, virtuous, and happy". Individuals had to play their part. In his Bristol monkey lecture, Saull encouraged mothers in the audience to teach their children only the "known facts" in order "to form the good, benevolent, and best character". In the Owenite/Enlightenment ideal, virtue was to be gained by obeying Nature's laws—science, in a word. The young mind, following nature, would be healthier, happier, and more sympathetic, rather than oppressed and fearful—it was Holbach's prescription, still being promoted in the latest Carlile edition, to search nature "for motives suitable to infuse into the human heart propensities truly useful to society".[14]

The infidel sects, carried away by the revelations of geology and physiology, could become as wildly messianic as the placarding Baptists outside their halls. For one believer,

> This education and training will effectually supersede the necessity for any human laws, opposed to Nature's laws; for Nature's laws will alone direct man, and insure his happiness. Judges, and all law-officers, prisons, and all punishments, will be useless, and will cease to exist.[15]

Millenarian Owenites were promiscuously mangling judicial and natural law, and, even then, envisaging the latter as an edict 'governing' matter.

This overriding concern to infuse morality alongside political dissidence, to keep the heart pure as the barricades rose, was universal in deist and anti-Christian halls. Each of Saull's mentors had emphasized as much. Some had tried, in their way, to put policy into practice. Carlile had set up a School of Free Discussion in 1829, based

---

12 TS, 27 Dec. 1833, 4; Crisis 3 (28 Dec. 1833): 144; (4 Jan. 1834): 150–51.
13 Crisis 3 (4 Jan. 1834): 150. Saull also helped later to set up a day school in Whitechapel (NMW 12 [9 Dec. 1844]: 192).
14 Holbach [Mirabaud] 1834, 1: 280.
15 NMW 2 (26 Dec. 1835): 65.

on "free, fair, and fearless" study of the works of Tom Paine, the atheist and philosophical anarchist William Godwin, and the utilitarian Jeremy Bentham, something he thought "may ultimately produce a lever to move the intellect of the earth". But at 6*d* a Sabbath, or 10*s* a quarter, attendance was expensive, even if it helped Carlile keep his own rational head above water.[16] Charlotte Street was continuing the tradition with a co-operative slant and a bit more cash. Even this school was 4*d* weekly for the under-8's, but lessons were on two nights a week and parents could pay half the fees in labour notes.[17] Over one hundred pupils attended this blame- and merit-free co-operative school. But mixing ages and sexes caused chaos, as did the cold winter without heating. Therefore, it was re-rationalized and, because it interrupted the Bazaar,[18] transferred to new premises in Millbank as the Westminster Rational School and General Scientific Institution.[19]

Saull stated that he was bequeathing his "valuable museum for the purpose of education", implying that children could examine "the works of nature" first-hand.[20] Like many Owenites, he saw geology as an integral part of a wider rational curriculum. The educational goal was a geological foundation on which Owen's environmental-conditioning superstructure could be erected. Some actually rendered this axiomatic, making geological revelation "indispensably necessary to all rational schools" in order to teach infants how they got here.[21] Thus spoke the pseudonymous "Student in Realities", a common correspondent in socialist weeklies. Under the running head "BEGIN BY THE HISTORY OF THE EARTH", the critique wanted schools to

> Begin by mineralogy, geology; by the *history of the earth before man's appearance thereon;*– make them acquainted with all substances in the same order in which they were successively generated in nature, and

---

16   *Lion* 3 (23 Jan. 1829): 122–24; costs: 3 (2 Jan. 1829): 30 (9 Jan. 1829): 64. Carlile had tried to establish a Sunday school at his Fleet Street premises, where adults and youths could unlearn the religious mischief they had been taught. That, too, had been 10*s* a quarter: *Lion* 1 (21 Mar. 1828): 361; (28 Mar. 1828): 385. It had closed in June 1828. McCalman 1988, 189–90.
17   *Crisis* 3 (4 Jan. 1834): 150.
18   *Crisis* 3 (11 Jan. 1834): 155; (22 Feb. 1834): 216.
19   It survived into 1835–36 (*NMW* 1 [20 June 1835]: 270) after which it became a Hall of Science, then Westminster Mechanics' Institution.
20   *Crisis* 3 (4 Jan. 1834): 150.
21   *NMW* 6 (5 Oct. 1839): 789–91.

> conclude with the tottering selfish politics, and *contending* psychological *systems* of occult knowledge, still called indiscriminately *"true religions"*.[22]

This was the trajectory of Saull's lectures. Such a geo-political hammer was deemed essential to chip away at the rival Rock of Ages, which resulted in it being adopted in Owenite curriculums far and wide. And the pick was put into the hands of both boys and girls. The London Working Men's Association proposed School Committees be set up nationally ("elected by universal suffrage of all the adult population, male and female") and that the teaching should integrate geology, but *they* saw the science restricted to "High Schools", that is, nine- to twelve-year-olds.[23] The Owenite communities, by contrast, taught it at all age levels. Stockport's and Rochdale's "Rational day schools" were soon setting exams for boys and girls in the "sciences of astronomy, geography, botany, geology, and physiology".[24] Practicals, too, were an important part. They might involve day trips: the youngsters would be driven in carts, the older boys and girls hiking, tricolour flags waving, cornets playing, and having reached some rocky outcrop they would listen to "addresses on geology and the features of the country".[25]

The London Owenite caucus, unimpressed by the increasingly watered-down Westminster Rational School experiment of infant-instruction, started another school in-house. Here, the curriculum took in the traditional subjects, "Writing and Arithmetic; English Grammar and Composition", and the sciences: chemistry, geology, physiology, and astronomy. Only now there was an addition to the list, showing that the hottest of contemporary topics was galvanizing the Owenite classroom: "Electricity".[26]

Long before science was professionalized and constrained around standardized academic and accreditation procedures, and before state-regulated curriculums set subject norms, the cultic niches provided havens not only for dissident sciences but contemporary rages. So, it is no surprise to find the latest, the Owenite star T. Symmonds Mackintosh's "Electrical Theory of the Universe", firing up the school children.

---

22 Student in Realities 1836–37, 254–6. This was extracted in the *NMW*.
23 *NMW* 4 (6 Jan. 1838): 82.
24 *NMW* 11 (17 Sept. 1842): 99; (17 Dec. 1842): 203.
25 *NMW* 12 (22 Jul. 1843): 32.
26 *NMW* 7 (30 May 1840): 1262–3.

Although his grandiose unifying scheme had first been published in the *Mechanics' Magazine* in 1835, "Mack", as he was called, was already 'one-of-us'. He was a former weaver, blooded as another Carlile shop stalwart,[27] an inventor, promoter, engineer, who had come to embrace Owenism passionately. Not surprisingly, having sustained trial by fire in Carlile's shop, he was deeply anti-clerical although clearly deistic. To him, the solar system was a vast electrical machine: the Sun, a giant conductor crackling with electricity, bathed and held the planets in an electrical field. The earth was suspended in an electric medium, like a brass ball held on silk next to a generating machine. Such a spectacular debunking of Newton's gravitation caused huge controversy, making it clap-trap to some, and breathtaking to others. But it satisfied the socialist need for an all-encompassing, home-spun, anti-Newtonian, anti-occult theory of everything. With spectacular shows the talk of the town, as Iwan Morus has so entertainingly demonstrated in *Frankenstein's Children*, Mack's electrical gadgetry satisfied the Owenite mood. Mackintosh frequented the usual haunts, an "animated speaker, with a faculty for vivid and humorous scientific illustration".[28] He told them at Watson's Mechanics' Hall of Science, in a six-lecture series in 1836, running concurrently with Saull's lectures, that electricity was the most potent agent in the physical world, and "the ultimate source of motion".[29] He could be found, too, at other Saull venues, at the Mutual Instruction Society, Great Tower-street, and the Rational Institution in Curtain Road. And being 'one of us' there was decided chauvinism in the *New Moral World*'s re-printing the lectures.[30] Birmingham's main seller of the "Great Unstamped", James Guest, publisher of the *New Moral World*, got out threepenny numbers of the *Electrical Theory* in 1838, making it the first book publication. At this juncture, "Mack" became a social missionary, taking his theories to local branches in Manchester, Salford, Oldham, and Liverpool, where he drew enthusiastic audiences.[31]

---

27   Anon 1858, 62–63. The theory was first serialized in *MM* 24 (3 Oct. 1835): 11–13. Mackintosh's patents included cooling and condenser improvements for the steam engine.
28   Holyoake 1906, 1: 235.
29   *MM* 26 (22 Oct. 1836): 48. He was talking here again in 1838: *NMW* 4 (6 Jan. 1838): 85.
30   *NMW* 3 (20 May 1837): 239; 4 (28 Oct. 1837): 5; *PM* 1 (13 May 1837): 232.
31   Morus 1998, 135–9; *NMW* 5 (24 Nov. 1838): 80.

"Electricity" entered the socialist-school curriculum because it ticked all the boxes. This was the hottest science, providing demystified explanations of origins and actions, and was moreover invented by a social missionary. This home-grown Electrical Theory explained everything, from the motion of planets to the growth of plants. Electricity could possibly even generate life, if it was true, as Sir Richard Phillips reported in 1837, that a voltaic battery continuously charging a piece of Vesuvius rock had produced a horde of tiny bristly 'insects'.[32]

Humans, anatomically wired with current-sensitive nerves, were themselves seen as electric machines. At least, it was more empowering to look to this sort of knowable and controllable physics than to trace causes "to sources beyond our knowledge and above our control".[33] And, since the sun and people were running down on charge, Mackintosh argued that it was better to seek "bread now rather than cake tomorrow", in Morus's words. "Time therefore, for revolution on earth since there was no hope of heaven."[34] 'Mack' would push the moral in talks. He toed the Owen line that humans shaped by circumstance should not be judged, and he attacked Christianity for downplaying man's social duty to man in the here-and-now in favour of the Kingdom to come.[35]

The children were in safe, rational, and self-empowering hands. And they were being crammed with a double dose of geology, as 'Mack' used it extensively to support his all-encompassing theory. Perhaps, though, they were perturbed to find that the Earth had once had five moons, but that four with depleted charges had crashed into it, their wrecks causing the major mass extinctions through history. And the remaining moon, with its decaying orbit, would share the same fate. The apocalypse was not Millennial, but Lunar.[36] Even Saull might have been staggered at that.

32　*Annals of Electricity* 1 (Apr. 1837): 242–44. Richard Phillips had visited Andrew Crosse in 1836 to inspect his electrical apparatus which would generate these 'insects' (*MM* 26 [8 Oct. 1836]: 13–14. Mackintosh was already citing Crosse: *MM* 26 (22 Oct. 1836): 48. Colonel Macerone had his own galvano-electric theory and discussed its use in generating "animalized rudiments" (Macerone [Maceroni] 1837, 19; Macerone [Maceroni] 1848, 1: 143, 362, 412–13). For the context of Crosse's supposed electrical biogenesis see J. A. Secord 1989; Morus 1998, 110; 2011, pt. 2.
33　Mackintosh 1846, 361.
34　Morus 2011, 78–79.
35　Mackintosh [1840], 96.
36　*NMW* 4 (28 Oct. 1837): 6; Mackintosh 1846, 155–56, 224, 228.

13. A Purpose-built Museum — 1835

Saull's museum and lectures were a major means of propagating the geological naturalism that lay behind the Owenites' alternative education. Hence his offers to "open the great book of nature to your view" personally at Aldersgate Street, which he made to every audience. These included the Owenites at the Institution of the Industrious Classes, the Rotundanists buoyed up by "science and liberty", union activists, *Mechanics' Magazine* readers, provincial co-operators, French *savans*, and so on. All were invited, "rich and poor", women and children equally with the men.[1] And here they were to have a hands-on introduction to the hundreds of sequenced fossils in his gallery and hear them described "as facts much to[o] hard for the parsons".[2]

Only now there were more than hundreds of fossils, many more. Sir Richard Phillips talked of "ten thousand". By 1835, when Sir Richard was commending Saull's museum in his *Million of Facts*, the exhibition had grown huge. Saull's astronomical mentor and Jacobin compatriot was one to know, as a welcome visitor who could spot expensive fossils of the "highest interest".[3] A *Mining Journal* reporter called the fossil repository "very extensive, consisting of many thousand specimens, and I found it particularly rich in the department of *fossils*, of which it possesses a very perfect and valuable series. The whole is arranged stratigraphically, in a very instructive and judicious manner". The collection was clearly a visually stunning display, being laid out in expensive "glass cases,

---

1   *TS*, 27 Dec. 1833, 4; *MM* 19 (25 May 1833): 117–18; *Crisis* 3 (5 Oct. 1833): 38; (28 Dec. 1833): 144; (4 Jan. 1834): 150; *Isis* 1 (3 Mar. 1832): 59–60; *NS*, 18 Sept. 1841; Saull 1836, 30; *Bulletin de la Société Géologique de France* 15 (1835): 67.
2   *NS*, 31 Oct. 1846, 3.
3   R. Phillips 1835, 293.

[which] have the advantage of being easily commanded by the eye, and are much more conveniently seen than by the ordinary arrangement in drawers." Saull had been on a phenomenal spending spree through the early thirties and had built up one of the country's premier private collections. His long purse from wine profits meant that the museum had now become "Immense". It was bursting at the seams in Aldersgate Street. The *Mining Journal* complained that the "extensive series of well preserved *diluvial bones* completes the collection", but that their crowding now ruined the overall effect, only to be told by Saull that this would soon "be obviated".[4] He was about to expand.

Not only was the warehouse overflowing, but courtyard stables were hardly a salubrious siting for an exhibition assuming national prominence. So, to accommodate the burgeoning exhibits, in 1835 he substantially rebuilt the complex to house both the museum and wine depot. Although no contemporary description survives, the new building must have been large. The sales advert for the Champion Commercial Hotel, which bought the site in later Victorian times, lists a "substantially-built Warehouse in the rear, four storeys in height, being No. 6 Falcon-street, conveniently occupying an area of 4,700 feet".[5] We do know that the assemblage now occupied two floors, with an upper and lower gallery. The complex with its "stratigraphically arranged" exhibits was then re-launched, with British and French press adverts appearing from October 1835.[6] Coincidentally (or not), the wine business partnership of W. D. Saull, J. Castle, and T. Saull was dissolved in February 1835, with Castle leaving.[7] Perhaps this gave Saull greater control and freer rein to rebuild and devote more space to display his fossil assets.

He certainly needed the shelf space, given the sudden influx of gigantic slabs, some of which required multiple workmen to move. Newly identified giant reptile fossils were being washed out of Wealden rocks on the Isle of Wight. Saull was quickly on the spot, becoming a

---

4   Mining Journal and Commercial Gazette 1 (7 Nov. 1835): 83. "Immense": *National Standard* 3 (18 Jan. 1834): 44–45.
5   *The Era*, 19 Mar. 1887.
6   London and Edinburgh Philosophical Magazine 7 (3rd Series) (Nov. 1835): 431; *MNH* 8 (Dec. 1835): 679–80: *Bulletin de la Société Géologique de France*, tome 7 (1835–36): 49.
7   *Courier*, 14 Feb. 1835, 4; *TS*, 14 Feb. 1835, 4; *MC*, 14 Feb. 1835. 1.

frequent steamboat visitor to the island, and, by November 1835, he had stocked the museum with "many portions of *saurian animals*",[8] probably from this source. A year later, as Hugh Torrens points out, Saull and the Spitalfields silk manufacturer Thomas Field Gibson were arranging to exhibit parts of an *Iguanodon* femur from the Isle of Wight, "nearly as large as Mr. Mantell's", at the Geological Society. Saull, Mantell, and Gibson were evidently well acquainted by 1835,[9] and Gibson's country house at Sandown on the Isle of Wight seems to have been their base.

Saull's deep pocket was now stretching to foreign fossils. After the sale of Big Bone Lick mastodon bones from Kentucky by Stevens's auction room in 1836 (with tusks or jaws selling for twelve or thirteen guineas[10]), the tourist guides reported seeing some in Aldersgate Street. The *Stranger's Intellectual Guide to London* found, among Saull's "extensive and valuable" mastodon collection, "a scapula, which must have measured, when perfect, five feet long by three feet wide; with vertebrae fourteen inches in diameter".[11] Others spotted fossil horse teeth from the Kentucky site.[12] With proliferating exhibits and so many hefty fossils, the museum was apparently expanded yet again in December 1838. A syndicated review of the refurbishment now appeared in periodicals as disparate as the *Gardeners' Gazette* and *Court Gazette*:

> The extensive and valuable collection of Mr. Saull, F.G.S., in Aldersgate-street, was, on Thursday [27 December 1838], after receiving a great many additions, re-opened to the public. It is satisfactory to add, that the wishes of the liberal and benevolent proprietor were gratified by a very numerous attendance. There is not, perhaps, so perfect a school of geology in the metropolis, or one to which the attention of the young geologist may more advantageously be directed than to the one in question.[13]

---

8   *Mining Journal and Commercial Gazette* 1 (7 Nov. 1835): 83. On travel to the Isle of Wight and the Portsmouth steamboats: M. Freeman 2004. How important these cliffs were becoming was shown by the Royal Polytechnic Institution displaying a huge scale model of them in 1843: *Royal Polytechnic Institution Catalogue* 1843, 85. The following year they were eulogized in Loudon 1844.
9   Torrens 2014, 670; J. A. Cooper 2010, 63.
10 Mantell 1846.
11 A. Booth 1839, 122; *Morning Post*, 31 Dec. 1841. On Big Bone Lick see Matijasic (1988), and H. D. Rogers 1835.
12 Karkeek 1841c, 701.
13 *Gardeners' Gazette*, 29 Dec. 1838, 827; *Court Gazette*, 29 Dec. 1838, 614–15.

The bulging, purpose-built museum drew more praise from the press. By 1841, Saull's exhibits vastly exceeded 10,000, making this the "principal museum of geology in London".[14] Within a decade of its founding, the collection was being hailed as "one of the first in the metropolis for number of specimens and the excellence of their arrangement".[15] And thus it was trailed extensively in the newspaper listings for the capital's holiday attractions, suggesting that many visitors must have been given a Christmas or Easter conducted tour. Some scientific visitors were now mooting that it could actually be "the largest private collection of fossil remains in the kingdom".[16] In truth, one wonders whether the selling point, "the largest private...", repeated in so many press reports and tourist guides, did not emanate from within the museum itself. Was it a self-aggrandizing selling-point raised as a lure?[17] Another draw was the continual influx of new fossils. The arrival of shipments was announced by press releases, which in turn increased visitor numbers. Typical was the *Morning Post*'s note on the "numerous attendance" after the museum received one particular batch of "valuable additions".[18]

At some point in the thirties these additions were truly spectacular, notably the trove of *Iguanodon* bones. Mantell had been the reptile's discoverer. He had first unearthed a thigh bone twenty-three inches in circumference from Tilgate Forest in Sussex, christened it *Iguanodon*, and scaled up using a lizard model to suggest a reptile "from sixty to a hundred feet in length!" Its teeth were massive, as were the vertebrae and claws, and it appeared to have a horn on its snout.[19] By 1839, it was reported that Saull had "The largest known collection of the bones of iguanodon, from the Isle of Wight, consisting of humeri, numerous vertebrae and ribs of this stupendous animal, whose claw alone must have measured seven inches."[20] He also had a huge, intact, and complete sacrum (made up of the fused vertebrae of the pelvic region, that part of

---

14 *Courier*, 12 Apr. 1841, 3.
15 *Courier*, 27 Dec. 1841, 1.
16 Karkeek 1841a, 73; 1841b, 175.
17 Again, it was called "the largest private Geological collection in the United Kingdom" in *NS*, 31 Oct. 1846, 3.
18 *Morning Post*, 31 Dec. 1841.
19 Mantell 1831, 184; Norman 1993.
20 A. Booth 1839, 122; *Morning Post*, 31 Dec. 1841. In addition, he possessed an ilium (Mantell and Melville 1849, 293; Mantell 1851, 270).

the spine supporting the hip girdle).²¹ It was unique, nothing like it was known in any other collection, or in any other reptile.

Tellingly, there was one expert who possibly did not realize the importance of this low-life museum until late in the day—the grave young Anglican Richard Owen, the new Hunterian Professor at the Royal College of Surgeons. This is surprising on two counts. Firstly, press reports in 1839–41, if nothing else, should have alerted the rising star of palaeontology to this horde. The Peelite *Morning Post*'s report of acquisitions in December 1841 (which reiterated that Saull had "the largest known collection of bones of the Iguanodon, from the Isle of Wight"), would have left the Peelite Owen in no doubt of the museum's importance.²² Geology primers themselves talked of the enormous size of the bones in Aldersgate Street.²³ And secondly, Owen, the darling of the Oxford divines, had actually been tasked by their British Association for the Advancement of Science to draw up a report on British Fossil Reptiles. He was on the lookout for saurian fossils, and had consequently "ransacked" every "collection, public and private". What he had not found was a good *Iguanodon* sacrum. Mantell had been working on a partial skeleton from a Maidstone quarry in 1834, but the blasting had fragmented the pieces, and there was next to nothing of the spine in this crucial sacral region.²⁴ Only *after* delivering his verbal report to the BAAS, somewhere between September 1841 and April 1842,²⁵ did Owen

---

21 This sacrum was found on Brook Point (now Hanover Point), Isle of Wight: Mantell 1847, 319. It came into the museum some time between late 1836 and 1840 (Torrens 2014, 671). Karkeek (1841a, 72) saw "several pelvis" in Saull's museum when he visited in 1840.
22 *Morning Post*, 31 Dec. 1841.
23 G. F. Richardson 1842, 402.
24 Norman 1993.
25 Torrens 1992, 1997, 2014, 671. See also Dear 1986; Desmond 1979. Richard Owen 1841 [1842], 127–41. While it is clear that Owen doctored his paper to include the "Dinosauria" after he gave his verbal report to the BAAS, we do not know when he first visited Saull's museum. It was evidently before 1839, for, in that year, Owen mentioned having seen the collection, while searching for ichthyosaurs and plesiosaurs to complete the first part of his "Report on British Fossil Reptiles" (Richard Owen 1840, 44). So presumably he returned and spotted the *Iguanodon* in late 1841–early 1842, when he came looking specifically for these fossils while writing up his second report. What makes this likely is that Owen (1841) read a paper on *Cetiosaurus* on 30 June 1841, yet it did not mention Saull's Wealden specimens, which suggests that Owen had also yet to see the *Cetiosaurus* in Saull's collection. This reinforces the idea that he returned looking particularly for Wealden cetiosaurs and iguanodons late in 1841 or early 1842.

finally catch up with the prize specimen in the "well-stored museum of J. [sic] Devonshire Saull". The mistake is telling: the fact that the grave Church-and-Queen anatomist did not know the correct initial suggests a social distancing. In fact, I doubt there was much beyond formal contact between the haughty Tory and the irreligious socialist. Owen might even have been reluctant to step into the socialist cess-pit, knowing all that it stood for. The explicitly anti-transmutationist and anti-materialist conclusions to so many of Owen's major papers at this time suggest that personal intimacy would have been very difficult, with contacts kept at a formal level.[26]

Richard Owen was an exceptional and prolific fossil anatomist. He revealed that this sacrum was composed of five fused vertebrae, not the two typical of living reptiles. This sturdy, fused bone, supporting the pelvic girdle, became central to his reconstruction of these fossil giants. Rather than Mantell's long, stupendous lizards, Owen shortened the iguanodons and megalosaurs by making them stand erect like huge mammals. Then he separated them off as a unique group of advanced reptiles, and in 1842 he christened them 'dinosaurs'.

Press interest suggest that Saull's own strange views were becoming better known. Hence the syndicated *Gardeners' Gazette* puff for Christmas visitors to London in 1838: it describes a

> very rich and rare museum of specimens, placed in a commodious and well-lighted building, erected for their reception at the back of his premises. The great variety of specimens are displayed to the utmost advantage, from being arranged in the order of the deposition of stratas— from where there was almost apparent chaos, through the gradations of animated species, to the period when the earth became a fit receptacle for man. The museum is open to the public every Thursday morning, at eleven, when Mr. Saull attends his visitors, and describes the gradations, according to his arrangement of the numerous specimens.[27]

"Gradation" was a keyword, used by Sir Richard Phillips in the 1832 essay republished by Saull. It was caused, in Phillips's understanding, by the "progressive evolutions" of life,[28] which Saull's infidel circle saw as an unfolding or growth from 'lower' to 'higher' animals and plants.

---

26 Desmond 1985b, 1989.
27 *Gardeners' Gazette*, 22 Dec. 1838, 810–11; *Court Gazette*, 22 Dec. 1838, 604.
28 R. Phillips 1832a, 51–53, 70.

Indeed "gradation" would shortly move from keyword to buzzword, standing for something akin to what would later be called "evolution" (a transmutation of one form into another). It might have meant that in the *Gazette* story. Certainly, the reporter returned after Christmas and filed a more revealing account. Saull obviously explained his views, perhaps monkey-to-man and all, but the perplexed hack was not about to pass them on to innocent readers:

> Apart from all theory, the arrangement is a very satisfactory one, taking the successive depositions of the earth's strata in the order in which they occur in nature; bearing evident marks of a succession of agencies before the world was fitted for the comfort and reception of man. There is no science which leads to such speculation as geology, or in which the mind may so soon lose itself in its bewilderment, and Mr. Saull's theory is as compatible and consistent as that of any other. The various agencies may have been millions of years at work, and the gradations have been produced by many successive alterations of place and structure, without impugning the common received records of the formation of the world.

Despite sidestepping "Mr. Saull's theory", the journalist left a positive impression of Saull's dioramic display of advancing ancient life, even if he was a bit befuddled:

> The fossil remains of animals are historically curious, from the period when the immense saurians reigned the undisputed masters of the seas, to where the earth became fitted for living animals [mammals]. Here again we perceive the huge mammoth, the elephant, and rhinoceros, full three times their present natural size. In the collection of fossil organic remains of our own country, Mr. Saull has shown a most praiseworthy and persevering industry. The cabinet of animal remains is, perhaps, the most extensive of any in this country; and shows that the elephant, rhinoceros, hippopotamus, tiger, hyaena, and boar, were its inhabitants; whilst the very numerous specimens of tropical vegetation, the cocoa, the tamarind, the coffee, and various spices, seem likewise to prove, that they once flourished here. Mr. Saull attends his visitors personally every Thursday morning, at eleven o'clock.[29]

The hack was not alone in scrambling these novel vistas, so new and confusing. Some religious rationalizations of these "hideous"

---

29   *Gardeners' Gazette*, 29 Dec. 1838, 827; *Court Gazette*, 29 Dec. 1838, 614–15.

creatures, supposedly living among pre-Adamite men, could miss the chronological mark completely:

> Why may we not suppose that some of the hideous animals, which the comparative anatomists have found for us, may have been more suitable to the very corrupt state of the ante-diluvian population, than to ourselves, corrupt as we are; and been suffered to co-exist in the way of correction and punishment, to annoy, plague, harass, and alarm, those sinful, incorrigible generations of men?[30]

Then there were the visitors who *did* grasp the chronology. The museum was now large and important enough for famous faces to be seen, geological and otherwise. One of Saull's closer geological confidants was another outsider, Gideon Mantell. His books and chauvinistic talks on his *Iguanodon* and the other saurians "of a most appalling magnitude" did so much to promote the idea of an "age of reptiles". It was becoming apparent that monstrous lizard-like creatures had once dominated the earth. They were, in Mantell's words, "the Lords of the Creation"—even if this "romantic doctrine" was too alien and unorthodox for many and dismissed as more "infidel" nonsense by scripturalists.[31] Saull and Mantell had exchanged fossils and visits through the 1830s, constantly checking one another's new exhibits. Mantell would breakfast at Saull's, and take tea or sleep overnight when he was in town. They would make long-haul visits together to see other collections—up to Stratford, for instance, to examine elephant fossils from Ifield (there were "many fine elephantine remains" in Saull's collection, according to Mantell[32]). Or to visit quarries, such as the commercial quarries in Maidstone, Kent, which were exposing more *Iguanodon* bones.[33]

There was clearly a camaraderie between Saull and Mantell, which was shown as Saull brought his priceless fossils to illustrate Mantell's London lectures on Britain's great reptile past.[34] Not only did Saull allow Mantell to take casts of his prize *Iguanodon* sacrum, but to chisel away some covering rock. Inside, Mantell claimed to have exposed six fused

---

30   Nares 1834, 158–60.
31   Mantell 1831; disbelief in an "age of reptiles": *Christian Observer*, July 1839, 400–401.
32   Mantell 1838, 1:131; J. A. Cooper 2010, 63.
33   D. R. Dean 1999, 132–34.
34   *Literary Gazette*, 1669 (Jan. 1849): 24.

vertebrae, not Richard Owen's five. Mantell was careful to publish the first illustration of this increasingly important fossil as a riposte to the haughty Owen, whom he had come to hate.[35] A rather snippy Owen then got his own back by publishing his own plates of Saull's sacrum.[36] Personal rivalries were now pushing Saull's fossils to the fore.

Far less familiarity was shown by Mantell's *bête noire*, the pious Richard Owen, whose contacts with the infidel republican were undoubtedly restricted to professional visits. The ambitious Owen abominated transmutation. Patronized by Oxford divines, Owen proved his worth by a masterly study of adult chimpanzee anatomy in 1836. This emphasized their bestial, heavy-jawed physiognomy and, by so doing, distanced apes from men to discredit the transmutationists' claim that we came from monkeys. And look again at the underlying worth of Owen's ideological *coup de grâce* in 1842, the 'dinosaur'. That an indicted blasphemer actually *owned* the prime specimen on which Owen based his new "Dinosauria" could only have re-emphasized the materialist threat for the young Anglican. He now used Saull's fossils among others to reshape Mantell's long, sprawling, lizard-like monster into an upright-standing, rhinoceros-like, crown of the reptilian creation. This was again used to scotch any belief in the inexorable, upward, self-propelling progress of life, his point being that reptiles had degenerated from their majestic dinosaurian heights.[37] Owen was poised to become the premier British palaeontologist, the "English Cuvier", and he was shortly to start monographing his way through the *History of British Fossil Reptiles* (ultimately collected into a monumental four-volume set). Publishing the descriptions of Saull's fossils was mutually beneficial: Owen had a ready source of fossils, while the museum acquired kudos,

---

35 D. R. Dean 1999, 132–4; J. A. Cooper 2010, 126, 131, 153; Mantell and Melville 1849, 275–76, 300 Plate XXVI; Mantell 1851, 303. Mantell's illustration of Saull's sacrum is also reported in *Literary Gazette* 1681 (Apr. 1849): 259.

36 Richard Owen 1854, 11, 13–14, tab. 3. This sacrum from Saull's museum is now BMNH 37685.

37 Saull's flagrantly anti-Christian 'evolutionism' thus raises the possibility of a second target aimed at by Owen at the end of his BAAS paper. Owen had explicitly fingered his rival, the radical transmutationist at University College London, Robert Edmond Grant: Desmond, 1979; 1989, 321–27; Torrens 1997. Richard Owen 1841 [1842], 196–204. But Saull's own obnoxious materialism could have further spurred Owen on to draw damning anti-transmutatory conclusions in his paper.

the more so, Saull "rejoiced", when Owen turned up new species.[38] But there must always have been an elephant in the room, even if Victorian decorum rendered it invisible. Any personal contact between the Carlilean atheist and haughty Anglican must have been mediated by the fossils, with "Mr. Saull's theory" and its socialist underpinnings studiously avoided.

Saull's fossil trove provided rich pickings for Owen. In Aldersgate Street, Owen found more evidence of the reptile fauna of ancient tropical Britain. From the Isle of Wight's south coast came sea-rolled vertebral fragments of a narrow-snouted, gavial-like crocodile (*Streptospondylus*), and a new species of the recently-discovered colossal saurian, called by Owen evocatively *Cetiosaurus* ("whale reptile"),[39] as well as the remains of toes, hip, and spine of the *Iguanodon*. These were enigmatic creatures, and quite startling for many. No one was better equipped than Owen to make sense of the fragments, or to show how distinct many of the reconstructed reptiles were from anything on the planet today. In Aldersgate Street, he also distinguished bits of *Megalosaurus* ("giant saurian") from Oxford's rocks, and the vertebrae of a twelve-foot constricting snake (*Palaeophis*) from the later London Clay deposits of the Isle of Sheppey, at the mouth of the Thames.[40]

As Owen and Mantell grew increasingly antagonistic and tussled over interpretations of the *Iguanodon* sacrum, the gainer was Saull's open museum. Mantell, piquing the public's interest with his popular books, announced that "As Mr. Saull, with great liberality, throws his museum open to visitors every Thursday after mid-day, this unique fossil can be seen by any person interested" in the conflicting views of these giants.[41]

It meant that the museum was a port of call for the geological gentry as much as the rough-hewn radicals. As the proprietor, Saull mixed with the learned elite, and in return he was invited to *soirées*. He must

---

38   W. D. Saull to Richard Owen, British Museum (Natural History), Owen Collection, 23: ff. 112–15.

39   On its first discovery by navvies working on the London and Birmingham railway in 1836–37, whose managers undoubtedly sent him specimens, see Richard Owen 1841; M. Freeman 2001, 59.

40   Richard Owen 1841 [1842], 92, 94, 109, 127, 129–30, 135, 141, 180; *Streptospondylus* is also reported in Richard Owen 1842. Owen's Royal College of Surgeons had a plaster cast of the *Iguanodon* toe bone in Saull's collection: Royal College of Surgeons 1854, 29–30. Rieppel 2012, 2015, on the use of fossil casts.

41   Mantell 1851, 269.

have been the only socialist and atheist, once indicted as part of Robert Taylor's nest of "vermin", to move among the swells at Charles Lyell's parties. He was known to be at one in May 1835, the same month that he was making stirring speeches at Robert Owen's Institution demanding the repatriation of the Tolpuddle Martyrs (the Dorchester labourers transported for taking an oath not to accept a wage cut.)[42] Yet, here he was, rubbing shoulders with those staunch Tory fossil fish enthusiasts Lord Cole and Sir Philip Egerton.[43] Esoteric fossils were an unexpected arbitrating medium in the 1830s.

Even higher up the social ladder, the aristocrats who were mercilessly slandered in the pauper press made an appearance in his museum. Enter the Conde de Montemolin, exiled pretender to the Spanish Bourbon throne and "now located, we hope comfortably, in this island of refuge for all distressed notabilities".[44] Accompanied by his suite, the Chevalier de Berard and Colonel Garcimartin, the "illustrious Prince" toured the Library of the British Museum and the Royal Mint "where every preparation had been made for his reception by the principal officers of the different departments".[45] The *Times*'s deferential attitude towards the "Prince" and the red-carpet treatment meted out by the big national institutions, suggests that Saull's museum, also on his itinerary, was now in the big league. The irony of a staunch republican entertaining royalty, even if it was not British, was probably not lost on "Citizen" Saull—not least because he shared the radical hatred of the "loathsome Bourbons".[46] This, after all, was the Saull who had once threatened the King with the fate of Charles I, causing a national uproar and censure in the *Times*.[47] Anyway, the Carlist pretender graced his warehouse with a visit. The entourage included a reporter from the unyielding Tory *Standard* (London's best-selling evening paper, founded to support the Iron Duke of Wellington's efforts to block reform):

> The Illustrious prince examined attentively the arrangements of the strata and the numerous fossils contained in that select collection. The

---

42 *TS*, 20 May 1835, 4.
43 Morrell 2005, 137.
44 *Chambers's Papers for the People* 4 (1854): 24–25.
45 *Times*, 16 Jan. 1847, 5.
46 *Destructive* 1 (1833–34): 334.
47 *Times*, 25 Dec. 1834, 2; 26 Dec. 1834, 2; *TS*, 26 Dec. 1834, 2; 1 Jan. 1835, 4.

Conde listened with great interest to the explanations given by Mr. Saull, which occupied full two hours, and expressed himself much gratified with his visit.[48]

Polite etiquette probably covered what he really thought of the republican's materialist take on life. Court circular events (the visit was also reported in *The Lady's Newspaper*[49]) had rarely seemed so incongruous. One wonders whether the affable Saull muted his levelling monkey talk for the royal ears. Or did he take advantage of the occasion?

Saull's contacts with geology's urban gentry were probably limited. In the 1830s he delivered no papers to the Geological Society, remaining essentially a merchant onlooker. And although he subscribed to the *Reports of the British Association for the Advancement of Science*, and attended the peripatetic jamborees, which visited a different city each year, he exhibited but once, and that only a drawing of his fossil hippopotamus teeth (in Edinburgh, 1834).[50] His 'geological' venues of choice remained the Co-Operative Institutes and Halls of Science, where the lecture profits were funnelled off to Robert Owen's "Missionary Society" or the jailed street vendors.[51] And with the rational education of the underclass his goal, his target audience remained radical sympathizers, even if he was preaching to the converted.

---

48 *Standard*, 5 July 1847.
49 *The Lady's Newspaper*, 10 July 1847, 29.
50 *Report of the Fourth Meeting of the British Association for the Advancement of Science; Held At Edinburgh in 1834* (London, Murray, 1835), xlvii; *Edinburgh New Philosophical Journal* 17 (Oct. 1834): 430: *Literary Gazette*, 922 (Sept. 1834): 637; Richard Owen 1846, 410 for details on the "fine portions of the under jaw, and several detached teeth of the *Hippopotamus major* from the post-pliocene freshwater beds at Alconbury, near Huntingdon" in Saull's cabinet.
51 *PMG*, 24 Nov. 1832; *Crisis* 1 (8 Dec. 1832), 159; *Lancashire and Yorkshire Co-Operator* ns no. 10 (nd [Oct.1832]): 23.

14. Satires on Saull

It was not the geological gentry that took on Saull, it was a fellow socialist. Saull was dogged more than ever by the Rev. James Elishama Smith, the Owenite insider, who was transforming from outrageous Antichrist into a sober peddler of middle-class fodder.

Smith's *Crisis* had reached its own crisis in 1834. The failing Labour Exchange and disorganized unions had caused a slump in sales. Smith, moreover, was falling out with Robert Owen, particularly over Smith's support for the Grand National Consolidated Trades Union.[1] The 'Social Father' had actually voted down a resolution by co-operators to support a 5s minimum daily wage, infuriating trades' union organs like *The Agitator*,[2] and the militants told him in no uncertain terms to keep out of union affairs. Owen looked more to social regeneration and a rebalancing of relations between masters and men, not to strikes and confrontation, and he grew exasperated by the increasing class warfare. Saull, more radical, was sensitive to the turbulent events and supported strikers and the locked-out.[3] "Brother Saull" spoke at Trades' Union anniversaries and invited unionists to the museum.[4] But he never deserted Owen. Smith did. He left in August 1834, and the *Crisis* folded in acrimony. Smith took the publisher of the *Crisis*, B. D. Cousins, with him and they planned a series of new ventures, starting with the *Shepherd* (1834–38)— from which he would take his future soubriquet, "Shepherd Smith". Smith's transformation seemed at first sight astonishing: a reverse transmogrification, apparently completed without any emotional wrench. From Antichrist and millenarian Owenite he remade himself

---

1 Saville 1971, 129-38.
2 *The Agitator, and Political Anatomist* (1831): 8, in HO 64/19, f. 138 (Dec. 1831).
3 *TS*, 25 Dec. 1833, 4; *PMG*, 28 Dec. 1833; *People's Conservative* [*Destructive*] 1 (18 Jan. 1834): 402; *Pioneer* 1 (28 Dec. 1833): 135–36; Saville 1971, 136–38.
4 *TS*, 22 Apr. 1835, 2.

into a purveyor of (almost) wholesome family fare. He had fairly "done with the Infidels".[5] Now he helped pioneer a new form of family journals, and, with their wider appeal, his penny weeklies took Saull-baiting mainstream.

The *Shepherd* touted a milk-and-water universalism, soft science in snippets, softer socialism, literature, drama, and poetry. The page-one leaders were tellingly called "The System of Nature". Here, the regenerate took on "the infidel, the materialist, and the atheist".[6] Obviously the monkey on Smith's back continued teasing and biting—it seems that Smith could never quite escape the beast. Issue 5 scratched the itch with an article on "The Man Monkey"

> There are oddities in the world, who, being sadly puzzled with the subject of the origin of man, surmount the obstacle all at once, as they conjecture, by supposing him to be a civilized monkey; but like the Indians who support the earth on an elephant, and the elephant on a tortoise, they are left in the lurch after all their heroism in leaping over the ditch.

Monkey origins were rubbish, given that every species was an "original", and talk of "an effort of nature" was absurd. This was a rehash of his *Crisis* critique of Saull's speech, suggesting that perhaps the *Shepherd* was not such a jump after all.[7] But it was Smith's next venture, the *Penny Satirist* (1837–46), which mocked Saull before the widest audience. 40,000 sales a week were not unknown for this sense-and-skit periodical, catering mostly to the self-improving poor (hence its common paper and coarse woodcuts). Not that the Countess of Leiningen, Queen Victoria's sister-in-law, had not been spotted "with the *Penny Satirist* in her hand", so its reach could extend upwards.[8] The rag joined John Cleave's *Penny Gazette of Variety and Amusement*, itself a smorgasbord of fiction, farce, anecdote, moralizing, science, and street politics, with a rough political caricature on the title page as a lure—mid-brow entertainment for the family in short—and reaching a swathe of the artisanal and shopkeeping classes "who desired not study but amusement".[9]

---

5   Saville 1971, 138.
6   *Shepherd* 1 (1 Aug. 1835): 386. McCalman 1992, 64.
7   *Shepherd* 1 (27 Sept. 1834): 40; 2 (15 Feb. 1837): 33–35.
8   W. A. Smith 1892, 167–8; Latham 1999, 126. Maidment 2013 on the vitality of these cheap comic woodcuts.
9   J. F. C. Harrison 1961, 30.

The editor's moral injunctions were hardly different from any other family paper, and the goal was to Christianize and Owenize society by getting the conditions right to encourage good behaviour. Saull's radical causes might be championed, but never his materialism, and Smith baulked at putting baboon blood in our veins. Democracy must extend to barbers, but pushing it to barbary apes was the beginning of a joke. Evidently, readers were familiar enough with Saull, his radicalism, and his monkey-man, for these jibes to be run at his expense. In fact, unless you knew who Saull was, the *Satirist*'s British Association for the Advancement of Science jest would make no sense:

> Mr. Saull read an ingenious essay, to prove that the baboon is the original form of the human species, and expressed his hope that the day would arrive when the whole of the monkey species would be entitled to the elective franchise. Universal suffrage would not be complete without it. Lord Brougham said, that if this enfranchisement took place in Mr. Saull's day, he hoped that Mr. S. would be chosen as the first representative of the new elective body.[10]

That "Mr. Saull has employed his geological learning to the noblest endeavours— ... the amelioration of the political condition of his species, and the eradication of error and superstition from the mind" was laudable in Smith's eyes.[11] But the deed was dirtied by Saull's making man a hairless ape, a statement so outrageous that it had to be parodied. Smith lost no opportunity to make a monkey of the man. Facts would rather "convince the most obdurate, that man was originally an ass, and not a monkey or baboon, as Mr. Saull, the matter and motion philosopher insinuates".[12] Even papers which simply advertised Saull's lectures now took a lashing, most notably that "WHITE headed Beldame of Shoe Lane", the *Morning Herald*, a long-standing daily rival to the *Times*. It was independent of party, growing perhaps a bit Conservative, but hiring the *True Sun*'s old editor John Bell—a "popular democrat, demagogue, and republican"—provided the counterbalance. The *Penny Satirist* lambasted the "old lady" and her "sand-blind, feeble-eyed, and spectacled subscribers" for smuggling radicalism into a Tory rag for the sake of profit:

---

10   *PS*, 23 Sept. 1837.
11   *Crisis* 3 (5 Oct. 1833), 36–39.
12   *PS*, 4 July 1840.

> The love of gold in the old woman is stronger than the love of principle, and therefore, she is induced at times—in order to make herself useful and amiable to all parties—to patronise not only Radicalism and Chartism, but even Atheism itself. She regularly, every Easter, and sometimes oftener, publicly recommends Mr. Saull's lectures on Geology, in which the lecturer's chief aim is to throw discredit on the Mosaic account of creation, and from thence on the whole Bible itself!![13]

Ultimately Smith turned against an autodidactic geology completely. When Cousins started yet another penny periodical, *Franklin's Miscellany* (1838–39), he ran a "Letters on Science" column. This was Smith's penny-a-liner contribution, under the name "Mercury". The erstwhile arch-infidel—the fiercest, remember, that the spy had ever heard—now rather lamely, and perhaps disingenuously, used his new outlet to attack sceptical geology for leading humanity astray:

> Many who have got a smattering of phrenology and geology ... set themselves up as liberals and *savans*, with such airs of Rationalism, that one would imagine that they had unriddled all the mysteries of Nature, and dived into the deepest arcana of physical knowledge; whereas, the fact is, they know nothing positive. Their liberalism consists merely in an abjuration of some old ancestral notions about religion and politics, and the substitution of some few crudities in their stead, accompanied by a farrago of dry facts and detail, from which no active and useful principle of living truth can be deduced.[14]

What was once applauded as Saull's effort to ameliorate conditions and remove superstitions was now dismissed as junking a few "ancestral notions". And what was once praised as geology's liberating power was now derided as a "farrago of dry facts" bereft of social meaning. From having worked hand-in-hand at the Labour Exchange, materialist Saull and millennialist Smith were now at loggerheads.

Saull continued to champion a rival Enlightenment equation of well being with material 'naturalness'—giving the dry facts meaning. And since, for him, "geology will tend, more than any other portion of natural philosophy, to direct our reasoning in its proper path", it was to the rocks that artisans must look for salvation. Or rather to Saull's museum, where the fossils were laid out to illustrate the moral of the

---

13   *PS*, 23 July 1842; [James Grant] 1837, 2: 32.
14   *Franklin's Miscellany* 1 (17 Nov. 1838): 388; J. E. Smith 1853, 39.

myriad creatures that lived before humans. Hence his continual offers, along the lines of this early one in the *Mechanics' Magazine*:

> I hereby offer my extensive museum of geology, containing many thousand specimens (illustrating the various changes and productions on the crust of the earth), to the inspection and observation of all those of both sexes who feel an interest in the acquisition of this branch of knowledge ... I shall feel great pleasure in going over my collection with them ... and this I frankly offer, without pecuniary fee or reward, for my full and ample reward will be in the delightful sensations that are always experienced when developing and elucidating truth.[15]

"Truth" was its own reward, but it was an Owenite truth resting on material foundations, and virtue resulted in obedience to this evolving nature. This explains the title of many of his 1830s lectures in freethought dives, such as "Geology in reference to Human Nature" or the influence of science "in Forming the Character of the Future Generations of Mankind". Artisans could hear these at Owen's Institution, the Tower Street Mutual Instruction Society, or the Finsbury Mutual Instruction Society. The venues in turn would promote his museum ("filled with fine geological specimens"). One wonders, in fact, whether Smith's spoofs did not actually increase attendance. Mechanics were now asking where they could see Saull's wonders, and the venue managers would duly send them along to his Thursday open sessions.[16]

Geological and Judicial Law

The antagonisms of the old comrades were now beginning to run wider. Where Shepherd Smith thought throwing the poor onto their own resources under the New Poor Law would be fine, as long as the clergy and land owners were treated likewise and made to work for their tithes and rents,[17] Saull was uncompromising in his hatred of the Act. To Owenites, the wretched workhouses were a sign of society having gone

---

15 Saull 1833c.
16 *PM* 2 (20 Jan. 1838): 200. Saull's London lectures and venues are listed in the *Penny Mechanic* and *New Moral World*. He also took his "extremely interesting lecture on geology, in connection with the social improvement of the people" to the provinces, lecturing, for example, at the Social Institution in Salford (*NMW* [23 Sept. 1837]: 387).
17 *Shepherd* 1 (2 May 1835): 288.

off the rails and needing realignment. The Whigs made the workhouses execrable in order to keep all but the most incorrigible or indigent out, forcing the rest to compete in the marketplace. This saved the government money, while the increased competition worked to lower wages, benefiting employers, as Saull recognized.[18] The Whigs—unlike the Owenites—saw poverty as a function of character, and the poor were declared "deserving" or "undeserving" according to their prudence or industry.[19] But the result was that the sickly and old suffered terribly. To Owenite journals, Saull relayed heart-rending stories of the new law which "carries such desolation and distress through the land", stories which upended liberal explanations of indigence that resulted from low moral character. They characteristically redirected blame away from the victim and on to social injustice. He told of sad sights from his native village in Northampton: of a widow, her son transported, living in rags, and not expected to survive winter. And now "under the new unfeeling poor law her usual small allowance from the parish was stopped." The law leaves "poor forlorn widows, who, surely of all persons, most require assistance, from being deprived of support through the death of their husbands. Professing Christians, where is your consistency, your honour, or justice?"[20] Through his blasphemous specs, Saull saw the problem as partly a Christian one.

The workhouses were going up in the later thirties, and Saull reacted angrily against this "cruel treatment of the poor". He attended rallies attacking the new poor law, which was "iniquitous in its nature, and oppressive in its operation", and he supported radical MPs in their forlorn efforts to repeal the law.[21]

Not that he was unusual in this; not, at least, in London. Here, says David Green, the myriad (local government) vestries "operated almost as if they were separate ratepayer republics". From 1831, the franchise in these vestries had potentially been open to all resident ratepayers, women included, and the higher percentage of artisans able to vote

---

18   *Times*, 28 Feb. 1837, 6. R. Dean 1995 on Owenite attitudes to Malthusian solutions.
19   Claeys 2000, 10.
20   *NMW* 4 (23 Jun. 1838): 278–80. Breton 2016 on such accounts of the poor that dismissed bourgeois explanations based on character.
21   *Times*, 28 Feb. 1837, 6; 2 May 1839, 5; *Courier*, 27 Feb. 1837, 3; *Charter*, 5 May 1839, 226.

meant the vestries took a democratic, and anti-poor law, turn.[22] The result was that the vestries drove London's parliamentary radicalism. But agitating at both vestry and parliamentary-level meant more committee work for Saull, especially for the Metropolitan Association for the Repeal of the New Poor Law, formed after a meeting at the Freemasons' Tavern in February 1838. (The *Patriot*—a paper run by the evangelical Independents that railed against geological infidelity and considered Owenism a malignant depravity—derided the society's members as "chiefly the votaries of Saint Monday", that is, layabouts, and added that "Placards of a very inflammatory description had previously been posted on the walls; but the meeting went off like a damp fire-work".[23]) The "cruel" workhouses were lambasted as "Bastiles"—after the hated prison destroyed in the French Revolution—and when *The Book of the Bastiles* (1841) came out attacking them, Saull gave it a puff.[24]

Since application of the Poor Law Amendment Act was not mandatory, he offered to "lend both his purse and personal exertion" in any fight to keep the "obnoxious and abhorrent" law from applying in his own Aldersgate vestry, or indeed to the City of London generally. He saw it as simply unnecessary. For one thing it hardly saved money. He had been auditor of his parish accounts for 16–18 years (as he reported in 1837), where the poor rates had been reduced to only a shilling in the pound, which was still sufficient to support the local paupers. He even declared he would refuse to pay his rates if the commissioners set up the system in his parish. Saull examined the workhouses while travelling the country on business and reported that many aged labourers, "poor old creatures, upwards of seventy years of age, had declared that they would sooner perish in the streets than go into one of the new union workhouses".[25]

An optimistic palaeontology, rightly viewed, pointed to a more correct political path and dictated action. From his mentor, Sir Richard Phillips, Saull had taken the pregnant notion of the 'pabulum'. This was, in effect, the prepared substrate on which plants lived, the soil

---

22   Green 2010, 82–93.
23   *Patriot*, 22 Feb. 1838, 124, cf. *MC*, 20 Feb. 1838; *London Dispatch and Peoples Political and Social Reformer*, 25 Feb. 1838. 6.
24   Baxter 1841, vii.
25   *TS*, 25 Apr. 1836, 1; 28 Feb. 1837, 1; *Times*, 17 Feb. 1837, 6; 28 Feb. 1837, 6.

and nutrients. On the early Earth, the basal granite rocks could sustain no life, but heat and erosion started the trituration process, and the cyclical immersion and tidal erosion as the poles swung provided the means of "restoring an exhausted world and improving it". "Soil thus becomes a more and more refined pabulum in every revolution", in Phillips's words, by means of which "the strata [could] prove the gradual evolution of all things". Saull had a theory to work by. He told the Salford socialists in 1837 that the ground for life was prepared in advance, and that there would never be more mouths in each geological epoch than the prepared pabulum could support:

> that calcareous matter was necessary to the production of life. It might be termed its pabulum—and hence, it was observable, that no animated existences were discoverable till an abundance of it had been provided for their sustenance.[26]

A fuller report, of a geology lecture at Leeds in 1840, fleshes this out more. Saull started with the formation of the earth's crust, beginning

> with the primitive rocks in which no remains of life are discoverable, up to the time when nature having been gradually readied for this event, man makes his appearance. He [Saull] dwelt particularly upon the fact that no animal, of any description, is ever found in existence, until an abundant supply of the means for maintaining that existence has been previously provided; and that from the felspar, in which organic remains are first discovered, and in which calcareous matter, the grand supporter of life, is also first discovered in the ratio of about 2 or 3 per cent., up to the present time, we find that this pabulum of life is constantly on the increase, and, as a consequence, animated and organised beings more numerous and prolific.[27]

It was the modification of an old idea, going back to Holbachian notions of ancient abundance, that the regenerating earth would always provide. This news was "greeted with cordial and frequently repeated cheering", not, perhaps, for the arcana of ancient saurians, but because of the science's anti-workhouse, anti-Malthusian moral. The ground was prepared by weathering as a result of cyclical astronomical events, increasing the copiousness with each turn. Here Saull was, in 1840,

---

26 Saull 1837; R. Phillips 1832a, 47–48.
27 *NMW* 8 (18 Jul. 1840): 37.

at the height of the economic depression, with starvation and mass agitation, pointing out that England obviously had taken a wrong political turn. Pessimistic Malthusian predictions took no account of geology's proof of planetary provisions, or its "law of progress". Were men to recognize these, they would "conduct themselves in accordance with the bountiful arrangements of nature"—meaning the rich would share with the poor—because, as fossil life showed, the Earth "has provided abundant resources for the enjoyment of all animated beings." And understanding ancestral life's materialist cause would "annihilate those unnatural feelings produced and perpetuated by ignorance"—the time wasted by superstitious reverence—and our "best feelings will be called forth in sympathy with general humanity, and, as a necessary consequence, all must advance in a much greater accelerated ratio". So it was back to the rocks "to direct our reasoning in its proper path".[28]

The 'pabulum' had been provided, but the hunger and poverty persisted. People were "surrounded on all sides by abundance", Saull said in 1837, "but starved, like Tantalus, in the midst of it, solely in consequence of the irrational institutions" foisted on them.[29] And one glaring 'irrational' institution stood far above the rest—the iniquitous Corn Law, pushed by the farmers and aristocratic landowners for personal profit, which made bread expensive and edged the poor towards starvation.

Saull became a major anti-corn-law activist shortly after opening his rebuilt museum. The bad harvest of 1836 had caused a hike in the price of corn. The taxes on foreign grain imports kept domestic prices and landowners' profits high, even while industrialization and urban growth made lower bread prices essential. As the attacks on protectionism grew, Saull joined the clutch of Radical MPs on the Committee of the London Anti-Corn-Law Association (founded 1836), and so began a decade of activity against the "Bread Tax".[30] The Association demanded the total repeal of the Corn Law, which favoured the landowning interest of the political elite. Reduction would cheapen bread for the manufacturing

---

28 Saull 1853, vii.
29 Saull 1837.
30 *Examiner*, Dec. 1836, 814; *Shipping Gazette*, 14 Dec. 1836, 1; *MC*, 22 Dec.1836, 1; *TS*, 22 Dec. 1836, 1; 6 Mar. 1837, 1. Prentice 1853, 1:49–50. The Radical MPs were Thomas Wakley, Benjamin Hawes, Joseph Hume, and Thomas Duncombe.

poor.³¹ But removal of price guarantees threatened the farmers, who excoriated the "bare-faced lies" of the liberal press and fingered this new Association—which would sometimes meet in the Mechanics' Hall of Science³²—as the centre of the "iniquity and mischief".³³ As palaeontology pointed out the problem, so corn-law removal became another solution: Saull's science and politics were blending into a seamless stream of activism across all fronts.

Whatever the cause of poverty in the face of plenty, a rational geological education would ready the mechanics for the New Jerusalem when the political situation was redressed. "Brother Saull" repeated it again at the Anniversary Dinner of the Trades' Unions in 1835: if "The people, the only true source of legitimate power" was the toast, then training was the key, and "he would be proud to exhibit" his museum to any comrade to show what could be done.³⁴ He would shortly move from Chartist lectures on "the social and political condition of the country" to advocacy of his geology museum with the same ease.³⁵ Like a spinning top, the alternating educational geology and radical politics blurred into a bigger progressive picture.

Infidelity: Geological and Matrimonial

Convincing the middling ranks that Owenism would lead to regeneration, or that culture is responsible for crime, not criminals,³⁶ or that co-operation would lead to harmony, would never be easy. Convincing the religious was harder still. Conservative critics always

---

31 Even at the risk of wage lowering, which is why the "cotton lords" were in favour (Prothero 1979, 220).
32 *Charter*, 1 May 1840, 16; *Commercial Daily List*, 5 Mar. 1840, 1.
33 *British Farmers Magazine* 1 (Apr. 1837): 355–56. Saull also had a role in the subsequent Metropolitan Anti-Corn-Law-Association, a branch of the Anti-Corn-Law League formed in 1840: *MC*, 25 Feb. 1840; *Charter*, 1 Mar. 1840, 3; *Constitutionalist*, 1 Mar. 1840. 4; *Examiner*, 1 Mar. 1840. On his proposal of a petition to be drawn up by local London groups: *MC*, 11 Apr. 1843. This association would also meet in the Mechanics' Hall of Science: *MC*, 12 Apr. 1843. Saull also joined in the City of London free-trade agitation as the corn laws became blamed for scarcity in the hungry forties: *MC*, 16 Dec. 1845; *Standard*, 16 Dec. 1845; *Atlas*, 20 Dec. 1845, 817.
34 *TS*, 22 Apr. 1835, 2.
35 *NS*, 18 Sept. 1841; 9 Oct. 1841.
36 Saull 1838b.

returned to the infidel, counter-cultural independence of socialism. A preacher at the evangelical London City Mission, which targeted Owenism, was aghast: "Never before did men calmly and openly unite together, organize institutions, frame laws, and employ missionaries to overturn the constitution of society, destroy the social relations, abolish marriage, and blot out from the mind the belief and love of the one living and true God."[37]

While Labour Exchanges were anathema to many critics, who slanderously spoofed co-operators for exchanging their wives and scowled at them for indoctrinating the unwashed, it was the touchiest sacrament, marriage, that generated the biggest backlash. In fact, Owen's marriage proposals produced more apoplexy than his irreligion.

"Of all the sources of evils in human life, under existing arrangements, marriage, according to popular notions and as now solemnized, is one of the most considerable, if not the chief." That was Owen speaking in his *Lectures on an Entire New State of Society*, delivered in Saull's Albion Hall in 1831. To replace the state/religious coercion and legalization of a husband's ownership of his wife "'for better and for worse,' (the absurd phrase used on this momentous occasion, to express the nature of their bondage)", Owen proposed something shocking to a society whose evangelical laces were straitening. This "solemnizing" of an indissoluble bond, dressed up by the priesthood into a self-serving sacrament, was "a species of private property in persons of the most objectionable character, and without the removal of which, private property in riches cannot be abandoned in any society".[38] The slaves, let alone the slaves of the slaves (women), had to be liberated for the coercive capitalist and religious straightjackets to be removed. The Owenites were challenging the church's authority, not only over the sacrament of marriage, but of baptism and death too, and taking control of these rituals.

Female emancipation was imperative for co-operators such as William Thompson, to end "domestic slavery".[39] Women had to be equally educated in economic and scientific knowledge to enable them to become joint possessors "of the world's wealth, and an equal partaker

---

37 Quoted by Topham 2022, 366.
38 Robert Owen 1830, 76, 80.
39 W. Thompson 1824, 298–99.

in all the delights which flow from mental and moral culture".[40] This had long been a Utopian dream. The paternalist Owen was equally worried, as Barbara Taylor has shown, by the nuclear family as "a key source of competitive ideology, as well as the main institution responsible for the transmission of private property"—as a den of selfishness which looked only to its own advantage and ignored its neighbours. Part of his solution was a probationary marriage period following an Owenite civil ceremony. This would ensure the couple's compatibility and compliance with co-operative ideals. Following a failure, the union could be dissolved after a cooling-off period.[41] Owen worked up the details in his *Lectures on the Marriages of the Priesthood in the Old Immoral World* (1835). Even if, as Eileen Yeo says, "Owen's prescriptions would not jar the modern ear", the idea of cheap marriage, quick divorce, and no priestly interloper brought the Victorian roof down. Never had an issue generated so much acrimony, whipped up by the clergy.

> By 1840 the Lectures had been torn apart in dozens of anti-Socialist publications; quoted aloud in endless public debates; denounced in pulpits from Canterbury Cathedral to the Primitive Methodist chapels of Belper; banned from many public bookstalls; and on one occasion publicly burnt by an opponent with a flair for the dramatic. 'Let no man, let no woman especially, dare to become a Socialist without first reading these ten lectures ...' *The Evangelical Magazine* warned.[42]

The consequences even shocked some co-operators. William Lovett said that it "was like the bursting of a bomb-shell".[43] Others accepted the 'rationality' of it, if not the practice. Many accepted the practice, and it worked well: although it turned out that the Owenite ceremony was just as ritualized, with organ, choir, a social sermon, and a wedding breakfast.[44] How many availed themselves of a quick divorce, though, is unknown. William Thompson went further to suggest everyone should enter his commune single. His relatives thought it a sign of his insanity and accordingly challenged his will when he died. Tittle tattle even had George Petrie's mental collapse and death (in 1836) brought on

---

40 Southwell [1840], 20.
41 Robert Owen 1830, 75–84. B. Taylor 1983, 39; Frow and Frow 1989, ch. 7.
42 B. Taylor 1983, 183–84; Yeo 1971, 101–02.
43 Lovett 1920, 1: 51.
44 Yeo 1971, 102.

"by his wife's enthusiastic acceptance of Owenite marriage doctrine".[45] To evangelicals, for whom hearth and home were sacrosanct, Owen's outrage cut to the very heart of the family. Some reprobates did not help the cause, notably the piano finisher-turned-infidel-bookshop-owner Charles Southwell (see Chapter 18), who echoed Owen's belief that marriage was "simply a law framed by priests and legislators to maintain their power".[46] He rather justified the critics' accusations: the thirty-third child of his father and his third wife, a servant girl, he himself had embraced a live-for-today "licentiousness", marrying one adulterous girl, then living with her aunt, then another married woman who managed his finances, and he finally kissed and told all in his lurid *Confessions*.[47] Southwell might have been the exception, but this was where the orthodox saw it leading.

The issue was exacerbated when a reporter's notes of Owen's 1834 lectures in Charlotte Street were republished as *Marriage System of the New Moral World* in 1838. According to Edward Royle, the notes were sensationalized, which did not help. Marriage was again denounced as "a Satanic device of the Priesthood to ... keep mankind with their slavish superstitions, and to render them subservient".[48] An apoplectic *Fraser's Magazine* managed to invoke jingoism, xenophobia, and revolution into its critique of these "horrible abominations", which devolve ultimately into "indiscriminate prostitution". Was this "an attempt at transplantation into English and consecrated soil of the unholy impurities with which Hindooism and Mahommedanism are rife, and which Jean Jacques Rousseau, and other kindred spirits, bequeathed as their chief legacies to after generations"?[49]

It was too easy to paint Owenites with street-arab morals or to equate such delinquency in racist terms with the 'savagery' of the 'lower' orders.[50] With Anglicans unable to penetrate the rookeries (the "modern Sodom"[51]) to make marriage the inviolable sacrament demanded by

---

45 Chase 1988, 158 n33; Lovett 1920, 1: 51.
46 Southwell [1840], 21.
47 Southwell 1850, leading *The Young Man's Magazine* (1854): 76, to roundly condemn his "shameful immoralities". The "Confessions" were begun in Southwell's *Lancashire Beacon*, no 7 (1849): 49.
48 Robert Owen 1838, 7; 1839; Royle, 1974, p. 62; B. Taylor 1983, ch. 6.
49 *Fraser's Magazine*, 21 (Jan.-June 1840): 689–90.
50 Qureshi 2011, 21.
51 Duncombe 1848, 19.

prudish society, co-habitation was the norm here. The flash costers in their shabby velveteen coats had a "lively dislike" of the missionaries and preferred to (literally) shack up together in their teens.[52] Most barrow boys who poured onto St Giles streets to flog the poor their fruit and fish were 'illegitimate', not one in ten of their parents were "tucked up" (married). So perhaps it was no surprise that one fanatical anti-socialist, John Brindley, would attack Owenites on stage using "the low ribald slang of the costermonger".[53] Fear of the ghettoed "Sodomites" could be used to suggest the direction of Owenite travel, towards social degradation and ignominy.

It was too much for the incredulous *Quarterly Review*. Socialism was "a wide-spreading moral plague". In Owenite Halls, sedition and irreligion were seen as a piece, but this attack on the sacraments was the final straw. The review blamed the Newport Chartist uprising in 1839 on such an infidel 'education', which released the wide-eyed underclass from religious restraint. It "teaches the poor to read without accompanying that gift with such moral and religious instruction as may regulate and purify the use of it". The traitors were perverted by "those infamous and seditious publications which are everywhere corrupting our population". As a sign of this religious abandon, it pointed to Owen "and—we hardly know how, with decency, to express the monstrous proposition—the abolition of that restrictive engagement which we call *marriage*, but which Mr. Owen stigmatizes as 'an accursed thing,' 'an unnatural crime,' 'a satanic device.'" It was clear that "the man who could even imagine, and, still worse, publish such abominations, must be insane". Here was the "wickedness and folly of Socialism". Misrepresenting Owen as proposing "a licensed system of adultery"[54] was a deliberate attack on his respectability, and it worked. As so often in conservative critiques, secular learning was made the seed bed of sedition, rational rearrangement the harbinger of moral doom—and, in Saull's case, irreligious evolution would bring out the beast.

---

52 Chesney 1970, 51, his reworking of Mayhew's *London Labour and the London Poor*. Street-patois journalism was designed to shock the class voyeurs, so the cadgers' houses in St Giles with their lax sexual arrangements were vividly portrayed (Duncombe 1848, 16–19; Beames 1852, 130, 203).
53 *NMW* 6 (2 Nov. 1839), 857.
54 [Croker] 1839–40, 304.

Saull apparently never questioned his own conjugal arrangements. He had married Elizabeth, *née* Weedon (1789–1860), in 1808, and she remained his "dear wife" for life. But that did not stop him from also denigrating solemnized marriages. He viewed these inflexible Christian sacraments just like the rest. So, in Charlotte Street in 1833, after Owen reaffirmed that affection, not legality, should be the guiding rule, something which requires a trial period to assess, Saull, in support, talked of his Continental experience. With some "animation" he "observed, that in Prussia and Holland marriages are made and dissolved by the magistrates alone, at the due notice and request of the parties". Even in geology lectures "in connection with the social and moral improvement of the people", he would wind up with comparisons of the clergy and moral state of the populace in France and England—to the detriment of the latter—before finishing on their respective "marriage laws".[55]

It could only have made his infidel geology more suspect in orthodox eyes. By their fruits should poisonous philosophies be known was the *Quarterly*'s attitude. After all, here was the sort of archetypal educationalist it really hated: 'immoral', shown by his support of Owen's marriage views, irreligious, proven by his indictment for blasphemy, and seditious. The latter would be cemented by Saull's now supporting the condemned leader of the Newport Chartist uprising, the former tailor, indeed former magistrate and Mayor, and brilliant spokesperson for the movement, John Frost. Saull was among the "friends" of Frost, a group led by Bronterre O'Brien and Feargus O'Connor, who campaigned and subscribed to meet his legal costs. Most importantly, they organized public shows of solidarity to counter "the poison infused by the daily press into the public mind" about Frost.[56] Although Frost was convicted of high treason, and sentenced to be hanged, drawn, and quartered, the appeals and protests helped to get a commutation, and he was transported to Van Dieman's Land. Still Saull (and Mrs. Saull) continued with the rest to campaign for his repatriation (a third of the petitioners were women).[57] The group got Mrs. Frost an annuity and worked to

---

55 Saull 1837; *Crisis* 2 (11 May 1833): 144.
56 *Charter*, 15 Dec. 1839, 741; 5 Jan. 1840, 792; 12 Jan. 1840, 888 *CPG*, 21 Dec. 1839, 2; *The Odd Fellow*, 11 Jan. 1840; Lovett 1920, 1: 208.
57 Dinwiddy 1992, 406. D. Thompson 1984, 81, actually doubts that the petitioners had as much influence as the Chartist-sympathising Lord Chief Justice, who recommended mercy.

release her estate—in fact, from the subscription lists it looked like Saull was putting money in monthly.[58]

The equation of infidelity (geological or matrimonial) with sedition harped on by the press was not wildly misplaced—although whether the guilt was by association, or whether infidelity, sedition, and materialist geology were constitutive of a larger radical world view was the real question. Whichever, the Tory policing strategy was simply to point out where secular education would lead: sedition, blasphemy, and the breakdown of sacred marriage bonds.

Other personal traits could be used to denigrate socialist science: extremist attitudes which were considered character 'flaws' by *John Bull*'s red-blooded followers. Vegetarianism attracted an eclectic bunch as some radicals extended their sympathy to enslaved farm animals. So many around Saull embedded vegetarianism into their moral strategy for social regeneration that one wonders about Saull himself. Many deists experimented—his mentor Sir Richard Phillips had not tasted meat since he was twelve, and detractors poked fun at his strict 'Pythagorean diet'.[59] Carlile was another, and Saull was his benefactor, so did it rub off? William Thompson, Roland Detrosier, Robert Dale Owen, Julian Hibbert, all abhorred the killing of animals, or at least eating their flesh.[60] They had a scunner against the 'Roast Beef of Olde Englande', with its traditional gentrified taint, not to mention the tainted gentry and their blood sports. John Gale Jones had managed to get an anti-cruelty petition to the Lords in his fight against blood sports.[61]

But the issue was a complicated one. Although Paine's *Age of Reason* saw "cruelty to animals" as a "violation of moral duty",[62] feelings were confused by the intrusion of the hated evangelical societies. These were not only for the 'suppression of vice', locking up Carlile and his crew, but also against cruelty to animals. While Carlile portrayed animal cruelty as a Christian vice,[63] evangelicals also campaigning against it conflicted the issue. Many radicals correctly saw the evangelical do-gooders as

---

58   NS, 10 Oct. 1840; 20 Mar. 1841; 17 Apr. 1841; 24 Apr. 1841; 18 Sept. 1841; 30 Apr. 1842; *National Association Gazette* 1 (1842); Maccoby 1935, 208–11.
59   *Crisis* 4 (2 Aug. 1834): 13; R. Phillips *ODNB*.
60   J. F. C. Harrison 1987; Gleadle 2003, 202; Leopold 1940, 72.
61   *Newcastle Courant*, 26 Mar. 1825.
62   Carlile read this into his trial proceedings: Carlile 1822, 79; Conway 1892, 2: 103.
63   *Republican* 13 (30 June1826): 816.

singling out the avocations of the poor (cock fighting, badger baiting), while letting the gentlemanly fox-hunters off scot-free, which led to more cross-currents of confusion. So even if the oppressive vices were occasionally condemned as a job lot—"We pity the oppressed, we sympathise with the slave, we justly condemn cruelty to animals"[64]— in truth cruelty and vegetarianism did not figure constitutively in campaigns.

In the later forties, as Owenism disintegrated into freethought, educationalist institutions, and spiritualist communes, vegetarianism seems to have become the prerogative of the sacred socialists.[65] On the materialist side of the new fence, young Brummie socialist and atheist George Jacob Holyoake had "vainly tried to rise to the level of vegetarianism", in the words of his biographer. But the attempt degenerated into ribaldry as he later knocked "the foible so prevalent among our vegetarian friends, of complacently imagining that the imbibing of peculiar food endows them with unusual purity and intellectuality".[66] So vegetarianism was far from *de rigueur*, even if Saull was surrounded by it. If the "Mr. Saul" said by the *True Sun* to have opposed the building of an abattoir in Islington is our man (which is far from certain), then that is the most that can be said for him.[67]

Although obituarists talked of Saull's "frugal habits", we do not know what those habits were.[68] Vegetarianism being a form of physical puritanism, it was frequently associated with other morally-regenerative stances, particularly temperance. Only in the 1840s did temperance begin to mean teetotalism, but both were common among Saull's peers.[69] Again, they were attractive to Carlile, Hetherington, Allen Davenport,

---

64 *Reasoner* 17 (1 Oct. 1854): 218 quoting the London kindergarten teacher Madame Ronge.
65 James Pierrepont Greaves, Charles Lane, and A. Bronson Alcott were strict vegetarians.
66 *UR*, 27 Oct. 1847, 96, bound with *Reasoner* 3 (1847); McCabe 1908, 1: 91.
67 *TS*, 14 Mar. 1834, 3.
68 *JBAA* 12 (1856): 186–87.
69 J. F. C. Harrison 1987; 1967, 206; Cole [1944], 76. Place 1834, pointed out that pubs were often the only place where single working men could find companionship and amusement. Lack of alternatives explains the rise of Owenite tea festivals and radical coffee houses. B. Harrison 1994, ch. 5, on the parallel rise of the evangelical teetotal movement.

William Thompson, Baume, Robert Dale Owen, and other Owenites.[70] "GIN and JESUS" might have been the curse of the "bamboozled" classes, as the *Yahoo* had it.[71] But boozing was blamed for different evils by the rival parties. In the Owenites' alternative society, where festivals and tea parties were to rival pub culture, drunkenness was perceived, not in capitalist-management terms as an impediment to worker productivity, so much as destructive to socialist community relations.[72] Hence, they kept drunks out of the Labour Exchange. Yet, in practice, Owenite halls were no different from prim mechanics' institutions, some of which had actually grown out of temperance societies, and whose governors linked drink with promiscuity, improvidence, and absenteeism.[73]

Monthly Co-operative tea parties were in full swing by 1832–33 to cover the Exchange's rent, and Halls of Science in the 1840s continued the tea-party tradition where alcohol was barred.[74] This moral aspect of Owenism put Saull the dealer in wines and brandy on the spot. If a "drunkard sells his soul, children, and country at an election for a glass of gin", and if (as John Finch, a social missionary credited with founding some seventy teetotal societies, reported in the *Crisis*):

> Gentlemen boast of having alcohol (wine) enough in their cellars to poison 1,000 men, and merchants calling themselves moderate drinkers are not ashamed to acknowledge that they drink 21 glasses of the best French brandy per week. 10 millions of pounds are spent in wine, 20 millions in spirits, and 22 millions in ale, porter, &c. annually in this kingdom; and nearly all the wine, a great share of the ardent spirits, and no small part of the ale is consumed by the higher and middle classes; thus affording a most pernicious example to all below them[75]

then Saull was left between a rock and a hard place. But there was, he reported, no drunkenness in the French countryside, where wine was cheap.[76] So, for him, it was an urban proletarian problem, involving

---

70 *PMG*, 23 July 1831, 22 (B. Warden); Leopold 1940, 76 (R. D. Owen); Cooter 2006 (Baume); Thomas Cooper 1849; Barker n.d. [1938], 51 (Hetherington); Wiener 1983, 60 (Carlile).
71 [Watts] 1830, xxvi.
72 Yeo 1971, 95.
73 C. Turner 1980, 344.
74 Faucher 1969, 17.
75 *Crisis* 4 (17 May 1834): 43; J. F. C. Harrison 1969, 122–24.
76 Saull 1837.

exploitation, misery, and poverty. Saull, it seems, could live with his trade because of his Robin Hood attitude: selling expensive French imports to the cognac-imbibing gentry while pouring the proceeds into pauper education. That is not the only incongruity. It was the final irony that Saull, one of London's major wine and brandy importers, was a known donor to temperance societies.[77]

It was this underwriting role that seemed to be his redemption. Saull, clubbable and sociable, was still the perennial chairman and treasurer of untold causes. His brandy largesse was spread lavishly into every radical cause, as was his financial and organizational expertise. Whether it was the Co-Operative Building Society in 1839, or the Third National Trades Conference in 1845, or organizing funds for the families of killed Chartists,[78] the man made wealthy by the wine business was in demand to manage the cash.

---

77 *Monthly Notices of the Royal Astronomical Society* 16 (Feb. 1856): 90.
78 NS, 27 Oct. 1849; NMW 6 (14 Sept. 1839), 752; 13 (16 Aug. 1845): 486.

15. Martyrs, Churches, and Vestries

By far Saull's most famous redemptive campaign was for the repatriation of the 'Tolpuddle Martyrs'. This was extraordinarily emotive episode. The farm labourers in the tiny Dorset village had been transported for resisting a shilling wage-cut and swearing a union oath. As their case became a *cause célèbre*, Saull achieved his highest public profile as an organizer. He acted as one of the lieutenants to the firebrand Thomas Wakley, as well as Treasurer of the Dorchester Committee from 1835. If you wanted to buy tickets for the benefit concert at the Royal Victoria Theatre (a new drama with the actors giving their services free, all in support of the wives and children), you contacted Saull. If you wanted to attend the Dorchester Committee's public dinners at White Conduit House, Thomas Wakley presiding (with Saull sitting in when Wakley left), you paid your 3s to him, or 1s for just the Ball afterwards. It was a huge undertaking, juggling a torrent of little sums, the £5 made up from Spitalfields weavers' pennies, the umpteen receipts, dispersements, audited balances, in short hundreds of accumulated pounds to fund the repatriation campaign or keep the martyrs' fatherless families afloat.[1]

Wakley, a medical journalist who founded the campaigning *Lancet*, was the newly-elected doctrinaire radical MP for the enormous new Finsbury constituency (created as a result of the Reform Bill), with its third of a million inhabitants. He was voted in by the shopkeepers, whose trade was among the working classes, so he represented, in effect, the poorer communities in Parliament. Conservatives derided

---

1   TS, 16 Sept. 1836, 5; 25 Apr. 1837, 2; 30 Apr. 1837, 1. *London Dispatch and People's Political and Social Reformer*, 24 Sept. 1836; 13 Nov. 1836; 30 Apr. 1837. *Spitalfields Weavers' Journal* 1 (3 Oct. 1837): 24. For another theatrical benefit, at the Royal Pavilion Theatre, see *London Mercury*, 4 Dec. 1836, 8.

him with racist sneers as "the *honourable* (!) member who represents the Jew clothes-venders of Finsbury".² Wakley passionately pleaded the convicts' case in the House, while Saull organized union support and fundraising for the families,³ and all this as he was rebuilding and re-launching his museum.

Wakley's emotive speeches on the exiles' plight reduced the House of Commons to tears: "The great paunch-bellied, whiskered fellows were to be seen sobbing in all directions".⁴ Saull, for a moment, could be just as rousing. The transportation had exposed the threat tactic, as Saull (not above issuing threats himself) told trades' unionists: the magistracy had

> a deeper scheme—it had wished to intimidate the many by the example made of the few. The rising spirit of liberty was to be crushed by the blow—(Never, never.) And let it not be disguised unless the people roused themselves and imitated their brethren in France, some dreadful attack must be anticipated.⁵

Ultimately Wakley's affecting oratory, backed by radical clamour, got the prisoners their pardons—and in 1837 their repatriation as well at Her Majesty's expense.⁶ So that left the public procession through London to organize,⁷ as well as the *dénouement*: buying the men small farms with the £600 surplus, which would protect them from further harassment by the magistrates.⁸ The reforming *Morning Chronicle* gave an upbeat account of the day-long procession on 3 April 1838 to celebrate this success—the "dense" cheering crowds, twenty-four thousand it reckoned, the six thousand trades people with their "splendid banners"

---

2   *London Medical Gazette* 15 (17 Jan. 1835): 562; Desmond 1989, 156.
3   *NMW* 1 (14 Mar. 1835): 160; 21 Mar. 1835, 168; *PMG*, 4 Apr. 1835; *TS*, 22 Apr. 1835, 2; 4 May 1835, 1; 12 May 1835, 1; 20 May 1835, 4; 16 June 1835, 1; 8 July 1835, 2, 6.
4   *NS*, 21 Apr. 1838.
5   *TS*, 22 Apr. 1835, 2.
6   *London Mercury*, 16 July 1837, 6.
7   *TS*, 15 July 1837, 8.
8   *CPG*, 31 Mar. 1838, 2; *NS*, 31 Mar. 1838; *Charter*, 14 Apr. 1839, 184; 13 Oct. 1839, 606. On the Crown and Anchor meeting (chaired by Saull) and subsequent meetings, to discuss buying farms, which they did in Chipping Ongar, Essex, and Harlow: *London Dispatch and People's Political and Social Reformer*, 13 May 1838; 26 Aug. 1838. 6; 1 Sept. 1839; *Bell's Life in London and Sporting Chronicle*, 13 May 1838; *Champion*, 14 May 1838, 11–13; *NS*, 9 June 1838; *CPG*, 8 Sept. 1838. The committee was wound up in 1842: *National Association Gazette*, 30 Apr. 1842, 140.

snaking one after the other out of Kennington Common on their way to Oxford Street and beyond, the bands striking up "See the conquering hero comes" as they passed the Home Office, just to rub it in. The five Dorchester men sat royally in an open landau carriage drawn by four horses. The *Chronicle* noted the good humour and best behaviour, with the readied Bow Street officers being told to stand down in consequence. The event ended with a gala dinner in a thousand-seater tent, where Saull sat in till Wakley, the chairman, arrived.[9]

Saull's stock was rising but only among radicals. Contrast that sympathetic view with the derision of the *Morning Post*, which belittled the procession for its conservative readers. It is a fascinating counter-account, and we only learn of Saull's real standing from its attempt to stoke up hostility. The day was miserable, pelting with rain, and what few banners there were ended up in tatters. The "mob", only two or three thousand at the start, had "poured forth from courts, alleys, and 'back-slums'". "A more ruffianly set" could not be imagined. The snide asides piled up, feeding the prejudice: "Nearly a third of them were dressed in flannel jackets, like those worn by bricklayers' men, and a vast number had traces of their Sunday frays, in blackened eyes and swollen faces." Hints of drunken sprawls were to litter the report to the end. Expectation was met with deflation: "There were none of the 'Liberty or Death' banners" of former parades, as if that would at least have roused the passions of its Tory readers. The "rabble-rout" seemed "very apathetic about the matter". "Indeed a large proportion of the crowd showed by their remarks that they regarded the whole foolery as most ridiculous." The sneering then fell on Saull:

> The order of march was first, a few men, mounted on dray horses, to clear the way; then the trades in their order, next the "Dorchester Committee," followed by the landau and its contents, and closed by a miscellaneous rabble. We had almost forgotten an important personage who figured in the printed programme as "the Treasurer of the Committee," and who was to follow immediately after the landau "in his carriage." We looked long and anxiously for the "carriage," and at last discovered a "one horse chaise," in which were seated the "treasurer" aforesaid, who is named Saull ... !

---

9   MC, 17 Apr. 1838; *Patriot*, 19 Apr. 1838, 250.

Frugality and modesty obviously cut no ice, but it shows how important the object of derision was—riding directly behind the Dorset men in his little chaise. By the time the procession reached the banquet tent at White Conduit House, the sodden "labourers" were covered in mud and "a great many of the men were much intoxicated." The sour note continued with the feast, and "whether the gin or the air had whetted their appetites", the ravenous hordes rushed at the food: "tables were overturned, women screamed, and men swore, and blows were exchanged with frightful rapidity", confirming a genteel audience's view of the visceral proclivities of the sons of toil. Here a host of "mobocrats" made the usual "blasphemous exhibition" in their speeches, while the overwhelmed Dorchester "labourers" showed themselves to be "men wholly illiterate". Then came the parody of dropped 'h's, when one speaker was reported as saying that "sooner than [being] torn from his home by them there miscreants he would have suffered a dagger to be plunged in this here art".

But it was again the *Post*'s view of Saull's speechifying that is revealing: it was "of that kind of oratory 'which the learned call rigmarole.'" To a Tory, his words to the "congregated rabble", pleading for more funds (to buy farms) and calls for a show of hands of those who had no vote— "when instantly a whole forest of paws was exhibited"—was incoherent. In fact, given a radical's tacit knowledge, the speech made perfect sense: Saull said that democracy would have prevented the Tolpuddle abuse in the first place. The *Post* depicted the "ladies" (a common slur using quote marks) attending the ball afterwards, and the effects of further "raw gin" into the night were left to the imagination of its readers.[10] It was a clever exercise to reinforce just about every genteel prejudice. But

---

10   *Morning Post*, 17 Apr. 1838; for more denigration, see *Berrow's Worcester Journal*, 19 Apr. 1838. Political slants dominated such reports, so by contrast the *Globe* applauded the "good order, decorum, and respectability" of the crowd, as reported by *NS*, 21 Apr. 1838. This was the general view: *Bell's Life in London and Sporting Chronicle*, 22 Apr. 1838. Experience had taught them to expect the slanders, and the people "seemed resolved that no act of theirs during the day should afford a pretext for abuse or ribaldry to their enemies": *Champion and Weekly Herald*, 23 Apr. 1838. This source estimated the numbers at 80–100,000. The supposedly illiterate George Loveless, the most senior of the convicts, was an affable lay preacher. His 4*d* pamphlet, *The Victims of Whiggery* (1838)—published from Cleave's, Hetherington's and Watson's premises—had gone through eight editions and sold 12,000 copies by January 1838.

from this negative we can extract the positive—we can see how central Saull was inside the organization.

The Dorchester Committee might have seen Saull in highest profile, but it was not an isolated instance. He seemed to be a treasurer here, there, and everywhere. For example, he had barely begun to contemplate buying farms for the families, when he was chairing meetings and acting treasurer to the defence committee of another victim, the Rev. Joseph Rayner Stephens.[11] A Methodist preacher (although separated from the Connection because he advocated the disestablishment of the Church), Stephens had swayed men across Yorkshire, Lancashire, and the Midlands with his stirring sermons. He preached from open carts in town squares on the "misery" of factory workers and the iniquity of child labour. As for the Poor Law workhouse, it was "so abhorrent" that Stephens told his listeners to tar and feather the guardians and send them back to London. The mill owners were already sacking Stephens's supporters, now the Bow Street runners finished the job. In 1838, he was charged in Manchester with using "violent and inflammatory language" to incite the crowd of "evil-disposed and disorderly persons". Witnesses said he was naming the poor-law guardians and mill owners, and quoting Scripture—"Vengeance is mine"—while telling onlookers to "get their guns and pikes". (Apparently the sales of pikes did rise after each Stephens speech.)[12] Despite a campaign in Manchester, and a Saull-led defence fund in London, the authorities cracked down harshly. Even the bails and sureties before the trial ran to £4600, and, at Chester Assizes in August 1839, the Rev. Stephens was given an eighteen-month sentence.[13]

It is a wonder Saull found time to run his business and promote his science while juggling so many political balls. When not geology lecturing, visiting Paris with Owen, attacking the poor laws, campaigning, working to get George de Lacy Evans (the General just returned from leading the British Legion volunteers against the Carlist insurrection in Spain) re-elected radical MP for Westminster,[14] or

---

11 *The Operative*, 14 Apr. 1839; 21 Apr. 1839; *Champion and Weekly Herald*, 14 Apr. 1839, 5; 28 Apr. 1839, 4; *CGV*, 27 Apr. 1839, 2; *Charter*, 21 Apr. 1839, 200.
12 *MC*, 31 Dec. 1838; Holyoake 1881, 16, 27–28, 47–56.
13 *Times*, 4 May 1839, 6; *NS*, 11 May 1839; Holyoake 1881, 146, 172–74.
14 *MC*, 28 June 1837; *TS*, 28 June 1837, 6.

decrying the milk-and-water reforms which left the mass of the people untouched (at the Metropolitan Parliamentary Reform Association),[15] he was devoting his energies to corporation reform. On this Saull was indefatigable: he surfaced more in the daily press campaigning for City democracy in the thirties and forties than on any other issue. As an Aldersgate street ward elector of "common councilmen" of the City of London, and an officer—an auditor of the council's books—he had long been at the heart of affairs.[16] He was on the Committee to aid the Corporate Reform commissioners. No Guildhall meeting passed with the Livery of London pressuring the Whigs on reform without Saull's Committee work.[17] Nor was he less forceful in fighting the 'tyrannical power of Court of Aldermen' for refusing to admit an "Infidel"—Michael Scales of Portsoken Ward—who had twice been elected but kept out.[18]

Saull was, perhaps rather incongruously, a member of that droll body of extreme liberals, the "Ancient and Honourable Lumber Troop", by all accounts an uproarious dining club which met in Dr Johnson's House and once included Hogarth as a Trooper.[19] And, on one occasion, Saull had the old chaps see off a prospective Tory MP who stumbled into their group trying to canvas support to represent the City.[20] In this instance, the Tory was trounced at the polls by the East India merchant and reformer William Crawford, who did have Saull's backing. The City returned four MPs and the four candidates Saull supported were all reformers—and, given the long leftward drift of the City, they were all elected. Not, of course, that they were not grilled in advance on key issues. One of the founders of the new London University, George Grote, then struggling to write his *History of Greece*, was keen to see the ballot, triennial parliaments, and Church reform brought in, and the taxes on

---

15 *Daily News*, 28 Apr. 1849.
16 *Courier*, 26 June 1832, 3; *Atlas*, 1 July 1832, 421; *Times*, 26 June 1832, 3; *MC*, 26 June 1832; 25 June 1834; 25 June 1834; *Standard*, 25 June 1834, 1; *Royal Kalendar*, 1836, 297; 1838, 297; *TS*, 26 June 1834, 6; 28 June 1834, 3.
17 *Times*, 20 Sept. 1831, 3; 30 Sept. 1831, 3; *MC*, 28 Sept. 1833, 1.
18 Carlile claimed that Scales had been barred because he was an "Infidel": *Prompter* 1 (21 May 1831): 462. *Morning Post*, 22 Dec. 1832; *TS*, 6 July 1833; *MC*, 21 Sept. 1833.
19 James Grant 1838, 89.
20 The candidate was Francis Kemble (Beaven 1908, 1: 283, 294), who was obviously unaware that he was stepping into the lion's den: *TS*, 8 Aug. 1833, 2; 9 Aug. 1833, 2. There was a sort of freemasonry to the Troop, so I suspect that Saull was using it for business purposes.

knowledge and tithes thrown out. But Saull still pressured him on the iniquity of Dissenters being forced to pay church rates before finally giving his endorsement.[21]

As an anti-clerical Owenite and tormentor of his local vicar, Saull had long railed against church rates, levied for the upkeep of local Anglican churches. London, awash with "washed & unwashed Radicals", might have been godless in the eyes of Oxford and Cambridge clerics, but spires still dominated its skyline. It had an astonishing 400 churches and chapels.[22] Saull could waft away his local vicar's attempt at conversion with astro-geological confidence, but behind his contempt for a state-paid priesthood was real ire. Like all blasphemy-radicals, he fulminated against the "fat livings" of many a London incumbent, these "plundering oppressors" of the working poor.[23] Rectories could be in a bishop's or aristocrat's gift. Ones such as St Botolph's in Bishopsgate were pulling in £2500 a year. Cripplegate's vicar received £2,300 in tithes, church rates, and so on, and he was not even resident.[24]

The disparity between clerical wealth and parishioner poverty explains the rise of subversive rags like Cleave's *Slap at the Church* (1832) with it visceral laceration of that "destructive species of black slug called PARSONS".[25] This paper was so in-your-face that even when it morphed into the more sheepishly-clothed *Church Examiner* (1832) it was still prosecuted.[26] While the penny blasphemies remained the most colourfully vulgar, opposition to tithes and church rates spread through the whole Dissenting community, and since Dissenters by 1833 for the first time outnumbered Anglicans in the country, resistance

---

21 *British Traveller And Commercial And Law Gazette*, 24 Nov. 1832, 1. For Saull's endorsements: *TS*, 31 Dec. 1834, 2; *MC*, 31 Dec. 1834. In 1846 Saull was one of the merchants on the City's "Liberal Registration Association", which was designed to update the electoral register to maximise enfranchisement: *Daily News*, 23 Dec. 1846, 1.

22 *Cosmopolite*, 29 Sept. 1832, in HO 64/18 f. 652; "washed" quoted by J. A. Secord 2000, 267.

23 Hetherington 1830; [1832], 14; Saull 1828a.

24 *Cosmopolite*, 22 Sept. 1832, in HO 64/18, f. 667; *The Church Examiner and Ecclesiastical Record*, 1 Sept. 1832.

25 *A Slap at the Church*, 12 May 1832, in HO 64/11, f. 418. Saull pointed out that "The archbishop of Paris received £860 per annum, the archbishop of Canterbury, £25000!" And the French clergy made a greater effort to promote the "welfare of their flocks" so that "the lowest paid was best served", which was designed to appeal in an age of 'cheap government': Saull 1837.

26 *Church Examiner and Ecclesiastical Record*, 15 Sept. 1832, in HO 64/18, f. 384.

was actually widespread. Saull told his vestry meeting that there was a "gross injustice of imposing such a tax upon dissenters", and that he "would never, so long as he could raise his voice or his hand against it, consent to a church-rate [loud applause]". He helped to block it in his Aldersgate parish, where it was made a voluntary contribution rather than compulsory levy.[27]

Saull is rarely reported speaking on anything but radical politics at his vestry meetings.[28] Aldersgate ward politics were just as lively. Each year at the September elections of Common Councilmen, Saull would ask tricky question or propose reformist resolutions, almost all of which were carried. Thus in 1834 and 1837 he insisted on knowing the candidates' political views, which had previously not been considered important.[29] In 1836, he questioned the exorbitance of City spending on the King's domestic servants and wanted to know how prospective councilmen stood on the £500 set aside to build churches, which were irrelevant to City trade.[30] He did not operate alone. The Aldersgate ward radicals ran in tandem. So Saull would propose a council candidate (in 1834 and again 1837, T. Alcock, sometime spelled Allcock), who "was a sincere reformer, a friend to triennial Parliaments, household suffrage, and vote by ballot."[31] And Alcock would then successfully second Saull's resolution to ban the "abominable" anti-Catholic oath, which had to be pledged by prospective Common Councilmen. (This holdover from the years before Catholic emancipation, which was still barring Catholics from council offices, sat uneasily with the reformists' demands for religious toleration.[32]) And both men took part in the "spirited" denunciations on the new Police Bill in 1839, which radicals saw designed more to suppress discontent than prevent crime. It increased

---

27 *MC*, 15 June 1840.
28 The only exception I can find referred to a vestry meeting where he complained that the new river company was not supplying enough water to Aldersgate: *TS*, 19 Sept. 1835, 3.
29 *MC*, 23 Dec. 1834; 27 Sept. 1837; 29 Sept. 1837; *TS*, 23 Dec. 1834, 8; *Times*, 29 Sept. 1837, 3.
30 *Patriot*, 22 Dec. 1836, 565; *Baldwin's London Weekly Journal*, 24 Dec. 1836, 4.
31 *MC*, 23 Dec. 1834; *TS*, 23 Dec. 1834, 8. On Saull and Alcock see also *TS*, 22 Dec. 1835, 8; *MC*, 27 Sept. 1837; 29 Sept. 1837; *Times*, 29 Sept. 1837, 3.
32 *TS*, 26 Dec. 1835, 4.

the number of offences and police powers, and allowed constables to close establishments on the Sabbath.³³

Even inside the broader coalition making up the "City of London Corporation Reform Association"—designed to make the City more fiscally prudent by rooting out the vestiges of corruption—the radicals, including Alcock and Saull, pushed to increase the ballot at both the City and local levels.³⁴ The City's diluted response to corporation reform led to more radical petitioning, and attacks on representatives whose corporation power displayed itself in "ostentation, jobbing, and gluttony". Saull, the abstemious Owenite, actually resigned as a City auditor in the mid-thirties, because he was "so ashamed of the large sums of money which were voted away for eating and drinking".³⁵

Others in the City could be counted on. Perhaps the most interesting was fellow municipal reformer Henry Bradshaw Fearon, who mixed religious and political radicalism. Fearon was a Freethinking Christian. This tiny sect denied Christ's divinity and the doctrine of the Fall, and Fearon swallowed Volney and Holbach like the rest. And even though Saull's materialists never understood the sect's faith in Scripture, it did not stop their joint activism.³⁶ Like Saull and Alcott in Aldersgate, Fearon was a reformer in Farringdon ward. All of these men backed one another at Guildhall meetings, for example in agitating for the Lords to pass the Corporation Reform Bill.³⁷ Fearon was also in the liquor trade, a brewer and owner of Thompson and Fearon's gin palace on Holborn Hill, where a tot could be gulped on the trot (there were no tables or chairs in the main bar). A noggin of "Fearon's best" was "served by young women dressed up like the BELLE LIMONADIERE of a Paris Coffee-house, and the establishment in all its parts is nearly as fine as VEREY's or the CAFE DE PARIS".³⁸ This was supposedly the largest gin-shop in England, possibly the world, with its profits again funding the movement—and,

---

33   *Times*, 18 Mar. 1839, 5; Goodway 1982, 103–05.
34   *MC*, 19 Jan. 1839. See also *Courier*, 9 Nov. 1838, 4; *MC*, 16 Nov. 1838; 23 Nov. 1838; 12 Jan. 1839; 21 Feb. 1839; 18 Apr. 1840; *Patriot*, 26 Nov. 1838, 766; *Times*, 1 Dec. 1838, 6.
35   *MC*, 4 Apr. 1839; 20 Mar. 1840.
36   *GM* 143 (1828): 507–12. Henry Hetherington and Saull's solicitor William Henry Ashurst were also members of the sect. McCalman 1988, ch. 4, for the sect's history.
37   *TS*, 30 Sept. 1835, 2; also 13 Aug. 1835, 8; 28 Aug. 28 1835, 2.
38   *The Man*, 8 Dec. 1833, 176; Grant 1838, 223; J. White 2007, 283.

like Saull's depot, it *was* profitable. As a Freethinking Christian, Fearon shows an alternative scientific direction that radicalism could take. He used Lawrence's *Lectures on Man* to make life and mind depend on bodily organization. Thus, body *and* mind were material, and with Fearon using Scripture to make body and soul one, the soul was itself so much matter, and mortal. In fact, every "beast of the earth" was a living soul.[39] The soul goes the way of all flesh, and it is only by a later resurrection of the dead that eternal life will begin. Thus Fearon, the mental materialist, ran with the anti-clerical radicals. In fact, they were all part of the same "mares'-nest" to a bemused *Fraser's Magazine*, which was reduced to ribaldry. Since Fearon, "this infidel high priest of a Spirit Temple", had made his money in spirits and was looking to retire, he was obviously suffering "remorse for the souls and bodies he has been thereby the instrument of ruining", and therefore was inclined "to disbelieve the existence of either".[40] More orthodox medical journalists just sighed that mental materialism "can scarcely make a single convert in an age so enlightened as the present". But it got a better hearing in London's back-street medical schools,[41] and Fearon became a powerful ally in ward politics.

So radical were many City electors that one purveyor of "devilish poison", James Harmer, a Spitalfields-weaver's son turned solicitor and then Alderman or chief officer of Farringdon ward—the man who presented Saull's petition against his blasphemy charge to the Court of Common Council in 1828—narrowly failed to become Mayor in 1835.[42] His failure reflected less his role as the "Thieves' Attorney-General": he was what would later be called a civil rights lawyer, specializing in wrong committals. It was more the content of Harmer's huge-selling *Weekly Dispatch* which drew the ire of the rival *Times* and the Tory aristocracy's favourite *The Age*, that mainstay of the gentlemen's clubs. What radicals saw as the *Dispatch*'s dare-devil attacks on religious and political abuses, they saw as "blasphemy, disloyalty, and immorality", and not without justification. The *Dispatch*'s "foetid and ... loathsome"

---

39 Fearon 1833, 53.
40 *Fraser's Magazine* 9 (Apr. 1834): 424–34.
41 Desmond 1989, ch. 4; *London Medical and Surgical Journal* ns 4 (25 Jan. 1834): 819–23.
42 *TS*, 30 Sept. 1835, 1. He ran again in 1840.

letters penned by "Publicola" (John Williams, the turncoat scion of a Tory family who skewered princes and priests in his coruscating epistles) showed him to be "the greatest single foe of the Church in the country".[43] One onlooker fingered the *Dispatch* as the only paper (above-ground, that is) "which openly advocated Atheism".[44] It helped to make what Shepherd Smith slated as the "blackguard paper"[45] into the best-selling weekly in the country. It was shifting 60,000 copies every Sunday in 1840. The *Dispatch*'s announcement that "There is no more moral depravity in being an infidel than in being a clergyman" left *The Age* incandescent. In the pubs, by contrast, Publicola's letters were voraciously devoured: huge numbers of *Dispatches* ended up in the gin joints and coffee houses, despite being an expensive paper (8½d before the repeal of the newspaper tax in 1836, 6d afterwards). *The Age* could claim that the "beer-shop thieves' dens are filled with ruffians, whose principal incitement to crime is in the columns" of Alderman Harmer's paper. The *Times* and *Age* hammered away at the *Dispatch*'s insults to the sovereign, religion and what Publicola called "that bloody and beastly book" (the Bible). It was enough to frighten the electors into placing the mayoralty into safer hands. But only just.[46]

---

43   Maccoby 1935, 420.
44   Grant 1871–72, 3: 42. And at least one respectable Mechanics' Institute, Gloucester's, cancelled its subscription accordingly: C. Turner 1980, 264.
45   W. A. Smith 1892, 210.
46   *The Age*, 27 Sept. 1840, 308, 309; Bourne 1887, 2: 101–02; Maccoby 1935, 416; J. Williams 1840.

16. Lease-holder of the New Moral World

In the 1830s, Saull operated at the highest Owenite echelon. He had become financially indispensable. Robert Owen's London home for twenty years was his eleven-room town house, No. 4, Crescent Place, Burton Crescent, near the new London University. The American Henry Darwin Rogers, who had alternated geology lectures with Saull at Owen's Institution, had made himself "comfortable" here while lodging with Owen.[1] But it was not actually Owen's house. It was Saull's. He owned it and presumably leased it to Owen.[2]

Owen additionally hired cheap "dilapidated" rooms in the adjoining Burton Street,[3] and, in 1835, he moved his lecturing operations there. This was to be the centre of his newly-formed "Association of All Classes of All Nations"—a title not designed to appease the *Poor Man's Guardian* class warriors—which was to "effect an entire change in the Character and Condition of Mankind". The old immoral world was to be rejected—hence the title of the Association's penny weekly paper in November 1834, to replace the collapsed *Crisis*: the *New Moral World*. Ever the wag, Holyoake remarked that Owen's opening speech in 1835 occupied two entire numbers of the *New Moral World*, "long enough to weary both worlds at once".[4]

This new Burton Street venue saw Saull in 1836 give two lectures on "Geology in reference to Human Nature". With so many applying

---

1   Gerstner 1994, 24.
2   In 1852, after Owen moved, Saull advertised it at a rent of £50 per year: *Reasoner* 13 (16 June 1852): 288; 15 (28 Sept. 1853): 208.
3   Ron Dobie to Robert Owen, 18 Jan. 1831, ROC/4/25/1, Owen Collection 2011, Co-Operative Heritage Trust Archive, Manchester.
4   Holyoake 1906, 1: 137.

to hear about the geological portents of mankind's moral destiny, these were then turned into a longer course. Saull directed audiences to good books on geology, which he himself consulted "for their concurrence with his views in some cases, as for their opposition in others". The opposition was obvious, for none would dare condone any sort of continuous, uninterrupted, filiation of fossil life. The most outstanding source was the monumental work of the moment: the Rev. William Buckland's two-volume *Geology and Mineralogy Considered with Reference to Natural Theology*, hot off the press in 1836. One of a series of Bridgewater Treatises, it was typeset, printed, and priced by Pickering, "a publisher of taste" to the gentry, in the mistaken belief that it was a theological work.[5] Conservative texts had always managed to fire up a reaction in radical readers.[6] Saull's judicious sifting was a good example. Acceptable were Buckland's deep-time chronology, loss of Moses (the Oxford divine had long done with the "Days" of creation and now recanted his belief in a geological Flood), and progressing diversity and periodic appearance of "higher" types of fossil life.[7] Jettisoned would have been the Reverend's designful explanations with their backdoor to Providence, as well as talk of "the direct agency of Creative Interference" to produce them.[8] Saull told socialists that the volumes were "of great importance in the settlement of leading truths in Geology", while "striking at the root of certain mysterious traditions".[9] He was trying to push Buckland into an uncomfortable role as a fifth-columnist.

At Owen's yearly birthday celebrations, the rational entertainment preceded the dancing. As at bourgeois *soirées* and *conversaziones*, curios were exhibited as talking points. In Burton Street, splendid lithographic drawings would line the walls; on tables sat phrenological busts, while large electrical instruments would be set in motion.[10] These pre-dance

---

5   Topham 1998, 242; Topham 2022.
6   Rose 2002, 39.
7   Rudwick 2008, 426 et seq.
8   W. Buckland 1836, 1: 586; Hilton 2000, 187, for an understanding of Buckland's "succession of separate dispensations" within a law-based Creation.
9   *NMW* 3 (10 Dec. 1836) 53.
10 *NMW* 3 (20 May 1837): 235. J. F. C. Harrison 1969, 223 on the Owenites' fondness for dancing; Morrell 1976, 137. Morus 2010a, b on the choreography, and Morus 1998 on the electrical entertainments of the moment. Socialist conversaziones mirrored those in polite society, with their "learned lectures and dazzling displays", scientific curios, and organ music: Alberti 2003.

performances were as delicately choreographed as in any scientific auditorium—a mix of showmanship and technical wizardry, stage-managed to proclaim a rival authority for a compatible Owenite nature: electrical machines to sustain Mackintosh's self-regulating cosmos, fossils to establish trust in a self-regulating evolution. And, for that, Saull's ancient saurians and crinoids were invariably on display, with Saull on hand to give impromptu explanations.[11] His specimens could be show-stoppers. Big was always best—and his museum had the biggest and the best. Since these were given priority by travel guides, we know most about them: the gigantic reptile *Iguanodon* was estimated by Mantell to have reached seventy feet long, but specimens in Saull's cabinet appeared to have come from even larger individuals. His fossils were simply "of enormous size", and what better to amaze the guests than its seven-inch claw? Not only was the marine reptile *Ichthyosaurus platyodon*—a dolphin-shaped saurian—the largest known from the rocks of Lyme Regis and Gloucestershire, but Saull had the largest known vertebra, which measured almost eight inches. Then there were the tusks of mastodons, a "full three times" the size of an elephant's, or else a huge, showy ammonite, *Ceratites nodosus*, a "museum quality" specimen from Hanover, highly ornamented with elaborate wavy sutures. Rarity too was a draw. At events, Saull would show scarce fossil palms[12] or tree ferns from the British coal mines, perhaps even the trunk of *Sigillaria saullii* that Adolphe Brongniart had named after him.

Burton Street could be a transformative experience for inquisitive youngsters just setting out, none more so than the future 'Darwinian' evolutionist Alfred Russel Wallace. Old and celebrated, Wallace, in *My Life* (1905), recalled leaving school at fourteen and joining his apprentice-carpenter brother in London for a month or two "early in 1837":

> our evenings were most frequently spent at what was then termed a "Hall of Science", situated in John Street, Tottenham Court Road ... Here we sometimes heard lectures on Owen's doctrines, or on the principles of secularism or agnosticism, as it is now called; at other times we read

---

11 NMW 3 (20 May 1837): 235.
12 *Crisis* 1 (5 Jan. 1833): 174; *Ceratites*: Spath 1934, 477; tusks: *Gardeners Gazette*, 29 Dec. 1838, 827; *Court Gazette*, 29 Dec. 1838, 614–15; A. Booth 1939, 122; *platyodon* vertebra: Lydekker 1889a, 94, 97, 101–02; claw: G. F. Richardson 1842, 402; A. Booth 1839, 122.

papers or books, or played draughts, dominoes, or bagatelle, and coffee was also supplied to any who wished for it. It was here that I first made acquaintance with Owen's writings ... I also received my first knowledge of the arguments of skeptics, and read among other books Paine's "Age of Reason"... I have a recollection of having once heard him [Owen] give a short address at this "Hall of Science", and that I was struck by his tall spare figure, very lofty head, and highly benevolent countenance and mode of speaking.[13]

The flame was sparked here, and Wallace became a torch-bearing socialist and future land nationalizer. "Here", writes Jim Moore, "Wallace picked up the political values that stayed with him more or less for life: human nature is perfectible through education and changed environments; all humans are equal partners in progress."[14] But *where*, exactly? What *was* the venue? Seventy-year-old memories are notoriously flaky, and anachronisms abound in Wallace's recount. If it really was "early in 1837", then the institution was in Burton Street. Owen himself was around in January and February 1837,[15] so Wallace could have heard him then. There is a possible alternative though—a daughter institution that actually was in John Street: at No. 49, the Community Friendly Society (1836–39). By 1837, the labour exchanges and the co-ops had nearly all collapsed, and this was a rare survivor. It was small, only thirty-three members, yet it had a grocery store, paid out sick and unemployment benefits, and, more to the point for an inquisitive fourteen-year-old, held Sunday meetings with lectures, and it celebrated its anniversary each April with dancing and singing.[16] Moore himself, however, favours the probability that Wallace attended the John Street Institution (founded 1840, at No. 23) in the forties but mis-remembered the year. Whatever the venue, the impressionable youngster was imbibing the flagrantly anti-Christian ethos which encouraged materialist explanations of nature and Saull's evolutionary ascent of life. The strong cultural relativism here would mark Wallace's

---

13   A. R. Wallace 1905, 1: 79, 87, 89, 104.
14   J. R. Moore 1997, 301; G. Jones 2002, 74. Jim Moore (pers. comm.) thinks Wallace probably saw Owen in 1844, giving the reasons in his forthcoming study of the young Wallace.
15   *NMW* 3 (7 Jan. 1837): 85.
16   *NMW* 3 (22 April 1837): 202; (6 May 1837): 220–21; Garnett 1972, 145–46.

social thought, just as the environmental conditioning and emphasis on perfectibility would shape his evolutionary optimism.

Owenism had always been a broad church, with the materialists making up one wing. Saull's deism of the 1820s had become atheism by the thirties. The fracture points inside the Owenian fellowship were already evident by 1838 as the sacred socialists started pulling away from Saull's materialists. The split widened as Saull injected big money into educational communities. To bypass Church- and Dissent-administered education for the co-operators' children, Saull had long argued for a demystified, science-based 'rational schooling'. Nothing but an emancipationist programme would counter the gentry's efforts to put the children into livery.[17] However his transcendental friends were looking to a more mystical, holistic, ascetic communitarianism, and schooling for them would require a spiritual dimension.

The issue exploded in 1838 when Saull received £100, subsequently upped to £1000, from an anonymous donor, with instructions that the Owenite community rent land in order to establish an Educational Friendly Society. There were few strings, except that Saull was to be Treasurer and new recruits to the community were to be sought immediately.[18] Middle-class philanthropy always raised the spectre of loss of independence, but, given that there were few conditions, a *New Moral World* editorial was in favour. Not so a long-standing Saull colleague, the vegetarian transcendentalist Charles Lane. He was an acolyte of the Pestalozzian James Pierrepont Greaves, and (like Saull and Hetherington) had been a radical in the London Mechanics' Institution, where he introduced Pestalozzian teaching techniques, encouraging invention and mutual-instruction classes. All three men had been further radicalized in the reform years and had worked inside the Metropolitan Political Union.[19] But Lane's growing transcendentalism meant a complete rejection of Saull's money offer: only the "submissive harmonious concurrence of humans" can lead to true co-operation, and accepting middle-class cash would "ruinously fetter its operations".

---

17   *Crisis* 3 (4 Jan. 1834): 150.
18   *NMW* 4 (20 Jan. 1838): 100; (17 Feb. 1838): 131–32; *NS*, 3 Mar. 1838. Johnson 1979, 98–99, on the fear of middle-class control.
19   Flexner 2014, 3, 32, 159, 161; J. F. C. Harrison 1969, 128–29; Armytage 1961, 138, 173–78, 182–83.

Money, he argued, can hardly buy the "FEELING of universal fraternity, of which community of property is the SOCIAL FORM". The editor of the *New Moral World* was more materialistically inclined (in all senses) and responded: "But the real truth is, that the principal bond of union, the primary agent in the production of real or apparent *spiritual* accordance, is the abundance and quality of the *material* foundation on which it rests."[20] Donations were now coming in from others quarters, £250 from a "very aged" socialist, £500 more when Pierre Baume toured Bradford as a representative of the new venture. By May 1838, there was £2500 in the pot.[21]

The freethinking Owenites carried the day. The sectarian split became total as Lane left to found the Alcott House community and boarding school in Ham, Surrey. Here, individual regeneration rather than social engineering would become the new goal—a sort of spiritual enlightenment gained by abstinence, celibacy, and raw carrots—a monastic retreat from the dehumanized industrial world where men grew long hair and beards.[22] Not much sympathy was shown by the other side. The social missionary Lloyd Jones complained that communitarianism here had become "a receptacle for all moral and intellectual delinquents—empty-headed young men bordering on idiocy, babblers and quibblers, long-haired, bearded and vegetarians, etc".[23] The cold and raw carrots drove many out, and Lane himself sailed to America in 1842 to found the short-lived Utopian Fruitlands community near Harvard University.[24] It was a foretaste of the Owenite splintering to come as the materialists became more vociferously atheistical.

Saull's group continued in their uncompromising anti-spiritual stance. They hosted debates on the divine inspiration of the Bible, in which up-and-coming social missionaries would take on young evangelicals, fresh out of divinity school. These were lively and always big crowd-pullers. A number of ambitious tyros on both sides cut their teeth of these lions-den sessions. For example, Saull chaired one before Christmas 1838, in which a Trinity College, Dublin graduate,

---

20 *NMW* 4 (10 Mar. 1838): 155–57; (17 Mar. 1838): 163–64, 168.
21 *Proceedings of the Third Congress of the Association of all Classes of all Nations*, 1838, 39; *NMW* 4 (31 Mar. 1838): 183; (14 Apr. 1838): 197; Holyoake 1906, 1: 145.
22 Frost 1880, 41–48; Latham 1999, 20–21; Hardy 1979, 60–61.
23 Claeys 2002, 261; McCabe 1908, 1: 113.
24 R. Francis 2010.

the Rev. Joseph Baylee, took on Holyoake's fellow student fresh out of Birmingham Mechanics' Institute, Frederick Hollick. (Even Holyoake stood in awe of the 25-year-old, who had "the brightest mind of any student in the Mechanics' Institution."[25]) The protagonists used the Social Institution as their career spring-board in 1838. The debate might have been lively, but it was not particularly enlightening. After Saull failed to stop Baylee from kicking off with a hymn and prayer, it settled into the standard bicker on Moses' plagiarism and the bishops' politicking to get certain gospels accepted as inspired.[26] But such public exchanges allowed both men to sharpen their rhetorical strategies. The protagonists would both become doctors, Baylee of divinity, Hollick of medicine (or so he claimed).[27] Baylee, a regular at such theatrical confrontations,[28] would go on to found St Aidan's Theological College, Birkenhead, in the forties. Hollick sailed to America in 1842 to gain fame for popularizing esoteric medical lore, as befitted an Owenite.

---

25   Holyoake 1892, 1:49, 60.
26   Baylee and Hollick, 1839.
27   Hollick's "M.D." is problematic. From 1840 to early 1842, he was successively a social missionary in Liverpool, Edinburgh, and Birmingham. In August 1840, while in Liverpool, he gave four popular lectures on teeth (despite a lack of books), followed by a short course on human physiology. We get an idea of his (lack of) qualifications from a branch report:
"What! a Social Missionary,—a mere stripling, a youth of two-and-twenty ... presuming to step into a path hitherto trod only by the learned of the faculty! to teach a theme monopolised by colleges, and carefully kept within their time-honoured walls as something too sacred and precious for vulgar minds. These Socialists ... have upset the parsons, and are now tilting with the doctors. Who shall stay them before the world is turned upside down? Mr. Hollick delivers these lectures in the Hall of Science, and commands a larger audience than can be found to attend any other Institution in Liverpool; they are given in a scientific, popular, and practical manner, and so clearly enunciated, that every one may understand without difficulty [NMW 8 (26 Sept. 1840): 203]."
His movements show him in Edinburgh inaugurating the new Clyde Street Hall of Science and giving the same lectures in December 1840. These seem to have been his sole experience in 'medical' lecturing before emigrating to America in Spring 1842. In December 1843 he was at the new Social Institution in New York, and in 1844 press reports of his popular medical lectures here referred to him as Dr Hollick, while the title page of his *Origin of Life* has him "Frederick Hollick, M.D.", so it remains to be seen how he acquired this qualification. With licensing laws absent in most American states, and diploma mills not uncommon, it was not hard to obtain one. Like many Owenites, he was actually suspicious of medical power and merely exchanged a missionary platform in social engineering for a New York platform in anatomical popularizing.
28   NMW 6 (7 Dec. 1839): 940 passim; Larsen 2004, 106.

"Dr" Hollick—like other socialists and Chartists who published self-help science books[29]—went on to champion the democratization of anatomical knowledge. This was in spite of clerical distaste for anatomy and dissection and contemporary fears for the moral consequences of physiological knowledge among the "corruptible" poor.[30] Owenite teaching strategies were designed to demystify medicine as much as religion. This could mean breaking the doctor's financial hold on poorer patients and, with it, his moral influence, which too often perpetuated middle-class mores.[31] No more strenuous plebification of knowledge was to be found than Hollick's. He extended Owenite ideals of female liberation and marriage as a loving contract to pioneer books naturalizing sexual health and pleasure. The title of his first said it all: *The Origin of Life: A Popular Treatise on the Philosophy and Physiology of Reproduction, In Plants and Animals, Including The Details of Human Generation with a Full Description of the Male and Female Organs* (1845). This would have been difficult to publish in England. Even in New York, the revelation of Latin-guarded lore (which was how he presented it[32]) upset public propriety, and he was prosecuted for obscenity. But the book still passed through twenty editions in three years. For his part, the Rev. Joseph Baylee would see in print, among many books, the obligatory *Genesis and Geology; The Holy Word of God Defended from its Assailants* (1857).[33] It was a sign of the later times that Baylee eventually took a parish (Sheepscombe in Gloucestershire), where the sacred socialist Greaves's closest disciple,

---

29   For instance, William Lovett (1851, viii–xix), who concurred with Hollick that mankind's moral deficiency, exhibited in his "class dominations", showed him to be "defectively taught" and in need of a levelling anatomical education. Lovett superintended a day school for the London branch of the National Charter Association in 1846, which taught the secular sciences and other "improving" subjects. With encouragement from George Combe and John Elliotson (Lovett 1920, 2: 370–73, also 326–27, 384–89), the self-taught Lovett went on to publish an *Elementary Anatomy and Physiology*, which was well received as a plebeian teaching aid.
30   *Lancet* 1 (1 Jan. 1831): 470–72.
31   R. D. Owen 1839, 8–10.
32   Hollick 1848, xv-xvi; *UR*, 2 June 1847, 53–54, bound with *Reasoner* 3 (1847); J. F. C. Harrison 1987, 211. The Owenites' fascination with the creative aspects of electricity was taken further by Hollick, who used it to cure neuralgia, paralysis, and rheumatism.
33   Baylee's exegesis relied on an unrecorded length of time between God creating the world and then re-ordering its chaotic state in the Six Days. During this time, geological processes could occur in the supposedly chaotic undersea world (Baylee 1857, 8–13).

Georgina Welch, now rather orthodox, patronized and supported his church.[34]

The People's Charter

At the end of the thirties and into the hungry forties, there was an angry cast to poor urban life. Reform for the middling sort had palpably failed the working class. The economic depression begun in 1836 had reached a peak, leading to hunger, frustration, and increasingly violent agitation, including the abortive Chartist uprising in 1839. In the first Birmingham council elections, in 1838, the Tories were "mangled and minced". The radicals took power, and the Mayor's court even included a Charter signatory. With the Birmingham Political Union in control, and the local police force yet to organize, Chartists in May 1839 moved their national Convention to the city—to finalize their national petition for male suffrage—away from the heavily-policed capital. But sixty Metropolitan policemen were bussed in from London, and their attempts, backed by the military, to stop banned meetings in the Bull Ring, led to riots in July. The insurgency was fierce: some Chartists were armed, a neighbour of Holyoake's had his nose chopped off, and ten policemen were hospitalized. Lovett was arrested for printing placards condemning the magistrates and was sent to Warwick gaol.

Birmingham also beckoned the socialists. The printing of the *New Moral World* was shifted there, and the Central Board set up in the town. Holyoake and Hollick were made paid social missionaries by the local branch. In May 1839 the fifty countrywide Owenite branches convened the longest socialist congress to date (sixteen days) in the town.[35] At this tense time, the socialists distributed half-a-million tracts locally, discussing everything from co-operation to female emancipation and the abolition of traditional marriage. Fear of the Chartists led to a sparse meeting of the British Association for the Advancement of Science in Birmingham that August. Many of the savants were simply frightened away. In the event, the government commuted the death sentences on three Bull Ring rioters, which dampened the insurgency, but the town

---

34 Latham 1999, 148.
35 Maccoby 1935, 194–98; Hovell 1918, 156; Holyoake 1892, 1: ch. 5, ch. 17; Holyoake 1906, 1: 127; Fraser 1979, 88–89; Royle 1974, 50, 62.

was still a "feverish quiet" with peace ensured by "men in green and men in red, police staves and cavalry sabres".[36]

Birmingham was the citadel of English Jacobinism, or so said the *Courier*.[37] It was also young George Jacob Holyoake's town. Holyoake was one of the new-breed recruits, aggressively pushing the secular implications of Owenism. Extremely talented, he had come out of Birmingham Mechanics' Institute, where he had studied science and won the maths prize, and had ended up on the committee conducting classes.[38] The socialists having set up in Allison Street, he sat their lectures too[39] and he heard Robert Owen speak there on 15 June 1836. By 1839, he was a social missionary, and, in 1840, actually wrote a book on Euclid for use in schools and Owenite halls. But despite this science interest, he, like Hetherington before him, now emerged arguing that such intellectual pursuits were secondary to the political struggle. One can understand it, given the horror of the economic depression, which peaked in 1839–42. But he still had little sense of materialist science's longer-term ideological pay-off.[40] He knew that geology was all the rage, and, already displaying that brilliantly humorous streak of his, he talked of "Saurian" Tories as long extinct, if they but knew it, and the age clamouring for "Geology, and Gaslights".[41] However, it was not long before he too was forced into the fray, denying orthodox claims that geology supported true religion.[42] Then came the obligatory recommendation that autodidacts visit Saull's "excellent Museum", so vast now, and "so rich in curiosities", that enumerating the exhibits would require "converting [his report] into a Catalogue".[43]

Shock waves from the 1839 socialist congress in Birmingham rippled through polite society. The Bishop of Exeter stood in the Lords to

---

36 Morrell and Thackray 1984, 257, 321–22, 326–27; 1981, 252.
37 *Courier*, 26 Dec. 1836, 4.
38 George Jacob Holyoake, "Brief Notes of Lectures" (1838–1839), MS, passim and "Log Book" No. 1, MS, Bishopsgate Institute, London. Holyoake 1892, 1: 142; McCabe 1908, 1: 48.
39 *NMW* 2 (19 Mar. 1836): 168; Holyoake "Brief Notes," passim.
40 *NMW* 9 (6 Feb. 1841): 88. However, he never failed to respond when provincials wanted a lecturer to demolish some itinerant's geology-supports-genesis line (*NMW* 10 [26 Mar. 1842]: 311).
41 *NMW* 10 (9 Oct. 1841): 114.
42 *NMW* 10 (26 Mar. 1842): 311.
43 *Reasoner* 1 (6 Aug. 1846): 159.

blame William Pare, the superintendent-registrar of births, marriages, and deaths in the town, and the local socialist vice-president, who had officiated in Congress debates on the Bible as "a tissue of lies". It was enough to have the bishop declare him unfit to hold public office, and Pare was forced to resign. "Was this Christianity"? asked Saull speaking in Pare's defence "doing as they would be done by?"[44] Of course, worse could have befallen them. The clergy, magistracy, and manufacturers were a formidable foe. It was not unknown for the bigger bosses to employ drunken gangs to storm socialist meetings, and, in one terrible instance, to try to kill the lecturer.[45]

Saull seems fairly unique in standing at the intersection of all three of these Birmingham congresses: Socialist, Chartist, and British Association for the Advancement of Science. But it is not actually known whether he attended any of them.[46] He *might* have been at the BAAS meeting, for the reason that the comparative anatomist Richard Owen was reading the first part of his "Report on British Fossil Reptiles",[47] to prepare which he had visited Saull's museum. And Saull *did* take a steam vessel over to Boulogne that September with a BAAS delegation. There they had a week-long joint meeting with the Société Géologique de France, of which Saull was also a member. At one of the sessions, Saull acted as liaison officer, and inducted a new member into the French Société.[48]

It is questionable whether Saull had much to do with the Chartist Convention, though, save the odd pound put into its collection, or donations to Frost's defence fund.[49] But having his feet in the Owenite camp did not prevent some sympathy or stop him lecturing the Chartists at their Hall in the Old Bailey. The first talk was on 16 September 1841, when he delivered "a very excellent and instructive lecture" on the

---

44 *NMW* 7 (20 June 1840): 1322; (8 Feb. 1840): 1093–94.
45 *NMW* 7 (9 Feb. 1840): 1176; Buchanan 1840b, 407.
46 His whereabouts can only be pinpointed for one summer day: he was in London, at a Numismatic Society meeting, on 18 July, a week after the violent Bull Ring meetings, which he was unlikely to have attended anyway (*Proceedings of the Numismatic Society*, 1838–39, 351). His absence is indicated by the *Northern Star*'s lament, commenting on Lovett's arrest: "Where were George Rogers and Mr. Saul [sic], that they did not immediately repair to Birmingham, to give bail for the glorious Lovett"? (*NS*, 13 July 1839).
47 Richard Owen 1840, 43–44.
48 *MC*, 14 Sept. 1839; *Bulletin de la Société Géologique de France* 10 (1839): 385–86, 431.
49 *Charter*, 24 Mar. 1839; *Operative*, 24 Mar. 1839; *NS*, 13 July 1839.

"social and political condition of the country." Local Chartism was now in uproar—a fracture in the movement had led physical-force Chartists to condemn the moderate "Lovettites" for setting up a broader coalition, a "National Association".[50] Being a moral-force Owenite, Saull had feared he was stepping into a lion's den. Saull's lecture referenced an ugly incident a few nights earlier, on 7 September, when, at his Mechanics' Hall of Science, Watson had excoriated a physical-force extremist who had called for Saull's long-time friend Lovett to be assassinated.[51] It cast a pall over the Old Bailey lecture, and Saull had "expected to meet with much opposition". In the event, the talk went off well, probably because the physical-force activists were off listening to their "patriotic champion" Feargus O'Connor, fresh out of prison and in town that night. So Saull set up a quarterly series of lectures in their hall, beginning on geology, and, naturally, he invited the Old Bailey Chartists to his museum.[52]

Of all the Chartists, Saull's heart was closest to Lovett, Watson, Hetherington, and the other "traitors, assassins, and spies", as they were branded by the wilder insurgents.[53] 'Knowledge Chartists' was a kinder appellation for Lovett's activists. But even that was supposed to be a slur on their talk of self-improvement and rational education, and on their demands for a penny subscription to set up schools and libraries. These ideals, though, remained dear to Saull's heart. The Knowledge Chartists took over a hall capable of holding 1,000 people in High Holborn, which was to become the National Association headquarters, and Saull personally assured Lovett he would attend the opening in July 1842.[54] Here they planned to house libraries and schools, with children taught by day, and adults receiving lectures at night on "physical, moral, and political science".[55]

Lovett's goal was to ready the populace for power, as was Saull's. But the differences in temper and intent between Owenism and Chartism

---

50 Stack 1999, 1028.
51 Goodway 1982, 42; *NS*, 11 Sept. 1841; Lovett 1920, 2: 254–64; Wiener 1989, 87.
52 *NS*, 18 Sept. 1841; 9 Oct. 1841.
53 Goodway 1982, 42; *NS*, 11 Sept. 1841; Wiener 1989, 86.
54 W. D. Saull to W. Lovett, 13 July 1842, British Library MSS Catalogue Add. 78161, f. 162.
55 *Nonconformist*, 27 July 1842, 515; *National Association Gazette* 1 (30 July 1842): 243–44.

generally were immense. As J. F. C. Harrison once put it, Owen's propaganda machine was unequalled. "In the peak years 1839-41 two and a half million tracts were distributed: 1450 lectures were delivered in a year, and Sunday lectures were attended by up to 50,000 weekly."[56] This was a literary onslaught, whereas Chartists went for vast open-air rallies demanding industrial action to gain their demands: suffrage, private ballot, annual parliaments, and so on. Owenism at times looked to be devolving into an educational movement, and such would Saull have it. Despite his odd lecture to Chartists on "just government", support for Lovett, or treasurer's role holding funds for the families of Chartists who had died in jail from cholera,[57] there was little apparent contact.

All of this makes it strange what that left-wing insurrectionist, "the 'Marat' of 1839",[58] George Julian Harney, learned in Scotland. Harney was one of the youngest members of the 1839 Chartist National Convention (just twenty-two, but he already had form: as Hetherington's shop-boy he had been imprisoned twice for hawking the *Poor Man's Guardian* while still in his teens). After the Convention, he made a confidence-boosting tour of Chartist communities in the North and Scotland. He reached Kinross, twenty-five miles north of Edinburgh in February 1841. Here, the small band of local Chartists had taken steps towards building a meeting hall, which "will enable them to laugh at the petty tyranny of the idiotic, knavish 'respectables.'" Further, Harney heard, they were planning to ask Saull to stand as their Radical MP "in opposition to the Whig tool and placeman Admiral Adam. A resolution was passed at the meeting, inviting Mr. Saull to explain his views and principles upon public subjects, particularly as to the People's Charter."[59] If the offer did reach Saull, he never stood. He was never a Chartist, even if he supported Lovett. We do not even know how he viewed the "Charter-Socialists", who advocated a republic based on the Charter's democratic principles but with socialist institutions. In truth, he was more the suit

---

56 J. F. C. Harrison 1969, 31.
57 NS, 3 Dec. 1842; 28 Jan. 1843, 8; 27 Oct. 1849. He was joined as treasurer by George Rogers.
58 Epstein 1994, 19–20; Claeys 1987, 160. There was no love lost between Lovett and Harney: Lovett (1920, 1: 207) deplored the dagger-brandishing Harney's "insane and foolish conduct".
59 NS, 13 Feb. 1841; 26 June 1841.

behind the scenes, the financier and organizer, his 'rigmarole' speeches lacking Wakley's fiery lustre. It was just as well. Chartism was vilified in the press and candidates were guaranteed a bad reception, as Harney himself found out on standing at Tiverton later. With few ten-pound householders daring to publicly declare for a Chartist (there being no secret ballot), he did not pick up a single vote.[60]

---

60 W. E. Adams 1903, 1: 223.

PART III

1840s

ATHEISTS AND ABORIGINES

17. Halls of Science

Saull's home remained London Owenism. But that home kept shifting. The geographical base seemed always in a state of flux, and the peripatetic Association of All Classes of All Nations moved again in October 1837 to 69 Great Queen Street, Lincoln's Inn Fields. Here the stalwarts began to experience "a very material change for the better". London membership soared, a day school was established, and lectures became solidly attended. In the latest "commodious" premises, Saull was presumably picking up more listeners.[1] He was supporting the Association with small annual donations, and the AACAN was itself thriving in its countrywide branches, judging by the numbers of converts during the depression. The Birmingham Congress could count on one hundred thousand members by 1839.

Countrywide control was maintained by a hierarchical chain of command, a tight organizational structure which sat comfortably with Owen's paternalist attitudes. England, Scotland, and Ireland were divided into fourteen missionary districts, with an itinerant lecturer appointed to each. The bigger the district, the more it was subdivided into branches, anything up to ten, each with its own local lecturers. The lot reported to a District Board of Directors, who paid the local missionaries some £80 to £100 a year. The District Board itself reported up the chain to the Central Board, which superintended the whole and reported to Congress. At the height of the agitation, 1839–42, the *New Moral World* was circulating 40,000 copies a week at 2*d* a number.[2] Even the *Westminster Review* conceded that Owen's co-operative, labour-exchanging, anti-capitalist values seemed to be the creed of a great

---

1 *NMW* 4 (28 Oct. 1837), 5; (31 Dec. 1837): 84. Donations: 4 (6 Jan. 1838): 85; 5 (1 Dec. 1838): 96.
2 Robert Owen *ODNB*.

portion of the working classes and supported by surprising numbers of the professional and middle-classes.³

The flow of funds and need for giant halls of science in each district prompted the Great Queen Street caucus to found the London Co-operative Building Society in 1839, with Saull as a Trustee. It was to open access to funds, to ease borrowing for the huge capital outlay needed for these big venues. They should be "commodious buildings, containing lecture rooms capable of accommodating about one thousand persons each, with committee rooms, reading room, library, shop for the sale of publications, baths, and other conveniences."⁴ It was stipulated that the halls were to provide platforms for lectures on "scientific, literary, theological, moral, social, political" topics. These lectures and open debates, by encouraging the usual challenging discussions—something that made the halls exciting and distinguished them from mechanics' institutions—were designed to promote community goals.

The first project off the ground was a yet another new institution in London, for the central "A1" Branch—the branches being ranked numerically to avoid confusion. The building chosen on John Street (off Goodge Street, Tottenham Court Road) had a massive 2300 square foot lecture hall, and, with the gallery, it could accommodate up to 1300 people. A large organ was expressly built for the rostrum, so essential for the social festivals. Across the hall was a tea room where the monthly festivals could be held, and the rest of the building was transformed into a reading room, library, dressing apartments and kitchen, all with gas lighting and hot running water.⁵ This finally promised a long-term home. But the whole operation came at no small cost (£3,000), and Saull sank £200 into the building society to help it along.⁶ This was initially a loan, but on 25 December 1840 he donated it as a "Christmas Box", requiring only interest to be paid yearly.⁷

---

3   Claeys 1987, 164; Garnet 1972, 157; *NMW*, 7 (20 Apr. 1839): 404.
4   *NMW* 6 (14 Sept. 1839): 752.
5   *NMW* 7 (28 Mar. 1840): 1205; Royle 1998, 104.
6   *NMW* 7 (23 May 1840): 1243.
7   *NMW* 9 (9 Jan. 1841): 24, 28. After 1846 the London Co-Operative Building Society started a drive to pay off the outstanding £1200 debts on the hall and it bought out Saull's £10 per annum annuity: *The Age*, 11 Nov. 1848, 252; *Reasoner* 10 (1 Jan. 1851): 252; *Leader* 2 (4 Jan. 1851): 19.

Robert Owen presided over the opening on 23 February 1840. At the new institution, four lectures were to be delivered weekly—two on the social system, two on science. "Scientific Classes" too were planned, on chemistry, geology, astronomy, and mathematics.[8] Elementary instruction would include more sciences, physiology and electricity, as well as writing, grammar and composition. And the syllabus received a morbid bonus within weeks in 1840, by way of an impromptu clinical pathology demonstration. The emaciated Secretary of the A1 branch died of consumption, that great killer of middle-aged Victorian clerks who suffered from poor housing and sanitation.[9] Among the Owenite corpus, even the corpse could benefit the community. The Secretary had willed his body for public dissection at John Street, hoping that the cause of his death would be elucidated as his lungs were opened, so that others might be guided to a healthier lifestyle. This was proof that they were serious about teaching the people how the human body worked, instruction rarely available elsewhere outside of the guarded medical portals. A hundred local socialists (including women) tested their ideological mettle by attending.[10] This was one instance, perhaps, when even Saull's geology seemed less pernicious to prim Christians, who thought this shocking exposure of the great unwashed to mortal flesh would only encourage materialism.[11]

A couple of weeks later, Saull gave a series of three lectures on the rise of fossil life and the emergence of rational man, who was evidently not afraid to look inside his own body for explanations.[12] Saull was a mainstay of the Social Institution (or Literary and Scientific Institution, as it was also called). He was visibly front and centre, presiding yearly over the "Family Party and Ball" at the anniversary celebrations,[13] but he also worked behind the scenes to sort out extensions of the lease.[14] John Street became the focal point of London Owenism, not only for lectures and propaganda (anti-capitalist, anti-poor law, and anti-clerical),

---

8   *NMW* 7 (28 Mar. 1840): 1205; (23 May 1840): 1244.
9   G. Anderson 1976, 18.
10 *NMW* 7 (23 May 1840): 1344.
11 Nash 1995a, 161–62; R. Richardson 1989 on dissection and the taboos surrounding it.
12 *NMW* 7 (6 June 1840): 1290; (13 June 1840):1310.
13 For example, *NMW* 10 (19 Feb. 1842): 272; (12 Mar. 1842): 296; 11 (27 Apr. 1843): 284.
14 *Reasoner* 5 (22 Nov. 1848): 411.

but also for the life-affirming rituals usually entrusted to Christian ministers. Marriages were solemnized here, replete with choir. Babies were ceremoniously named after heroes in the pantheon—a sort of socialist imitation of Catholic confirmation with its appropriation of saints. Owen, the 'Social Father', would induct them into the rational community at his Sunday meetings.[15] Holyoake chose a euphonic "Mazzini Truelove" for his six-month old (after the Italian revolutionary and the Institution's secretary Edward Truelove).[16] Congresses would occasionally be held here, issuing "proclamations, manifestoes, and addresses to her Majesty". ("If the Queen preserved them," laughed Holyoake, "she must have left a fine collection."[17]) And when others failed to find a home, the institution's doors were always open: so the Chartists the day after the great Kennington Common meeting in 1848 "crowded into the John Street Institution" when every other venue was closed to them.[18]

During the depths of the economic depression, Owenite anti-capitalist alternatives were doing a booming business. Halls were now springing up all over the country. A thousand guests at Owen's seventieth birthday celebrations in June 1841—with Saull officiating—heard the latest:

> they had now Halls of Science in Manchester built at an expense of £7000, another in Liverpool which cost nearly as much; and they had halls in Halifax, in Huddersfield, and many other places, built at an expense of £30,000.[19]

These were all "Halls of Science"—referring both to the science of society, and to the physical sciences, those motors of 'rational' enquiry. At a deeper level the two were inextricably linked, for Saull's scientism would ultimately have social relations explained by physiological principles.[20] These were the first Halls of Science so-called since Watson's City Road "Mechanics' Hall of Science" in 1834, now rebranded Owenite Branch 16

---

15 Yeo 1971, 101.
16 Truelove's book shop next door sold works by "Owen, Fourier, Godwin, Voltaire, Paine, Volney, Mirabaud, Carlile, (not Carlisle) Southwell", and of course "Holyoake": *Reasoner* 7 (14 Nov. 1849): 305–07.
17 Holyoake 1906, 1: 129.
18 Holyoake 1905, 1:76; McCabe 1908, 1:134.
19 *NMW* 9 (12 June 1841): 374–77.
20 J. F. C. Harrison 1969, 78.

(Saull, with proprietorial fondness, was still lecturing here, on geology, astronomy, and the "Influence of Scientific Inquiry"[21]).

Some of the new Halls in industrial towns were monsters. Manchester's held 3000 people. The entire cost had been met by the savings of artisans, who ploughed their little profits into a Hall of Science Building Association. By 1840 the huge working-class support meant that it was relatively easy to raise the cash from £1 shares, paid in weekly instalments, from workers' wages.[22] Manchester's hall was an immense building, "the finest and most spacious in the town".[23] As well as hosting public lectures, it held evening classes every night and a Sunday School to teach the sciences. It was proud, self-assertive, and successful in the contested Sabbath lecturing space. But their Sunday concerts were what stood out, rousing oratorios with a 100-piece band and chorus performing Handel and Haydn, interspersed with social hymns, and "trumpet parts" to give "one general thrill of rapture". Enthralled Mancunian crowds would then sit through Robert Buchanan's intermission lecture, which trashed Christ's miracles and made a rational world seem miraculous in its potential.[24] Uplifting classical music was often used in bourgeois establishments to 'refine' the masses, or, at least, keep them out of music halls.[25] This was true in some provincial mechanics' institutions patronized by the clergy and gentry.[26] But in the Halls of Science the "Grand Oratorios" functioned as recruiting drives, and to give the intermission blasphemy lecture an almost sacred aura.

These big, self-assertive branches would put a strain on Owen's patrician hold over the movement. They put an even bigger strain on the provincial mechanics' institutions, run by wealthy elites with social safety in mind, and the clergy in support. These kept labourers out of the boardroom, and politics and theology off the syllabus. By contrast they were boring, and the socialists criticized "the excessively individualistic and technological orientation" of the mechanics' institutions, which

---

21 NMW 4 (6 Jan. 1838): 85; 5 (26 Jan. 1839): 224; 6 (26 Oct. 1839): 848.
22 Yeo 1971, 92.
23 Faucher 1844, 25; A. Black 1955.
24 NMW 8 (5 Dec. 1840): 368; (18 Dec. 1840): 400; 9 (16 Jan. 1841): 40; 10 (4 Dec. 1841): 184.
25 H. Cunningham 1980, 61.
26 Tylecote 1957, 273; Inkster 1976, 281; J. F. C. Harrison 1961, 64, 70–71.

simply helped workbench hands "'get on' in their jobs as rational competitive atoms", while shunning the screaming social, political, and moral questions of the day.[27] Socialist halls were not honing better machine operators, but morally and rationally rounded men and women for a new egalitarian Jerusalem.

Little has been written on the provincial differences among these Halls or the hostility they met. The flurry of building activity nationwide was documented just as cursorily by co-operation's first home-grown historian, Holyoake:

> A Hall of Science was erected in Rockingham Street, Sheffield, in 1839: a commodious and handsome building for the time. Mr. Joseph Smith had erected the first at Salford, less pretentious, but a pleasant structure, costing £850, and capable of holding six hundred persons. The Liverpool Hall, a building of mark for those days, cost £5,000...More than £22,000 was spent in one year in securing "Social Institutions" ... [28]

Stiff opposition was encountered in many localities. In Bristol, rioters rammed the doors of the huge 4000-capacity Hall with a cart, and fifty or sixty burst in with sticks to ransack the place. Their attempts to burn it down only failed because the gas cocks were turned off.[29] At Manchester, an arsonist tried to destroy the unfinished building.[30] The city's clergy and employers both decried this "hideous form of infidelity". The pressure was fairly relentless from the first week of opening, when the local clergyman took the stewards to court for charging admission for Sabbath lectures, which, under an old anti-sedition act, was illegal for all but licensed places of religious worship.[31]

Sermon-inspired hatreds could produce quite dangerous situations in the provinces. The buildings were an "abomination",[32] ideologically not architecturally, and their entrances were placarded continually by irate Christians. In the more locked-down towns, attempts by Owenites to enter the halls could mean running the gauntlet. When that "greyheaded panderer to immorality" Robert Owen arrived in the Potteries,

---

27 Yeo 1971, 90; Shapin and Barnes 1977.
28 Holyoake 1906, 1: 187.
29 *Weekly Chronicle*, 28 Feb. 1841, 7; size: *Nonconformist*, 27 July 1842, 516.
30 *Patriot*, 20 Apr. 1840, 253; *Weekly Chronicle*, 26 Apr. 1840, 2.
31 *Patriot*, 13 July 1840, 493; Royle 1974, 66–67; Podmore 1907, 2: 522–24; A. Black 1955, 42–44.
32 *Argus*, 12 Jan. 1840, 2.

at the Burslem Hall of Science, "a miserable hovel" in the eyes of the *Staffordshire Gazette*, he faced an "infuriated, fanatical, and drunken mob," fired up by the clergy and inflammatory handbills denouncing his "Poison". Plied with free drink and egged on by the authorities, the mob attacked Owen's coach and he was frog-marched away by the clergy-led crowd and held in a house for two hours. The mob caught the pacifist Alexander Campbell, "the most fatherly-minded of all the missionaries",[33] a gentle man moving towards sacred socialism, beat him up, and then "bonnetted" him (covered him in clay), to the delight of the local conservative press.[34]

These hate incidents reached a zenith after 1840. The high church Tory Bishop of Exeter in the Lords lambasted Owen's views, not least on the "blessed state", marriage—which he declared "not fit for your Lordships, or any decent person to hear"—but also Owen's blasphemous denial of immortality and revelation. As heinous in the Bishop's eyes was the Owenite "doctrine of the irresponsibility of man", which undercut Christianity to the core. If humans were products of circumstance, and therefore blameless, whence the heavenly rewards and hellish punishments? And with those disappearing threats and promises went the earthly power of the Church. Worse, it would let the thieves off in this life. He had heard of a social missionary in Liverpool arguing that a reprobate convicted of manslaughter "ought not to be punished, because—he could not help doing what he did!" Saull had said as much himself. He used his native Northampton experience to drive the point home. His homily related the story of a local boy, uneducated, running wild because his starving mother was widowed. The youngster, thus lacking in education and guidance, started stealing food and ended up being transported. Where was the guilt when "gross neglect" was to blame? But Saull failed to persuade the locals that the wretch was "more sinned against than sinning".[35] Similar mitigating claims based on a belief that circumstance shaped character outraged the Bishop. It struck "at the foundation of all law, human and divine".

---

33 Holyoake 1906, 1: 211, 236; J. F. C. Harrison 1969, 128.
34 Buchanan 1840a, 142–6; *Weekly Chronicle*, 5 July 1840, 2.
35 *NMW* 4 (23 June 1838): 278–80.

This made the Owenites not only immoral, but "an illegal society" and the Halls of Science had to be shut down.[36]

Though the Bishop failed to get Owen prosecuted, the clergy and manufacturers became pro-active. Some factory bosses threatened to sack socialists.[37] Attempts were made to "depopulate" the Halls, as the Church set up rival "Parochial Libraries" under "the patronage of the Clergy—carefully ... supplied with books, suited to open both natural and revealed truth". They were to instil "holy principles and habits" in order to "consecrate" science—to make it point to a higher truth and thus halt the spread of infidel contagion. An apocalyptic image in the *Church Magazine* hinted at the blood and terror in store, with the execrable Halls "sapping the very foundations of society, and threatening the evils of an 'Age of Science,' not inferior to those which in the last century attended a boasted 'Age of Reason.'"[38] Angry letters venting frustration filled the Tory papers: "Why are those dens of blasphemy—Social halls of Science (Science!!!) permitted to exist in every quarter of the metropolis ... sending out their missionaries to brutalise—to demonise our countrymen ... Sir, this should be stopped at once."[39] With over 30,000 attending Sabbath lectures in the Halls of Science nationwide, the clergy were called upon to do more to "suppress this monstrous heresy".[40]

In this hysterical climate, the more outrageous the anecdote, the more drooling the Tory press. One of the more querulous socialist proprietors of Manchester Zoological Society, a Dr Hulley, was reported by the *Argus* to have mooted "the animals *receiving the sacrament*" on the Sabbath. A local joke became a national scandal. Such "disgustingly profane ribaldry" had its parallel only in the acts "of the fiends who flourished in the 'reign of terror'".[41] When a sermon-drilled populace feared it would have its throats cut by socialists, amicable dialogue was all but impossible. Even mechanics' institutions were suffering the backlash. As Lord Brougham heard in 1839: "the clergy are now openly

---

36 *Mirror of Parliament* 1 (1840); 312–31; *Atlas*, 25 Jan. 1840, 51.
37 A. Black 1955, 42–44; Buchanan 1840a, 139; *NMW* 8 (4 July 1840): 4.
38 *Church Magazine* 6 (Feb. 1844): 54–56.
39 *The Age*, 4 Sept. 1842, 5. *NMW* 8 (15 Aug. 1840): 105, on *The Age*'s "ravings" about socialism's success in the manufacturing districts and "the church" being "the sacred source and unsullied sanctuary of Conservatism".
40 *British Magazine* 20 (July 1841): 65–66; *Union* 1 (1 Dec. 1842): 368.
41 *Argus*, 29 Nov. 1840, 700. On attempts to hush the socialist Hulley up in the society: *Manchester Times and Gazette*, 12 Dec. 1840.

hostile to Mech. Inst^s., as the seed beds of infidelity, & are founding Societies for the diffusion of Religious & Useful Knowledge under the patronage of the Bishops".[42] Teachers complained that children would learn their Bible-tract disseminating techniques in Sunday School; then they would be lured over to John Street with its promise of exciting science and use these techniques to distribute infidel propaganda. The religious teachers' solution in 1840 was to set up a rival "Senior Scholars' Institute" for advanced science and sacred studies in Red Lion Square. It was an overt effort to hold on to their Sunday School graduates.[43]

Ironically, where the clergy and squirearchy did stop the Owenites building, the local socialists sometimes resorted to buying old chapels instead. In Glasgow they managed to acquire a large parish church, and, in Birmingham, socialists bought the Lawrence Street Chapel and rebranded it the Hall of Science. In a more clerically-deregulated London, the socialists and Metropolitan Churches Committee actually vied for building sites. With radical vestries opposing the clergy, and the socialists able to stump up the huge sums quickly, they would often win the bid to build on church-designated plots.[44] Sometimes provincial efforts to thwart the clerical-blocking badly backfired. In Huddersfield, they put the Hall building funds under a spinner's name, because he was not known to be a socialist, but the spinner thought the £130 better than his meagre wage and promptly absconded to America with it.[45]

Saull toured the new Halls in July 1840. From lecturing the socialists in Leeds—where nearly four hundred heard his two-hour exposition on geology—he migrated over to the Huddersfield and Manchester Halls of Science. By now, a tacit acceptance of progressive 'evolution' was beginning to pervade the socialist halls. In 1840, it received a further boost from the import of French St Simonian ideas on the "New Genesis", with its reworking of Bible verses to include monstrous fossil beasts and "hideous hippopotami" born and raised by the "God of

---

42 T. Coates to H. Brougham, 27 Sept. 1839, Brougham Correspondence 95, University College London.
43 *Evangelical Repository* 1 (Oct. 1840): 209–10.
44 *Patriot*, 4 Nov. 1841, 742; A. Black 1955, 43; Simon 1960. 239–40.
45 *Courier*, 22 June 1840, 3; *Patriot*, 29 June 1840, 450. On this Hall: A. Brooke, "Huddersfield Hall of Science."

Progress".⁴⁶ As a metaphoric expropriation of palaeontology it might have struck a chord with the Francophile Saull. But his leaner geology still came unfashionably backed by astronomical causes. Using a globe, Saull explained what he tellingly called the "pregnant history" of the strata-embedded life, looking to the cyclical climatic changes as "the causes of the great varieties of fossil remains". It was driven by the poles swinging towards the equator, "and vice versa; and this not once, but many times." He drew both anti-Malthusian and Owenite morals. The first from the fact that the "pabulum of life is constantly on the increase, and, as a consequence, animated and organised beings [become] more numerous and prolific"—in short, nature provides an ever-increasing abundance of food allowing an ever-increasing number of mouths. And he drew the final Owenite conclusion for an age collapsing into violence

> that gradual progression is the universal law of nature. It is found equally developed in planets, vegetables, and animals. True philosophy will apply it to society; and, avoiding all violent and reckless changes on the one hand, avoid with equal care a stand-still policy on the other.

The Leeds socialists knew that "Mr. Saull holds some peculiar notions upon the subject, which we believe are not yet acquiesced in by some of his brothers in the science".⁴⁷ That was something of an understatement. However hot the topic for branch socialists, it still burned the fingers of the *Penny Satirist*. Smith remained on Saull's tail, blinded by incomprehension, still claiming that Saull's ancestor was an ass, not a monkey. Saull's "witchcraft" was a devil's brew of "delusion and vain imagination" stirred with "infidel bigotry", said the *Penny Satirist*, and so much superstition in its own right.⁴⁸ Nor was his museum of more worth. Forget the poor biblical "enthusiast, who adores the relic of some pious saint", the curator's fanaticism "is, in many respects, less reasonable, less honourable". The acquisitiveness and "worship" of such fossil relics in their shrines was only another case of "idolatry".⁴⁹

Saull, it seems, had fostered Smith's growing belief that science itself was soulless when it was not pedantic. The literary world was

---

46 *NMW* 8 (26 Dec. 1840): 40. It was introduced by gentlemanly communist Goodwyn Barmby.
47 *NMW* 8 (18 July 1840): 37; On Leeds: Morrell 1985.
48 *PS*, 4 July 1840; 1 Oct. 1842.
49 J. E. Smith 1873 [1848], 1: 310.

becoming contemptuous of these "dull and prosy" men of science. See the "geologist collecting petrified dung, and insects, and shells, to prove that which cannot be proved ... Poor simpleton!" and then looking "as full of scientific pride as a ginger-beer glass when the puff is running over."[50] Gone was the revolutionary day when "matter—the dross of existence" could be found "seated on the throne of God himself", and when man had lost his self-respect. Nor could he regain his standing, so long as he imagined himself the son of "a shaved, and untailed, and cultivated monkey, as Mr. Saull" would have him.[51]

Saull's guest lectures of 1840 were among the last he was to deliver on geology. Not that he was put off by the critics. But with younger activists rising to the challenge of life's origin, he could move on to a complementary area, the final stage of the evolutionary drama: mankind's rise from savagery.

Relatives

The countrywide Halls of Science were among the few places that could host Saull's sort of infidel performance outside of London. This is what rankled the Anglican clergy: their loss of town control. The Halls provided venues for undesirables who might otherwise be kept from holding meetings by being denied rooms. Take "the bigoted town of Northampton", as the *New Moral World* called it, at least before the "banner of Socialism" was unfurled.[52] Saull was a native: he had been baptized at nearby Byfield, and he still owned a property there.[53] His relatives lived on the outskirts, so he was often in the region. The cabinets of other fossil collectors contained Northampton specimens donated by Saull, an *Ammonites elegans* from Bugbrook, close by the town, or a fish from King's Cliff (in the northern corner of Northamptonshire), suggesting that he was also buying or prospecting in the county.[54]

In August 1840, Saull had moved on from Manchester to lecture the socialists in Northampton on "the Present and Future prospects

---

50   *PS*, 1 Oct. 1842.
51   *Family Herald* 2 (26 Oct. 1844): 394.
52   *NMW* 5 (13 Apr. 1839): 395.
53   William Devonshire Saull, Will, Public Record Office, National Archives. *NMW* 4 (23 June 1838): 278–80.
54   Delair 1985.

of Society" and the French peasantry.⁵⁵ Presumably he visited his kin in nearby villages. These local parishes were tightly controlled by the clergy. John Saull (one of his two nephews there) owned a pub, the "Admiral Rodney" on the High Street of Long Buckby, north-west of Northampton. The village was, in Holyoake's (not-unbiased) eyes, "an intellectual desert, where priests rule and freedom is dead". The previous landlord had been warned by the vicar of Buckby that he risked his licence if he leased his "club room" for freethought lectures. The minister, the Rev. Richard Gardner, a member of the Church Missionary Society,⁵⁶ was happy to send the Word to heathen Africa but evidently not to let in infidel missionaries from darkest London. The new landlord John Saull now offered Holyoake the same backroom to speak, despite the same threat. It led to clerical anger, but whether John Saull's subsequent bankruptcy was due to his pub being blacklisted is not known.⁵⁷

These threats against landlords, and the barring of venues, explains why the Northampton Chartists looked "with more than common delight upon that feature of our times ... we mean the erection of Trades' Halls and Halls of Science":

> we must ... have our Halls of Science, wherein to discuss our grievances, and to advocate our political rights. It is well known ... that all the public buildings are monopolized by the privileged classes; and the very sacred edifices of the priests are too holy, to have their doors unfolded for free enquiry. This being the state of things ... it becomes the Working Classes immediately to commence business for themselves in politics, and especially to become their own priests, and to erect for themselves temples, dedicated to truth and free enquiry.⁵⁸

Nor were the mechanics' institutions themselves happy about the Halls, for they saw their own clientele draining away. A report on the institutes'

---

55 *NMW* 8 (22 Aug. 1840): 124.
56 *Church Missionary Record* 13 (July 1842): 173.
57 *UR*, 3 Feb. 1847, 20; *Reasoner* 2 (17 Feb. 1847): 69–70; (3 Mar. 1847): 106–07; (17 Mar. 1847): 134; (21 Apr. 1847): 219. Bankruptcy: *The Jurist*, 13 Pt 2 (1849): 198; *Patriot*, 24 May 1849, 336. Both of Saull's nephews were sympathetic to radical freethought. One (unnamed) turned up at a Utilitarian Society meeting in the City Road Hall of Science, which Holyoake and W. D. Saull addressed (*UR*, 6 Oct. 1847, 89). William Saull (the other nephew) was Chairman of the Working Men's Association in Long Buckby (*CGV*, 3 Nov. 1838, 3).
58 *CGV*, 27 Mar. 1841, 4.

decline and takeover by clerks was commissioned in 1839. Its author told Henry Brougham that the mechanics' institutions'

> omission of politics & religion (the two things which men think most useful objects of speculation) keeps away all the sober, industrious & reflecting of the handicraftsmen, who resort to Socialism & go through severe moral discipline & privations to become members of the Community of Socialists.[59]

The report itself was equally damning. It blamed the "systematic exclusion" of politics and theology "in almost every shape" for driving men to the socialist Halls.[60] By 1841, this failure seemed to be threatening the existence of the institutes themselves. The socialists "zealously diffuse their opinions far and wide; they have erected halls, and established places of meeting in which they discourse to thousands; they invite persons of adverse opinions to listen to and freely discuss the expositions of their principles":

> they have lectures on the sciences [it referenced the calibre of the John Street science lecturers], they have music, and in some cases other classes, and they add to these the occasional attraction of tea-parties, accompanied by dancing. The number of members of Socialist Institutions in London is much smaller than that of members of Mechanics' Institutions, but the attendance at their lectures, discussions, and festive meetings, is much greater ...

The reason was that the Halls encouraged free-for-all debates on religion and politics; they were lively, vocal, entertaining, and liberating. The exclusion of debate and the "right of free inquiry" left the mechanics' institutions, "if not distasteful, at least uninteresting".[61] Participation was a large part of the Halls' success.

But, even here, there is an irony. In 1840, just as the report was going to press, the John Street Institution, under Owen's tight control, *abolished* the free-for-alls after Sunday lectures. It seemed to be encouraging "conceited and ambitious persons to offer opposition for mere display", flamboyant individualism, against the socialist ethos. From now on,

---

59   T. Coates to H. Brougham, 27 Sept. 1839, Brougham Correspondence 95, University College London.
60   Coates 1841, 24.
61   Coates 1841, 29–30; Shapin and Barnes 1977.

pertinent questions had to be delivered in writing. Time was set apart to answer them "and proper arrangements made for the maintenance of the strictest order".[62] Many baulked at Owen's undemocratic chaperoning, and other Halls, notably Manchester with its worker control, remained much freer.[63] The same went for other London venues, where boisterous cross questioning was allowed to continue.

But the best lures for interested artisans were the theatrical debates staged between socialist missionaries (slammed as the "missionaries of Satan"[64]) and ministers of the cloth. Saull chaired a typical thrust-and-parry session in 1842, this time at his old Carlilean stamping ground, the Rotunda—now re-branded the South London Hall of Science (and enrolled as Branch 53). The place was as lively as ever: one advert showed where its priorities lay:

> BRANCH 53 of the Rational Society, SOUTH LONDON HALL of SCIENCE. BLACKFRIARS ROAD.—
>
> On Sunday Evening Mr A. CAMPBELL will LECTURE on "The Distress in the Manufacturing Districts and the only apparent Remedy," commencing at Half-past Seven.
>
> On Monday Mr. MANSFIELD will conclude his course of LECTURES on GEOLOGY; to be followed on the two succeeding Mondays by Mr. BUCHANAN in "An Inquiry into the Creation and Fall of Man and the Deluge, with reference to Geology and Astronomy".
>
> On Tuesday a SOCIAL FESTIVAL. Single Tickets 1s, double 1s 6d.[65]

Here Saull was right at home, and, in September 1842, he chaired a two-day debate on the "Disadvantages of Christianity", between a local lecturer and Independent minister.[66] Such set pieces were the big draws and acted as recruiting spectacles for the cause—this was socialism showing its Rotundanist teeth.

Tory histrionics over such blasphemous goings-on had only increased through the decade. Outside the socialist venue would be a placard

---

62 *NMW* 7 (23 May 1840): 1243.
63 The lectures, festivals, and refreshments in the northern halls were cheaper too: Yeo 1971, 88, 94–95.
64 *NMW* 5 (25 May 1839): 490; A. J. Booth 1869, 199.
65 *NMW* 11 (6 Aug. 1842): 42.
66 *NMW* 11 (10 Sept. 1842): 90.

advertising the latest lecture. "The bible no revelation" was one, causing upright citizens to gasp, especially when they saw that the lecturer was "a person calling herself Mrs. Martin." More irate letters to the *Standard* (the city's top-selling evening rag, with half-a-million sales):

> Who this female blasphemer may be, is a matter of no moment; but it is lamentable that English women, if such they are, can be found capable of so disgracing their sex and their country, and if it is necessary to impose a legal restraint upon the sale of arsenic and other deleterious drugs, it is surely the duty of the magistrates in whose district this nuisance is situated, to endeavour to prevent the diffusion of a moral poison ... [67]

The incisive, witty Emma Martin turned out to be the biggest draw of all. A mother of three, once a "zealous" Baptist of fierce Calvinistic persuasion, she had recently converted just as zealously to Owenism. The economic exploitation, social degradation, and debased state of women in Christian society convinced her that freethinking socialism was a more moral way forward.[68] That is not how others saw it. She had deserted her husband and deserted her Saviour. Appalled at women being "seduced" by socialism, *Fraser's Magazine* pictured the "coteries of courtesans" suffering the "horrible abominations" of Owen's trial-and-error marriage system. Only "under the Cross" would women find true salvation.[69] But Martin knew her Bible backwards and became a star attraction, not least in exposing the patriarchal iniquities of the Church. Just as Saull's monkey-man made milksop socialists queasy, so did Martin's extreme feminist freethought. And with her social realignment came a contingent scientific shift. Like Saull and others, she would soon be arguing that the history of life resembles the history of the individual, a natural growth, with mankind "but an improvement upon the lower animals" and a "new product of Nature's increasing power".[70]

No disputant so celebrated had appeared since Richard Carlile, it was said.[71] So it was fitting that Mrs Martin saw the old lion off. Carlile, fifty-two, his asthma made worse by London's belching chimneys with

---

67 *Standard*, 20 Dec. 1842.
68 B. Taylor 1983, 130ff; Frow and Frow 1989, 86–87.
69 *Fraser's Magazine* 21 (Jan-June 1840): 689–90. For a study of the "fallen" woman, literally "seduced", but as applicable to the metaphorical "seduction" by Owenism causing an equal loss of 'character', see A. Anderson 1993.
70 J. A. Secord 2000, 314–16; E. Martin [1844].
71 *NS*, 22 Nov. 1851.

their pall of black soot, died on 10 February 1843.[72] Defying prejudice to the end, he wanted his corpse to go to his own hero, the surgeon William Lawrence, to benefit mankind. But Lawrence, cowed since a court had declared his *Lectures on Man* blasphemous, never took on the task. Instead, the body went to St Thomas's Hospital, where the dissector pointed out Carlile's uncivilized sloping face but big heart, while the accompanying oration commended Carlile's action in donating his body to science. But Carlile's notoriety only exacerbated what many saw as a desecration, anatomical dissection. As it was, dissecting rooms were associated with executed felons, grave robbery, and posthumous punishment of poor-house victims. Carlile added blasphemous atheism to the list. But better dismemberment than rotting in hallowed ground, had been his view.[73] The grave was a sanctuary of silent repose till Judgement Day for many Christians, and while putting Carlile on the slab with the murderers might have been fitting, it did not lesson the Christian opposition. Even the anodyne oration at St Thomas's over the "miserable 'atheist'" elicited "abhorrence and disgust".[74] When the body was finally released for a Kensal Green burial, one radical generation paid tribute to the other. No one was better suited to deliver the eulogy at the City Road Hall of Science than Emma Martin herself.[75]

Mrs Martin points up another advantage of the Halls over Mechanics' Institutions: their family orientation and female friendliness. Women were not to be veiled and shuffled off into the gallery. Against the chauvinist claim that educated women would lead to better domestic economy and a more comfortable working man's home, feminists, male and female, wanted women's intellectual liberation proper. But as Barbara Taylor in *Eve and the New Jerusalem* admits, rearranging the "social

---

72   *Examiner*, Feb. 1843, 88; burial: *NMW* 11 (25 Feb. 1843): 284; Wiener 1983, 259–60; Holyoake 1849a, 27. The autopsy revealed a cerebral haemorrhage causing paralysis during an asthma attack: *London Medical Gazette* n.s. 1 (24 Feb. 1843): 781–85.
73   *Lion*, 3 (20 Mar. 1829): 353–59. Many freethinkers left their bodies to anatomy schools to set an unChristian example when corpses were so desperately needed. R. Richardson 1989.
74   *Medical Times* 7 (25 Mar. 1843): 419. Because some were looking for pathological symptoms to explain Carlile's aberrant atheism, it is no coincidence that the report published in the conservative *Medical Gazette* began by racially profiling Carlile's skull, which was put on the degraded level of an American Indian (*London Medical Gazette* n.s. 1 [24 Feb. 1843]: 782).
75   *NMW* 11 (18 Mar. 1843): 308; *CPG*, 8 Apr. 1843, 3.

relations of the New World" within the cramped sphere of tacit Victorian expectations was "obviously going to run into practical limitations". Nevertheless, women had greater opportunities in the branches.[76] They were to be welcomed at lectures, and as lecturers, and, at the festival, to the high table. This was a real worry for the clergy. When Owen visited Leeds Hall of Science, posters went up "affectionately warning young women who wish to maintain a good reputation, to keep away from socialist meetings".[77] The increasing egalitarian ethos, and co-operative Lamarckian demand that women be educated alongside men, meant a hospitable environment in the Halls.[78] As the ubiquitous chair of meetings, Saull would open sessions with the greeting: "Social friends of both sexes," rather unnecessarily rubbing in the point.[79]

Saull also actively encouraged women to hear his museum talks on our fossil past and co-educational future.[80] Primarily the offer was aimed at the spouses and girl friends of the fustian-jacketed mechanics, but they came, high as well as low—even the genteel readers of the *Lady's Magazine*. Quite what the *Lady's* reporter made of his speech over the fossils is not known. Cryptically, he or she politely quoted the old proverb "Many men of many minds", meaning there were all manner of "views of the cause of the wonderful changes which have taken place and are still going on upon the surface of this earth", Saull's odd view among them.[81]

In the late 1830s and early 1840s, his free museum was well promoted in the artisan press, from the *Penny Mechanic* to *Cleave's Penny Gazette of Variety and Amusement*, and name checked in geology books.[82] But it was even more widely trailed in the dailies. It figured in all the syndicated listings of free Christmas and Easter recreations in the 1840s, for families who were visiting London for the festivities.[83] The *Courier* at Easter 1841

---

76 B. Taylor 1983, 234–37.
77 *Leeds Weekly Chronicle*, 22 Dec. 1839, 3.
78 Yeo 1971, 96–97; W. Thompson 1826b.
79 *NMW* 13 (11 Jan. 1845): 229–31.
80 *MM* 19 (25 May 1833): 117–18.
81 *Lady's Magazine and Museum* 3 (Nov. 1833): 297.
82 G. F. Richardson 1842, 80, 368, 386, esp. 402; Mantell 1844, 1: 135: 2: 780, 838–39, 902–903. *PM* 2 (20 Jan. 1838): 200 passim; *CPG*, 15 Apr. 1843; *Literary World* 3 (1840): 166.
83 For example, *Standard*, 25 Dec. 1838; 25 Dec. 1840; *Morning Post*, 18 Apr. 1840; 26 Dec. 1843; *Courier*, 18 Apr. 1840, 4; *Morning Post*, 14 Apr. 1843; *Argus*, 15 Apr. 1843,

made it the top geology museum in London for the huge number of fossils and "the excellence of their arrangement".[84] And the *Morning Post* that Christmas saw it as "quite unique, both for objects of rarity and beauty".[85] For a fuller account of the contents, there were the London visitors' handbooks. Booth's *Stranger's Intellectual Guide to London for 1839-40* had it vying "with any private Museum of a similar nature in the kingdom". In Aldersgate Street you could see beautiful ferns from the coal measures, valuable pear-shaped sea lilies, ammonites, the biggest collection of bones of the "stupendous" *Iguanodon*, and remains of gigantic American mastodons. Every Thursday tourists would get a walk-through on geology. They were told how coal contributed to our comfort, how knowledge of the rocks helped the agriculturalist understand soil and the surveyor plan house foundations. So that

> even in a pecuniary point of view, Geology may be advantageous to the mere speculator; but its study raises us above mere mercenary considerations, in showing us that no animated being came into existence until preparation had been made for its reception [the reporter had clearly listened to Saull]; thus proving the great laws, founded on the purest benevolence, which regulate the universe.[86]

The "purest benevolence" was Booth's euphemism for Saull's quasi-teleological idea of the 'pabulum' preceding new life. This was the cue for Saull's imagining of the strata—from oldest to youngest—housing ancient sea lilies or coal age ferns, creatures from the golden age of reptiles, then the elephants and rhinos that once roamed Britain. And so to the shocking origins of human savages and their gradual civilization. There was no fall from grace here, no Edenic idyll, but a steady rise over aeons as the earth's elliptical shifts made conditions suitable, leaving 'Nature' to do the rest. Geology was presented as an optimistic philosophy of ascension, deliberately juxtaposed to Genesis's fall of man and "theology of degradation".[87]

Nature's productive powers might have made "Satirist" Smith's eyes swivel, but materialist assumptions like Saull's remained *de rigueur*

6; *The Era*, 16 Apr. 1843.
84 *Courier*, 12 Apr. 1841, 3; 27 Dec. 1841, 1.
85 *Morning Post*, 31 Dec. 1841.
86 A. Booth 1839, 15, 121–22.
87 Kenny 2007, 370.

among the extremists. Sectarian splits in the Owenite community would now throw up even more overt atheists, paving the way for the fullest discussion in Britain yet of life's material progress. For Shepherd Smith things were about to get far worse.

18. The Atheist Breakaway

Robert Owen's response to clerical attacks in 1839–1840 had massive sectarian repercussions. To streamline the management of the burgeoning community, the 1839 Birmingham Congress agreed to unite the propagandizing "Association of All Classes of All Nations" with the fund-collecting "National Community Friendly Society" into "something more wonderful still" (in Holyoake's facetious phrase): The Universal Community Society of Rational Religionists.[1] The operative was *"Rational Religionists"*, which was a contradiction in terms to Holyoake's way of thinking.[2] But this fall-back position was a subterfuge. It was to get round the prosecutions initiated by the Manchester clergy, who had taken the Hall's gate keepers to court for charging on the door for Sabbath lectures, which was illegal, except in churches. So Owen had the Halls registered with the Bishop's Court as a place of worship for the sect of Protestants, called "Rational Religionists".[3] But the attempt to thwart the Sabbath noose only snared them tighter. The Bishop's Court then insisted that local missionaries take an oath that they really were Christians: that they believed the Scriptures to be the revealed word of God. In Manchester and Bristol, lecturers who had preached diametrically opposite doctrines, after a lot of soul-searching and defiant manoeuvres, finally bit their lips, took the oath, and perjured themselves to keep their Halls open.[4]

The Manchester missionary who knuckled under was Robert Buchanan (giving him the soubriquet "Rev.-swear-at-last"). But it was to little avail. The day after his talk at the Hall of Science "on 'Geology, and the Mosaic Account of World-Making,' with dioramic illustrations",

---

1   Holyoake 1875, 1: 193; *Union*, 1 (1 Dec. 1842): 367.
2   Holyoake 1892, 1: 134.
3   Royle 1974, 66–67.
4   Podmore 1907, 2: 534–36; Royle 1974, 66–68; Holyoake 1906, 1: 158–62.

Town Mission preachers and hecklers flooded the theatre, and the altercation stopped proceedings.⁵ Then, on inaugurating the new hall of science in Whitehaven Cumberland, in 1842, Buchanan was besieged by a mob, the hall was destroyed and he "ran for it; but was hunted, caught, and very roughly treated". The mob, containing many women, went on to burn the shops and houses belonging to the socialists.⁶ The beating finally forced Buchanan to abandon peripatetic Owenite lecturing altogether.

The cumbersome "Universal Community Society of Rational Religionists" slimmed its title down to the Rational Society in 1842, with Saull, as usual, one of the Society's auditors.⁷ The Halls of Science continued to operate under this legal protection as licenced sites of religious observance, and, for four years, Owenism as a mass movement peaked under these strange conditions. By now, sixty-two branches were enrolled, and 50,000 people attended Sunday lectures weekly.⁸ And while the Bishop of Exeter had thought to stamp out socialism, his outburst in the Lords actually resulted in 50,000 copies of Owen's reply being sold.⁹

In fact, tract production was massively stepped up, thanks to the London Tract Society (based in John Street), whose meetings Saull would chair. The Society managed to shift increasing numbers, from 50,000 in 1840, to 140,000 in 1841–42 (in the same period the *New Moral World* increased its circulation sixfold).¹⁰ They were cranked out with a labour-intensive hand press, even as the editor of the *New Moral World* dreamed of the day when they had "steam-driven cylinder machines throwing out printed sheets by the Million".¹¹ Still, in a pamphleteering age, the socialists almost held their own against the flood of religious

---

5   *NMW* 10 (14 Aug. 1841): 56.
6   *NMW* 10 (29 Jan. 1842): 247–48. B. Taylor (1983, 189) points out how often women—the religious pulse of the family—were the hecklers.
7   *NMW* 11 (10 June 1843): 418; 12 (8 June 1844): 401.
8   Podmore 1907, 2: 469.
9   A. J. Booth 1869, 204.
10 *NMW* 11 (1 Oct. 1842): 116; (22 Oct. 1842): 139. 1841–42 was the peak: 12 (21 Oct. 1843): 136; (4 Nov. 1843): 150; 13 (12 Oct. 1844): 126; (19 Oct. 1844): 136. The London Tract Society developed provincial branches and changed its name to the Rational Tract Society in 1842.
11 *NMW* 13 (11 Jan. 1845): 229–31; also 13 (5 Oct. 1844): 116; (24 May 1845): 387. The tracts were dispatched to the branches at 4*d* a dozen.

tracts, anti-corn-law pamphlets, and Chartist flyers. Pamphlets were sent to all the commercial towns, with the manufacturing districts getting the lion's share, to be distributed by the local branches. Emigrants were encouraged to pack them in their trunks, and they were translated and sent to European cities. The proselytizing of their anti-capitalist, culturally-deterministic, regenerative message reached its peak in 1841. Religious evangelists complained that the John Street depot was not only pinching their Bible-distribution techniques, but their pupils as well, as one Sunday School graduate was spotted working there as a secretary.[12] The tract distributors were also copying religious foot-in-the-door techniques: socialists were encouraged to lend pamphlets and call back for them later to engage the reader. So successful was the Tract Society that it was soon holding its own festivals and tea parties, with Saull in the vice-chair below Owen, or chairing himself.

All the while there was simmering anger at Owen's registering the socialists as "Rational Religionists" and encouraging missionaries to take the oath—"be-reverended", in firebrand Charles Southwell's words.[13] Others equally refused to play "the whore to the priests," as Southwell's Bristol colleague William Chilton said.[14] Since Southwell was the highest-profile defector and started a more militant trend with extreme scientific repercussions, it will pay to look at him in greater detail.

Charles Southwell was prodigiously talented, highly opinionated, and socially irresponsible (in a non-Owenite sense). He had started as an unpaid Lambeth lecturer, delivering a hundred and fifty talks in 1839 in his spare time. His punchy rhetoric on socialism, marriage, capitalism, or Creation attracted huge audiences. Up to a thousand turned up on Kennington Common each Sunday to hear his soapbox harangues on the uninspired Bible.[15] For a piano finisher, he was astonishingly literate. Quips and quotes would effortlessly roll off his tongue. A melodramatic delivery helped, but then the Thespian trod

---

12 *Evangelical Repository* 1 (Oct. 1840): 209–10. Fyfe 2004 on evangelical tract production.
13 *OR* 1 (4 Dec. 1841): 33.
14 W. Chilton to G. J. Holyoake, 26 Dec. 1841, Holyoake Collection no. 22, Co-operative Union, Manchester; Royle 1974, 68.
15 *NMW* 6 (10 Aug. 1839): 665; (7 Sept. 1839): 733; (21 Sept. 1839): 763; (2 Nov. 1839): 861.

the boards in the theatre as well as the park, and he happily deployed "subversive Shakespeareanisms" in his political repartee. As Marsh says, "Shakespeare gave gloss and heritage to atheistical materialism".[16] But it took a toll, so Southwell swapped his day job at the piano-forte factory for paid Owenite lecturing. He was assessed on his knowledge and tested on his speaking skills and allowed to put "S.M." (Social Missionary) after his name.[17]

London loved Southwell, and he was massively in demand. His talks on marriage were interspersed with Saull's on science at the City Road Mechanics' Hall, while his lectures on "Drama" followed Saull's on "Geology" at John Street. In fact, he was a star attraction at the A1 Branch, exhibiting a huge talent for ancient science and classical literature as much as biblical exegetics.[18] Southwell was Carlile redivivus, eager, chafing at the bit, aggressively atheistic. He had even taken on an ageing anti-socialist Carlile at Lambeth in a marathon two-night session late in 1839, in a hall "crowded to suffocation" with hundreds unable to get in.[19] Young Southwell might have been wittier and nimbler, and whether or not he could outflank the old fox he certainly had his uncompromising style. It quickly showed. When he guest lectured in Dover in October 1840, the local press were startled by his "violent manner" in debunking religion. There was no disguising his virulent tongue. The Bible was so many "cunningly-devised fables", Christianity so many "wild absurdities" which "taught man to murder and to do all those things which were against the first principles of our nature".[20]

But Southwell flexing his atheistic muscles was Southwell snubbing the new softly-softly approach of the Central Board. He was losing sympathy for the spineless Board, which was growing as "self-complacent" as "a bearded Methodist Conference".[21] The crunch came when the missionaries had to swear the oath as "Rational Religionists". Right into 1841, he sympathized with the "poor fellows who had large families" and who needed their lecture income, so he refused to

---

16 Marsh 1998, 111.
17 *NMW* 9 (6 June 1841): 351.
18 *NMW* 6 (26 Oct. 1839): 848; 7 (16 May 1840): 1213; (6 June 1840): 1286; (13 June 1840): 1310; (20 June 1840): 1320.
19 *NMW* 6 (2 Nov. 1839): 861; (14 Dec. 1839): 957.
20 *NMW* 8 (17 Oct. 1840): 252.
21 Southwell 1850, 60.

condemn them, even if he would "rather fall dead on the platform, than take the oath in question".[22] Kowtowing was never his way. But as the year wore on, it would rankle more and more, and oath-taking would only exacerbate other grievances. By then, he had been re-assigned to Birmingham (he moved in November 1840). Here he helped raise £800 to move the headquarters from a small room in Well Lane to the Southcottian Lawrence Street Chapel, which was capable of holding a thousand.

On 13 June 1841, he moved again, to become a lecturer stationed at Bristol Hall of Science, where he opened up the discussion classes to the public and democratized their proceedings, creating elected Presidents and Secretaries on a three-monthly rotation.[23] The paternal, undemocratic aspects of London A1 Owenism galled him, as it did others, and he started moving Bristol in new directions. There was now a hint that nothing was off the table in these classes. His new co-worker, a young compositor *"with* brains", William Chilton, added that they were "imitating the Eclectics", who believed that "that no one man"—he was pointing at Owen—or "system ever yet contained within themselves all truth".[24] These were intimations of a world beyond Owen, and that it would arrive sooner than expected.

At the end of his three-month stint in Bristol, Southwell had had enough. He returned to London and announced he was resigning as a Social Missionary. It caused a sensation. Even though his valedictory lecture on 26 September 1841 to explain his decision was given at a few hours' notice, it was to a packed hall. By all accounts, it was the most forceful and funny off-the-cuff talk he had ever given, with the Owenite greybeards taking the brunt. Once Owenism had been the "very poetry of politics", but that was before socialism had been "churched, shorn of its consistency", and its preachers "reverended". It had lost its way after the Bishop had forced a "shuffling, equivocating, white-feather policy".[25] Stopping impromptu questions after lectures for fear that they introduce controversy was short-sighted. Encouraging poor socialists to take the

---

22 *NMW* 9 (14 May 1841): 351.
23 *NMW* 8 (12 Dec. 1840): 381; 9 (22 May 1841): 316; 10 (3 July 1841): 7; (11 Sept. 1841): 86.
24 *NMW* (11 Sept. 1841): 86; Southwell 1850, 65.
25 *OR* 1 (4 Dec. 1841): 33–34; (8 Jan. 1842): 58; (5 Mar. 1842): 90–91; (9 Apr. 1842): 131; *NMW* 10 (23 Oct. 1841): 134–35.

oath as Rational Religionists was cringe-making. "Religion is the blight upon the fair harvest of reason", which made this a "miserable, truckling, unprincipled policy". In Southwell's view, they should have stood their ground. He was straining at the leash. Like other democrats, he hated Owen's strangling of branch democracy, and his patriarchal hold.[26] He hated the rampant "idolatry" of "our dear father". This fawning, with Owen succumbing to the "poison of flattery", was a stumbling block to developing a more progressive stance. He wrote off the *New Moral World* as a blinkered party organ, whose editorial policy was designed to stop the more adventurous from rocking the boat. His shaft was not surprising, for the *NMW* editor had refused to run his farewell address or even his letter explaining his resignation.

A Spate of Atheist Prints

The "thing was damned", Southwell said, defecting. He took Chilton with him, the sharp twenty-six-year-old compositor on the *Bristol Mercury*. Compositors were elite artisans. Literacy marked them out, they had to be able to read fast and hammer metal type accurately—meaning they often had to interpret copy, which necessitated spelling and grammatical skills.[27] They had even been known to suggest improvements to authors themselves.[28] Chilton was among the most incisive. He was also another doctrinaire infidel—in fact Holyoake called him "the only absolute atheist I have known".[29] In Chilton's words, he was not prepared to "coquette" with the priests. Or with the Central Board; the malcontents pricked Owen, the *New Moral World*, "a *disgrace* to *our* society", and the "weak stomachs" of flunkies.[30] Now, by adopting an aggressive stance, they helped spawn the first independent and overtly atheist literature. Southwell's *Oracle of Reason* (1841–43) premiered on 6 November 1841. It proudly declared on its masthead that "WE WAR NOT WITH THE CHURCH,

---

26 Claeys 2002, 253–54; Royle 1974, 69; Southwell 1850, xiv.
27 P. Duffy 2000.
28 Compositors even suggested a better title for one of Charles Darwin's books: Desmond and Moore 1991, 547.
29 Holyoake 1892, 1: 142. Chilton also had Chartist sympathies: he was a delegate to the Birmingham Chartist conference in 1842: *Evening Star*, 28 Dec. 1842, 3; Royle 1974, 68–72; Chilton *ODNB*.
30 *OR* 1 (16 Apr. 1842): 142.

BUT THE ALTAR; NOT WITH FORMS OF WORSHIP, BUT WORSHIP ITSELF; NOT WITH THE ATTRIBUTES, BUT THE EXISTENCE, OF DEITY."[31]

In its train came a welter of confrontationist prints. Owen's truckling had opened up the deeper fault lines, and uncompromising atheists were prying the crack open. Within weeks of the *Oracle* appeared the *Atheist and Republican* (1841–42), by another fallen star, Frederick Hollick. These circulating missionaries had long circled around one another. Hollick had been made district missionary in Birmingham when Southwell left for Bristol, although his flock complained that he was not advancing socialism but his own agenda.[32] His collective had three numbers of the *Atheist* out by 18 December 1841, and nine issues in June 1842, by which time Hollick himself had sailed to America.[33] There followed another atheist print, the half-penny *Blasphemer*, which appeared on 1 January 1842, but, like the *Atheist and Republican*, "after burning for a while, [it] flickered and died".[34]

At this point, Saull's long-time friend Henry Hetherington weighed in, with yet another atheist paper. In February 1842, he started the *Free-Thinker's Information for the People* (1842–43). It mimicked Chambers's respectable *Information for the People*, but junked the twee and anodyne and substituted the subjects on which "all such publishers are studiously silent", namely, a debunking of prophecies, miracles, supernaturalism, and gospels. It also had lashings of Hindu mythology and Pagan philosophy, all picked for their anti-clerical impact.[35]

It was old-style confrontation appropriate to Hetherington's purpose. Saull's geological-perfection principle was as secondary as Owen's "beatific scenes" awaiting us in the new moral world. The object was to change society so the servant could sit at the same table as his master. Hetherington remained less interested in perfecting man than in removing the "curse" of the dispossessed.[36] Central to that curse was the village clergyman, educated among the gentry at Oxford University

---

31 *OR* 1 (12 Feb. 1842): 67; (6 Nov. 1841): 1. J. A. Secord 2000, 307; Rectenwald 2013, 235–36; Mullen 1992; Desmond 1987.
32 *NMW* 7 (20 June 1840): 1320; 9 (12 June 1841): 367; 10 (21 May 1842): 371.
33 *NMW* 10 (18 Dec. 1841): 200; 10 (25 June 1842): 424; Royle 1974, 75.
34 *NMW* 10 (1 Jan. 1842): 216; *OR* 2 (4 Feb. 1843): 62.
35 *FTI* 1 (1842); *NMW* 10 (12 Feb. 1842): 264; Royle 1974, 75.
36 *PMG*, 4 Apr. 1835; 12 Sept. 1835. Claeys 2002, 175–230. Claeys points out that Hetherington also doubted that home colonies would ever thrive without competition. By contrast, Saull stayed with Owen on this point.

or at Cambridge. His village presence with the magistrate and squire ensured compliance among commoners; as such, the clergy were seen as the policing agents. Hetherington's anti-clericalism never dimmed. It hardened further when he was charged with selling a "blasphemous" set of pamphlets—the Manchester socialist C. J. Haslam's ridiculing *Letters to the Clergy*—and given another jail term, four months in 1841. Saull chaired a meeting in John Street to petition Parliament for his release. But no one was sanguine of success, when the jailing was clearly for mixing attacks on clerical extortion with demands for equal rights for the poor.[37]

Fresh out of jail, and angry, Hetherington commissioned articles for his *Free-Thinker's Information*. Despite his previous reluctance to use science, Hetherington now ran sermons in stone to illegitimate the clergy's spiritual sanction, but in an unexpected way. The opening article, on the "Mosaic Account of the Creation and Fall of Man", used geological immensity to contradict Genesis. It was penned by twenty-year-old Thomas Frost, a Croydon-bred printing apprentice, yet to fully grasp the Owenite doctrine of the "influence of circumstances" and trying his hand at writing.[38] It rehashed the story of those giants which so fascinated his generation: the exotic "Ichthyosaurus and Plesiosaurus, two gigantic sea reptiles ... the monstrous Iguanodon, the remains of which have been found sixty or seventy feet long", and so on. They had been mainly amphibious, Frost assumed, thriving in a torrid, oceanic world. Only two fossils of contemporary "marsupial animals" were known, real rarities, whose bones from their island homes must have washed down the rivers to be luckily preserved in the sediment. When dry land predominated, so did the mammals, and Palaeotheriums and mastodons became the new "lords of creation".[39] It was a simple story to score simple points. The earth had passed through untold aeons, an immensity of time beyond the ken of humans, let alone Genesis. And the *Iguanodon* and *Ichthyosaurus* were extinct, each genus had vanished, every species and every individual. The rock strata were ledgers of the dead. Mortality did not begin with Adam's fall; in nature's mausoleum, it had been recorded since time immemorial.

---

37 Hetherington 1840; *NS*, 27 Feb. 1841.
38 Frost 1880, 15; 1886, 40; Frost *ODNB*.
39 Frost 1842, 6–7; Desmond 1984 on the controversy over these first fossil mammals.

The interesting thing about Frost's piece, and possibly the reason why Hetherington published it, was that there was no perfectibility, the bedrock of Saull's geological Owenism. Saull swore that nothing but "Socialism fully carried out can meet or remedy the manifold evils that afflict mankind", and that a democratic education would put society back on the path to progress. He expected clergymen to fall in, and echoed Owen in believing that even "the highest ranks of society" would eventually follow suit.[40] Hetherington had long scoffed at the idea of the wealthy voluntarily relinquishing power and profit. The "designing knaves", using the Church "to perpetuate their plunderings", could no more join a co-operative than socialists could attain Utopia through tea parties.[41] He had no need of Saull's "universal law" of fossil progression, still being heavily promoted to underwrite the march to the millennium.[42] So, where Saull had commandeered the directionalist fossil record of the Oxford don, the Rev. William Buckland, Hetherington published Frost's article in the *Free-Thinker's Information*, which appropriated the discordant science of Charles Lyell.

Great authorities were needed to command respect. No matter that geology's gentrified exponents were otherwise anathematized by activists: by the canons of the age, the men of science were seen as rationally constrained by nature, which gave their scientific voice its validity. Because Genesis was directional—a miraculous sequence of creations culminating in Adam and Eve—Frost had Lyell offer a conflicting image. At £2 7s, six week's pay for a Dorchester Labourer, Lyell's three-volume *Principles of Geology* was aimed at wealthy readers. Genteel book buyers expected the conventional pieties. And since Lyell was to suggest that nothing stronger than today's climatic and volcanic forces were needed to change past landscapes, he was careful to assuage readers' fears that such did not apply to ancient species as well. Any hint that past life had been altered by everyday causes would have been morally reckless. What sort of delinquent would brutalize man by making him a better sort of ape? Lyell himself was revolted at the fantasy of a blood line imperilling man's immortal soul. So he crafted *Principles* to avoid any imputation of bad taste or judgement. He undermined

---

40   *NMW* 7 (20 June 1840): 1319–25.
41   Hetherington [1832], vi; *FTI* 1 (1842–43): 245–51.
42   *NMW* 8 (18 July 1840): 37.

talk of life's inexorable rise, undercutting potential evidence for the transmutation from 'lower' to 'higher' forms. The fossils pointed to no continuous upward direction, and any evidence to the contrary was an artefact of preservation.[43] This part of Frost's article was what attracted Hetherington: Lyell's denial of the "progressive development of life from simple types" in the old strata "to completer developments" in the later rocks.

In Frost's paraphrase, life was "complex and complete" from the first. There was no "gradual development", nothing to correspond to the Genesis of the Sabbath sermon. Armoured fish were turning up in ancient Scottish rocks, and scales and footprints possibly told of tortoises at the same time. Even the odd reptile had now been found in strata as deep as the coal seams. Mammals were absent from these early deposits only because of the odds against the preservation of their remains. At the time they had probably been living on scattered islands in the wide oceans, and their carcases had not been preserved because they required estuarine sediments for entombing. But who knew what would eventually turn up? The first fossil monkeys had unexpectedly been found in the late 1830s. So perhaps ancient humans were awaiting discovery. This left a twin-pronged conclusion: that complex life was unimaginably ancient, and that "Millions of years are inadequate" to explain its entombing formations. Therefore, geology could provide no comfort to Genesis, whether of Six Days or Six Thousand Years. But the lack of progression also proved that there had been no natural trajectory towards Heaven or the Millennium.[44]

By the time the *Free-Thinker's Information for the People* came out, Charles Southwell was in prison. Southwell had been an idiosyncratic Owenite, a fellow traveller who had not travelled very far. Having bridled at Owen's patriarchy and his truckling to Christians, he had founded the *Oracle of Reason* with a more confrontationist aim. It was an "exclusively ATHEISTICAL print",[45] whose calculated crudity raised the expected storm. The opening inflammatory articles were Southwell's, refuting God's existence and undercutting the clerical props of the Anglican state. The crudity peaked in the fourth number. Southwell's

---

43 J. A. Secord 1997, xxx–xxxv: Corsi 1978; Bartholomew 1973; Ospovat 1977.
44 Frost 1842, 6–7.
45 *OR* 1 (1842): Preface ii.

sexual frankness was uncommon for his day. It seems to have bordered on obsession, judging by his shocking *Confessions*. But coupled with his earthy delight in "old and rude" English, it could re-craft the Pentateuch as a depraved Shakespearean tragedy. The dramatist, who had "smelt the lamps", could not resist playing the Old Testament as "a history of lust, sodomies, wholesale slaughtering, and horrible depravity".[46] But the crudity cut much deeper. Where Owenites generally displayed a cultural generosity, he was a racist. He seemed at times to anticipate the race warrior Dr Robert Knox in dispensing stereotypical judgements (hence modern Greeks were a "pirating, lying people"; Egyptians "a degenerate race", and so forth). And like imperial phrenologists he could place foreign heads beyond educational redemption: the small brain of the "Carib" or "stunted, dwarfish Laplander" rendered them immune to Owenite benevolence.[47]

Southwell's was a toxic combination. He rejected a Christian 'soul' encompassing all races; therefore he lambasted all talk of Adam and Eve as common parents of all races; and he rejected Owen's forgiving cultural relativism. The result was a rarely seen moral drift and racism. When he lashed that "revoltingly odious Jew production, called BIBLE", he was, as others have pointed out, cashing in on the prevalent "anti-Semitism as an alienation tactic".[48] Southwell confessed that he meant the "Jew Book" diatribe to cause outrage "and, with that view, used terms the most offensive I was able to use."[49] The old hands hated it. Carlile loathed the "splutter and clatter" of his socialist enemy and cancelled his subscription. Even the dagger-brandishing Julian Harney was saddened by the *Oracle*'s "ribaldry and disgusting language"—although, like Carlile, he contributed to Southwell's prison fund. One wonders whether Saull was so disgusted. Here, after all, was the activist who had financed Taylor's theatrical assaults on the "Jewish vampire", and had supported Smith's Antichrist—and he too routinely supported

---

46 *OR* 1 (1842): 25; Marsh 1998, 111–14; Royle 1974, 76; Mullen 1992.
47 Southwell 1840, 2–5, 10. Southwell had access to Charles White's *An Account of the Regular Gradation in Man* (1799), which posited separate origins for the 'lower' black and 'higher' white races (*OR* 1 (1840): 5–6). Stenhouse 2005, examines how Southwell's venom turned on Maori Christians after he emigrated to New Zealand in 1856, how he rejected amalgamation, or the notion that Maori could be civilized, and how finally he embraced genocide as an option.
48 Marsh 1998, 113; J. A. Secord 2000, 312–13; *OR* 1 (1842): 25.
49 Southwell 1850, 66.

Southwell's fighting fund. Others put Southwell's language down to an expletive tit-for-tat, the "same Billingsgate abuse" used in "Christian attacks [on] the Infidel".⁵⁰

Shrieking insults did indeed fly both ways. For Evangelicals, this atheist profanity was a warning that Satan's work was almost done. The *Oracle* fulfilled the prophecy of the sixth vial in Revelation—it was the "Unclean Spirit from the Mouth of the Dragon [the Devil]", spreading "its filthy slime over Christendom".⁵¹ But society was asymmetrical, Christianity was the law of the land,⁵² and Southwell was handed a year's jail term for blasphemy.⁵³

The sentence was a foregone conclusion, given the "somewhat rampant piety of the times", said the *Satirist*.⁵⁴ Large crowds demonstrated at the Halls of Science: 2,000 in Bristol, 1,400 in John Street, with Hetherington doing the branch rounds, starting petitions and collecting funds. Southwell's defence committee made sure that a faithful transcript of the trial, including his ten-hour defence speech, was published as part of the propaganda.⁵⁵ But the Owenite Central Board showed little sympathy, even if it deplored the state's "fierce intolerance". It found itself suddenly sensitive to "violent attacks upon the opinions or prejudices of our fellow-beings".⁵⁶

Young Turks sprang into action all over the country. A twenty-two-year-old in Glasgow, Robert Cooper, started a collection.⁵⁷ Cooper was a shooting star, destined to shine as brightly as Holyoake for a time. He was to the cause born: his father had been on the platform at the Peterloo massacre, and Robert was from the first generation to come out

---

50 Royle 1974, 75. G. J. Harney to G. J. Holyoake, 17 Nov. 1843, Holyoake Corresp. No. 102, Bishopsgate Institute; R. Carlile to G. J. Holyoake, 16 Oct. 1842, ibid., No. 79.
51 Bickersteth 1843, 8, 21–22.
52 Southwell (*Investigator* [1843]: 71) claimed that the witch-burning Sir Matthew Hale invented that "silly sentence" about Christianity being part of the law of the land, "so often quoted as infallible wisdom, by the judges", as a pretext for crushing those who disrespected Christianity.
53 Southwell 1842, 1.
54 *Satirist or the Censor of the Times*, 23 Jan. 1842, 27.
55 *Nonconformist*, 19 Jan. 1842, 43; *NMW* 10 (22 Jan. 1842): 239; *NS*, 15 Jan. 1842; *CPG*, 5 Feb. 1842. The trial was well reported in the press with long coverage in the *Bristol Mercury*, 15, 22 Jan. 1842, *Weekly Chronicle* 22 Jan. 1842, 4; and *NS*, 22 Jan. 1842, with précis in many London and regional papers.
56 *NMW* 10 (11 Dec. 1841): 191; (25 Dec. 1841): 208.
57 *NMW* 10 (26 Feb. 1842): 280.

of Salford co-operative school, where he had heard Owen.[58] Evangelical spies smeared him as an "effeminate and affected-looking man",[59] but he drew adoring crowds in the North. Like Southwell he had a functional approach to science: what holed and sank the biblical Ark worked. As an example, he held up the French anatomist Pierre Flourens, who found a skin layer in black people that was "altogether wanting" in whites, from which Flourens concluded that the races were "essentially and specifically distinct". Cooper extended this to argue that they "must have originally sprung from *perfectly separate stocks*" to contradict the Adam and Eve story.[60] There was no suggestion that Cooper was following Southwell's path, even if racists were to make Flourens's findings part of the pro-slavery ideology in the 1850s and 1860s. Cooper stayed loyal to Owen and did not indulge in racial slurs. He had a Rousseauean respect for ancient Confucians and moral 'savages'[61] and used Flourens only with irreligious intent. But it does emphasize again how artisan atheism in the 1840s could open up potentially dangerous channels.

Cooper's defence money joined the rest. And although the pleas on Southwell's behalf to the Central Board fell on deaf ears, it could not stop a benefit concert at John Street. The urbane Saull did his bit: he chipped in, not much, ten shillings here, six shillings there.[62] Whether this was out of duty, or real sympathy, was a moot point, since atheist breakaways obviously put Owen's ally on the spot.

---

58 *LI* 2 (May 1855): 28–29.
59 *Monthly Christian Spectator* 2 (Dec. 1852): 718; Royle 1974, 89.
60 R. Cooper 1846, 158–59. The substance (pp. 361–66) of Flourens' article (*Annales des Sciences Naturelles* 10–Zool. [Dec. 1838], 357–66) was translated as "On the Natural History of Man" in the *Edinburgh New Philosophical Journal* (27 [Oct. 1839]: 351–58). Since Cooper was stationed at Kirkaldy (1842), Glasgow (1842), and Edinburgh (1843, 1845), this was probably his source. As a challenging speaker who drew Edinburgh students, he was actually invited in 1843 to attend lectures at the university. Those of "Professor Millar", he recalled, "on Practical Anatomy were of eminent service to me". Possibly this was the new (1842) professor of surgery James Miller, whose historical talks on Pictorial Anatomy discussed religious art. As a Free Church advocate who wrote Christian tracts for labourers (*Edinburgh Medical Journal* 10 [July 1864], 92–96), Miller would certainly have found Cooper a challenge. This gave Cooper "access to libraries in Edinburgh inferior to none" (*LI* 2 [Dec. 1855]: 30). Such penetration of a higher learning institution was unprecedented among socialist missionaries.
61 R. Cooper 1846, 193–97.
62 *OR* 1 (25 June 1842): 224; (2 July 1842): 232; *NMW* 10 (5 Feb. 1842): 256.

The Blending of Life and Society

Four issues of the *Oracle of Reason* were all Southwell printed before the authorities pounced. The legacy, though, was evident. Besides a flagship series of articles "Is There a God?" (to establish the paper's atheist credentials), another series on "Symbol Worship" (on ancient Hindu and Egyptian religions, to decentralize Moses), there was a more cryptic third series. He initiated it with a teaser: an evocative illustration—the first in the *Oracle*—placed at its head: a near-naked, axe wielding "Fossil Man". This accompanied the "Theory of Regular Gradation" series, which could occupy a quarter of each eight-page issue, and, continued by Chilton, would run to 48 articles across two volumes. All told, the series stretched to 80,000 words (the size of a book), and that before it spilled over into subsequent publications.

"Regular Gradation" sounds innocuous, but these were inflammatory words. They translate into something like serial transmutation: "the *blending* of one animal into another, the growing out of, or changing of, one form into another", recorded in the rocks as a gradual progression of life forms from simple to complex. After untold aeons this resulted in mankind, "an animal so long in coming to perfection".[63] A similar concept was shortly to be called "Development", and, by 1870, the word "Evolution" was taking over, although conceptually very different by that point. Different, not least, in its usage: if nothing else, this overtly atheistic palaeontology in the 1840s left no doubt that the mutation of life was now a constitutive part of anti-clerical propaganda.

As schismatic freethinkers, Southwell and Chilton wielded disquieting doctrines like transmutation to assail the Church rather than support socialism or advance science. "Gentlemen, the learned counsel told you that I wished to reduce man to a level with the brutes", said Southwell at his trial.[64] But while counsel was thinking of Southwell's attack on mankind's immortal attributes, the literal brute-levelling came in "Regular Gradation". Whither the "dignity of the soul", said Southwell in his talk on the tailed ancestry of man, and what use the divine, "were it proved that his flock were, after all, but the fiftieth cousins of sheep".[65]

---

63 Chilton, OR 1 (2 July 1842): 228; (14 Oct. 1843): 347.
64 Southwell 1842, 30.
65 Southwell, OR 1 [6 Nov. 1841]: 6.

This picture gave cynical satisfaction to the insurgent atheists. As hard-bitten editors, struggling through the 1840s' famine and depression, they were slapping down hauteur. The "ridiculous conceits" of the "more *nice-than-wise*" class took a lashing: a class puffed up with its supposed superiority, whose "false delicacy" about mankind's naked animality shielded it from the sordid truth.[66]

How had Southwell got to this point? In two years, he had moved by leaps and bounds to attack human conceits at their core. As his Owenism loosened and atheism strengthened, his science became more extreme. At Lambeth in 1839, he had glossed "Man, in relation to other Animals", to show that mind as well as body "changes under the influence of external circumstances", but simply to set up the malleability of the human brain in phrenological terms, and its susceptibility to Owenite education.[67] By 1841, as the missionary rattled the bars of his Owenite cage in Birmingham and Bristol, he had moved on to "the true meaning of the book of Genesis", or "Life, Death—the Genesis account of the Creation".[68] In palaeontology, he now saw true potential.

What seems to have catalyzed his leap was an article with the baited title "L'Homme Fossile" in the popular French *Magasin Universel*. This capitalized on a spectacular find. In the 1830s, fossil human skulls, bones, and worked tools were turned up alongside extinct animal remains in caves near Liége. No one knew how old they were or even if they differed from modern skulls. But it was enough for the radical French geologist Pierre Boitard, in 1838, to title his article "L'Homme Fossile" and open with an ape-savage illustration. Despite its heading, the piece actually romped through the whole of fossil zoology, emphasizing the graduated rise of life. The title was a lure, and readers only encountered this fossil human on the last page.[69] But Southwell saw the potential. He cribbed Boitard's illustrations, pasting them into "Regular Gradation", including the monkey-faced "fossil man", and clothed the article's framework of gradual complexification with hardcore atheist apparel. Configured thus, the "Theory of Regular Gradation" would blossom into a full-blown naturalistic vision, cosmic in scope. Admittedly it

---

66 Southwell, *OR* 1 (13 Nov. 1842): "6" [13].
67 *NMW* 6 (14 Sept. 1839): 752; (12 Oct. 1839): 807.
68 *NMW* 9 (8 May 1841): 296; 10 (21 July 1841): 30.
69 Boitard 1838, 240. Rudwick 2008, 412–16; Grayson 1983, 6ff; Riper 1993, 61–63.

moved in a very ramshackle way from planetary formation and life's chemical origin through to an ape ancestry for mankind. And nothing was beyond its evolutionary scope, for, in Chilton's words, unbiased philosophy must admit "that the inherent properties of 'dull matter,' as some *bright* portions ... have designated it," are "sufficient to produce all the varied, complicated, and beautiful phenomena of the universe".[70]

Southwell was still looking imperiously *down*—from the Caucasian crowning heights, from the destination which life aspired to. Hence the "fossil man" he pictured was "man undeveloped".[71] No longer monkey, he was not yet man, but was gazing upwards. The aspect "higher" and "lower" dominated the natural world no less than the social. And, although Southwell and Chilton chafed at the latter, they tacitly accepted the former (as did almost all men of science in the 1840s). They saw mankind as nature perfected—hence the *serial* transmutation. Life was trundling up towards its apotheosis.

On Southwell's jailing, Chilton—who said he had held similar views for "several years"—took over the "Regular Gradation" series, beginning on 19 February 1842. He immediately introduced more modern sources.[72] He plundered geological and medical tomes, quoting verbatim passages about ascending fossil sequences. As Southwell had said, kicking off the series, fossil animals were "in a state of continual flux"[73] and changed gradually into more complex forms. Chilton explained that they did so because the "life-producing and life-sustaining" environment of each age increased, as Saull had argued from Phillips's postulates, and this resulted in an expanding number of varieties and their more modern appearance. As ever, he resorted to the Owenite-Lamarckian stand-by of environmental modification, according to which the developing species "either accommodated themselves to the different circumstances, or became extinct".[74]

---

70 Chilton, OR 1 (19 Feb. 1842): 77.
71 Southwell, OR 1 (20 Nov. 1841): 21.
72 Chilton, OR 1 (19 Feb. 1842): 77. His sources included geologists Henry De la Beche, Charles Lyell, John Phillips, and William Buckland, and comparative anatomists Robert Edmond Grant, Richard Owen, W. B. Carpenter, Thomas King, and George Newport, although Chambers' *Information for the People* seems to have been the stock source.
73 Southwell, OR 1 (20 Nov. 1841): 21.
74 Chilton, OR 1 (11 June 1842): 204–08.

Wealth and Power

No freethinker had covered the inflammatory subject so exhaustively before or exploited such up-to-date sources. If this kind of thinking was suffusing the pauper presses, Chilton knew why so few gentlemen savants were prepared to believe it. Carlile, in typical 1820s' language, had seen the squirearchy of science "crouched to the established tyrannies of Kingcraft and Priestcraft".[75] Chilton now tore into state-sanctioned knowledge likewise, looking at it through class spectacles. The *Oracle* slated authorized science as "a matter of traffic and trade amongst the *savants*, and the higher classes", with polite geologists looking to support from "right reverends, right honourables, &c., in fine, on those who are interested in keeping up the usual common-place *go* in society".[76] Chilton was not so naive to believe that his materialist science would turn minds, which it could only ever have done in his alley audience. If nothing else, as Secord says, elite science grounded its political authority in expert factual knowledge.[77] The gentlemen argued that this was arrived at by a royal road of inductive reasoning based on lengthy observation and time-consuming travels, something which put it beyond the reach of women and the working classes. It was the standing of the wealthy 'experts' which gave official science its imprimatur. As Steven Shapin puts it, the claim of true knowledge was assessed according to criteria of personal competence. This relied on trust, which itself was socially generated. And in an age when gentlemen savants were barely distinguishable from their social peers,[78] those assessments were largely class based, meaning the socialist "scum" were denied a hearing as untrustworthy fanatics.

It was the intractable class nature of the interpretive authority, backed by wealth and rival clerical power, that Chilton was up against. Elite gentlemen were not going to change their minds. But Chilton was not talking *to* them, nor were they listening. The closest the dons and divines came to his trash was to tut-tut about the show trials reported in

---

75   Carlile 1821, 101.
76   OR, 29 July 1843, 261; Chilton 1842.
77   J. A. Secord 2000, 312–13.
78   Shapin 1990.

their morning's *Times*. Chilton was actually persuading his angry alley followers of the bias of bourgeois science backed by wealth.

Wealth bought the old order time to indulge. As an old revolutionary (and Saull neighbour) said, "literary pursuits, like law suits, are beyond the means of those who cannot command means to enforce their claims."[79] Chilton was gazing at another world. He could not have entered the manor's front door, let alone have sat at the same table as a geological savant. It was the tradesman's entrance for him. Though in a higher trade, he probably earned about 48s a week,[80] so much small change for the rich. On a tight wage, he could only dream of the sums attracted by the geological gentry. The Rev. William Buckland was paid £1000 out of the Earl of Bridgewater's estate to write his *Geology and Mineralogy* (1836)—the book bastardized by Saull and Chilton—on the stipulation that it exhorted "The Power Wisdom and Goodness of God as Manifested in the Creation".[81] As "Regular Gradation" was running, Buckland was chivvying Prime Minister Sir Robert Peel to give his devout Anglican and anti-transmutationist protégé Richard Owen a Civil List pension, a top-up of £300 a year for life. Owen was awarded it after his religious respectability and scientific potential were vouched safe. It was a Peelite prop which would allow a lifestyle to match his scientific rank.[82] Respectability brought its own reward.

What was said of other talented activists would have applied to Chilton, that "had he been less poor he would have been more famous".[83] Although the *bon mot* is not as profound as it seems, because it was poverty-inducing inequality that turned Chilton into an activist grinding a fossil axe. Looking at some of the obscenely rich geologists of his world explains why. Take the "King of Siluria" Roderick Impey Murchison, former military man turned geological imperialist, who followed his Silurian System into Russia, annexing the rocks as it were. He could afford to be conceited. He once lost £10,000 on a single railway speculation, which gave some inkling of his fortune. As a haughty Tory, with a mansion in Belgrave Square, he had a ludicrous "thirst for

---

79   Allen Davenport's words: *Reasoner* 2 (16 Dec. 1846): 18.
80   *LMR* 1 (13 Nov. 1824): 28.
81   Topham 1992.
82   Desmond 1989, 354–57.
83   *Reasoner* 2 (16 Dec.1846): 18.

honours" that was duly sated by the Russian Tsar. Though a religious doubter, unconvinced of Christ's divinity, still the old soldier declared he would "stoutly fight for the Church, as a great and essential moral engine".[84] The Anglican proprieties were kept up along with the appearances. And what glittering appearances: Murchison's Belgravia mansion saw 700-strong *soirées*, and a young John Ruskin attending one in 1842 was agog: "rooms all pale grey & gold—magnificent cornices—with arabesques like those of Pompeii in colour, furniture all dark crimson damask silk & gold ... at least four footmen playing shuttlecock with peoples names up the stairs." Even he admitted this was "coming it rather strong".[85]

So Chilton's question to his poor readers came down to this: if, as he believed, theology and science were "natural enemies", why had they entered into a "hollow conspiracy" during the "fashionable reign of the Bridgewater treatises"? The likes of Murchison provided the answer. To the unemancipated and disenfranchised, the self-constituted guardians of knowledge were using pro-Christian, anti-democratic science to sustain their wealth and rank. For another *Oracle* editor this was why they "ignominiously betray their trust" as "expounders of truth".[86] To share in the rewards from the exploitation of labour, they were prostituting their science, or such was the radical view on the street. And the rewards were great: the pauper presses were quick to point to the clergy having a financial stake in the status quo: the £9 million collected in tithes and taxes that would be forfeit by secularizing society and disestablishing the church.[87] But that would be nothing to their losses if democracy followed. Not that there was a chance: Lyell, Murchison, Sedgwick, Buckland, all had a horror of being swamped by the underclass. Lyell spoke for all when he said that good breeding, superior education, and independent station made proper leaders, not a popular vote.[88]

A gruff Dalesman, the Rev. Adam Sedgwick, had cracked the Cambrian system, but he, too, conflated the natural and social strata,

---

84 Geikie 1875, 1: 263; Stafford 1989, 6, 15, 190, 209; J. A. Secord 1982.
85 J. A. Secord 1986b, 123.
86 *Movement* 1 (1 June 1844): 196–97.
87 Hetherington [1832].
88 Lyell 1849, 1: 33.

no less than Saull or the *Oracle* activists had done. Preaching to the coal-blackened colliers, he reminded the "rabble" of the providential aspect of the coal industry and their own beneficial relations to mine-owners and capitalists, superimposing the moral, economic, and geological orders to gird up class barriers. Even ignoring the atheists' onslaughts on the Christian faith, their very language was threatening to him: the concept of blending "lower" into "higher" was abhorrent. Sedgwick might have been a Whig and mildly reformist, but the preservation of rank, with all its moral attributes, character, dignity, and ornament, was still of paramount importance to him. Even if he did "wish that the barriers between man and man, between rank and rank, should not be harsh, and high, and thorny; but rather that they should be a kind of sunk fence", it still had to be "sufficient to draw lines of demarcation" between them. Blending was impossible; this was God's ditch. As in society, so it was in nature. God's "elevation of the Fauna of successive periods" was by "creative additions". Successive groups were introduced—Fishes, Reptiles, Mammals, Man—each with "an organic perfection corresponding to their exalted rank in Nature's kingdom". The suggestion by blaspheming democrats that the "lower" could push itself up and transmute into a "higher" rank was anathema for blurring these God-given boundaries.[89]

What Chilton saw as conscious exploitation was a far more tacit, nuanced, and complex situation. In a pre-professional age, the elite practitioners were invariably Oxford and Cambridge divines, or trained by them, or else wealthy ex-lawyers, military officers, or medical men. In entry ledgers, under "Occupation", they would write "Gentleman", to distinguish them from the 'lower' orders. That is, they were financially independent, with the time and wherewithal to indulge their passion for the rocks. They could afford to buy and write expensive tomes, and stump up exorbitant society fees. By controlling the learned societies and publications, they became the self-declared arbiters of content and taste. Many had a dual calling: Buckland would become dean of Westminster, Sedgwick added to his Cambridge professorship a prebendary at Norwich Cathedral, which he hoped would net him £600 annually for

---

89   Clark and Hughes 1890, 1: 515–16; 2: 47, 189; Morrell and Thackray 1981, 31–32, 127.

two months' attendance.[90] Others would be knighted—Davy, De la Beche, Murchison, and Lyell. Even that young anti-radical and anti-materialist medical professor Richard Owen—already entertaining Prince Albert in his Hunterian Museum and arranging for his own portrait to be hung at the Prime Minister's country house—was to be offered one. All would court their Royal Highnesses. Owen, having been fitted by the palace tailor for the necessary cocked-hat and "elegant attire", would go on to teach the royal children at Buckingham Palace.[91] Lyell was on intimate terms with the Prince Consort at Balmoral.[92] These scientific gentlemen had more than a stake in the status quo. They were closely knitted into the power structure at a personal level.

For science to succeed at an institutional level, it needed titled and royal patrons. A "By Appointment"-status conferred conviction and gravitas; it helped elevate its ranking alongside the proper professions. And the improving aristocracy, as guardians of morals, manners, and Mammon, saw it as part of their public calling to officiate at these learned bodies. Duty might only mean an honorary station, but the trickle-down effect was palpable. Thus the British Museum was run like a rotten borough. The clergy and nobility considered it a show-piece for the nation's treasures, not necessarily somewhere to advance knowledge. It was revealed in their *noblesse oblige*. Appointments, in the gift of the trustees, headed by the Archbishop of Canterbury, were restricted to safe men of science, while librarians and functionaries were enlisted "from the inferior departments of the church and public offices".[93] The Zoological Society was top-heavy with noblemen, who were intent on turning the Zoo into a game park, which promised delicacies for the gentleman's table. The aristocracy conferred prestige; they also attracted patronage, meaning the Zoological Society quickly acquired its royal charter. They could underwrite its success and negotiate face-to-face with government ministers, especially if they required land (as in the Zoological's case in Regent's Park).[94]

---

90 Clark and Hughes 1890, 1: 435.
91 Desmond 1989, 358; Rev. R. S. Owen 1894, 1: 246–47, 353–55; 2: 98.
92 K. M. Lyell 1881, 2: 156–58.
93 *Hansard* 1836, 31: 308–12; Desmond 1989, 145.
94 Desmond 1985a, 226ff.

Premises too were in the administration's gift, and it was the government which granted the Geological Society its spacious apartments on the site of a former palace. Somerset House, on the busy Strand, was a "magnificent" quadrangular edifice of solid granite, a pile that spoke of permanence and security. With its exquisite statues "consisting of the arms of the British empire, supported by the Genius of England, and Fame sounding her trumpet", the state building would confer patriotic prestige on the Geologicals. It too had royal connections as Queen Caroline's former town residence. Moving here put the geologists alongside the prestigious Royal Society, Society of Antiquaries, and the Royal Academy with its fashionable art exhibitions. But it also put them beside something more sinister. The imposing block was actually a government administrative centre, the income tax and audit office, and, to the horror of campaigners, home to the hated Poor Law Commissioners. The Geological governors sat under the same roof as the "Tyrant of Somerset House", responsible for incarcerating the old and unfortunate—while the cry on the street was for "freedom from the despicable bondage of the 'lickspittle' despots of Somerset-house and Downing-street!"[95]

The Geological's apartments provided a traditional gentleman's club facilities, with reading and lounging rooms.[96] The same geological squires effectively controlled the British Association for the Advancement of Science. Here again, faced with agitation for "fierce democracy", they rendered it a peripatetic vehicle of calm knowledge, and the odd radical recalcitrant who tried to rock the boat quickly found his avenues for advancement blocked off.[97] The scientific barons were committed to a pyramidal social structure propping up a wealthy intellectual elite. At most, they promoted gentle reforms as a panacea for working-class discontent. For them, scientific truth had a very different moral dimension. It encompassed responsibility and social stability, and spoke loudly against radical redistribution. It tacitly underpinned 'creation'

---

95 Baxter 1841, 36; Young 1960, 50; Bartlett 1852, 176–78; Brady 1838, 77; G. N. Wright 1837, 2: 671; Cruchley [1831], 28. Political ties between science and state were also ensured by the huge back-bench presence of Members of Parliament in all the learned societies. In the Zoological Society, nine per cent of members were MPs (Desmond 1985a).
96 Rudwick 1985, 23.
97 Morrell and Thackray 1981, 302, also ch. 1, and 245–56.

and subordinated matter to a guiding Will, whereupon 'duty' could be dictated by Church authority.

This was what an impoverished Chilton was raging at in 1842, during the death and starvation wrought by the economic depression. Where Owenites saw the scientific bosses acting according to their station, Chilton attributed base motives. They were traitors to true science, he seethed in "The Cowardice and Dishonesty of Scientific Men". They lacked the "honesty" to come clean about its materialism, opting to lay it at the Christian altar to preserve their privileges. The knights of science were in league with the political and clerical masters, finding it in their "interests to keep us in this position". "This is the unkindest cut of all; coming as it does, from those who should pour the balm of hope upon the despairing and wounded spirit; instead of which, They smile, and murder us while they smile!"[98]

The scientific Eucharist was handled like "contraband goods," religiously cloaked and kept among the cognoscenti "lest the trade and tithes of the priest be injured".[99] That religious profiteering amounted to "Nine millions of money", seethed Hetherington, "the greater part of which is paid to lazy luxurious bishops, the younger scions of the aristocracy, or to deans, chapters, deacons, vicars, rectors, &c., &c., most of whom are non-resident, fox-hunting, dissipated, immoral, and unprincipled".[100] The activists were convincing the dispossessed that a self-transmuting nature sanctioned social action against this enemy. Out went the priests' "puerile" notion of "creation".[101] No more could life be conjured up at the beginning, out of nothing, than continually, through geological time by a Deity. Instead, the militants promoted an image of spontaneously emerging and self-rising animals and plants. The idea of a ceaselessly tinkering God was laughable—fit only for that joke by Saull's fellow financier Julian Hibbert: "It must be dull work to be eternally trundling a wheel-barrow, and perhaps hard work too for an incorporeal Being."[102] The strata showed lowly species growing into complex ones. But no Almighty craftsman would have worked this way.

---

98 Chilton 1842, 194.
99 Carlile 1821, 111, 120.
100 *FTI* 1 (1842–43): 251.
101 Chilton, *OR* 2 (14 Oct. 1843): 347.
102 Hibbert 1828, Appendix 3: 7; *Investigator* (1843): 26.

To push the point home, Chilton used a shop-floor analogy: "we do not find a coach-maker, when he has to build a nobleman's carriage, begin by making a mud cart or pair of trucks."[103] Life had built on itself, pulled itself up by its own bootstraps—a perfect artisanal metaphor broadcast by umpteen autodidacts. This self-striving world of life, powered from below, legitimated democratic change. Nature did not need divine sanction, any more than the social atoms needed patrician permission. Sovereignty rested with them: that was their democratic mandate.

The *Oracle* hardliners chipped away at Christianity's defence of creationist miracles and revelation, but they also hammered hard at "design". The 1840s saw newer, sophisticated approaches to "design", based on the archetypal plans linking various animal groups. The most prominent was in another Bridgewater book with a £1000 payoff, *Animal and Vegetable Physiology Considered with Reference to Natural Theology* by P. M. Roget (of thesaurus fame).[104] But the *Oracle* protagonists took aim at a much softer target. That was Archdeacon Paley's by now decrepit argument underpinning other Bridgewater Treatises, that God's existence and benevolence could be deduced from the perfect fit of each species to its niche. Such a degree of planning showed foresight, therefore there must be a caring Planner. These old "proofs of design" were "sadly hacknied", said Southwell; parsons learn the argument "from Paley, Paley stole it from Condillac, and where *he* got it from is not of much consequence". Others forgot the argument and went for the jugular, with the ultimate ad hominem, that Paley was "the greatest drunkard and debauchee of his time". And the argument itself left great scope for facetiousness. That eyes were made to see was as silly "as to say that stones were *made* to break heads, legs were *made* to wear stockings, or sheep were *made* to have their throats cut".[105] So said Southwell, continuing his series "Is There a God?" while sitting in jail.

Not only was there no "design" but, given the unemployment and starvation in the depression, the hubris of Paley's "happy" nature seemed outrageous. Far from seeing nature teem with delighted existence, the

---

103 Chilton, *OR* 1 (11 June 1842): 206.
104 On the history of this book, and how Roget was domesticating (cribbing, insiders said) the radical anatomy of Robert Edmond Grant at London University, see Desmond 1989, 222ff.
105 Southwell, *OR* 1 (19 Mar. 1842): 109, 111; (8 Jan. 1842), 61.

struggling Chilton swore that "all nature cries aloud" against such nonsense. Only parsons living luxuriously on tithes wrenched from the down-trodden masses could fancy nature as a hymn to God's goodness. This was "worse than ridiculous", it was a "vilely pernicious teaching".[106] Exposing Paley's "happy" world—where "all is for the best" with everything in its proper place—was intended to bring the back-broken poor aboard. Cynical *Oracle* activists, looking from below, exposed its dark underbelly. Why had not the squires' deity designed "less suffering and more enjoyment, less hypocrisy and more sincerity, fewer rapes, frauds, pious and impious butcheries?"[107]

Chilton's "Regular Gradation" series fought on many fronts. As a result, it quickly began to lose coherence. It interspersed attacks on Genesis and design with descriptions of fossil life and digressions on anatomy. Eventually Chilton lost his way in the arcana of comparative anatomy: for nineteen issues he trudged through the organs and tissues of the animal kingdom, lifting whole sections from a medical student's compendium.[108] Readers complained. They could not see the relevance of undigested comparative anatomy, and they had a point. These illegal prints, bought from street sellers dodging the authorities in working-class neighbourhoods, were violent, angry, and served an immediate purpose. Tolerance only extended to science so long as it had meaning for the struggle. The series had gone off target and some called for it to be scrapped. To pave the way for popular sovereignty, knowledge had to function; the more esoteric it was, or bogged down in minutiae, the more useless. It needed to be simple, demystified, in a word (and an ugly one) "unintellectualistic".[109] The complaints led to apologies for the "uninteresting and unpopular manner" of the digressions.[110]

But lack of coherence had a more mundane cause too. Chilton was beset with difficulties—accidents, police raids, and the imprisonment of his fellow editors all helped to break the narrative thread. At one point, five of his friends were in prison, and Chilton was campaigning on their behalf, raising bail, attending court, lecturing, and writing

---

106 Chilton, *OR* 2 (11 Nov. 1843): 379.
107 Southwell, *OR* 1 (7 May 1842): 165.
108 He was cannibalizing Evers 1838, a hundred-page digest based on the latest works of R. E. Grant, R. B. Todd, P. M. Roget, and others.
109 Johnson 1979, 84–85, 94.
110 Chilton, *OR* 1 (15 Oct. 1842): 356; 2 (24 June 1843): 220.

thirty to forty letters a week.[111] All of this had to be squeezed into spare time outside of his ten-hour working day as a compositor. His hectic life shows that even the practise of writing could be very different for a man in his position. A gentleman's wealth bought him the leisure to read and write, often (as in Charles Darwin's case) with an amanuensis who would make a fair copy, which would go off to the publisher—and then, in Darwin's case again, he would doctor the proofs at colossal cost, forcing the printers to re-set the type.[112] For a scrimping, rushed compositor, for whom time really was money, there was no such luxury. Chilton would have to camp at his works for two or three weeks at a time. "My life was a continual race; I had not proper time to eat, to sleep, and certainly not to think." Some articles were actually set straight to type on the frame, which explains the series' fractured nature. Nor was the series financially rewarding, or the *Oracle* financially viable. Only Chilton's pay cut to subsidize the publisher and a cash float from a John Street insider (we do not know who) kept the paper solvent.[113] But it left Chilton in poverty.

Composed on the fly, his pieces had a searing tone which told of a militant who thought on his feet. That in itself led to a certain serendipity. Impromptu modifications and digressions could be dropped in weekly. For instance, the appearance of Hetherington's *Free-Thinker's Information*, which denied any "progressive development" to give Genesis the lie direct, caused Chilton to take evasive action. He argued that fossil families in "*each* stratum" might show a simple to complex gradation. Because life, like Owenite man, accommodated to conditions, environmental changes in one period might encourage an extended progression. In the next stratum, simple life would re-appear and start *its* journey upwards. No longer was ascent unilinear or straightforward, even if there was an aggregate increase in complexity. With this, Chilton explicitly ditched a "continuous, *uninterrupted* chain of progression",[114] and adopted a more complex image. Although it was never spelt out. Nor was his new image necessarily a genealogical tree. He possibly had in mind a "hundred"

---

111 Chilton 1847.
112 Desmond and Moore 1991, 322, 325, 596.
113 Chilton 1847. For the stress of other newspaper compositors, see W. E. Adams 1903, 2: ch. 38.
114 Chilton, *OR* 1 (30 Apr. 1842): 159.

parallel lineages, all springing from simple, spontaneously-originated ancestral stock, and each reaching a different level.[115]

The lurching series was also contingent on the literature that fell his way. A disinterred monkey's jaw bone was noticed to show that long-missing fossils could unexpectedly turn up.[116] Elsewhere, he exploited old Jamaica lobby books and cannibalized passages which suggested that humans were so many separate species.[117] Ideas and illustrations would turn up in the *Oracle* undigested and disconnected, thrown in with little commentary. How best to introduce Lamarck and his notion of chimpanzees standing erect, freeing their hands, converting warning cries into speech, and emerging as men? Turn Lyell upside down—and that is what Chilton did through five instalments. He simply imported verbatim passages from Lyell's refutation of Lamarck in *Principles of Geology* and stripped out each and every caveat to leave a positive image.[118]

Human origins were demystified for naked politico-religious reasons. Like Saull's Owenites, the *Oracle* splinter group was antagonistic to any notion of an "immortal principle"[119] that would put humans under Divine obligation and legitimize a powerful priesthood. The series was *for* the downtrodden—to show them that the elite puffed "men up with the absurd notion that they are an anomaly among animated existences"[120] as an excuse to police the poor. It was a defence against clerical protagonists who still asked, as they had in Carlile's younger day: "how can you account for natural phenomena without a god?"

The use of such tactics showed its heritage in the Owenites' policy of engaging Christians in public debate. These familiar spectacles gave the *Oracle* series its structure and dialectical value. Hence Chilton's conclusion:

> If atheists can show that matter may make a man ... theists will waive all other objects to materialism. The object of this series of articles ... was to show the reasonableness of the belief that matter can make men

---

115 As envisaged by Hodge 1972.
116 Chilton, *OR* 1 (2 July 1842): 229.
117 Chilton, *OR* 2 (22 July 1843): 253.
118 Chilton, *OR* 2 (12 Aug. 1843) 279, and subsequent issues.
119 Chilton, *OR* 1 (26 Feb. 1842): 83.
120 Southwell, *OR* 1 (13 Nov. 1841): 5.

and women, and every other natural phenomena [sic] —unassisted, undirected, and uncontrolled.[121]

The militants, like their Enlightenment heroes, had faith that the "Augean stable of religion, fouled and polluted by human blood and misery, will yet be swept with the flood of science."[122] Nor did they doubt transmutation's serviceability, and the unaided progression of life became a cornerstone of their strategy for social and political betterment. With this enormous "Theory of Regular Gradation" series stretching across two years, rambling and fragmented though it was, Chilton had provided a major asset which redefined the science of emergent organic change for the republican, deist, and socialist market.

Saull's geology lecturing—pale by comparison—was effectively rendered redundant, and the furore surrounding the *Oracle of Reason* put his merchant position even further in jeopardy. Whether as a consequence or not, in the 1840s he would switch to a study of the last stage of the human ascent, local British aborigines. This too was more suitable given his growing involvement in the London archaeological community.

---

121 Chilton, *OR* 2 (11 Nov. 1843): 379–80.
122 Chilton, *OR* 1 (9 Apr. 1842): 135.

19. Backlash

The *Oracle of Reason* was run by a tiny disparate 'collective' of mobile artisans. Having been peripatetic social missionaries, they were used to re-location, and easily shifted their editorial offices from Bristol to London to Edinburgh, while keeping a continuous flow of issues. They were proud, too. Unlike many previous illegal rags, they prominently displayed their names on the title page, as well as purposely signing articles. Of course, driven by anger to use the most offensive language—a deliberate provocation—they set themselves up for a fall.

The viciousness of the *Oracle*'s railings against a priestly-constrained science matched the mood countrywide. Starvation, strikes, and misery marked the depth of the economic depression. By mid-August 1842, the Lancashire mills were grinding to a halt. A mass turn-out by weavers, striking against a continual reduction in wages—supported by the hatters and miners—led to city-centre demonstrations of 10,000 or more in Ashton, Oldham, and Manchester. Hundreds of special constables were sworn in and, despite a huge military presence, the mills and police stations were attacked. Mill towns looked as if they were "in a state of siege or civil war".[1] Hundreds were imprisoned. One Chartist leader was arrested in bed, days later, and charged with "riotously" assembling at the Manchester Hall of Science. In his house, police found a rifle, pistols, gunpowder, shot flasks, and percussion caps—weaponry that the *Church and State Gazette* highlighted to intimate that this was to be an armed rebellion, not that there was any evidence of it on the street. The guns, as always, were on the state's side: the protesters faced dragoons, rifles, and, in one demonstration, a cannon was actually aimed

---

1   *NMW* 11 (20 Aug. 1842): 65.

at them. The disturbances spread to the outlying regions and seditious placards were soon seen in Holyoake's Birmingham.[2]

This inflamed period was a turning point for Saull himself. He was fast approaching sixty. And, unlike the young tearaways, he remained a staunch Owen ally. He now gave up guest lectures on geology (perhaps leaving it to these young bloods) and moved into primeval archaeology. Ostensibly, he gave up political meetings as well, although in 1842 he did make a concession. Called back because of the "most awful distress" in the country, he chaired meetings in February at John Street and the Finsbury Social Institution. Given the lay-offs in the economic downturn, Owenites wanted 'home colonization' communities set up: self-sustaining, state-backed, self-reliant, and not run to assuage capitalist greed. The mass redundancies and strikes resulted in so many "miserable starving creatures in all our streets and towns". Saull's socialists inveighed against the bosses: the "industrious population ... are sunk into the most abject state of wretchedness, and are left at the will of their casual employers, to perish in the streets and cellars of our towns and cities without even food, clothing, or comfort..."[3]

The price of standing on principle at this time was tragically evident in Holyoake's case. When Southwell was jailed for blasphemy—to shield society from godless "confusion and crime", in the prosecutor's words[4]—Chilton stepped in as the *Oracle*'s sub-editor, and he invited Holyoake to become the figurehead editor. Holyoake accepted the poisoned chalice. Walking ninety miles from his home town Birmingham to Bristol to see Southwell, he stopped on 24 May 1841 to give a lecture at Cheltenham Mechanics' Institution. It was familiar territory. As a sixteen-bob-a-week social missionary, he had already addressed the mechanics here

---

2   *Church And State Gazette*, 26 Aug. 1842, 442; *Evening Star*, 22 Aug. 1842, 1; 25 Aug. 1842, 2; *Weekly Chronicle*, 20 Aug. 1842, 1; *Nonconformist*, 17 Aug. 1842, 563. The weaponry belonged to the Manchester smith Alexander Hutchinson, who blended Chartism and trades' unionism: Chase 2000, 177–78, 186; Webb 1920, 207–08.

3   *NMW* 10 (19 Feb. 1842): 267–72. There were a number of possible reasons why Saull dropped political attendances. Besides age, there was Harmony's development (see below), and with the commencement of the millennium at Tytherly, such old-world activities might have seemed redundant. Nor would this have been so odd, with Owen now sacking all the social missionaries as no longer necessary. Or, with Saull about to devote more energy to the archaeology of savage Britain, perhaps that was his rationale.

4   Southwell 1842, 16.

in February. His gist then was economic liberation before the luxury of intellectual gratification, which marked his shift away from what he saw as the bourgeois socialist emphasis on polite education. (Even as he spoke, the *New Moral World* was advising on classes ranging from elocution to astronomy.[5]) The poor dying of starvation or committing suicide dictated his priorities:

> Botany was offered as a delightful science, and so it was; but there was little gratification in knowing the structure of an ear of corn when bread could not be got to eat. Geology stood in the same rank. If all of you knew as much of coal mines as Dr. Buckland, could you get coal here this winter for less than 1s. 6d. per cwt? If so, geology might be useful to you. (Cheers.) With most persons religion, or rather a certain intensity of faith, was deemed essential, and was sought to be connected with every system of education. But will faith fill empty cupboards? (Cries of "No no")[6]

It was Southwell's trial and treatment that turned Holyoake towards atheism. Returning to Cheltenham on 24 May, he had a much harder anti-clerical message. With the Church costing twenty million a year, he considered "that the people were too poor to have a religion".[7] In the same way, the grave Hetherington was railing against "the holy trades' unions"—the church commissioners—who were "very successful in keeping up the rate of wages" for the 18,000 clergy, though unemployment was rampant all around them.[8] But Holyoake had the flippant edge. Given the depression, he suggested that the priests be put on half pay like the subalterns, especially as their god was a fiction. This outraged the local clergy. In August 1842, in a sensational trial, he was convicted of denying God's existence ("with improper levity"). The *Times*'s reporter described him as a "thin miserable-looking lad" and ridiculed him for his "prosy, incoherent and absurd harangue" delivered in a "shrill discordant voice" with an incomprehensible 'Brummie'

---

5 *NMW* 9 (13 Feb. 1841): 91–92.
6 *NMW* 9 (6 Feb. 1841): 88.
7 *Bristol Mercury*, 11 June 1842; *Derby Mercury*, 15 June 1842; *OR* 1 (11 June 1842): 202; Holyoake 1842.
8 *FTI* 1 (1842–43): 249–50.

accent.[9] He was given six months, and some were sorry that they could not "send you and Owen ... to the stake instead of to Gloucester gaol."[10]

While Holyoake was in prison, the rotten underbelly of Paley's happy world was exposed again. Given a copy of Paley's *Natural Theology* by a magistrate, Holyoake responded in typical fashion by writing *Paley Refuted in His Own Words*, which remained a back-catalogue pamphlet in pauper bookshops for a generation.[11] The *Oracle*'s editors might have sought respectability in martyrdom, but it came at a cost. Holyoake received a black-edged letter two months into his term. The small sum that benefactors had collected for him, he posted home to buy his two-year old daughter Madeline a winter coat. Instead, it bought her a coffin. The family had survived on hand-outs, pitiably few in the depression, and she had succumbed to a fever aggravated by cold and malnutrition—a death the poor knew only too well.

Atheism was rare in the wider society, and certainly atheism that asserted itself in times of grief. Just how rare was demonstrated by Madeline's funeral. Her mother Eleanor stood firm on no chapel and no "priestly mummeries", which flummoxed the cemetery officials. As such, Madeline was apparently the first to be laid in Birmingham cemetery with no minister or sacraments, just a tearful but wholly un-religious farewell.[12]

The *New Moral World* distanced itself from Holyoake, insisting that his theological provocations could only retard the socialist cause.[13] Saull, who had been threatened himself under the blasphemy laws, may have had more sympathy, at least to the extent of dipping into his pocket. When Hetherington, Watson, and the *Oracle* men set up the Anti-Persecution Union to fight these court cases and campaign for free speech, Saull donated.[14] But the blowback for Owenites of these show

---

9   *Times*, 17 Aug. 1842, 7; 18 Aug. 1842, 7; *Morning Post*, 17 Aug. 1843; 08–18; McCabe 1908, 1: 74–75.
10 Holyoake 1850, 12. Royle 1974, 74.
11 Holyoake [1847]. Published in August 1844, it had passed through six editions by 1866. Goss 1908, 3, lists the counter-refutations.
12 OR 1 (22 Oct. 1842): 368; 2 (15 Apr. 1843):143; McCabe 1908, 1: 86. Holyoake 1850, 74–78, left a heart-rending account and he always looked back on this episode with "mute terror".
13 NMW 10 (18 June 1842): 414.
14 OR 1 (17 Dec. 1842): 432; *Evening Star*, 7 Dec. 1842, 3; Royle 1974, 82; 1976, 54–57; Barker 1938, 40, on Hetherington's lead role in setting up the Anti-Persecution

trials was considerable. The police actually quizzed Holyoake in jail about Owen's influence, and whether it was Owen who had turned him into an atheist.[15]

Not only was there a surge in atheist prints at this time, but the Tory press was screaming about the proliferating alleyway shops spreading this "filthy and deadly nuisance". All noticed the rise of these new dens, flouting their "odious and corrupting books",[16] these "storehouses of all that is vile and nauseous thrust[ing] forward their unblushing fronts, soliciting the attention of passers by".[17] None could understand why these "execrable *fomites* of impiety and impurity [are] permitted to infect our metropolis" by a "Christian Government".[18] The tirades of the *Standard*, *Argus*, and *John Bull* were unremitting against these "disgusting depots" trading in their "vile merchandise". The patriotic press wanted the *Oracle*'s "filthy" vendors rounded up and new powers for magistrates to shut down these "emporiums of obscene prints".[19] Not least it would clean the cities and prevent future "OWENS and CARLILES" from trading in "sedition, sensuality, and atheism".[20]

What particularly frightened the family patriarchs was the effect of this unclean knowledge on impressionable women. Women were active in Owen's campaigns, just as they had been in Carlile's and Taylor's. The Owenites had always mirrored a religious sect, with Owen the pontiff, who required three-months' training and probation for his apostles, before the ordinands could march off with their *New Moral World* bibles. To become a socialist, an old activist said, was to be "born again".[21] The tea parties and festivals were a substitute for church socials, which were the glue holding the community together. They provided the camaraderie and the week's focal point for many families. The difference was that, like the millenarians, and unlike the Anglicans, socialist women could equally be ordained. Emma Martin and Margaret Chappellsmith were paid social missionaries.

    Union. Holyoake (or his wife) received 10s a week from APU the during his incarceration (Holyoake 1850, 75).
15 Holyoake 1850, 11–12.
16 *Standard*, 21 Mar. 1843.
17 *Argus*, 10 Dec. 1842, 8–9.
18 *Standard*, 21 Mar. 1843.
19 *Argus*, 18 Mar. 1843, 9.
20 *John Bull*, 4 Feb. 1843, 72–73.
21 B. Taylor 1983, 122.

Merely taking to the socialist stump was insurrectionary in an age of sexual conservativism driven by the new evangelical ideology, when women lost their voices in the church.[22] Women activists have been rightfully restored to the arena by recent feminist historians. They remind us of how active women were in the public sphere—not merely fighting for workers' rights, better employment conditions, or in union activity, but as blasphemy and anti-clerical activists. Emma Martin had been at the 1839 socialist congress and her infidel lectures on Owen's marriage system, divorce, and woman as property struck a deep chord in an age when women had little legal recourse and no financial status in marriage.[23]

These campaigners were not simply acolytes. There were plenty of those too: the flock of adoring fashionables fascinated by the Rev. Robert Taylor's debonair debauchery and the Rev. James Smith's dark theatricality. This fearful flocking extended into the Owenite period. *The Christian Lady's Magazine* could lament that ungodly Sabbath lectures could "draw a crowded audience of *women* to listen to what ought to kindle the most burning indignation in every female bosom."[24] Like Saull's geology lectures, equally commended to women, the "Theory of Regular Gradation" apparently excited great interest, "more especially with the female portion of our readers".[25] This moral contagion was hugely worrying to the patriarchs, who held women to their rightful place—the hearth and home.

The evangelicals' sexual conservativism reinforced the ideal of position and place, and, with it, profession. The geological divine, the Rev. Adam Sedgwick, a bachelor, spelled it out bluntly: women excelled "in every thing which forms, not merely the grace and ornament, but is the cementing principle and bond of all that is most exalted and delightful in society", but the "ascent up the hill of science is rugged and thorny, and ill-fitted for the drapery of a petticoat".[26] A professional

---

22 B. Taylor 1983, 124–28; Frow and Frow 1989, 101–06.
23 Frow and Frow 1989, viii, 85; Keane 2006.
24 B. Taylor 1983, 137.
25 *OR* 1 (19 Feb. 1842): 77.
26 *Edinburgh Review* 82 (July 1845): 4; J. A. Secord 2000, 243. This is not to suggest that there were not equally chauvinistic atheists. Some reacted badly to Owenite dictums about "laws fettering female genius" and saw no more than Sedgwick in the ability of the "petticoat" (*OR* 2 [14 Jan. 1843]: 25). But they were rare.

exclusion order went up on science and politics: women were neither to be tainted nor tempted by what they might there find. The *Oracle*'s "poisonous mischief", by appealing to women, was a disgrace to "Christian England".[27] It was up to the fathers and brothers to cast out "the unclean thing" and protect the chaste maidens, lest "the purity of the daughters of our land should be contaminated by the sight of such publications".[28]

The patriotic harangues led to plain-clothes police scouring the metropolis for dens of blasphemous prints.[29] The *Penny Satirist* saw the double standard, with the detectives turning a blind eye to the rich and raiding the pauper shops.[30] It was not the two-pound blasphemies but the penny ones the authorities found so dangerous—the books that spoke to the hungry masses angrily peering through the gates at the gentrified opulence. The *Sun* made Cicero's point: that the wealthy readers of infidelity were "generally *particeps criminis* with [the clergy] in the plunder of the working classes". The rich did not need "superstition" to keep them in line, but Christianity

> is absolutely necessary to keep the common people in subjection. "To the poor the gospel is preached," because the rich neither need nor believe it; and without such preaching, the probability is, that there would not long be any poor to preach to.[31]

The same "conspiracy" was thought to act against pauper science. While the Owenites and atheists are jailed, "no one dreams of prosecuting a patronised Professor", such as the Queen's favourite Charles Lyell, who calculated one hundred thousand years for the Mississippi to deposit its delta muds, which equally contradicted Moses. It seemed that "blasphemy only belongs to the lower orders".[32]

---

27 *Standard*, 14 Dec. 1842.
28 *Argus*, 18 Mar. 1843, 9.
29 *Morning Post*, 5 Jan. 1843, 2; *John Bull*, 7 Jan. 1843, 3.
30 *PS*, 21 Jan. 1843, 1.
31 *Sun* quoted in *NMW* 8 (26 Dec. 1840): 409.
32 *Reasoner* 1 (30 Sept. 1846): 244–45.

Another Street of Shame

The backlash peaked late in 1842 when the *third* editor of the *Oracle* set up in one of London's busy walkways, at 8 Holywell Street, behind the Strand. Here, among the old clothes shops and obscene print shops, the fearless Scot and former soldier Thomas Paterson, who had assisted Holyoake's mission in Sheffield, picked up the baton. His eight-foot window, festooned with two-foot posters, emblazoned with extracts in large letters from the "Jew Book" article, attracted "mobs of the lower classes", "enjoying the ridicule with which the monster who owns the shop has attempted to clothe the divine founder of our religion". So started Paterson's campaign of deliberate provocation. Crowds blocked the narrow street, "hooting and shouting" at the red-rag placards. Angry passers-by called for the proprietor "to be taken out and burned".[33]

Within days, the "filthy low street" had become the capital's scandal, judging by the press. It was screaming with outrage, "teeming with letters to editors, to ministers, and to bishops, backed by leading articles without end, complaining bitterly of the nuisance".[34] The shop was among "the most abandoned and blasphemous repositories of crime and infamy" of any street "inhabited by Christians".[35] Twice in one week, the "den of blasphemy" was attacked, its windows smashed and the offending placards on "That revolting, odious Jew production" snatched—once by the Vice Chancellor's son. The *Standard*, *Morning Post*, and *John Bull* egged on the attackers to rid the city of this "moral pestilence".[36]

All called for the Society for the Suppression of Vice to act, or the Home Office, or local magistrates.[37] "Day after day, month after month, the same exhibition is kept up" in one of the capital's most densely-peopled

---

33 Paterson [1843], v, 11, 13, 17, 21. *John Bull*, 10 Dec. 1842, 595, on the street. McCalman 1988, 205, 217–21, on the 'smut' also coming out of Holywell Street, with its pornographic printers running under a maze of aliases. William Dugdale (alias "H. Smith") worked at no. 37. The *Oracle* editors were shortly to relocate to no. 40.

34 *Argus*, 28 Jan. 1843, 9.

35 *Age*, 6 Nov. 1842, 5.

36 *Standard*, 14 Dec. 1842; *Morning Post*, 14, 17, 21, 22 Dec. 1842; *John Bull*, 17 Dec. 1842, 607. Also *Times*, 14 Dec. 1842, 6; *MC*, 14, 21, 24 Dec. 1842; *Argus*, 17 Dec. 1842, 3; *Examiner*, 17, 24 Dec. 1842; *Era*, 18, 24 Dec. 1842; *Spectator*, 24 Dec. 1842, 1227; *Lloyd's Illustrated London Newspaper*, 25 Dec. 1842.

37 *The Age*, 6 Nov. 1842.

lanes, "offering the same unhealthy excitement to the ignorant, the untaught, and the depraved", said the *Times*. If "Christianity is the law of the land", why has Government not acted?[38] A letter-writer to the *Standard* said that Paterson's shop window advertises "'atheism for the million,' as he terms his 'Oracle of Reason',' and "Two editors in gaol and the third ready." Why not oblige him?[39] Paterson had deliberately forced the issue, and the Church-and-Queen reaction against this "sinful exhibition"[40] had its effect.

The police were continually called to disturbances, and ended up colluding with the crowd.[41] Four summonses were ignored by Paterson.[42] But God had not been checkmated by Satan, or at least the Jewish Jehovah would have his revenge. Paterson was finally "driven from his lair"[43] by his landlord, a "Jew", as the press pointedly noted. The "Jew" and "atheist" had somewhat similar signification as untrusted 'others' in these papers. Both had restricted civil rights; Jews, too, were "infidels", unbelievers; they could not hold municipal office or a Parliamentary seat and were viewed prejudicially by many Christians. But the "Jew" equally hated the *Oracle* atheists, because it was largely the Old Testament, not Christ's ministry, they were reviling. (A Jewish protestor had once smashed Carlile's window when it displayed an offensive cartoon of Jehovah.[44]) Now it was a Jewish landlord who turfed Paterson out.

Paterson was finally charged at Bow Street magistrates court "with exhibiting to view a profane paper in a thoroughfare".[45] *John Bull* was not alone in pointing out that Paterson's defence in his trial was "more abominable" than his offence.[46] He read into the court record "the most

---

38 *Times*, 23 Dec. 1842, 4.
39 *Standard*, 20 Dec. 1842.
40 *Morning Post*, 26 Dec. 1842.
41 Paterson [1843], 13–14, 17.
42 *John Bull*, 19 Dec. 1842, 612; 26 Dec. 1842, 624; *Observer*, 18, 19 Dec. 1842; *Morning Post*, 19, 26 Dec. 1842; *Times*, 19 Dec. 1842, 7; *Standard*, 19, 26 Dec. 1842, 1; *MC*, 26 Dec. 1842.
43 *Essex Standard*, 30 Dec. 1842, 1; *John Bull*, 31 Dec. 1831, 627. Paterson duly moved to Wych Street nearby (*Court Gazette*, 28 Jan. 1843, 53). A few months after Paterson was evicted, Hetherington moved his business to 40 Holywell Street, selling the same subversive literature (*PS*, 12 Aug. 1843, 2).
44 Paterson [1843], 58.
45 Paterson [1843], 3.
46 *John Bull*, 28 Jan. 1843, 64.

horrible and disgusting expressions ever uttered by human being"[47], a common tactic in an attempt to get them re-broadcast by the dailies. A clergyman pleaded with the *Times* not to give this publicity to the "work of Satan".[48] And when the *Standard* did repeat "the vile blasphemies" to shock the pious into action, it appalled the *Times* and *John Bull* even more. The anger reached a peak when Paterson was fined a mere 40s with costs on each of the charges. But, in the event, he preferred martyrdom and accepted a month's imprisonment in lieu.[49]

This press furore led to the government crackdown. The trouble was, all were now tarred with the same brush: the atheist Paterson was considered symptomatic of the socialist "supporters of the *New Moral World*." The "licentious and seditious trash" put out by "advocates of Socialism, sedition, and Infidelity" was shown to be indictable.[50] The harassment now extended to socialist institutions. In January 1843, plain-clothes police raided their libraries and coffee shops, confiscating all illicit pamphlets. The Rotunda was targeted, "now one of the halls for the propagation of the doctrines of Robert Owen", as well as Saull's favourite haunt, the City Road Mechanics' Hall of Science.[51] *John Bull* considered the halls of science even more dangerous than the Holywell shop because of their theatres, where "the notorious ROBERT OWEN preaches the fearful doctrine of Socialism and atheism", where "women—yes, women!—declaim upon the folly of religion and the sinfulness of marriage!"[52]

In this alarmist atmosphere, the *Age* responded to a report showing that half of the population could not read or write by exclaiming: "No wonder that every Socialist chapel, and chartist Lecture room is thronged by wretched men and women, who have no other intellectual amusement than to gulp the blasphemous and treasonable garbage served up to them ..." The wonder was that these demagogues turning "sweeps into politicians, and pot-boys into regicides", had not "led on

---

47 *Patriot*, 30 Jan. 1843, 68, 72.
48 *Times*, 28 Dec. 1842, 6.
49 *Standard*, 28, 30 Jan. 1843; *Times*, 31 Jan. 1843, 4; *Morning Post*, 28 Jan. 1843; *Morning Herald*, 28 Jan. 1843; *Era*, 29 Jan. 1843; *John Bull*, 2 Feb. 1843, 73.
50 *Argus*, 28 Jan. 1843, 9.
51 *Morning Post*, 5 Jan. 1843, 2; *John Bull*, 7 Jan. 1843, 3.
52 *John Bull*, 5 Aug. 1843, 490.

the brute multitude, after the laudable examples of their prototypes—
the DANTONS and ROBESPIERRES in 1793."⁵³

No distinction was made between the breakaway extremists and rump socialists. Despite the latter lying low, under Owen's orders, *John Bull* in 1843 reached a splenetic pitch in excoriating the "beastly depravities of Socialism, its hideous and disgusting depravity, its insane folly, and its blasphemous buffooneries".⁵⁴ The atheist schism had only ramped up the pressure on the beleaguered socialists, even though they had their heads down. It helped sow the seeds which would weaken socialism just as it had reached its zenith. With the hotheads splitting off, the socialist rump was left looking like a sheepish sect of pious un-professing Christians—at least in the eyes of Hetherington's new weekly, *The Odd Fellow*, set up to rival the *Penny Satirist* for working families who craved amusement. While *The Odd Fellow* praised the atheists' secession, and their exposure of socialist pusillanimity, it did think the *Oracle* went over-the-top in reacting to the Owenites' "moral cowardice".⁵⁵ In fact it believed that the *Oracle*'s foul mouth was damaging the cause. With hindsight Holyoake himself later admitted that, by splitting off, "we [atheists] weakened the force which held the recognised co-operative fort".⁵⁶ Chilton too came to regret the *Oracle*'s "sledge-hammer style". But, as he explained, "Mine was a war to the knife"; neither the editors nor vendors "obtained quarter at the hands of Christians, and I gave none". What resulted was an arms' race with the authorities, and a sort of seat-of-the-pants journalism: Chilton chose targets "as the impulse of the moment moved me", irrespective of the effect. And the effect of all that "coarseness, vulgarity, and even brutality", instead of convincing the "reasoning believer, or shaking the faith of a bigot," seemed on reflection more likely to "horrify the one, and madden the other".⁵⁷

---

53   *The Age*, 12 Mar. 1843, 5.
54   *John Bull*, 21 Oct. 1843, 668.
55   *The Odd Fellow*, 18 Dec. 1841.
56   Holyoake 1875, 1: 247. On *The Odd Fellow*: Holyoake1850, 6; J. F. C. Harrison 1961, 30; Linton 1894, 37–38.
57   Chilton 1847. The contingent aspect also shows in his reviving the "Regular Gradation" series intermittently in the follow-on periodical, Holyoake's *Movement* (1843–1845). Here he would comment on Sedgwick's exposure of the antiquated notions of Creation held by the Dean of York, or deconstruct the anonymous *Vestiges of the Natural History of Creation*.

Saull had been overshadowed, there was hardly much point in him continuing his public geology lectures, and it would be perilous to do so in this hysterical climate. Chilton's "Regular Gradation" series survived the raids and arrests, and he took the editor's chair at the *Oracle* in 1843 after Paterson was sent down. Like Saull, the *Oracle* editors had weaponized the new stratigraphy, but their bombardment was much more devastating. They used it to underscore a serial transmutation and, like Saull before them, a monkey ancestry for mankind. It had the same double liberating purpose. By circumventing "Creation", it undermined the miraculous props of Anglican aristocratic power; and, by envisaging life driven naturally 'upwards' by external pressure, it furnished an evolutionary model of social ascent powered from below: it legitimated the struggle for a secular republic.

For an illegal paper, the *Oracle* did not sell too badly, about 4,000 copies a week at first.[58] In London it was even hawked around both boys' and girls' schools, where it indulged a taste for danger and titillation.[59] And one imagines the outrage as copies were passed out to Protestants exiting their meetings—a case of their own tract tactics being turned against themselves.[60]

The *Oracle of Reason* finally terminated on 2 December 1843, dying on its feet rather than being drummed out of business. By then, the freed Southwell had started up his follow-on *Investigator!* (1843), which tried to put atheism on a more sound philosophical footing.[61] Holyoake, too, brought out a penny-ha'penny follow-up, the *Movement*, which took up the sceptical mantle from 1843 to 1845, in a marginally more measured way. Saull supported this with dribs and drabs—half-a-crown here, twelve and sixpence there.[62] But in truth his allegiance remained with Owenism, and, more accurately, with Robert Owen himself, who was now taking the movement in a very different direction.

---

58 *OR* 2 (1843): iii; Royle 1974, 74; cf. other unstamped papers, Hollis 1970, 118–19.
59 *Argus*, 18 Mar. 1843, 9.
60 *Derby Mercury*, 31 July 1844.
61 With its exegesis of Hume, Locke, Kant, and Spinoza, which was pretty ambitious for penny trash.
62 *Movement*, 1 (30 Oct. 1844): 408; 2 (8 Jan. 1845): 15; he also donated to the fund for a late *Oracle* and *Movement* stalwart, M. Q. Ryall: *NS*, 7 Mar. 1846, 5.

20. Peace and Harmony

Not only did Saull own the lease on Owen's London house,[1] but he was financially committed to Owen's grandiose projects. Holyoake grumbled that Owen was a "spendthrift when forwarding his own plans of human regeneration", so long as it was not his money.[2] What finally bankrupted the movement was not the atheist schism, but his practical (or impractical) plan for the Commencement of the Millennium, or "CM", as was carved over the entrance at Harmony Hall, the mansion at the centre of his new communitarian experiment at Tytherly, in Hampshire. Harmony looked from propagandist reports like a heavenly idyll. In fact, it was over-ambitious, mismanaged, and a crippling drain on resources. In short, the Millennium commenced and collapsed in about five years, taking the Central Board with it.

Millenarian optimism had provided the foundations, when the loyal Owenites took over the Tytherly estate on 1 October 1839, and re-started the calendar to mark the beginning of the new moral world. The branches supported it in cash and kind, and fifty-seven colonists settled in this new land. But things went awry from the first. Only nineteen remained by summer 1841—so few that local labourers had to be hired to gather in the harvest, draining the initial resources. Still, rich benefactors kept it afloat, and collections were taken nationwide, but the cash influx only caused more overstretch. Owen bumped up the number of colonists to 300.[3] A palatial mansion was built to house them—Harmony Hall, as it came to be called—accentuating, said Holyoake, the dis-harmony that

---

1   *Reasoner* 15 (28 Sept. 1853): 208.
2   *Movement* 1 (11 May 1844): 170.
3   *NMW* 11 (1 Apr. 1843): 319; (23 June 1843): 433: the 300 were intended to include some at a revamped farm, Rose Hill.

came to prevail.⁴ Money was lavished on this three-story pile, which was designed by Joseph Hansom (of Hansom Cabs fame) and built in 1841–42, to contain lecture rooms, a library, and classrooms, besides the usual dining rooms, bedrooms, and so forth.

Outlying farms—including one about a mile away that we are interested in, Rose Hill—were purchased, resulting in a total holding of a thousand acres. All in all, by mid-1842, the millennial venture had cost £19,000 in old money. At first there was much good will. Radical printers such as Watson supported the venture, shipping off the latest reprints of the classics, Volney, Holbach and the like, as they came off the presses. Thus, in 1843, Godwin's *Political Justice* arrived in the same packet as *Alphabet of Geology* at Harmony, by now a sort of radical British Library depository commanding new works.⁵

Schools at Harmony were started. They projected 200 pupils and £750 revenue a year from this alone, which was wishful thinking.⁶ Geology was on the curriculum from the first, and calls went out to the branches from Harmony's Governor for rocks and fossils to stock the museum.⁷ One could feel the flush of excitement as the children marched, waving their tricolour flags, up to the local Dean Hill to see how the world's ancient history was revealed in the geological strata.⁸ But while geology might have been useful in subverting Genesis, so far as explaining soil types and agriculture (and there was increasing emphasis on this in lectures⁹) it was time wasted: Harmony farming was a flop. Not that they could not get a crop in—good yields on bad soil could be got with systematic manuring, and they did manage to fill three barns with

---

4   Holyoake 1875, 1: 306. The Harmony colony was also called "Queenwood", because the manor had once belonged to Queen Philippa in the 14th century: Garnett 1972, 166.
5   *NMW* 12 (12 Aug. 1843): 53. The library had nearly 1,400 books by 1844: Garnett 1972, 202.
6   Armytage 1961, 164–66; Garnett 1972, 166–96; Podmore 1907, 2: 543–52; Hardy 1979, 54; Frost 1880, 18. By mid-1843 they were up to sixty-one pupils, but only thirty-five were fee paying, the rest being the children of residents. The governess of the infant school was a Quaker, and some members attended the local parish church, which suggests more openness than among the London cadres.
7   *NMW* 9 (1 May 1841): 282; (29 May 1841): 332; 11 (6 May 1843): 360; (13 May 1843): 368; (20 May 1843): 376.
8   *NMW* 12 (22 July 1843): 32.
9   *NMW* 4 (16 June 1838) 265–6; 6 (11 July 1839): 608; 11 (30 Oct. 1842): 147; 12 (11 Nov. 1843): 156–57.

wheat in 1844.[10] It was more that skilled industrial workers expecting a paradise had no experience of the land. Productivity was impaired because few understood the basics of spade culture.[11]

The merchant Saull might have been a metropolitan backer, with huge business interests, yet he never lost faith in community and commonality, even if at a rather theoretical level. He reiterated it at a John Street farewell to another loyal Owenite, the *New Moral World*'s former editor G. A. Fleming, about to take over as Governor of Harmony:

> The institution of individual property had been found to war against the best and highest interests of humanity; it was the origin of inequality, selfishness, poverty, strife, and all the accompanying vices of such a state of things. The economy of the new system was based on the doctrine of commonalty of property, and it would aim at making each individual habitually act upon the maxim of "all for each, and each for all" (cheers). The results of this system would be that the strong would support the weak, instead of crushing them as at present; it would establish over the world universal brotherhood, and reconcile the interests and the inclinations of each (cheers).[12]

Great hopes were vested in Harmony, making it a place of pilgrimage. The views were so delightful, said a visiting socialist, that "I have seen no spot that reminds me so much of the promised land".[13] The sylvan setting was about as far a cry as one could get from the press cacophony and police raids spreading from Holywell Street to London's socialist halls. Thus it was to this safe New Jerusalem that Saull now contemplated leaving his geology museum.[14] His connection went deep, financially at least. The deposit on the neighbouring Rose Hill estate, with its mansion and farm, had been put down by Owen as an agent for the Rational Society in May 1842.[15] The mansion was to be his new home. It was

---

10 *NMW* 13 (17 Aug. 1844): 61.
11 Garnett 1972, 169–171, 180, 197–200; Cole [1944], 34; cf. Hardy 1979, 56, for an upbeat assessment of the farms.
12 *NMW* 13 (11 Jan. 1845): 229. George Fleming was more interested in geologists exposing the exploitation of women as young as eight in the Lancashire mines, where they were used to heave coal waggons away from the hewn seams: *Union* 1 (1 Apr. 1842): 44.
13 *NMW* 12 (7 Oct. 1843): 119. These rosy reports of "happiness and concord" disguised the financial mess and carping: Bray 1841, 2: 609.
14 Holyoake 1906, 1: 190.
15 *NMW* 11 (23 July 1842): 26; (3 Sept. 1842): 81.

also, given Owen's insistence on wealthy patronage for the movement, to lodge grander visitors who wished to inspect the experiment but who might find Harmony too basic. This was to be Owen's showcase, therefore he wanted the garden and farm cultivated to "a high degree of perfection" to impress the socialites—hardly a priority for Harmony's productive classes.[16] What these elite supporters eventually found was "a very comfortable family residence",

> placed on a gentle eminence, in the midst of tastefully-disposed grounds, and commanding panoramic views of great extent and beauty. It is approached from the high turnpike road by a carriage drive with two entrance lodges, together with several park-like enclosures of arable and pasture land, interspersed with ornamental belts of plantation, and studded with timber, the whole lying in a ring fence, and containing sixty-two acres.[17]

To get Rose Hill to that salubrious state would cost £2,500. And Owen would need £10,000 for fitting out Harmony, stocking the farms, building the schools and so forth, money which the Rational Society could ill afford in 1842.[18] Owen's idea was actually for Board members to build their own houses at Rose Hill. "The estate itself was beautifully situated for that purpose ... and occupying a gentle elevation, it would be both a healthy and agreeable site." There was no shortage of vision at this point. It meant more fund-raising by the branches.[19] A salt-of-the-earth socialist staying at Harmony wandered over and gave a more realistic view of Rose Hill. "The gateway and lodge give us palpable signs of neglect: the place must have been uncared for during many months." But even he saw it

> passing into such hands, as will restore its former beauty, and under Providence, make it another Paradise. The mansion is not large, but

---

16 *NMW* 11 (6 Aug. 1842): 44. A Governor, the Unitarian iron merchant and teetotaller John Finch, put it bluntly: "As the diet of the members ... is plain and homely, perhaps our wealthy friends would find themselves more at home, by boarding at Rose Hill during their stay." Here they could be accommodated "very genteelly" (*NMW* 12 [13 Apr. 1844]: 335). 'Coarse' was how the workers described their food, and they criticized the culinary delights of Rose Hill: Garnett 1972, 183, 200.
17 In the words of the subsequent sales brochure: *Reasoner* 21 (21 Sept. 1856): 96.
18 *NMW* 11 (24 Dec. 1842): 208, (7 Jan. 1843): 223–24; (24 June 1843): 433.
19 *NMW* 11 (21 Jan. 1843): 240–41.

somewhat tastefully built, we enter the open door, pass from room to room ... we ascend the stairs, and thence to the flat roof, heavy with sheet-lead. The prospect on this hilly eminence, is most extensive and picturesque.[20]

This "delightful villa" at Rose Hill was fitted out by April 1843 and advertised in the *New Moral World* as a "Boarding Establishment." There was never any doubt that it was for Owen's "genteeler" parties who wished a comfortable view of the new community.[21] The Central Board re-located to Harmony Hall at the same time.[22] The optimism was unbounded. And it ranged widely: from thoughts of a vineyard, which would have been a cut above turnips, to a fully-fledged industrial college on-site, the first in a rapidly industrializing Britain—a 500-place technical school that would have been within the means of the trades.[23]

But the whole venture was proving a bottomless money pit. The Central Board continually needed to raise funds, and it looked to Saull's largesse in 1843. It mortgaged the Rose Hill estate to him on 9 October. He paid £2,900 (over a quarter of a million pounds in today's money) and took possession of the deeds. By that time, the board was cutting down its options and cancelled further plans to upgrade the estate, preferring to lease it on 6 November 1843 to a sympathizer, Thomas Marchant. Arrangements would remain: boarders would be accommodated, and the Central Board, and the Rational Society could retake possession at any point.[24]

Saull was to own this "elegant and COMMODIOUS Mansion"[25] through thick and thin for the rest of his life. But Owen's residence there lasted for little more than a year. The annual socialist congresses were held

---

20 *NMW* 11 (28 Jan. 1843): 245.
21 *NMW* 11 (22 Apr. 1843): 346; (29 Apr. 1843): 353, 356; (6 May 1843): 364.
22 *NMW* 12 (1 July 1843): 4.
23 *NMW* 11 (3 June 1843): 400; (10 June 1843): 416; 12 (29 July 1843): 36; (9 Sept. 1843): 82; (13 Apr. 1844): 335; Garnett 1972, 196. An industrial college, let alone a large one and meeting socialist objectives and benefiting the working classes, was much needed given the state's laissez-faire attitude to education and the Anglican universities' disregard for technology. Even University College London only managed to get professors of engineering established in the 1840s (Bellot 1929, 266).
24 Royle 1998, 213 n. 16, 235; *Reasoner* 21 (12 Oct. 1856): 115. In *Reasoner* 22 (29 Mar. 1857): 50, the late Saull's Trustees date the indenture of mortgage for Rose Hill as 28 December 1843. See also *NMW* 12 (25 May 1844): 378.
25 *NMW* 12 (4 May 1844): 359.

in Harmony from 1842 to the end, 1846, and they spent most of their time anguishing over its problems. Those were becoming legion. It was not the pilfering, presumably by the "stray ruffians" hired to work the land.[26] Something much more systemic was at fault. There had been a massive overspend on Harmony Hall, which in Holyoake's view was a "monument of ill-timed magnificence". Nothing was wanting "but utility, convenience, and economy".[27] A cash crisis had brought the project to its knees, and Saull, as an auditor of the Rational Society, must have been more worried than most.[28]

Building work was suspended in July 1842, and retrenchment called for. Cutbacks in the kitchen led almost to "a workhouse level of diet". Owen resigned, and Finch became governor. Building was re-started, but no lessons had been learned, judging by Harmony's "miniature railway for transporting meals from the kitchen". In 1843, Owen again resumed control. But Harmony continued to drain funds, and his extravagance led to dissensions in the branches. There was palpable anger at his failure to keep within budget.[29] Bills were not paid, "and credit was the agreeable but insidious canker-worm which ate up" the dream.[30] The branches were growing cynical, and appeals to them had diminishing results, which led to more cutbacks.

Then came perhaps the most short-sighted of all the decisions. The 1844 Congress dismissed the social missionaries, saving some thousands of pounds a year, on the principle that this community should come first in the new moral world, not contentious debate in the old. That resulted in the branches themselves losing local support, and they began shrinking. The cash flow became critical. By 1844, Harmony had £30,000 liabilities, and the Rational Society was approaching bankruptcy. That year, building work cost £3000, ten times the revenue from the branches. Not even parsimony could save it now, as colonists shed their broad-cloth uniforms and returned to the old immoral world. The school principal resigned on losing his salary and was replaced by an unpaid inmate.

---

26 Holyoake 1906, 1: 193. Thomas Cooper 1885, 118–20; *Movement* 1 (13 Nov. 1844): 417; Cole 1944, 34.
27 *Movement* 1 (6 Nov. 1844): 409.
28 *Report of the Proceedings of the Eighth Annual Congress of the Rational Society*, 1843, 208; NMW 12 (8 June 1844): 401.
29 Garnett 1972, 191–98.
30 Holyoake 1906, 1: 192–93.

The Halls of Science had taken the brunt, haemorrhaging cash, and the result was disastrous. They began to be sold in the slump. Liverpool's became a Concert Hall, and Huddersfield's was put up for sale in June 1844: the Methodists tried to buy it, if only out of poetic justice, before it passed to the Unitarians and eventually the Baptists.[31] Oldham's was let to teachers, and even the iconic Manchester Hall of Science would shortly become the City Music Hall. Onto its "black, dusty look of desertion"[32] was written the epitaph of the movement.

Harmony, the pilgrim's shrine, had lost its mystique, and the messiah his holy aura. Withering branch criticism of Owen's paternal arrogance and profligacy led to his withdrawing completely from Harmony in 1844, and, on 8 August, as he prepared to leave for America, the remaining Harmony schoolchildren trooped over to Rose Hill to say goodbye.[33] The children themselves departed shortly after.

Saull did his best. In May 1845, he dropped the interest on Rose Hill to help out Marchant and the Rational Society.[34] But the utopian experiment was finished. The Society decided to cut its losses and wind up the project in August 1845. Even then, the squabbling over how to proceed persisted. One branch wanted the estate disposed of quickly, the debts cleared, and "the proceeds [to] be invested in the National Funds in the names of W. D. Saul [*sic*], Esq., W. Pare, Esq., and Mr. Whittaker [John Street Institution], to be applied to cooperative purposes for the benefit of the subscribers to the funds of the late Society."[35] Even that was a tall order.

---

31 A. Brooke, "Huddersfield Hall of Science"; Garnett 1972, 199–202.
32 *Reasoner* 5 (12 July 1848): 107; (19 July 1848): 121.
33 *NMW* 13 (17 Aug. 1844): 61.
34 In 1845 Saull cut the interest from 5% to 3% of the money advanced, and released the Rational Society from all liability to the remaining 2%: *NMW* 13 (24 May 1845): 386; *Reasoner* 21 (21 Sept. 1856): 96. Of Saull's few surviving letters concerning Owen, most from this time centre on Rose Hill. W. D. Saull to Robert Owen, 16 Aug. 1845 (letter 1379), requests Owen's signature on an insurance policy. While the estate being conveyanced back to Saull is discussed in W. H. Ashurst (Owen's solicitor) to Robert Owen, 6 Oct. 1851 (letter 1955), 11 Nov. 1851 (letter 1973), all in The National Co-Operative Archive, Robert Owen collection, Manchester. From Christmas 1853 Saull re-let Rose Hill to Marchant for a term of twenty-one years, at a rent of £105 5s. per annum. The estate was worth £3,000 at the time of Saull's death: *Reasoner* 21 (21 Sept. 1856): 96.
35 *Reasoner* 1 (24 June 1846): 60; *Report of the Proceedings of the Eighth Annual Congress of the Rational Society*, 1843, 208; *NMW* 12 (8 June 1844): 401.

The final collapse saw tawdry infighting, as Finch and the trustees in 1846 evicted the governor from Harmony Hall (literally, onto the road). Since the Congress had been called, it had to be improvised in a tent, and ultimately it finished up in Saull's mansion, Rose Hill.[36] That was the last Owenite Congress. The trustees did pay off many debts. As the printing presses and type went,[37] so did the *New Moral World*, its last issue being on 23 August 1845. They tried to sell Harmony Hall as a lunatic asylum, which must have seemed appropriate to many. But it was soon disposed of more fittingly, to become Britain's first applied science school, Queenwood College. Small shareholders were never compensated, bigger investors were bankrupted, and those who had sold their houses and donated everything lost more than faith. There could have been no sorrier end of the Owenite saga.[38]

---

36 Podmore 1907, 575–76; Holyoake 1906, 1: 193; *Reasoner* 1 (8 July 1846): 82.
37 *NMW* 13 (23 Aug. 1845): 497.
38 Armytage 1961, 166–7; Holyoake 1906, 1: 193–95. As a vocational college, Queenwood would employ practical and laboratory work as well as theoretical science, in order to turn out "scientifically-trained farmers and engineering apprentices", not a bad educational outcome from an Owenite perspective. W. H. Brock 1996, xvii: 7; Barton 2018, 69. Here London's future leaders of science, including the physicist John Tyndall and chemist Edward Frankland, would teach in the late 1840s and 1850s, men who would go on themselves to shake religious orthodoxy.

21. Secularism and Salvage

With Harmony's collapse went any hope for Saull's museum bequest. Yet, Saull remained faithful to Owen, even though others would have nothing more to do with him after the liquidation fiasco.[1] Owen remained a stabilizing force, even as the atheist tearaways were prising the movement apart, and as Harmony, the missions, and the Halls of Science were collapsing. Saull's rump Owenites salvaged what they could. They tried to develop something more low-key and practicable, a National Land and Building Association (founded 1845). This was to use subscriptions to purchase freehold land, with the intention of building comfortable and healthy workers' houses in model villages.[2]

A skeletal Owenite machine rumbled on. Year in, year out, come May, Saull would sit in the chair at John Street to celebrate Owen's birthday, singing the *Marseilles*, praising the 1848 revolutions. He would listen to old Owen (who turned eighty in 1851) rebut the young Turks' claims that he was just a visionary, by pleading that he was a "practical man". And, in truth, for all the failures, when it came to secular and infant education and so much else, that was true.[3] Holyoake was usually there, a rising star and a sign that however weakened Owenism was structurally, a splinter movement could carry the torch.

---

1 *Reasoner* 5 (31 May 1848): 2–3.
2 *Reasoner* 1 (18 Nov. 1846): 301–02); *NMW* 13 (19 July 1845): 454; West 1920, 223–24. The West-London Central Anti-Enclosure Association pooled their funds into it: *National Reformer*, 3 Oct. 1846, 3. They bought 100 acres thirty miles from London to make a start. But the idea of renting did not appeal to many activists, there being no democratic community control (Frost 1880, 68).
3 *UR*, 19 May 1847, 50; *NS*, 22 May 1847; 20 May 1848; 18 May 1850; and so on, each May, with Saull chairing the birthday celebrations, through to Owen's eighty-third birthday: *Reasoner* 14 (1 Jun. 1853): 346; "practical": *Reasoner* 11 (28 May 1851): 20; *NS*, 24 May 1851. J. F. C. Harrison 1967 on this lasting claim regarding infant education.

Owen had tried to appease the Harmony-backing capitalists and defuse social hatreds by eschewing the old democratic radicalism and newer atheism. Only this, he believed, would allow his paternalist socialism to win through. The ruse backfired. Now Holyoake would give the lie to Owen's claim that anti-religious lectures were ruining the branches. Holyoake pointed out that atheist talks in Birmingham drew larger crowds than Owen's "sedate namby-pambyism" ever did.[4] Anyway, the freethinkers were now cut loose. A direct infidel link connected Carlile in the 1820s with Holyoake in the 1840s. It was Carlile's erstwhile shopman, the grave James Watson, now a leading infidel in his own right at the Mechanics' Hall of Science, who goaded Holyoake into getting the *Reasoner* off the ground. With trade slackening, partly from the Owenite collapse, Watson retrenched and moved his print shop to Queen's Head Passage, Paternoster Row. Pamphlets sat on shelves unsold, collecting dust, all for the want of an anti-Christian periodical to revive public appetite. He badgered Holyoake, who obligingly started the *Reasoner* on 3 June 1846,[5] and this long-lasting organ would carry the 'secularist' message of socialism into the post-1850s age of equipoise.

The *Reasoner*'s posthumous fame largely rested on its promotion of the neologism "secularism", used to describe the breakaway movement in 1851.[6] For many it seemed to simply swap the pejorative Christian 'infidelism' or negative 'atheism' for the neutral or positive 'secularism'. But these were not totally interchangeable terms: for example, a Muslim was also an 'infidel'—in Christian eyes—but not a secularist. Also, secularism could encompass people who were not atheists, those, for instance, opposed to the Church interfering in politics. The word 'scepticism' was no good: as Holyoake averred, "he was not sceptical—he was in no *doubt* about Christian error." 'Freethinking' did not remove the problem, because it also applied to certain Christian sects. Holyoake was to say that where freethought ends, secularism begins.[7] To him, it was not old wine in new bottles but a new vintage. It encompassed a morality resting on "material and social facts" rather than theology,

---

[4] *Movement* 1 (13 Nov. 1844): 419; Royle 1974, 52–53.
[5] Royle 1974, 92; Royle 1976, 61; Linton 1879, 83.
[6] *Reasoner* 11 (9 July 1851): 118, for the word's early usage, but not by Holyoake; Rectenwald 2013.
[7] Holyoake 1905, 1: 185; 2: 17; *Reasoner* 8 (20 Feb. 1850): 54.

making the secular itself "sacred". Embedded within his 'Secularism' was the precept of "ethical duty", or a code of conduct which owed nothing to revelation but allowed its adherents to detach the "truth of today" from the errors of yesterday.[8]

Even this was scarcely new but largely another iteration of the Holbachian ideal, and, as such, had been long advocated by radicals and socialists, including Saull. What *was* different—and warranted the re-branding—was the tone and tactics. Holyoake issued books in his "Cabinet of Reason" which were designed to fill the gap between "the dilettante Scepticism of gentlemen, and the undisciplined Rationalism of the poor".[9] It was this middling terrain occupied by "a hundred thousand sympathizers" that secularism was targeting. The word signalled a broadening: the movement spread its appeal to literary radicals already fleeing orthodox Christianity.[10] Holyoake was adapting Owen's outreach technique, encouraging not bourgeois capitalists to come over but bourgeois intellectuals. Detractors were not alone in seeing 'Secularism' provide a "respectable garb", which allowed Holyoake to manoeuvre among the intelligentsia undergoing their own 'crisis of faith'.[11] It was useful to a shrewd operator. But then Chilton had always recognized Holyoake as the "*pet*" of polite society.[12]

Secularism's slippery creed did not necessarily deny anything, and it equally offered little positive but the "Providence" of science and de-Christianized ethics. And because it emerged seamlessly out of socialist freethought, secularism remained overtly political, despite its indifference to the millennium (not a word to be found in the *Reasoner*). It continued the struggle for civil liberties, starting with a fight to get affirmations in place of Bible-based oaths, for so long a concern to Saull.[13] In short, it retained the political and moral message, the sort Saull had trumpeted in his lectures for a generation.

---

8   Holyoake 1892, 1: 254–55; 2: 292–93; Rectenwald 2013, 323–24; 2016, ch. 3; Marsh 1998, 124.
9   *Reasoner* 12 (19 Nov. 1851): 15; *Monthly Christian Spectator* 2 (Oct. 1852): 623; S. D. Collet 1855, 21.
10 Rectenwald 2013, 237–42; Royle 1974, 154; Ashton 2006, 8–9, 241; Nash 1995b, 124.
11 J. R. Moore 1990; Linton 1894, 163–64.
12 W. Chilton to G. J. Holyoake, 24 Dec. 1841, Holyoake Correspondence No. 22, Co-operative Union, Manchester.
13 Royle 1974, 4, 150–51.

The transition from the Harmony ethos, held by Owen and his wealthy backers, to Holyoake's Secularism with its broader appeal, was never abrupt. How smooth it was was shown by another financial backer who made the move. This was Owen's solicitor and fellow socialist, William Henry Ashurst. He was also Saull's solicitor, who dealt with his conveyancing of Rose Hill. Saull and Ashurst were long familiar: both attended Guildhall meetings on corporation reform, they jointly acted on deputations, both agitated against the church rates, and Ashurst actually refused to pay his. They stood on anti-corn-law platforms together,[14] and both were in the Metropolitan Parliamentary Reform Association in 1842.[15] This camaraderie would eventually be reflected in Ashurst's son representing his dying father at Saull's funeral.[16]

Political agreement on reform, suffrage, and Owenism overrode theological disparity. Saull was a more extreme anti-Christian. Ashurst was a lapsed Freethinking Christian, like his co-religionist Hetherington. But Ashurst's was a more gentle unbelief; he hated Hetherington's violent language. Ashurst's Muswell Hill home was an open radical salon. As the solicitor to the cause, he provided legal help in fighting the taxes on knowledge, and advising Holyoake during his 1842 blasphemy trial.[17] Being an Owenite, he was famous for exploring mitigating factors in court. By showing how circumstances might have helped induce a crime, he saved untold poor souls from transportation. Ashurst was one of the middle-class backers Owen was trying to keep onside during the Harmony years: Ashurst had actually devised the constitution of Owen's Home Colonization Society, which bankrolled the Harmony building.[18]

---

14 *TS*, 13 Aug. 1835, 8; 6 July 1836, 2; 22 Dec. 1836, 1; *MC*, 22 Dec. 1836, 1; *Atlas*, 20 Dec. 1845, 817; Ashurst *ODNB*.
15 Rowe 1870b, 71a, 129. Ashurst also helped Rowland Hill get his penny postage reform through in 1839. There had been a rash of petitions that year for a flat uniform postage rate, including one by Saull (*Journals of the House of Commons* 94 [12 July1839]: 437), although whether he was actuated by a desire to see newspaper and tract distribution streamlined, or working people able to afford mail, or even to benefit his business, which must have involved a huge mail accounting system, we do not know.
16 *Reasoner* 19 (13 May 1855): 55.
17 Holyoake 1892, 177, ch. 34; C. D. Collet 1933, 19, 84; J. F. C. Harrison 1969, 225; McCabe 1908, 1: 140.
18 Royle 1998, 79; 1974, 91; Holyoake 1906, 1: 191; 2: 600. Then Ashurst dealt with its winding up and even poured money into the collapsing *New Moral World* at the end.

In 1849, Ashurst bought the Owenite *Spirit of the Age* (run by Robert Buchanan), toned it down, stretched its appeal, and made it an organ of "unsectarian socialism", with Holyoake as editor.[19] It was this same humane, liberal attitude that lay behind the *Reasoner*. Ashurst funded the *Reasoner*, wrote for it, and helped shape its moderate stance and wider horizons. More than that, it seems that it was at Ashurst's suggestion that Holyoake adopted the word 'secularism' for this phoenix rising from the socialist ashes.[20]

For the mellowing merchant Saull, the slide across to the 'secularists' was just as easy. Holyoake had unbounded organizational flair and was an effective facilitator. By marshalling the rump of Carlilean-Owenites and focussing on state bias and Church privilege,[21] he allowed Saull and Ashurst to keep their Owenite credentials while aligning them with disadvantaged Dissenters on the one hand and literary young blades like G. H. Lewes on the other. Saull was as happy to support Holyoake's demand for tolerance and disestablishment as he had been to support Carlile's raspier calls a generation earlier. That Saull and Holyoake were close is obvious. For his part, Holyoake exploited Saull's pub-circuit connections, and, while on a lecture tour in 1847, Holyoake used nephew John Saull's "Admiral Rodney" pub in Long Buckby in Northamptonshire as a venue.[22] This nephew, on visiting London a few months later, turned up at Holyoake's new Utilitarian Society (founded 1846), and heard both Holyoake and uncle William Devonshire respond to talks on ancient mythology and the "Two-natured Christ of the Churches".[23] This latest Holyoake society was itself supported by Saull, who understood the new meaning of "Utilitarian". In politics, it had long meant rule for the people's benefit, but now religion, too, had to give account of itself. As Holyoake proclaimed: "We shall have sealed the work of intellectual reformation when we have written *cui bono* over the altar".[24]

---

19 Holyoake 1892, 1: ch. 34; Goss 1908, xxxvi, 67; McCabe 1908, 1: 146.
20 Royle 1974, 93, 154–55; McCabe 1908, 1: 160, 203.
21 Nash 1995b, 124.
22 *UR*, 3 Feb. 1847, 20; *Reasoner* 2 (17 Feb. 1847): 69–70; (3 Mar. 1847): 106–07.
23 *UR*, 6 Oct. 1847, 89.
24 *Reasoner*, 2 (2 Dec. 1846): 1; Royle 1974. 94–95.

Vestiges of Creation

Holyoake's Utilitarian Society met in the Mechanics' Hall of Science in City Road, so Saull was on familiar turf. He was a regular attendee. From the first lecture (April 1847) onwards, Saull chipped in during the discussions—live banter again being encouraged after the disastrous Owen ban.[25] These could be spirited follow-ups, recapturing the old excitement, judging by the reports of a "sharp discussion" at the first Utilitarian Society meeting or the "animation" after Holyoake's sermon on the "Moral Remains of Genesis". A varied group of "disputants" was starting to show up, indicating that the Society was attracting many from outside the socialist orbit.[26] Here, for example, Saull would meet Josiah Mason, the Birmingham pen manufacturer who went on to found Mason's College (now Birmingham University). Of course, there were the obligatory discussions on universal suffrage. But good debates were also had after Holyoake's lecture on "Knowledge without Books". This undoubtedly appealed to Saull, who, with his hands-on museum, agreed that observation "should precede Book learning".[27]

How much the young bloods had taken over from Saull was shown by the reaction to the uproar caused by the *Vestiges of the Natural History of Creation* (1844). This expensive gloss on progressive geology and comparative anatomy, domesticated and dressed up to appeal to middle-class readers bored with the latest crop of novels, piqued interest and promoted parlour guessing games from its anonymity. The slow continuous ascent of life was made acceptable for a fireside family readership, as Secord has demonstrated. The "vestiges" were fossils, their footsteps tracing out a path to mankind. The process was as natural as a foetus growing through childhood to adulthood. Life was maturing, and underlying it was a kind of lawful continuous creation. To

---

25 *UR*, 12 May 1847, 47. We can gauge Saull's activity here, from his name appearing in the *UR* (for 1847 alone) on 26 May, 16 June, 28 July, 1, 8 Sept., 6, 13 Oct., 10 Nov., and 15 Dec.

26 *UR*, 14 Apr. 1847, 39; 8 Sept. 1847, 81. Saull's fellow "disputants" might include Charles Savage (interested in comparative religion), George Hooper ("Eugene", an Oxford Classics scholar), and Jonathan Duncan, the Cambridge-educated currency reformer, fellow member of the Metropolitan Parliamentary Reform Association, and author of many works, from the blasphemy laws to the rights of property, capital, and labour.

27 *UR*, 28 Apr. 1847, 43.

mitigate any materialist imputation, the author carefully explained that God had set the ball rolling, and unleashed a 'natural law', considered a sort of edict running through nature, to do the rest. *Vestiges* was to prove another key piece of dissolvent literature, corroding the traditional glue of religious society. For many literati, already sceptical, it added fuel to the fire.

Most activists would have first heard of *Vestiges* through their trusted platform speakers. These came at the book from all sides. Chilton hammered away in the *Movement* and *Reasoner*, irked, it seems, by this godly whitewash of his "Regular Gradation" series. First off the mark, he framed the argument. He thought the book a feather bed, catching those already slipping from orthodoxy, and that these falling souls would ultimately land on materialist bedrock. After all, if God can be dispensed with on the planet's day-to-day running, why do we need Him at all? There was "nothing new in all this", Chilton said, except that *Vestiges* saw 'law' somehow pushing life on, whereas "materialists consider that ... the higher forms of existence are merely increased developments of the lower".[28] God's whim would now become Nature's for many a soul in crisis, and Chilton recognized that even the anodyne *Vestiges* would "startle many" from their "slumbers".

By contrast, Emma Martin took a softly-softly approach. Not for her any "Jew Book" diatribes, or Holywell Street provocations. Hers was a "tone of mellowed soberness", as befitted a former evangelical Baptist used to persuading by dialogue at the door.[29] Like most infidels, she already accepted a geological rise "from the most simple up to the most complex—from the Lily ... up to the man" as evidence that nature was "the maker, and not God". Even before *Vestiges*, she had published a pamphlet, which recast Chilton's strident voice as reassuring patter: it had a "Querist" chatting to a "Theist" and convincing him or her that "man is but an improvement upon the lower animals". But, at the end of the day, the difference was only in tone, not intent. She was as anti-clerical as the rest. Matter itself had the power, it did not need God's blessing. With no Almighty there was no moral authority for the priest, who thereby had no right to enforce any "law which nature has not taught".[30]

---

28 J. A. Secord 2000, 310ff; Chilton 1846, 1847a, 1847b.
29 J. A. Secord 2000, 314–16; *Movement* 1 (6 July 1844): 239.
30 E. Martin 1844, 4–7.

As a separated wife and mother, and an apostate, she was hated all the more for it. Within days of her finishing a three-lecture course on *Vestiges* at John Street in June 1846, a scurrilous pamphlet announcing her death from "the pangs of a guilty conscience" was circulating. It was the sort beloved by Christians, showing death-bed repentances, which had been stock fare since the days of Voltaire and Paine. Her fabricated "recantation" was now added to the list, and nicely illustrated with a picture of a lady weeping over a tomb.[31] For a dead penitent, Mrs Martin continued in lively fashion at John Street.

Even Holyoake got in on the act, although usually not one to puff science. He gave a one-off talk at the Mechanics' Hall of Science on the "Origin of Man" as envisaged by *Vestiges*. Probably, as Secord says, he saw the book "extending the constituency for freethought" as it drew in middle-class freethinkers.[32] But firebrands still abounded. Another who pitted *Vestiges* against scriptural literalists on the eternal question, whence came man?, was Robert Cooper. He turned up at the Utilitarian Society on his first trip to London, having been lecturing at Hull on "The Origin of the Earth, and the Origin of Man; or, the author of the 'Vestiges of Creation,' versus the author of the Pentateuch".[33] He was another future luminary. His *Infidel's Text-Book* was already in-press.[34] Of the lot of them, he was the one destined to take over the uncompromising mantle on the human origins question.

How many interested artisans actually laid hold of a copy of *Vestiges* is a moot point. Secord claims few had direct contact, even the cheaper editions falling mostly into liberal middle-class hands.[35] The half-crown 'people's edition' was the cheapest. It was, for example, sold by the socialist stalwart Edward Truelove. Truelove had been an Edgeware Road butcher who supplied the social community with their Christmas fare.[36] As A1 branch secretary, he had taken his family off to Harmony, only to see it collapse. Undaunted, he returned as Secretary at John Street

---

31 *Reasoner* 1 (8 July 1846): 92–93; (22 July 1846): 127. John Street course *Reasoner* 1 (3 June 1846): 15.
32 J. A. Secord 2000, 310, 314; *Movement* 2 (29 Jan. 1845): 40. The sort shortly to cluster around John Chapman's bourgeois radical publishing house at 142 Strand (Ashton 2006), to whom Holyoake would extend a hand.
33 *Reasoner* 1 (16 June 1846): 30; *UR*, 12 May 1847, 47; 26 May 1847, 51.
34 *Reasoner* 1 (3 June 1846): 16.
35 J. A. Secord 2000, 307.
36 *NMW* 10 (11 Dec. 1841): 192; 11 (17 Dec. 1842): 204.

in 1845. The institution was still going, despite the financial crash, even if they had to offer decorators "beautifying" the hall membership privilege in lieu of payment.[37] Truelove also set up an adjoining bookshop, at 22 John Street, with the 2/6 *Vestiges* top-billed in his ads. And for those who could not afford it, he sold a 4d abridgement.[38] There can be little doubt, therefore, that *Vestiges* ended up in the well-stocked John Street library.

But given the large number of radical coffee houses, discussion clubs, and social halls in the metropolis, the same might be said of their bookshelves. By 1840, there were an astonishing 1,600 coffee houses in London, many catering to artisans—this was the classic public sphere, where political consciousness was honed.[39] The trades, for instance, met in the Parthenium in St Martin's Lane, as did the Atheistical Society, while the Chartists preferred Huggett's in Lambeth, or Halliday's West Riding Coffee House in Holborn Hill. By contrast, the Christian Socialists met at the American Coffee House in Worship Street, while the Free Enquirers got together weekly at the Crown in Harrow Road. The list is endless, whatever your bent, there was a coffee house nearby for you. For freethinkers, there were also the Globe in Fleet Street, or Bailey's or the Hope Coffee House, both in Soho. For socialists, there were the Cambrian or Hudson's, both in Covent Garden. And atheists might fancy Southwell's Charlotte Street Institution, now gleefully taken over from the socialists and re-opened as the Paragon Hall and Coffee House.[40] Many had libraries and reading rooms, some actually setting aside specific nights for reading. And most had the resources to buy in the latest block-buster, especially when it suited their purpose. Since activists tended to cluster round one or other, they could undoubtedly have thumbed through a half-crown *Vestiges*, or even have afforded the 4d abridgement.

Then there were the bigger focal points of social activity. Lovett's National Association hall in Holborn had a reading room and a 700-volume library.[41] In fact, it was essential for the ubiquitous London socialist, Chartist, and freethinking groups to have a communal library

---

37 *UR*, 3 Mar. 1847, 28; Royle 1998 136.
38 *UR*, 30 June 1847, 62; J. A. Secord 2000, 306.
39 Simon 1960, 231; Royle 1974, 191; *OR* 1 (6 May 1843): 162.
40 Royle 1974, 89.
41 Stack 1999, 1028–29.

and reading room. At one of Saull's favourite's, the Finsbury Social Institution, in Goswell Road, they boasted proudly of "the addition, purchase, and loan of many valuable works", as well as "a coffee and reading room". While the Rotunda had a "constantly increasing library of literary and scientific works" in the 1840s.[42] In addition, there were the presses and radical book shops, the "geographic centers of the freethought movement"—social hubs which often encouraged reading: Holyoake's shop at 147 Fleet Street not only contained his printing press but set off a room specifically for readers.[43] The profusion of these radical reading places in London militates against interested artisans never having read a *Vestiges*. They might not have paid half-a-crown, but they could have picked it off any coffee shop shelf.

✱ ✱ ✱ ✱

Holyoake's extended hand to liberal Nonconformity and bourgeois radicals led to certain tensions and new accommodations. Queasy dialogues were started with receptive disputants.[44] Evening readings might include the Unitarian James Martineau's *Rationale of Religious Enquiry* (1836), a sensational work which argued that faith must not offend reason.[45] The Utilitarians were pushing beyond combativeness to get some perspective on fellow-travellers, and none intrigued them more than the scintillating young preacher George Dawson.[46] Here was an eclectic who pushed Protestant private judgement to its limits. Dawson was as happy talking at mechanics' institutions and Chartist halls as to his own Birmingham congregation. One handbill reported that

> Mr. Dawson's system is mainly Socialism, with an appendage of Christianity, and a slight admixture of Swedenborgianism, Mahometanism, and Rousseau-ism. His object hitherto has been to get as far as possible without the Church; and he is now struggling to get

---

42 *NMW* 11 (9 July 1842): 15; 13 (11 Mar. 1845): 287.
43 Mullen 1985, 226.
44 *Reasoner* 4 (18 Aug.1847): 457; Saull agreed that it was important to stretch a hand out: *UR*, 26 May 1847, 51.
45 *UR*, 8 Sept. 1847, 81. Saull took part in the animated discussions.
46 *UR*, 6 Oct. 1847, 89. That is not to say all disputants fell in line: the dry-as-dust 'Aliquis' (George Gwynne) uncompromisingly probed each outsider in turn, from the ethnologist Luke Burke to George Dawson.

outside of out—so that the most appropriate name for his religion is one partly of his own coining, namely, OUTSIDE-OF-OUT-ARIANISM.⁴⁷

They had to like a man who walked the 1848 Paris barricades with Emerson and urged free public libraries and secular schools. Such a nonconforming Nonconformist they could do business with.

All these meetings were reported in the house organ, the *Utilitarian Record*, appended weekly to the *Reasoner*. Together the paper took over the listings function and achieved a national sweep last reached by the *New Moral World*. Saull, as usual, stepped in with financial support. Where Holyoake asked in his "One-Thousand-Shilling-List" for a shilling from each of his 1,000 readers to recoup the paper's £50-a-year running costs (he got it, twice over), Saull pledged a sovereign yearly, and others followed suit.⁴⁸ In return came Holyoake's praise for Saull and "the strenuous opposition he ever gives to supernaturalism in the great name of science".⁴⁹ Given the Utilitarian's widening aegis, Holyoake was at last beginning to appreciate the use of science. To what extent Saull and Holyoake hobnobbed on geology is not recorded, although, at one Utilitarian meeting, Saull did relate "a geological anecdote of Robert Chambers".⁵⁰ Perhaps this was because Holyoake was in the process of issuing a third edition of *Paley Refuted in His Own Words*, newly dedicated to the Chambers brothers.⁵¹ Since nearly all tittle-tattle about Chambers at the moment concerned his presumed paternity of the *Vestiges of Creation*,⁵² it is possible that Saull had yet another smoking gun.

Out of justice, the *Reasoner* pointed visiting secularists to Saull's "excellent Geological Museum"—easy to find, just "a minute's walk from the General Post Office". Always it was "visitors" who were addressed, on the assumption that Londoners already knew the

---

47   *Reasoner* 6 (21 Feb. 1849): 117.
48   UR, 9 June 1847: 55; 22 Sept. 1847, 85; *Reasoner* 3 (31 July 1847): 400; and so on yearly. For Christian comments on the funding drive and Saull's contribution: *The Bible and the People for 1853*, n.s., 2: 7–13.
49   *Reasoner* 3 (31 July 1847): 400.
50   UR, 19 Jan. "1847" [1848], 15.
51   *Reasoner* 4 (5 Jan. "1847" [1848]): 83.
52   The *Vestiges'* author was still unknown. Chilton had heard through the grapevine—a leak from one of *Vestiges'* printers—that it was Robert Chambers: Royle 1976, 141–42; J. A. Secord 2000, 314; *Reasoner* 5 (22 Nov. 1848): 414.

museum well. This emporium had exploded in a decade, doubling in size. By now, the free museum was immensely "rich in curiosities" which were increasing weekly, with "Mr. Saul [sic] or an assistant"[53] on hand to explain their meaning. So much had accumulated that Saull's traveller and warehouseman, William Godfrey, doubled as the museum superintendent.[54] Welcoming a reporter from the Chartist *Northern Star* in 1846—a rag hated by the Tory press as a "pestilent publication" appealing to the "low and ignorant"[55]—Godfrey made an impression. He "conducts visitors with such thoroughly democratic urbanity, and explains the subject with such a graceful simplicity." Saull would also occasionally dispatch Godfrey to the Hunterian Museum with a fossil skull for Richard Owen to identify, rather treating the imperious Owen like a public servant.[56] But Saull's museum that greeted visitors in the 1840s was noticeably changing, as he shifted his focus on to the last stage in the rise towards Owenite man.

---

53 *Reasoner* 1 (6 Aug. 1846): 159.
54 Identified by the *Northern Star*, cross-matched with Saull's will, which bequeathed "To my Traveller and Warehouseman William Godfrey the sum of Three hundred pounds sterling": W. D. Saull Will, 31 Oct. 1855, Public Record Office, PROB 11/2215. William Godfrey was described by the *Northern Star*'s reporter in 1846 as "the author of the 'World's Catalogue of Geology'", although this has yet to be identified: *NS*, 31 Oct. 1846, 3.
55 *The Age*, 28 Aug. 1842, 4.
56 W. D. Saull to Richard Owen, 14 July 1851, British Museum (Natural History), Owen Collection, 23: ff. 112–15.

22. British Aborigines

In Mantell's regional museum, visitors were invited to imagine the past in their vicinity. Antiquity and locality linked the artefacts; the depth of time being less important, they allowed the antiquarian and geologist to rub shoulders. Saull was now to follow suit, but for different reasons, and with different results.

In the 1830s, the border between geology and archaeology was porous and unpatrolled. The very name of the uppermost geological deposits, *Diluvium*, indicating gravels left by the Flood, showed why they piqued antiquarian interest. Above these lay the alluvial silts and clays with their human remains. But the lines were blurring. In 1839, the Diluvium was re-named *Drift*, because it was now thought that this rocky debris was dropped by ocean icebergs, not stirred up by God's wrath. This broke the biblical time-marker. By this point, drift and alluvium were also being seen as products of the same natural causes: erosion and deposition.[1] Moreover, geologists by the early 1840s were turning up extinct mammal remains in cave drift deposits, alongside stone tools, even if the latter were dismissed at the time as remnants of later human burials.[2] As geologists pushed up from deep time, so antiquarians were drilling down through shallow time. This was especially evident from the 1840s as the Danish Three Age System of pre-Roman history—Stone, Bronze, and Iron Age—slowly began to take hold.[3] Collectors, like Mantell, now wandered across this porous border,

---

1   Rudwick 2008, ch. 13. A growing reciprocity was also evident. Geologists such as Lyell used classical temples familiar to gentlemen on their Grand Tour to gauge the rate of earth movement (Warwick 2017), while antiquarians used estimates of geological movement to judge the age of burial sites (Torrens 1998, 51). For a later attempt at precision dating the junction between the oldest human history and latest geological deposits, see Gold 2018.
2   Grayson 1983, 69–77.
3   Rowley-Conwy 2007.

©2024 Adrian Desmond, CC BY-NC 4.0   https://doi.org/10.11647/OBP.0393.22

amassing all things local, fossils *and* antiquities. But Saull would have a specific reason for trenching on the more recent past.

Saull was working in the years immediately preceding the arrival of this Danish chronology. Even so, Owenite imperatives led him equally to order the prehistoric past, to push back human pre-history, and to set up his own pre-Roman cultural stages, without biblical reference points. Apparently, he was one of the few London archaeologists interested in the pre-Roman Celtic past.[4] But, then, he was probably the only card-carrying socialist. It helped, too, that, by the late 1830s, interest in Celtic barrows was beginning to eclipse the focus on the Celtic language.[5] This favoured Saull the collector, because the resulting burial remains could be integrated into the museum's evolutionary display.

This porous interface meant that Saull's slide from rising fossil life in the 1830s to rising 'savage' life in the 1840s was easy. One was the continuation of the other, with an over-arching Owenite perfectibility doctrine knitting the lot into a whole, and an environmental determinism seamlessly running the process. The resulting museum narrative was a beguiling speculative sweep to compete with the best "corrosive fiction" and sensationalist broadsides favoured by working class audiences.[6] 'Savages' had been on the cards at least since Saull first encountered Davy's dream in *Consolations*. In his 1833 monkey-man talk, he envisaged "our ancestors" as originally "naked savages", establishing their ascendancy over the brutes "by the use of clubs, or other rude weapons".[7] From this point on, the ascent continued as an intellectual climb, and it was this human mental advancement that Saull now pursued. When he said in the Harmony years that he was giving up political meetings,[8] he probably meant that he was freeing up time to devote to this growing 'aboriginal' interest.

By the late 1830s, Saull's geology talks were bleeding off into 'primeval archaeology', as it was called, and given a narrowing, localized focus. The unshaven monkeys were making advances in civilization and mind, learning how to shave. He was setting the pattern for the

---

4 Rowley-Conwy 2007, ch. 4.
5 Morse 2005.
6 A. Buckland 2013, 64–66.
7 *Crisis* 3 (5 Oct. 1833): 37.
8 *NMW* 10 (19 Feb. 1842): 267.

next decade, which would largely be spent examining the successive ancient dwellings found during City demolition works. The oldest 'homes' discovered, "composed of mere sticks, or turf, mixed with the debris of the most simple culinary and other utensils", provided his material *entrée* into a study of the "aboriginal inhabitants of Britain". These people had lived "the precarious life of the wandering savage". They were the barbarians encountered by Caesar, and thought by him uncivilizable—the Romans viewing them, said Saull, as the Australian aborigines "are to-day, as compared with us."[9]

Such a reversal of perspective—having the civilized Romans look at 'us' as barbarians—was precisely the kind practised in socialist schools. Here, children were encouraged to view themselves from the outside, to understand how accidents of birth and education had given rise to their attitudes. It was designed to induce a moral relativism, to quash chauvinistic ideas that their own "national peculiarities" were "the standard of truth". They were taught to put themselves in another land (or, in Saull's case, another time) to see that we should have "escaped neither its peculiarities, nor its vices", indeed that "we might have been Cannibals or Hindoos, just as the circumstance of our birth should have placed us". By lessening "uncharitable or intolerant" attitudes,[10] Owenites were attacking growing racial supremacist ideology but also providing the means to unseat presentist views, by which standards of the modern age were used to judge the peoples of the past. Owenite cultural relativism made looking at our aboriginal roots a more egalitarian exercise. Saull saw no discrete stocks, no separate human species to be disparaged but, rather, humanity's rise as a co-operative endeavour. And just as 'we' had risen by dropping "Druidical superstition" through cultural exchanges with the Romans and Phoenicians, so would indigenous peoples rise in co-operation with us. There was nothing irreclaimable about Australian aborigines, any more than ancient British ones.[11]

---

9   Saull 1837.
10 R. D. Owen 1824, 47–48.
11 Saull 1837. Morse 2005, 35, on how historians came to associate the Celts with Druidical religion. But Saull, *qua* freethinker, constantly downplayed this priestly influence.

Reacting in part to the breakup of Owenism, and the rise in racial ethnology, Saull saw the move to aboriginal archaeology as an act of reclamation. And reclamation in a more material sense benefited his repository, which filled up with a new set of objects. The second phase of museum development mirrored his new emphasis in the 1840s, with incoming exhibits on archaeology, ethnology, and London prehistory. Ostensibly, there was nothing new in this. Almost all contemporary museums were omnivorous. Gideon Mantell's in Lewes also had its Roman pots and Sussex grave goods.[12] Likewise, Saull stored local Roman ware, English vessels, coins, and so on; and his choice artefacts, like Mantell's, ended up after his death in the British Museum.[13] Such mixed collections were the rule. Whether the Piccadilly Hall in London or the Ashmolean in Oxford, their contents ran from stuffed birds and strange fossils, to amphorae, coins, and tribal booty—curiosities that were literally that, curious.[14] In this respect, Aldersgate Street superficially resembled a miniature British Museum, which Cobbett had described as "the old curiosity-shop in Great Russell Street."[15] Actually, the fossils in the British Museum's North Gallery occupied only a fraction of the museum's portfolio. (In 1853 the keeper G. R. Waterhouse's 'inventory'—no inventory at all, but a tour of the interesting or typical non-invertebrate fossils—occupied a mere ten of the 270 pages of the museum's content *Synopsis*.[16]) Not only did Saull have more fossils, but the exhibits were not so much bric-a-brac, and their arrangement had an inner logic.

---

12 Walters 1908, 157, 159, 253, 269, 344, 365, 421; Cleevely and Chapman 1992, 354 n. 76. Mantell 1836, 37–40 for the list of antiquities in the upper back room of his museum, ranging from the pavement from Lewes Priory to funeral relics of South Downs tumuli, plus the usual swords, spears, skulls, and amulets. See also *Lancet* 2 (29 June 1839): 506–07, for his "interesting assemblage of antiquities, urns, vases, lachrymatories, celts, coins, &c. &c., British and Roman, collected in Sussex". Mantell's first publication was actually on the discovery of a Roman pavement and he never lost his interest in antiquities (A. Brook 2002).

13 Walters 1908, 324, 372, 435.

14 Yanni 1999, 21, 25–27; Pandora 2017.

15 Cowtan 1872, 64.

16 *Synopsis of Contents of British Museum* Sixtieth Edition (1853), although this was better than the four and a half pages in the 1842 *Synopsis*.

Reclamation

Saull's collecting, dictated by his political interest in ancient British life, was also constrained by pragmatic factors. Unaccountably, the City Corporation had no museum itself and saw no need to preserve London's antiquities. This despite pleas by the Romano-British expert Charles Roach Smith, who realized that many were being lost.[17] As an example, Smith noted that a small Roman altar, found during the excavations at Goldsmith's Hall and consigned to a rubbish heap, was only saved by the efforts of Saull and solicitor Edward Spencer, a fellow geologist, numismatist, and antiquarian, resulting in it being preserved in the Hall.[18] Just as strange, the Society of Antiquaries had no museum,[19] so artefacts of interest—and there were plenty as London's Roman wall was revealed during the metropolitan improvement works of the 1830s and 1840s—ended up with Saull in Aldersgate Street. But despite the emergency nature of this ad hoc preservation, it still served Saull's purpose. He was using primeval archaeology to take the geological story to modern humans. So, for all the eye-catching spears and skulls, there was less randomness than might be supposed, and more structure to fit his narrative of progression from pre-Roman aboriginals through Roman civilizers to commercial man.

What antiquarian artefacts were in Saull's museum in the mid-1840s? The travellers' guides in their surface scratching leave little clue. Booth's *Stranger's Intellectual Guide* (1839) mentions only a "good collection of Anglo-Roman remains", adding that they "throw much light upon the domestic habits and manners of the Romans during their residence in Britain, and have done much to illustrate the topography of ancient London."[20] In truth, burgeoning City works had thrown up huge

---

17   C. R. Smith 1854, iv.
18   C. R. Smith 1848, 1: 130, 134; 1859, 48. Even *Ainsworth's Magazine* 6 (Oct. 1844): 363, argued that Roman antiquities were not of "such immediate interest as those of later time". Spencer had long known Saull, having proposed him for the Geological Society in 1830.
19   A. Booth 1839, 15. DeCoursey (2013, 49–53) sees the society split between those studying texts and the more field-inclined monuments specialists. Sloppy financial management meant lack of funds, and those they had were earmarked, not for expensive preservation, but the library, which was recognised as superb. The society was recording rather than preserving.
20   A. Booth 1839, 15.

numbers of artefacts, many deposited with Saull. They were arriving from the mid-1830s as a result of the sewers being laid in Newgate Street. The excavations had gone down thirty feet, "underneath the whole of the foundations of the ancient cities". The navvies had revealed the "successive debris of the British, Roman, and later London"—and at the lowest depth, above the "diluvium" left by the ancient Thames, traces of "cinders and charcoal, the probable remnants of the destruction by fire of the rude wigwams or wooden huts, forming the first settlement of our British ancestors, where likewise a great quantity of human bones were found." Here was the beginning of Saull's evidence for ancient Britons. Above these were found

> Roman and Samian pottery [Samian was a very fine pottery made of Samian earth, and characteristic of Roman sites], consisting of vases, lachrymatories [tomb phials supposed to contain tears], amphorae [wine jugs], &c, many of which are in a fine state of preservation, retaining in legible characters the names of the makers. Coins of the Emperors Constantius, Constantinus Pius, Antoninus Pius, Nero, &c, a large quantity of vitrified tiles, &c.[21]

From the excavations for a new school in Honey-Lane Market came ancient human bones as well as Saxon coins. As a trustee of so many coins, Saull would take an active part in the foundling Numismatic Society (founded 1836) and become one of its scrutineers. In 1838, the President portrayed numismatics as that "branch of art" which was "the awakener of taste" in even the humblest (because everyone handled coins). As he did so, Saull was exhibiting flat, circular fossil nummulites from his museum at the Society because they so resembled coins (*nummulus* is Latin for a small coin), showing that nowhere was outside the reach of his fossil proselytizing. This caused the President to lapse into medieval panegyrics about this metaphoric anticipation, and to exclaim: "Nature herself would almost appear to have intended that numismatists should become the Honourable of the earth".[22] But it was below the coin-bearing layers that Saull's primary interest lay.

---

21 *New Monthly Magazine and Literary Journal*, 46 (Feb. 1836): 270–71; *Morning Post*, 6 Jan, 1836. Although London was the primary focus, the museum also housed Saull's Roman finds from his native Northampton (*JBAA* 4 [1849]: 396–97).

22 *Proceedings of the Numismatic Society* (1836–37): 89; (1837–38): 213, 250. As with fossils, so it was with coins; Saull often acted as a go-between. He would exhibit

How to assess the aboriginal inhabitants of this deeper past? The 1830s saw a more racially-inclined development of phrenology. The empire's expansion led to a new urgency in classifying its colonial subjects, often based on skull shapes. And, increasingly, phrenologists were imputing generalized cultural and psychological traits to these hierarchically-ranked 'races'. Such racist stereotyping, extending outwards, would, by the 1840s, also be extended *downwards*, into the Celtic deep past, as the degree of "savageness" of barrow skulls was used to define the relative age of the burials.[23] Already in the 1830s, a simplistic racial craniometry was being applied to Saull's collection. Thus, in a Cheapside excavation, in

> what is supposed to have been the ruins of a Human dwelling, was found a skull, now in Mr. Saull's collection, in a remarkably fine state of preservation, but which, phrenologically speaking, from the absence of the intellectual and great predominance of the animal organs, can give no exalted ideas of the moral character of the people to which the possessor belonged, the head being more like that of a Carib [indigenous West Indian people] than of one of the natives of modern Europe.[24]

Although Saull did not join the London Phrenological Society until 1844,[25] the museum's skulls were already having their bumps read by reporters in 1836 to show the 'savage' sloping-forehead of our ancestors. But there is no evidence that Saull himself had any great interest in phrenology. Indeed, the self-help science left many Owenites hopelessly conflicted. The worry was that it gave too little scope for "the modifying influence of external circumstances", which left it inadequate as "the

---

Northampton provincials' medals at the society: *Proceedings of the Numismatic Society* (1851–52): 20; *Numismatic Chronicle* 15 (Apr. 1852): 104–05.

23 Morse 2005, ch. 6; Goodrun 2016; Desmond and Moore 2009, ch. 2 on racial craniometry.

24 *New Monthly Magazine and Literary Journal* 46 (Feb. 1836): 270–71; *Morning Post*, 6 Jan. 1836. In *Archaeologia* 27 (1838): 150, only "a black wide-mouthed earthen pot" is mentioned from the Cheapside excavations. It, too, went into Saull's museum.

25 *Zoist* 2 (Apr. 1844): 30. His attitude towards phrenology is unrecorded, as it is towards that other self-help science mesmerism. According to Wiener (1983, 252), Saull attended demonstrations of mesmerism at the City Hall in Chancery Lane in 1841, given by William H. Halse, a self-proclaimed "Professor of Animal Magnetism" newly arrived from Torquay. These were arranged by Carlile, which might explain Saull's presence; as might the fact that Halse's galvanic experiments in revivifying drowned puppies were reported in the *NMW* (9 [27 Feb. 1841]: 132); Morus 1998, 144ff; 2011, 84–85; Winter 1998, 113, 369 n. 20.

basis of education and social and moral reform".[26] Despite the odd suggestion that Harmony residency should depend on a phrenological test of suitability, others agreed with William Godwin that the notion of inborn evil faculties, unchangeable, was "a libel upon our common nature". Owen himself ultimately discarded character-reading from bumps as an invidious restraint diluting the power of circumstance on the formation of character.[27] Yet the seemingly scientific measurement of skull-shapes—craniometry—remained beguiling. While Saull arranged his cultural artefacts in a "connected series of illustrations",[28] this probably meant in chronological order. We do not know whether he was himself using craniometry to produce a graduated sequence towards modern man. But it is telling that, in all of Saull's writings and reports, I cannot find a single mention of the word 'phrenology', so we have to be cautious with any craniometric imputation.

Saull's developing sequence, from savage Britons to civilized Romans, played a strategic role. Much of his work centred on the unearthing of London's Roman wall during the building of the French Protestant Church at the bottom of Aldersgate Street, close to his museum. Navvies uncovered the wall's foundations in December 1841, with Saull obviously on site. Freshly elected to the Society of Antiquaries (4 February 1841),[29] he made the wall the subject of his first (and only substantial) paper to the Society, in February 1842. He was qualified at this interface of geology and archaeology. He showed how a compacted flint base supported angular uncut blocks of Kentish ragstone (greensand) and ferruginous sandstone, probably brought in by Roman engineers from the Maidstone area. Outside of this defensive wall was a deep ditch containing Samian pottery, bones and horns of ruminants, as well as "handles of amphora, three glass lachrymatories, and an urn of a peculiar shape". All presumably ended up in his lower gallery. Saull's research helped give this wall its "celebrated" cachet, so that locals came to Aldersgate Street to see the remains of what the *Gentleman's Magazine*

---

26 *FTI* 1 (1842–43): 113.
27 Cooter 1984, 233; W. Godwin 1831, 370.
28 *New Monthly Magazine and Literary Journal* 46 (Feb. 1836): 270–71; *Morning Post*, 6 Jan. 1836.
29 *GM* 15 (Mar. 1841): 301.

assumed was the enlarged later Roman city of Londinium Augusta.[30] Since the wall was subsequently built over, Saull's was one of the few extant accounts and thus a constant source of reference.[31] With Saull's museum right on the spot, it was a port of call for anyone interested in Roman London.[32]

A Hotbed with a Heritage—Finsbury Social Institution

As with his geology, so Saull's antiquarian and political spheres were never totally discrete. Having started to rub in the contiguity between palaeo-evolution and aboriginal progression, he now made the antiquarian crossover more explicit. He extracted the moral of ancient history at another of his favourite radical haunts.

Aldersgate Street ran north into Goswell Road, and at the top was Finsbury Social Institution, with its compact 300-capacity lecture hall and coffee and reading rooms.[33] Finsbury had long been one of the most radical boroughs with its Spencean under-belly. The Spenceans had taken a harsh revolutionary line on agrarian democracy, demanding, in the aftermath of the French Revolution, that the land be reclaimed and apportioned.[34] One of them, Arthur Thistlewood, had responded to the

---

30 *GM* 22 (Nov. 1844): 506; *Illustrated London News* 1 (14 May 1842): 12, 16. Saull (1844) for his foundational paper. For the press coverage: *Times*, 3 Mar. 1842, 6; *Court Gazette*, 5 Mar. 1842, 1013; *GM*, 17 (Mar. 1842): 305. See also *Antiquarian and Architectural Yearbook for 1844* (1845), 81–82; *MC*, 4 Mar. 1842, 6. Saull's geological expertise came into play in other archaeological arenas, for instance, when analyzing cromlech granite engravings in Brittany (*Literary Gazette* 1624 [Mar. 1848]: 168).

31 More wall was shortly discovered (*GM* 19 [Jan. 1843]: 21–22), and a further 70 or 80 feet in the 1870s, which confirmed Saull's description (Price 1880, 20–21; 1881, 407–09).

32 Soon additions from other locations were added. The first indications of Roman habitation in West Smithfield—an urn containing the burnt bones of what Saull took to be a child, along with tell-tale Samian ware—also ended up at 15 Aldersgate Street: *GM* 19 (May 1843): 520.

33 *NMW* 12 (23 Dec. 1843): 208; capacity: *NS*, 18 Sept. 1847; *UR*, 8 Sept. 1847, 82.

34 Prothero 1979, 116–31; Chase 1988, 91, 117–20. Saull the urban merchant probably had little to do with agrarianism. About as far as he went was to help mitigate the plight of unemployed agricultural workers in the "Labourers' Friend Society" (founded 1832), which established allotments and cow pastures countrywide for the destitute, against the resistance of farmers and estate owners: *The Second General Report of the Committee of the Labourers' Friend Society*, 1833, 36; also *Third General Report*, 1834, 25; *The Labourers' Friend Magazine*, ns (Dec. 1836): following

bloody Peterloo Massacre in 1819, when peaceful demonstrators were cut down by the cavalry, by organizing the Cato Street conspiracy in 1820. The group planned to assassinate Cabinet ministers as a prelude to a general uprising. For his part, Thistlewood was hanged. But living in Goswell Road near to Saull was an old Spencean, his long-time friend Allen Davenport, now old and infirm.

Completely unschooled, Davenport had been successively a groom, soldier, and shoemaker before becoming an "out and out Spencean".[35] But he had moved on with the times, like Saull coming under the influence successively of Carlile, Taylor, and co-operation. He was also a radical bard, and, like all 'attic' poets, cripplingly poor. His "scientific and philosophical poem" *Urania* had been published by Watson in 1838, as a fund-raiser for the destitute old man. Urania was the muse of astronomy, so the poem's dedication to Saull was appropriate.[36] Davenport's poetic flights on "uncouth" man, making his debut on the earth, was a subject being fleshed out in more prosaic form by Saull. This proto-human, "Stood naked and alone in open space"

> Wherein no apples of temptation grew,
> No tree of knowledge met his longing view!
> He labor'd hard subsistence to obtain.
> And purchas'd days of joy with years of pain;
> So liv'd, so far'd the father of mankind.
> There tam'd wild animals & till'd the ground,
> And huts arose with moss and rushes crown'd.
> Thus Man created by his energies,
> Ere he enjoy'd his wretched paradise![37]

---

page 234; *NMW* 4 (23 Mar. 1838): 174–75. Not all were happy with this society. Some asked what right the rich had to patronize the poor by buying up and "letting out small portions of land", when labour exchanges were clearly the way to liberation (*PMG*, 10 Mar. 1832; *MC*, 25 Mar. 1833). Saull also supported the Agricultural Employment Institution (founded 1833) (*Royal Cornwall Gazette*, 30 Mar. 1833, 1).

35   *NS*, 5 Dec. 1846; Davenport 1845, 46–48; McCalman 1988, 193–94.

36   *NMW* 4 (11 Aug. 1838): 340. Saull and Davenport also frequented the Finsbury Mutual Instruction Society in Bunhill Row and the South Place Chapel. Saull lectured here, for example, on his Owenite theme: "The Influence of Scientific Knowledge in forming the Character of the Future Generations of Mankind" (*PM* 1 [29 July 1837]: 322; Davenport 1845, 71).

37   *The Man* 1 (28 July 1833): 32, extract called "The Origin of Man" from the unpublished "Urania". Janowitz 1998 on Davenport's "interventionist poetics".

Davenport had been running (well, limping) with Saull for a long time. They could be found together, in the old days, at the Optimist Chapel, itself in Finsbury, and the BAPCK, as well as at Owen's Labour Exchange. In later times, they met up at the Tower Street Mutual Instruction Society and, here, at Finsbury Social Institution.[38] Davenport was an agrarian polemicist, who wanted "the land, rivers, mines, coal-pits, &c," to be nationalized, and for all taxes to be paid out of the subsequent land rental, with the surplus to be returned to the people. Effectively, landed aristocratic wealth would be redistributed. He remained a popular draw and *The Origin of Man and the Progress of Society* (1846) comprised his talks critiquing private property.[39] He had gravitated to Chartism and, as President of the East London Democratic Association, had mentored the firebrand red republican Julian Harney—the "little man with the pen of a Marat".[40] Davenport was as one with Saull on freethought and universal secular education, but poverty now forced him to rely on whip-rounds arranged by Harney and Holyoake.

Another old Finsbury Spencean and friend of Saull's, George Petrie, had died in 1836. A plebeian bard himself, his lauded poem "Equality" remained pinned on a door in Saull's museum.[41] Other Petrie remains, more mortal than literary, ended up in Aldersgate Street, as we will see, suggesting that the radical galleries went far beyond traditional ammonites-and-amphora visitor attractions.

In short, Finsbury was a hotbed with a heritage, which put Saull at the centre of continuing agitation. Thomas Wakley was Finsbury's doctrinaire radical MP, for whom Saull would periodically deputize at meetings. In the 1840s, Finsbury remained one of the most active socialist branches (No. 16), with Davenport on its Council. Its members had a choice of meeting places: Watson's nearby Mechanics' Hall of Science—where Saull still lectured frequently—and now Finsbury Social Institution.

Finsbury Social Institution itself evolved with Saull's lectures. Owenite branch 16 had taken over the building in 6 Frederick Place,

---

38 *PM* 2 (5 Aug. 1837): 8; 24 Feb. 1838, 248.
39 Davenport *ODNB*; Davenport 1845, 67.
40 McCabe 1908, 42; Claeys 1987, 160; dagger: *Hansard Parliamentary Debates*, 3d. ser., 48 (1839), 33.
41 *NS*, 31 Oct. 1846, 3; Chase 1988, 160–61.

Goswell Road, during the Owenite boom in December 1840. As a local Hall of Science, it was not big; in fact,

> The lecture room is rather small, but is very well fitted up; and there is connected with it another room of equal size, well adapted for a coffee or refreshment room, with two kitchens, and a committee room. The whole forms a very complete little institution.[42]

In February 1841, a "festival"—a *conversazione* or *soirée* to the middle classes—had inaugurated the Institution. Its Sunday lectures on socialism attracted largely "mechanics and tradesmen".[43] Science was favoured from the outset, and Finsbury had a policy of running scientific talks weekly from 1844.

Bourgeois radicals were now welcomed as teachers. The medical practitioner in a Quaker's hat, Dr John Epps—whose phrenological work had long been interesting to co-operators[44]—talked on human physiology here in 1843–44.[45] Epps had his hand in many reforming pies, and could often be seen alongside Saull on committees.[46] As for radical sciences, Epps' latest interest got him nicknamed the "Homoeopathic Napoleon," for he had the stature "of the 'Little Corporal.'"[47] He was doing the rounds of the socialist halls, proselytizing phrenology and homoeopathy, and lecturing on human physiology, at a cheap rate (tuppence a lecture).[48] Finsbury's extensive sixteen-lecture course on physiology was about the biggest Epps delivered.

Owenite women were particularly active in Finsbury. There was a women-only mutual-instruction class, a woman on the Council, and another, Mary Jenneson, who was secretary of the branch and (almost

---

42 *NMW* 9 (27 Feb. 1841): 134; 8 (5 Dec. 1840): 368.
43 *NMW* 11 (29 Oct. 1842): 146–47.
44 *British Co-Operator* 1 (May 1830): 40 passim.
45 *NMW* 12 (16 Mar. 1844): 303. On Epps' medical radicalism: Desmond 1989, 166ff; J. F. C. Harrison 1987, 205.
46 Both had been on the Council of the National Political Union (*Destructive* (Hetherington), 1 [16 Feb. 1833]: 23); both were members of the Radical Club, and of the Metropolitan Parliamentary Reform Association in 1842–43 (Rowe 1970b, document nos. 71, 129), and both could be seen sitting on the stage at the opening of Lovett's Hall of the National Association (*National Association Gazette* 1 [30 July 1842]: 243).
47 Linton 1894, 160.
48 *NMW* 11 (5 Nov. 1842): 154.

uniquely) a delegate to the Owenite Congress.⁴⁹ Tellingly, come the 1848 revolution in France, a public assembly at Goswell Road sent an address to the "Citoyens François" applauding their "glorious accomplishments". It was signed off by "M. William Devonshire Saull, l'un des plus zélés partisans des *droits de la Femme*". This was enough for his fraternal greetings to be published in Eugenie Niboyet's pioneering feminist-socialist daily, *La Voix des Femmes*,⁵⁰ run exclusively by Parisian women and leading the call for women's enfranchisement.

Saull regularly gave cheap or free talks in the Finsbury Social Institution. What stands out is how many of them now spelled out the meaning of antiquities for freethinking socialism.⁵¹ We only have titles or a précis, but they are indicative. They revolved around what Mary Jenneson called Saull's "favourite antiquarian topic, 'The condition of the Ancient Britons during the Roman occupation of these islands'".⁵² Even Saull's levelling word "aborigine" for *Britain*'s "first inhabitants"⁵³ was itself shocking, given that the term was sneeringly associated in the public mind with those imperial 'throwbacks': the "wild and formidable"⁵⁴ African 'Caffres' and New Zealand Maoris.

---

49 Mary Ann Wiley married (1843–44) the tailor Charles Jenneson, himself on the pro-working-class wing of Owenism and a lecturer on the rights of women: *NMW* 10 (25 June 1842): unpaginated advert after p. 424, "Lectures at the Finsbury Social"; Frow and Frow 1989, 118 n. 24; Claeys 2002, 181. Charles Jenneson and Saull worked together to establish a non-sectarian, 2*d*-a-week Owenite day school in Whitechapel, *NMW* 12 (9 Dec. 1843): 192; (30 Dec. 1843): 215; *Movement* 1 (16 Dec. 1843): 8.
50 *La Voix des Femmes*, 27 Mar. 1848, 2; *UR*, 12 Apr. 1848, 39.
51 Some of Saull's slated lectures here are untitled, for example, *NMW* 12 (16 Mar. 1843): 303; 13 (21 June 1845): 426; *Reasoner* 1 (8 July 1846): 92. All of his titled lectures concern aborigines, except one, "On the Analysis of Opinion" in 1842, which covered the origin of prevailing "philosophical, political, and religious opinions": *NMW* 11 (29 Oct. 1842): 146–47; *NS*, 27 Feb. 1847; *UR*, 24 Feb. 1847, 26. Otherwise, his activities at Finsbury took in chairing a meeting to petition the Queen on the country's distress, backing Walter Cooper's stand on the wickedness of blasphemy laws, and collecting funds to see Owen off to America: *NMW* 10 (19 Feb. 1842): 271; 13 (13 Sept. 1844): 93–94; (5 Oct. 1844): 118.
52 *NMW* 13 (1 Mar. 1845): 287.
53 Saull 1845, 1.
54 Lindfors 1996. Even to have a humanitarian interest in modern "aborigines" could be written off as "mischievous and morbid sentimentalism", and there is some evidence that Thomas Hodgkin's sympathies (he was at that moment founding the Aborigines' Protection Society) helped lose him a Physician's post at Guy's Hospital in 1837: Kass and Kass 1988, 292, 377.

Some might have called his usage cynical. If it did not exploit, it certainly fitted in with the exotic peoples increasingly being exhibited in theatres, fairs, and music halls—peoples being marketed as 'savages' in the expanding imperial vernacular of the age.[55] London impresarios, by creating a clientele for viewing living "aborigines", could only have increased the audience for Saull's lower gallery. Not that Saull was the first to label the early Celts as aborigines or "savages".[56] But his usage was provocative and tailored to radical venues dedicated to cutting the plumed aristocracy down to size. As a piece of social reductionism, it sat in the Carlile-*Oracle* tradition of giving noblemen the same dirty roots as hod-bearers.[57] So his first Goswell talk, in February 1843, called "Customs and Manners", illustrated the aborigines' "history from the remotest antiquity, by the remains of their houses, furniture, dresses, implements, &c.", specifically to highlight the "changes which have taken place in the circumstances" *surrounding* "the inhabitants of these isles".[58] This circumlocution was meant to suggest that it was the Romans who changed the circumstances of the aboriginals they conquered. As such, it proved a test case of Owen's headlining maxim that "The Character of Man is formed for Him,—Not By Him", familiar on the masthead of the *Crisis* and *New Moral World*. Alter the conditions, and you alter the character, which is what the Romans did to civilize the aboriginals, and what Owenites were attempting to do to the Old Immoral World.

Notitia Britanniae

Now that debates following lectures had started up again, drawing the crowds, Saull made great use of them. One lesson he had in mind was stressed in these to-and-fro discussions. His talk on "British Antiquities" posed a question for the audience in September 1844: "Is the evidence of

---

55 Qureshi 2011.
56 This ethnographic analogy went all the way back to the seventeenth century, when reports of native peoples in America led to such "savages" becoming stand-ins for early Celts. The notion however was obnoxious to nationalist Celtic historians (Morse 2005, 17, 56).
57 The word "aborigine" does not appear, for example, in a parallel but contrasting work to Saull's *Notitia*, Akerman 1847. More conventional in structure, Akerman's tome described the types of ancient monuments rather than delineating a progressive trend. Saull's bent betrayed his Carlilean-Owenite heritage.
58 *NMW* 11 (11 Feb. 1843): 267.

Facts to be preferred to Written Testimony?", meaning should truth come from artefacts or sacred texts? And the next talk, in November, followed suit: "Will Antiquarian researches remove Traditionary Superstition?"[59] He was urging his listeners to treat "all the accounts descriptive of the earlier races of man" (read sacred and other texts) with "great suspicion" because of their unreliable hand-me-down nature. There was no contemporary written record, only word-of-mouth turning into untrustworthy folklore, which often ended with scribes "servilely copying one another, and repeating tales".[60] By contrast, an Owenite in a Pestalozzian object-teaching environment saw *artefacts* provide a spy hole into the past, from which more accurate historical insights might be had.

This was elaborated in Saull's short book in 1845, *Notitia Britanniae; Or An Enquiry Concerning the Localities, Habits, Condition, and Progressive Civilization of the Aborigines of Britain*. In it, he used a common aboriginal base for all peoples to let him oust fallen angels and racist demons alike. The Romans were the 'improvers' of their day, and Saull defended "the grand Roman plan of colonization" for the changes it affected in these aboriginal Britons.[61] *Notitia* was an expensive book, at 3s 6d, and obviously not aimed at plebeian socialists so much as wealthy antiquarians, among whom Saull was trying to establish his credentials. It was the fruit of three years spent visiting hut remains, tumuli, and barrows, as well as Roman villas and forts, and collating provincial accounts by private museum collectors, the guardians of so many relics. Much local lore, too, resided with the clerical antiquarians—it was the parsons, posted off to their rural diocese, who had the education and leisure to indulge a tastes for ancient civilization. Saull visited sites with one and all.

Ironically, it was a one-armed scriptural literalist who proved Saull's key source. The evangelical Scottish Presbyterian Dr George Young of Whitby was the last person one might imagine rambling amicably with the blasphemer, yet hut circles and fossils were a grand mediating point

---

59   *Movement* 1 (7 Sept. 1844): 328; (13 Nov. 1844): 424.
60   Saull 1845, 50; he was cleverly quoting from W. D. Cooley's new preface to *Larcher's Notes on Herodotus*, 1: 107, knowing that it applied *mutatis mutandis* to the Bible.
61   *NMW* 13 (1 Mar. 1845): 287.

and the two hunted happily together on the windswept moors. Dr Young (the 'Dr' was an honorary title from Miami College, Oxford, Ohio, in 1838, the year he published *Scriptural Geology*) was a mainstay of Whitby Literary and Philosophical Society, and a pastor who could publish tracts against infidelity and papers on ammonites with equal ease.[62] He had turned up the largest number of ancient hut remains in Yorkshire. Forty ancient British villages were to his credit, the huts signalled by circular depressions in the ground with stone surrounds. These beehive houses in their day were presumed to have had sod-packed walls and branch roofs. The inhabitants, according to Young (and Saull), were then on a cultural 'level' with the present day "Caffres" of South Africa.[63] At Harwood Dale, Saull and Young investigated fifty or sixty of these hut depressions, often characterized by charcoal remains in the centre where fires had been.

These 'primitive' hunter villages, successively occupied through the generations, provided Saull's baseline for his cultural levels. They were "rude abodes" with no signs of pottery or coins. Even in London, at the Cheapside sewage excavations, Saull found, on descending the shafts to the lowest point, similar concave remains of huts with central fireplaces, from an age when London was densely forested. Above this 'hunter' state was the next cultural level, the 'shepherd' society, with its fortified stations to hold the newly domesticated livestock. Saull could point to these in his native Northampton, in Long Buckby where his nephews lived. He visited another with a local vicar in Chipping Warden, although the biggest fortified complex, spread over 150 acres, occurred at Daventry.[64] This gave him two social 'strata', which sat at the base before the "momentous aera" ushered in by the Romans.

Coastal forts along the Channel were already "advanced in intelligence" before Caesar's arrival because of their contacts with Gaul, judging by the arms and crude money. This information came from Saull's "esteemed friend", the Devon antiquarian Captain Shortt. It had to be rather prised out of his texts, for W. T. P. Shortt's infuriating thickets

---

62 *Geological Curator* 7, no 7 (June 2002): 4–30; Cleevely 1974, 469 n. 48. Saull exchanged fossils and London Roman artefacts with Whitby museum, of which he was an honorary member: *Sixteenth Report Of The Whitby Literary And Philosophical Society*, 1838, 13; *Twentieth Report*, 1842, 15.
63 Saull 1845, 3–7.
64 Saull 1845, 13, 18–25.

of cataloguing detail and "discoursive" style belied his Oxford classical education. He had turned up coins in Exeter from Greek cities in Syria, Asia Minor, and Alexandria, and even an Isis bust with hieroglyphs, showing the extent of the early tin trade in Cornwall and the reach of Mediterranean trading vessels.[65] It was Caesar's arrival which extended this advanced cultural contact to the rest of the country.

Caesar found the inhabitants behind their bank and ditch hill forts, but the Romans brought these aboriginals down to the lush vales where they absorbed the "arts of civilization". The Romans introduced iron to replace the Celtic brass, drained the low lying "impassable swamps" and built the roads to establish wider communications. Again Saull, on home ground, described Daventry's Roman Road, twenty-feet wide, which was made of small stones with grouting. The locals learned from the Romans to cultivate and grind corn—allowing the next 'farming' phase. Log or board houses, cemented and tinted inside, replaced the old sod-and-branch huts. Villas were warmed by flues and hypocausts. Temples changed the "religious feelings" of the natives: glorious temples with tessellated pavements producing the effect of paintings. Saull's museum had some, found in Maiden Lane. Fine Samian cups and dishes were introduced, now made in Britain. And with decoration—of gods, musicians, hunting scenes and gladiators—came lettering, which was itself introduced to the natives. Saull's museum had some fifty pieces of this Samian pottery impressed with their makers' names. The art of stamping or coining money, with lettering again, was an innovation, as was glass, mode of dress, cremation, and urn burial.[66]

These social phases were the theme of *Notitia Britanniae*. Much of the information was culled from the knowledge of fellow antiquarian and private museum keeper, Charles Roach Smith. Roach Smith was a chemist in Finsbury, and the leading authority on Roman London. He was a passionate collector who descended the same shafts and examined the same excavations as Saull. They shared a similar serendipitous

---

65 Saull 1845, 26, 55–56. That Celts had advanced in civilization in Cornwall through commerce with Phoenicians was commonly accepted (Morse 2005, 90). Saull believed that Exeter was the site of a Phoenician colony, trading in tin, centuries before the arrival of the Greeks and Romans: Shortt n.d., iv. C. R. Smith 2015 [1886], 2: 257 on Shortt's disastrous prose.

66 Saull 1845, 26–48; on his digs in Northampton turning up Roman remains for his museum: *JBAA* 4 (Jan. 1849): 396–97.

approach, which relied on news of civic works and the navvies' good will. Road widening and sewer laying attracted them, and the dredging operations as the Thames was deepened at London Bridge, which revealed bronze statuettes and coins.[67] It was Roach Smith's dedication in scouring these public works that had already earned him the title "the Discoverer of Roman London".[68] The new sewerage shafts were particularly useful. The City had connected 11,200 houses (out of 16,200) to a City-wide sewerage system by 1852. This offered tremendous scope for antiquarians willing to descend shafts up to eighty feet deep during the building, and both Roach Smith and Saull took full advantage.[69]

In the end, it was up to these enthusiasts to store their sewer finds, at least until the laissez-faire state took a more interventionist interest. Roach Smith's huge cabinet would eventually become the foundation of the British Museum's Romano-British collection. Saull's, by contrast, had an ignominious fate.

When Roach Smith helped found the British Archaeological Association in 1843, Saull was on board immediately. He became a member ("Associates", they were called), later shared a seat with Roach Smith on the General Committee and attended the yearly congresses.[70] Roach Smith was influenced by the Comité des Arts et Monuments in Paris (an offshoot of a commission set up by Guizot, when Minister of Public Instruction), and he originally planned to emulate the Comité's series of illustrated works on France's heritage, to make a similar story of Britain's progress from "the earliest primeval period in which the first rude efforts of the hand of man might be traced, down to the latest

---

67 C. R. Smith 1854. They occasionally re-identified showmen's items. For instance, in 1848 Saull re-assigned a "Roman" harpoon dredged from the Thames as a modern whaler's: *Literary Gazette* 1657 (Oct. 1848): 700.

68 T. Wright 1845, 129. Thomas Wright was co-founder with Roach Smith of the British Archaeological Association, and Wright's chapter on the "Romans in London" in his *Archaeological Album* was based largely on Roach Smith's museum and publications.

69 *Archaeologia* 27 (1838): 140–51; 29 (1842): 145; *Literary Gazette* 1883 (Feb. 1853): 181; *MC*, 29 Jan. 1853. 5. J. White 2007, 50.

70 He attended from the first: *Times*, 16 Sept. 1844, 3; and yearly thereafter. Committee: *JBAA* 3 (1848): 133; *Lancaster Gazette*, 17 Aug. 1850, 4; *Nottinghamshire Guardian*, 31 July 1851; and in subsequent years. Here Roach Smith would often comment on Saull's papers, date his Roman findings, and identify the Roman stations subsequently mentioned in *Notitia*. This forum allowed great scope in understanding Roman Britain and its relation to 'primeval' archaeology.

division of the middle ages."[71] Although never carried through, it would have fitted Saull's agenda perfectly. Guizot had also instructed the Comité effectively to *preserve* French antiquities, and this emphasis on preservation was paramount in the Association.[72] This onus on saving, coming from the French, justified Saull's storage facility, which was now tilting heavily towards local Roman antiquities.

But still Saull had a deeper agenda. Even before *Notitia* was published, he was detailing his "primitive" to "pastoral" cultural sequence at the first annual meeting of the Association at Canterbury in 1844. In the "primaeval section" (it was divided into sections like the British Association for the Advancement of Science), presided over by Roach Smith and the geologist William Buckland among others, Saull described three Roman encampments near Dunstable, on the chalk Downs. One appeared to have been a "primitive" hill fort that had been extended later by the Romans, possibly as a forward observatory post, whence it became a "pastoral" camp.[73]

That transition was the novelty. This was not armchair archaeology but relied on legwork if not spadework[74]—Saull, like Roach Smith, was always on site and toured the country examining and collecting. But, to those indisposed to his philosophy, Saull's conclusions could be written off as armchair dilettantism. Sarcasm marked the *Athenaeum* review of *Notitia*, which excoriated the book from the first line: "Mr. Saull is one of the Pegge genus, but of an inferior species, since the latter did know something of what he was writing about." (A sly dig: the Rev. Dr Samuel Pegge was an eighteenth-century barrow specialist who "diligently collected the errors of his predecessors while adding another to the list".) The review went downhill from there, demanding Saull "prove who the 'Aborigines' of Britain were". There was widespread belief that the monuments based around these depressions were sepulchral, and

---

71 *JBAA* 2 (Jan. 1847): 302; *Archaeological Journal* 1 (1845): 71.
72 Individuals like Saull and Roach Smith remained the driving force. A Parliamentary Select Committee in 1841 did discuss the preservation of monuments, but only of "illustrious individuals" (Swenson 2013, 57). The state's hand was ineffectual compared to French government efforts.
73 Saull 1845, 54; *Times*, 16 Sept. 1844, 3.
74 For modern sympathetic ways of reimagining Victorian "armchair" prehistorians, see Sera-Shriar 2016; Barton 2022.

the *Athenaeum* reviewer doubted that the so-called "huts" were anything other than tombs.[75]

More resistance to the hut hypothesis came from the Archbishop of York's son, the Rev. Leveson Vernon Harcourt. He was a collector of lore to support, in his book's title, *The Doctrine of the Deluge; Vindicating the Scriptural Account from The Doubts which have recently been cast upon it by Geological Speculations* (1838). Two volumes, running to 1100 pages, proved a thousand times over that every ancient tradition was susceptible to "Arkite" reinterpretation (that is, pointing to the biblical Flood). The Flood waters were already receding from geology back into Sacred history, but Harcourt's work would be thrown in Saull's face. Harcourt's double-decker was overkill to many, with such attenuated evidence as to strain the patience of readers.[76] But it was his method that would have exasperated Saull. Harcourt side-stepped geology and amassed Pagan mythology, tapping "the memory" of the ancients "derived from their traditions, their superstitions, their monuments, and their usages", to show how Flood folk-lore was kept alive "till it was finally enlisted in the service of true religion", Christianity. Even the hill-top cromlechs and cairns were reinterpreted as monuments built by Noah's descendants. They commemorated a rejuvenated mankind's rise from "the purifying waters of the Deluge".[77] The mounds next to the depressions were sacred, for sacrifice and celebration, while the stone-sided pit-cavities were not houses, but water-holding tanks.[78] Looking at Harcourt's monster tome, one understands Saull's tactics, asking in Finsbury discussion forums whether "Facts" were not better guides than garbled Creationist "Testimony", his "facts" being artefacts, from visual fossil sequences to pot-shards.

---

75 *Athenaeum* 932, 6 Sept. 1845, 876 (the slashing review was by Samuel Astley Dunham); *Monthly Times*, 8 Sept. 1845, 7. On Pegge: *Archaeological Journal* 4 (1847): 30.

76 Even the reconciler Rev. Dr John Pye Smith (1839, 106)—so beloved of the *Patriot* (26 Apr. 1852, 270)— saw Harcourt "weakening an argument by an excess of amplification", while George Eliot thought he "rather shakes a weak position by weak arguments" (Kidd 2016, 14–15). By contrast, the appreciative *GM* 61 (Dec. 1841): 617–19, advised geologists to pay as much attention to this mass of testimony as they did to their physical evidence.

77 Harcourt 1838, 1: 9; 2: 469.

78 *GM* 40 (Aug. 1853): 183; (Oct. 1853): 389. This was Harcourt at the breakaway Archaeological Institute of Great Britain, with which Saull had nothing to do. See also Harcourt in *Sussex Archaeological Collections* 7 (1854): 32.

Saull's dry factual presentation might have chimed with the incipient positivism of those flashing young blades joining Holyoake's secular circle,[79] but his socialist implications went down badly with traditionalists. It was this twist at the end of *Notitia* that caused the public furore and atrocious reviews. Saull was warned by friends vetting the manuscript not to push these implications. But that was the whole point, the "portion of it, which I deem the most valuable".[80] At John Street, Saull had actually admitted to the socialists that he continued his connection with the learned societies precisely to extend Owenism into the bourgeois world.[81]

After the failures of the labour exchanges and co-operatives, and the loss of civic power bases in the Halls, it seemed that wealth and power would have to be redistributed voluntarily (at least in Owen's view), to produce a harmonious society. Although for those like Saull with a radical edge, the learned bourgeoisie could still be chivvied, and that was *Notitia*'s aim. From the opening talk of human cultural phases, "the hunter, (or rudest) state, the nomadic, shepherd, or pastoral state", proving that "man always has been—is now—and, by direct inference, ever will be, an advancing or progressive being", to the final lines, the evangelical Owenism stayed in. Those final lines might not have meant much to the archaeologists, but they repeated the aphorism on the mastheads of the *Crisis* and *New Moral World*, "If we cannot reconcile all opinions, let us endeavour to unite all hearts". Saull's aboriginal antiquarianism was used to point up "the universal law of nature and necessity" proved by geology,[82] that fossil and social progress must continue through "every gradation of mind" as society levels, equilibrates and perfects mankind.

Hence came Saull's call to scientific gentlemen to stop prostituting their talents. It was more muted than in Carlile and Chilton, but it was

---

79 Ashton 2006, 138. The rise from savagery was easily accommodated by Owenites themselves. Every socialist bookshelf would have had Minter Mogan's *Revolt of the Bees* (1826), which turned Manderville's fables on their head and chronicled the rise of the bees. As such it made a familiar allegory of "progression from a noble savagery through pastoral occupations, farming, and industry, to a fifth revolution pioneered by 'the wise bee' [Owen]", who would fairly redistribute wealth and knowledge: Armytage 1954, 1958.
80 Saull 1845, unpaginated "Introduction".
81 *Reasoner* 16 (5 Feb. 1854) Supplement, 97–98.
82 Saull 1845, 49, 57–58.

still there. The missionary Saull was calling for conversion. Despite integrating into learned bodies and attending elite *soirées*, Saull never really appreciated the depth to which a gentleman's gloss on science reflected his political, religious, and social beliefs, and that a scientific shift would prejudice these and thus his privileged position. Were a gentleman to step out of line, he would be immediately reminded that his *character* was at stake. And since, for a gentleman, knowledge without character was nothing, his authority would be shaken, his caste doubted. Character was the guarantor, it was the key chink that the Tory *Quarterly* looked to constantly when 'bad' science reared up. Threats to a gentleman's stature ensured conformity, as Lawrence and so many others discovered. Saull was calling for the Good and Great of science to act as social traitors. The freethinking future would be ushered in by 'unbiassed' men of science:

> Those only who are imbued with the love of science and philosophy, and who are consequently the disinterested advocates of free inquiry, have now ... momentous duties devolving on them: for to such minds appertain the execution of the task of supplanting the various antagonistic and conflicting opinions [that is, religion], which so materially tend to distract and mystify our common humanity; those alone who adhere to such principles can meet on common and neutral ground; for science recognizes none of the petty distinctions of sect, party, or persuasion; its effects on the mind being to establish universal philanthrophy [sic] in our communications with our fellow men, knowing, that the higher they advance in intelligence, the more perfect and enduring will be that congeniality of sentiment so much to be desired, and so worthy of their strenuous efforts for its accomplishment...

Such a coded Owenite request, to admit materialist implications, was doomed to fail because this would involve a total unpicking of a gentleman's social and religious standards, all of which were tacitly integrated into what became a block-box of belief in the truth of their science.[83] Hence the vehemence of the response.

The urbane *Gentleman's Magazine* in its apoplectic attacks on atheism, as the "delirium of a sick and suffering soul", pictured such godlessness as "spiritual leprosy" spread by Owenism, which itself sucks out every "patriotic conviction" of the heart. In the *Notitia*, it saw straight through

---

83 Latour 1987, 61.

Saull's anodyne snipe at retarding influences. The *Magazine* tried to retain its decorum, hoping that the author's intention was not "to say that the reign ... of the goddess of reason will supersede the great truths for our direction in time, and guidance to the mansions of eternity, to be found in the Bible." Were it the case, then this philosophy "of Voltaire and Rousseau, has been tried and found something worse than mere speculation. Take away the certainty of rewards and punishments which revealed religion announces, the social obligations are dissolved in an overwhelming flood of misery and crime".[84] Given that dusty antiquarian descriptions were so often dismissed as "dry, pedantic, and repulsive"; given, moreover, that the socialist Saull's approach was atypical at the time in stressing progressive transitions from aboriginal or 'primeval' to Roman, it is not surprising that polite readers found the results "curious" when not absurd.[85] But then Saull had come in at an idiosyncratic angle. Davy's dream had been fulfilled. The rise from savagery had been fleshed out in context-rich detail, with geological methods being used to locate archaeological remains at their correct developmental level,[86] all in aid of Owenite social ends.

Saull continued promoting this progressive social development at the more appreciative venues. But these were now changing rapidly. As a further sign of Owenism shrinking, the Finsbury branch of the Rational Society was re-launched as the Finsbury Literary and Mechanics' Institute in 1846. Wakley was to have chaired the inauguration, but Parliament kept him so Saull stepped in on 29 July. With Saull on the platform were figureheads of Owenism (Fleming), freethought (Holyoake), communism (Goodwyn Barmby), Christian radicalism and phrenology

---

84 *GM* 23 (Apr. 1845): 397–99; 35 (May 1851): 519–23.

85 *Leicester Chronicle*, 22 Feb. 1845, which contains an appreciative review, and commented on the "perfectly justifiable" geological approach to dating the stages; *Spectator* 18 (15 Feb. 1845): 162.

86 History transcended Saull's contingent Owenite meaning and used his first-hand descriptions as a resource, whether in the new anthropology of the 1860s, accepting the hut-circles of Young and Saull (*Journal of the Anthropological Society of London* 3 (1865) lxii), or later in the *Making of London* by Sir Lawrence Gomme (1912, 38), who uncritically quoted Saull's accounts of his London hut discoveries in the sewerage excavations. Saull's study of the immediate pre-Roman period, which would come to be called 'Late Celtic', though atypical for its day in that it tried to show sequential steps through to the Roman period, is now used as part of the backdrop from which the work on this transition by Augustus Lane Fox and Arthur Evans could be assessed (Hingley 2008, 294–95).

(Dr Epps), popular poetry (the former *Morning Chronicle* journalist Charles Mackay), as well as a trades-advocating ex-Unitarian minister (F. B. Barton), showing that, just as Saull was reaching out, so were the institutions.[87] He continued lecturing here on "The Earliest Histories of Man",[88] even as the Literary Institution underwent yet another relaunch under a new proprietor in 1847 and cast its net still wider in an effort to attract an audience. By now, the once-proud Owenite branch had shed its old mantle, as the new manager claimed that "it will be conducted upon principles entirely devoid of anything of a party or sectarian nature". The open-arms, clerk-receptive policy emphasized "comfort and convenience" and efforts to keep it "select, orderly, and respectable." Saull and the usual radical group were present at the re-opening, but there was no denying that the institutions had lost their Owenite exclusivity.[89]

With the loss of an ideologically-constrained base, Saull found himself buffeted by unexpected winds. Tensions at the Mechanics' Hall of Science were racked up by the 1847 intake. In came a new crop of acerbic freethinkers. At this point, Saull probably met Robert Cooper, and he certainly knew the ethnologist Luke Burke.[90] Holyoake's widening of his Utilitarian circle inevitably resulted in some discordant voices, but none more so than Burke's.

Luke Burke was a new-style 'ethnologist'. He wanted "ethnology" to be stripped not only of its "Hebrew chronology" but also its Christian obsession with the brotherhood of man, and what remained he would puff as a new science essentially untainted and data-driven.[91] 'Ethnology' for him, idiosyncratically, meant study of the "physical

---

87   *Reasoner* 1 (29 July 1846): 136; *NS*, 25 July 1846.

88   *UR*, 16 Feb. 1848, 24. He retraced the ground in "A Critical Examination of Ancient History" at Finsbury Hall, Bunhill Row, where the Finsbury radicals also met (*UR*, 13 Oct. 1847, 92); and the "Natural Law of Progress" at the newly formed and quickly forgotten Zetetic Society at his Mechanics' Hall of Science: *Reasoner* 1 (7 Oct. 1846): 256.

89   *UR*, 8 Sept. 1847, 82; *NS*, 18 Sept. 1847.

90   *UR*, 16 June 1847, 57; 6 Oct. 1847, 89; 13 Oct. 1847, 91. Another who came was well known to Saull, Walter Cooper, the Chartist tailor with Christian Socialist sympathies (and brother of the Chartist poet Thomas Cooper): 12 Jan. "1847" [1848], 13.

91   Desmond and Moore 2009, chs. 6–7, for the new attacks on J. C. Prichard's beliefs in the Adamic brotherhood of all mankind.

peculiarities of races",[92] in other words, what would later be called "Anthropology". As a deist keen to kick Moses out of science, he appealed to the Utilitarians. But by making the "races" unalterable, immune to any environmental modification, unchanged since the beginning, and by denying the "natural equality of men", his views clearly heralded the Victorian move from xenophobia to racism.[93] Unlike Saull, he had no truck with transmutation. "The primary differences are those which were established by the Creator at the origin of humanity," he announced in his *Ethnological Journal*. Therefore, utopian schemes of social improvement resting on the premise "That all men are of one genus, of one species, and of one family, brothers of the same blood, descended from one common father" were doomed. "Unity, equality, fraternity" to him were Christian chimaeras. Social revolutions based on them will fail because they ignore the "great and permanent diversities among mankind". This put him at loggerheads with Saull. Even worse, for Burke, some races were superior, and those "must be the rulers of the world."[94] A few activists, notably Southwell and Robert Cooper, found their own emphasis on discrete human stocks gaining strength from Burke's racial extremism. But it was abhorrent to Saull. By associating an environmentally-driven ascent from a common stock with Christianity's Adamic brotherhood, Burke was upping the ante. He might have been meeting the new imperial mood, but this was throwing the cat among the fat Owenite pigeons.

In widening secularism's remit to include racists, Holyoake was deepening the tensions. Burke's Utilitarian talk on 'savage' mythology might have piqued Saull's interest, given his aboriginal researches;

---

92 *Ethnological Journal* (June 1848): 3. That this was an early sign of a growing trend, note the parallel racial structuring that same month being promoted by Dr Robert Knox (E. Richards 1989; 2017, ch. 10; 2020, ch. 3) and sympathetically treated in the *Medical Times* (17 June 1848): 97, 114.
93 Lorimer 1978, 17 passim.
94 *Ethnological Journal* (June 1848): 5, 7, 29; (Mar. 1849): 470, 474. Burke had long attacked J. C. Prichard's environmentalism and Adamic brotherhood: e.g. *People's Phrenological Journal*, 2 (1844): 3, where Burke railed against those who believe circumstances have "converted fishes into reptiles, reptiles into quadrupeds, quadrupeds into monkeys, and monkeys into men; and, even at the present day, few persons can see any difficulty [because it might "harmonize with prevailing religious views"] in their blanching the negro, or blackening the Caucasian, in their converting the savage to civilization, and every civilized man into a philosopher."

even if Burke's "Demonstration of Deity" did not. Burke's indictment of converging ancestries as Christian spawn could only have outraged Saull. Burke's separate-origins pluralism sat better with Southern racists (Burke's "valued friends"[95]), whose science sustained the anti-black, pro-slavery ethos in ante-bellum America. And, as if to prove the point, Burke reviewed their works extensively in his *Ethnological Journal*.[96] Burke's views could hardly be avoided. In 1847, he was running courses at John Street and the City of London Mechanics' Institute in Gould Square, and emphasizing racial permanence.[97] But his anti-socialist, anti-environmentalism sat uneasily at the Utilitarian Society. This still had its Owen supporters, like Saull, and Burke was giving their social-amelioration policies and scientific environmentalism the lie direct. It was a sign that, as xenophobia hardened with imperial expansion into racism, and the sustaining Owenite community crumbled, a gradational blood-brother evolutionism based on the old Holbachian environmental sciences would lose its traction. The ground was being cleared. It now awaited the new Malthusian capitalist explanation of evolution to take on Burke, which the reclusive Charles Darwin still had under wraps.

---

95 *Ethnological Journal* (Feb. 1849): 438. The influential American pro-slavery pluralist, or what would shortly be called "polygenist", J. C. Nott, was actually fired by the "Gospel according to Luke Burke": Barnhardt 2005, 294–96; Desmond and Moore 2009, 168ff.
96 *Ethnological Journal* (Sept. 1848) 169ff.
97 *UR*, 31 Mar. 1847, 35, 36 et seq. and (26 May 1847): 52; *Howitt's Journal of Literature and Popular Progress* 2 (4 Sept. 1847): 160, for Gould Square.

23. Reforming Scientific Society

> Mr. Saull then alluded generally to the interest he had always taken in progress, religious, social, political, and scientific. He was a member of many learned societies, and he continued his connection with them in order to embrace every legitimate opportunity of advancing the principles he had at heart. He was now advanced in years, but his interest in the 'good old cause' was undiminished...
>
> Saull's talk reported in 1854, the year before his death.[1]

With the collapse of Owenism, and given his growing interest in our ancestral 'aborigines', Saull could now be found increasingly inside the antiquarian societies. The reforming of the old, corrupt Antiquaries (see below), and the rise of the Numismatic and Archaeological Societies, testified to the proliferation of artefacts as London was excavated to create the imperial city. Still more did it reflect the bourgeois influx in an industrializing age. Noisy reformers were joining, representing new trading and Dissenting interests, men who formed a liberal group of wealthy specialists. These new fellows were not "professionals", examined and accredited, certified as "experts", to be employed for their knowledge. They had yet to be split into 'professional' and 'amateur' status. But they were dedicated careerists. Even so, for them it was still a side line, and they were gainfully employed elsewhere, those who were not leisured "gentlemen": Saull was a wine merchant, Roach Smith a chemist, Edward Spencer a solicitor, and so on. The clergy's role was declining, although the Society of Antiquaries remained the vicars' club of choice.

The Owenite congresses having ceased, Saull effectively switched to the annual British Archaeological Association jamborees. He could be found at the Winchester Archaeological Congress in August 1845

---

1   *Reasoner* 16 (5 Feb. 1854) Supplement, 97–98.

talking on the development of the Saxon walls of Southampton.² As he slid across, his stock rose. At the third Association Congress in Gloucester (1846), he was one of the Secretaries. Here he exhibited Roman tiles, stamped with marks of the sixth and ninth legions,³ and he elaborated on remarks made in *Notitia* about early British villages on the moors near Sealing, in Yorkshire. That intellectual weekly, the *Literary Gazette*, which gave over huge space to learned society meetings, and now faithfully reported the Archaeological Association's congresses, positively purred over Saull. The Sealing speech sent the *Gazette* back to his *Notitia*, which "displays the zeal and research by which the writer has made himself so competent to handle this difficult inquiry". Saull was an "entertaining guide" to the "dark and distant questions involved in the gradual development of rude and savage men, primarily through Roman intercourse, into the beings of high intellect and refined civilization with which our island is now peopled." The rehabilitation must have been sweet.⁴

The talks give us a flavour of what must have been in the lower gallery of Saull's emporium: artefacts straddling the "Ancient British" and Roman divide, which, given its Owenite ambiance of progression and perfection, could be expected to be displayed to maximum effect. Again, at the November 1846 meeting of the Archaeological Association, Saull exhibited a late Roman urn, coin, and comb, all found in Godmanchester, while he also described the earlier earthworks in the area.⁵ By 1847, he was fully engaged with these societies. In this, one of his most productive antiquarian years, he: (1) discussed in depth the Roman roads at Dunstable, and the ancient British and Roman

---

2   *JBAA* 1 (Jan. 1846): 361; *John Bull*, 9 Aug. 1845; *Atlas*, 9 Aug. 1845, 502. The subject matter, not reported in these, *was* relayed in French journals: *Cahiers D'Instructions* (1846): 55; *Revue Archaeologique* (1845): 387. The *Revue Britannique* 5th ser., tome 27 (1845), 454, spoke of the "professor's" rich London museum.
3   *JBAA* 2 (Oct. 1846): 281.
4   *Literary Gazette* 1547 (Sept. 1846): 792; also 1539 (July 1846): 648, on the remains of an ancient British village on the moor near Sealing, Yorkshire; *GM* 26 (Oct. 1846): 407–12; *JBAA* 2 (Jan. 1847): 389–90. Saull discussed Scottish vitrified forts at this meeting. He dated them a little before the Roman period and suggested the locals used wood and kelp to 'vitrify' the walls, melting the material between the stones to fuse them together: *Literary Gazette* 1539 (July 1846): 649. On vitrified forts featuring in debates over Celtic pyrotechnical knowledge and the moral elevation of their designers: Ksiazkiewicz 2015.
5   *JBAA* 2 (Jan. 1847): 360; *Literary Gazette* 1560 (Dec. 1846): 1053.

settlements found alongside them;⁶ (2) defended his progressionist thesis by disputing that barrows were sepulchral rather than hut-based living spaces;⁷ and (3) talked on Roman mill-stones, important because corn-grinding was an innovation introduced into conquered Britain.⁸

When it came to the creaking Society of Antiquaries, Saull adopted a familiar ideological stance. Its meaning would have been obvious to anyone who knew him: the anti-clerical campaigner who attacked traditional mythologies encased in custom and law. Every socialist knew that there was no greater disrupter than geology—that deep-time disturber of revered chronologies. Ancient saurians and ruined worlds were dragging Victorians out of their parochial time frame. For some it was liberating, others cried in despair. Those "dreadful Hammers!", Ruskin wrote in 1851, "I hear the clink of them at the end of every cadence of the Bible verses".⁹ Saull believed that primeval archaeology now had the same devastating potential. When a speaker at the Society in 1847 argued "the necessity of collecting local legends" in order to preserve the "mythology of our forefathers at a very remote period of their history", Saull protested and "wished more attention were paid to facts, which he considered were of greater importance than traditions."¹⁰ Saull was promoting the object of secular Finsbury among the crusty Antiquaries.

---

6   GM 27 (Oct. 1847): 406; *Literary Gazette* 1571 (Feb. 1847): 174. In April he also announced the discovery, on the site of Roman Olenacum in Old Carlisle, of an altar stone inscribed to the goddess Bellona (the first such found in Britain) by the prefect of the local cavalry: *JBAA* 3 (Apr. 1847): 42; *Times*, 12 Apr. 1847; GM 27 (Oct. 1847): 594; *Literary Gazette* 1578 (Apr. 1847): 301. Roman roads and the British settlements alongside them were now stock subjects for Saull: when Archaeological Association members visited St Albans (the Roman Verulamium) in the autumn, he gave a talk on the London to St Albans Roman road (*Literary Gazette* 1602 [Oct. 1847]: 707).

7   On this point discussion now turned on a "primeval monument" with fifty stone-sided 'residences' at Ashbury, Berkshire, called "Wayland Smith's Cave", largely thought to have been a burial site, but which Saull stated (against opposition) was for the living, not the dead: *Literary Gazette* 1572 (Mar. 1847): 196; 1573 (Mar. 1847): 217, 221; GM 27 (Oct. 1847): 407; *Critic* 5 (Apr. 1847): 296. He was responding to the Numismatic Society founder John Yonge Akerman (*Archaeologia* 32 [1847]: 312–14), who thought the monument sepulchral. Akerman was a respected numismatist with a radical history to rival Saull's, having started out as Cobbett's secretary and Thomas Wakley's assistant (Sprigge 1897, 229).

8   *Literary Gazette* 1587 (June 1847): 447.
9   Quoted by D. R. Dean 1981, 123.
10 *Literary Gazette* 1573 (Mar. 1847): 217.

All of this explains why the museum threw "light upon the domestic habits and manners of the Romans".[11] Saull, at the growing heart of a modern empire, was looking at cultural imperialism in a positive light, as a helping hand of Owenite outreach. He was detailing the mechanism whereby a similar hand had been extended to benighted Britons in the dim past. Although, needless to say, to the travel guides the downstairs exhibits often looked a jumble. For example, John Timbs in his *Curiosities of London*:

> The *Antiquities*, principally excavated in the metropolis, consist of early British vases, Roman lamps and urns, amphorae, and dishes, tiles, bricks, and pavements, and fragments of Samian ware; also, a few Egyptian antiquities; and a cabinet of Greek, Roman, and early British coins ... Every article bears a descriptive label; and the localisation of the antiquities, some of which were dug up almost on the spot, renders these relics so many medals of our metropolitan civilisation.[12]

It seems that a mere fraction of Saull's Roman ware passed to the British Museum in the shambolic situation after his death: only a couple of fragments of bowl and part of a mortarium with its spout were worthy of note.[13] These surviving artefacts are therefore of little help in understanding the wealth of his exhibits. Also, some items exhibited by Saull at the societies—a sculptured thirteenth-century female head, or remnants of a Roman lamp found in Bishopsgate[14]—evidently belonged to provincial collectors, so we do not know whether they actually featured in Aldersgate Street.

The museum was evidently rich in coins, presumably dated sequentially through the Roman occupation. But while Saull diligently took part in the management of the Numismatic Society—as scrutineer from the late 1830s, auditor from the early 1840s, and Council member in 1844 and 1851–55—he did little else beyond chair meetings and exhibit

---

11 A. Booth 1839, 15.
12 Timbs 1855, 542.
13 Walters 1908, 324, 372, 435. The British Museum also purchased a seventeenth-century earthenware vessel, called Metropolitan slip-decorated ware—a coarse quality, red-clay vessel with the inscription "feare g[od]"—found in Princes Street (Hobson 1903, 109).
14 PSA 1 (1849): 222; JBAA 9 (Apr. 1853): 75.

the odd coin.¹⁵ It is easy to imagine that his primeval-to-pastoral theory simply found no scope for play here. But the crossover between coin collecting and political propagandism was visible in 1847. When a well-wisher donated some *ancient* coins to the *Reasoner*'s thousand-shilling fund (a float to keep the serial solvent), Saull put them to good use. He added others from his own collection so that "a little historical series may be made up". The set was to be sold, and the "purchaser afforded the means of using them in the study of ancient history," with the profits ploughed back into Holyoake's flagship journal.¹⁶

A Learned Joke

Of all the learned bodies, it was the ancient Society of Antiquaries that attracted Saull's reformist attention. Founded in 1707, it remained unreformed by the 1840s and was widely derided for its dilettantism. Saull was in some ways typical of the new influx: trading, lower middle-class, self-educated, using its forum for social leverage.¹⁷ Complaints about the "apathy and inactivity" of officials dogged the Society. A coalition of earthier reformers screamed about a "negligent" Council and its disregard for the conservation of finds in an age exploding with railway and sewerage diggings. As civic institutions reformed and democratized, it remained a rotten borough run by a Tory clique. An embarrassed *Literary Gazette* in 1846 called it "a laughing-stock".¹⁸

A storm tide of reform was sweeping over intellectual society. At the courtly Zoological Society, grubby and disenfranchised working zoologists were demanding a greater electoral role. Noble trustees at the British Museum were ignominiously subjected to a Select Committee probing their competence. And even the Royal Society was starting to move from an absolute to constitutional monarchy.¹⁹

---

15   Significantly, a silver medal of the executed king, Charles I, displayed while his fellow activist Dr John Lee was in the Chair: *Proceedings of the Numismatic Society* (1851–52): 20; *Numismatic Chronicle* 15 (Apr. 1852): 104–05.
16   *UR*, 15 Sept. 1847, 83.
17   DeCoursey 1997, 137, 158. There remained a residual prejudice against the "trade" taint. Roach Smith's own fellowship was resisted on this count (Hobley 1975, 329; Hingley 2007, 175).
18   *Literary Gazette* 1527 (Apr. 1846): 381.
19   Macleod 1983; McQuat 2001, 12; Desmond 1985a, 1989, 145–51.

These structural reforms reflected the wider political changes. In the 1830s, Parliament extended the franchise and granted rights to non-Anglicans, and municipal seats in the industrial regions were increasingly snatched by candidates (often Unitarians) with Dissenting backing. At the same time, London's learned bodies were invaded by Dissenting, mercantile, and professional groups making their own liberal demands. Medical reformers often led the way or, rather, the lobby representing the new class of General Practitioners. These GPs were educated in back-street anatomy schools and tended the poorer communities; for this they were derided by the hospital consultants as a "low-born, cell-bred, selfish, servile crew".[20] Just as Cobbett's *Political Register* had blasted "Old Corruption"—the traditional privileges of the aristocratic elite—so its medical mirror, Thomas Wakley's *Lancet*, led the GPs to attack the College of Surgeons' "self-perpetuating, tyrannical council".[21] The GP's campaign for rank-and-file rights partly paid off in 1843 when the College of Surgeons was rechartered. Councillors were no longer to hold seats for life or be self-electing. A new body of 300 Fellows, including some GP leaders, now had had the power to vote councillors on and off. Something similar occurred at the College of Physicians. Here, an oligarchic Council controlled London's lucrative hospital posts. The *Lancet* excoriated the College for its commitment to "the bigoted, Tory-engendering, law-established Church"—because it only admitted Fellows who had Oxford or Cambridge degrees, that is, wealthy Anglicans.[22] Under pressure, the Physicians too started reforming in the 1840s, finally admitting Dissenters to the Fellowship. Ultimately, the "medical aristocracy" had compromised just enough to defuse the situation—as Parliament had done. But there had been no concession to an "England revolutionized", or the universal suffrage demanded by the "democratic brawlers." With the Conservative Prime Minister Robert Peel in power in the 1840s, "moderate, practical" reforms had met the minimum liberal needs.[23]

---

20 A play on Pope's line: *Medico-Chirurgical Review* 17 (1 Oct. 1832): 574.
21 *Lancet*, 25 Sept. 1830, 4. On the reforms: Waddington 1984, ch. 3; Desmond 1989, chs. 4, 6; Underhill 1993.
22 *Lancet*, 19 May 1832, 219; J. F. Clarke 1874, 7; G. N. Clark 1964–72, 2: 702–12.
23 *London Medical Gazette* 29 (15 Oct. 1841): 117–20.

This was more or less the template for learned London. Squire-run societies were purged of their worst practices. By the time Saull's reformers were agitating against the bad management at the Society of Antiquaries, its unaccountability, corruption, and disdain for research, the Royal Society's own *ancien régime* attitudes were already changing. The Royal Society, that one-time "club for peers and dilettantes",[24] made its own compromises in 1847. In 1848, the President, the Marquess of Northampton, ten years in place, resigned, as did the Secretary (P. M. Roget, later of *Thesaurus* fame, twenty-one years in office and a time-server hated by the radicals). But with the Society still swamped by unproductive peers,[25] bad feelings continued to exist, as shown by *Punch*'s joke advert for a successor:

> WANTED, A NOBLEMAN who will undertake to dispense once a month, upon rather a liberal scale, tea, lemonade, and biscuits, for a large assembly. The company is select, and he will be allowed to mix with some of the greatest men in England.—Sealed tenders ... to be sent in to the Royal Society, marked "President." No scientific or literary man need apply.[26]

However, the fight to usurp control from aloof patricians, whose allegiances were to the land and the old order, and to steer policy towards more meritocratic and scientific ends, gained strength. The leaders, charged with Toryism, cronyism, and bad management, were constitutionally restrained. Rule by patronage was watered down as a more scientifically-qualified Council was formed. Committees were set up to vet papers for publication and recommend candidates for medals. Now a seat on the Council was to be the reward for active researchers and publishers. The result, in Roy MacLeod's words, was that "loyalties to Crown and Church were replaced by new contractual allegiances", and the Society would emerge with a new "image of philosophical integrity, public utility, open competition, and efficient administration".[27]

The campaign to get specialists into office at the aristocratically top-heavy Zoological Society were more chaotic. This is not surprising given that Sir Humphry Davy in the 1820s had originally envisaged

---

24 Berman 1975, 35.
25 Moxham and Fyfe 2022, 260, 272, point out that the noblemen had published nothing, while most had only a passing interest in science and joined out of duty.
26 *Punch* 14 (18 Mar. 1848), 111.
27 MacLeod 1983, 57–58.

the Zoological Gardens as a nobleman's game park. Here, ornamental fowl and exotic imports were to be bred to tempt a gentleman's palate. It appealed to the hunting-and-fishing squires, who were unrivalled in game management—they eagerly stocked the new zoo with llamas, kangaroos, and emus from their estates. The Gardens were set in the delightful promenading Regent's Park, putting it some distance, both in geography and ideology, from the Society's museum in Leicester Square. This was the largest zoology museum in England, with 460 feet of space, housing hundreds of mammals, thousands of birds, and tens of thousands of insects. Here the zoologists would study imperial imports, dissect the exotic cadavers, hold scientific meetings, and start a publishing programme. Lectures in the museum could be radical, including some on the way species changed, a science that was abhorrent to the noble managers. These zoologists, with their merchant, military, and East India Company contacts, had very different priorities. They criticized Council autocracy and its "raree"-show superficiality. This pro-science lobby got the gentry's game-breeding farm in Kingston closed down. The museum men argued that imports should be of scientific value. They wanted snails and snakes and the oddities of the moment, like the duck-billed platypus, not the tasty, or the plumed beauties that the fowling gentry had in mind.[28] But while these Fellows found their voice, they were never allowed to introduce a fiercer democracy. Even after a decade of demands only one concession was made to the reformers: the Vice-Presidents became electable. But the President, Treasurer, and Secretary placed themselves above the democratic fray. They still steered events, backing a Tory clique which got the fiercer critics voted off the Council in the turbulent 1835 elections. Back-bench grumblings persisted through 1836 when officials were caught giving Lord Stanley the zoo's ostriches as a gift. But a resolution by frustrated reformers about the Council's "irresponsible powers" was pointless.[29] By now Peelite Tories were firmly in control and the disappointed radicals started dropping out.

Amid this reforming ferment, that convivial gentleman's club, the Society of Antiquaries, top heavy with title and adornment, was itself hit by waves of dissent. The active members here, too, ran motions of

---

28 Desmond 1985a, 223–50; Åkerberg 2001, 84–89; Wheeler 1997.
29 Desmond 1985a, 200–11.

no confidence in the complacent and often absent management. The society had fallen "into a state of inefficiency and decline", according to the irate members' resolution in 1846. That was a polite way of putting it: the *Literary Gazette*, echoing the complaints of Roach Smith, Saull, and the medical practitioner Thomas Pettigrew, called it laughable. The *Gazette* listed the litany of managerial abuses, including the officers' use of funds for their own ends. Useless and absent officials told of a time-serving decrepitude. A decline in fellowships and finances said the same. It was a textbook gerontocracy: the treasurer had been in place twenty-five years. The grave Earl of Aberdeen had been President since 1812, and had long lost his interest in "Ancient rubbish". By 1846, he had absented himself from the previous sixty-six meetings. His absenteeism was understandable, given that he had held a Cabinet seat in every Tory administration since Wellington's government of 1828. Like so many aristocrats, he had assumed this figurehead position in the Society as part of his public "duty". He was still the Foreign Secretary in 1846 when he was finally, ignominiously, forced to resign his Antiquaries chair in the face of the clamour for an "efficient president", one who would see the post as an honour and attend its affairs.[30]

The incoming Vice Presidents included Samuel Wilberforce, the new Lord Bishop of Oxford. Though an energetic diocesan reformer and a paternalist who hated the evils of industrial society, he was a High Tory, and opposed to liberalism in all its forms, whether in church or science. He was just acquiring the soubriquet "Soapy Sam", a nickname the Darwinians would later hang on him with a vengeance. Another incoming Vice President was Sir Robert Inglis, a staunch, old fashioned High Tory who had a traditional view of the way society should be ordered, from the top down. He passionately defended Anglican privileges, resulting in Wakley slating him as a "sleek, oily, capon-lined man of God".[31] The hatred was mutual. The diehard Inglis led a rearguard action against Dissenting demands: he had resisted Parliamentary Reform, Catholic emancipation, the repeal of the Test and Corporation Acts, Church reform, and the Dissenters' call for the civil registration of marriages (until then a Church monopoly), denouncing it as the greatest attempt

---

30 *Literary Gazette* 1527 (Apr. 1846): 381; *PSA* 1 (1849): 129; "Ancient": Hingley 2007, 179.
31 *Lancet*, 27 Feb. 1841, 803.

to "to secularize the sacraments" since the Civil War.[32] Such patricians running the Antiquaries were hardly seen as improvements.

Inglis came with a reactionary track record. He had already fought to resist a leadership role for expertise at the British Museum. Here the Trustees were noblemen led by the Archbishop of Canterbury, who felt that wealth and rank qualified them to hold the nation's heritage in trust. Museums, for them, were to display ornaments and store treasures, not necessarily advance knowledge.[33] Titled officers could solicit patronage, being on 'hail fellow, well met' terms with government ministers, in a way impossible for the menial 'expert'. Just as at the Antiquaries, breeding was seen as a better qualification than researching. Against this closed world of hereditary privilege and Church sinecures, reformist groups of specialists, academics, medical radicals, and Dissenting teachers with their industrial backers were arguing for expertise on the Board, claiming that it was in the national interest. They were offering a counter-vision of a mobile, competitive, scientific society. In the 1835 Select Committee hearings on the British Museum, they argued that their lordships had neither the inclination nor competence to promote such goals.[34] But Inglis, representing the Trustees, refused to admit scientific "commoners". Such experts would be accountable to the new professional classes. They would bow, not to rank and wealth, but to talent and competition, and hold a meritocratic brief inimical to the hereditary principle. Inglis defended the track record of the Anglican Trustees, and their competence to run the national institution. He recognized that paid specialists with a meritocratic agenda posed a direct threat to Church-and-Crown authority. One fossilist (with a museum rivalling Saull's), J. S. Bowerbank, had the temerity to suggest that the British Museum should hire paid collectors, experts in evaluation, who could barter for exhibits. Such a "trading" taint was obnoxious to their lordships, who claimed it would degrade the museum.[35] The upshot

---

32 *Hansard* 1836, 32: 162; 1836, 34: 491; Hilton 2006, 382, 390, 431–32.
33 Gunther 1980, 75; Desmond 1989, 145.
34 *Hansard* 1836, 31: 308–12; *Report from the Select Committee on the Condition, Management and Affairs of the British Museum*, 1835, House of Commons Parliamentary Papers, 22, 27, 29, 30–31.
35 *Report from the Select Committee on British Museum*, 1836, House of Commons Parliamentary Papers, vi–vii, 73, 78–79, 118, 130–33; Gunther 1978, 84–85, 94–99; Desmond 1989, 145–51.

was that the landed interest quashed all idea of a 'specialist' board and left the stewardship safely in ennobled hands.

Inglis's taking a leading role at the Society of Antiquaries did not bode well for the reformers. Talk of the "great Rebellion", as Inglis slated the Civil War during the hearings, explains another radical exasperation. Among traditionalists, remembrance of the Civil War's most horrifying atrocity was still observed. In the loyalist calendar, 30 January was marked in black as the anniversary of Charles I's beheading. In bygone years, Tories would indulge their "superstitious veneration" by draping their rooms in black and fasting.[36] An anniversary sermon would be preached in Westminster Abbey, where the attendance waxed and waned according to the reactionary or reformist clamour of the age.[37] Fasting to expiate the country's sin was intended to keep alive a "sense of national guilt". It also initially served to vilify radicalism by pointing to its murderous consequences, but, increasingly, the regicide was being interpreted not as a political act but one of aberration in order to obscure its real cause.[38] At the royalist Antiquaries, the "Anniversary of the Martyrdom" was marked yearly by a ban on meetings. But bans, fasts, and prayers on 30 January remained a trigger for political opposition.

When the Tories suspended the Antiquaries on 30 January 1845, in observance of "the Fast of the death of King Charles I",[39] the radicals reacted angrily. Had not the republican Saull once reminded the reigning monarch of Charles's fate, outraging the *Times*?[40] To Saull, it was not martyrdom, it was royally deserved. He was far from averse to anniversary celebrations; ironically Tom Paine's birthday was a day earlier, 29 January, and Saull happily celebrated that, just as he did the French Revolution.[41] But he baulked at these loyalist observances. Another angry at this cancellation was Dr John Lee. He was a well-to-do 'advanced liberal', whose "weaknesses were very harmless", said the *Gentleman's Magazine* dismissively: teetotalism, women's suffrage,

---

36   Lord John Russell 1853, 3–4.
37   Emsley 2014, 54.
38   Vallance 2016.
39   *Proceedings of the Society of Antiquaries* 1 (1849): 70, 75–76.
40   *TS*, 25 Dec. 1834, 2; 26 Dec. 1834, 2; 1 Jan. 1835, 4; *MC*, 25 Dec. 1834; *Times*, 25 Dec. 1834, 2; 26 Dec. 1834, 2.
41   *UR*, 27 Jan. 1847, 18; *Reasoner* 2 (3 Feb. 1847): 60; *NS*, 6 Feb. 1847. As Epstein (1994, 152) says, radical anniversaries were counter-statements to these loyalist observances.

anti-smoking. Like Saull's, his principled stands were written off as eccentricities to reduce their import. Indeed, Lee's tea-drinking "Peace, Temperance, and Universal Brotherhood Festivals" on his grounds had all the hallmarks of Owenite festivals and were equally put down to his "peculiar views". Lee sat with Saull in every society—Astronomical, Geological, Archaeological, Chronological, and Numismatic—and now worked with him in the Antiquaries. He was another museum owner and fossilist, one whose catalogues survive, four volumes of them, covering the gamut, from Eastern antiquities to stuffed animals. (From these we get a glimpse of Saull's place in the exchange network. At least fourteen of Saull's fossils, duplicates possibly—from *Iguanodon* vertebrae to pecten shells, shale ferns to sponges—turn up in Lee's collection.[42]) It was Lee who introduced the Antiquaries motion that no 30 January suspension should take place again. There was nothing in the bye-laws to warrant it. The motion was backed by Saull but to what avail in a royalist stronghold is not known.[43]

The patrician council had run the Society of Antiquaries as their fiefdom. The managers were a self-electing "clique". Worst, for some reformers, was the Director, Albert Way. He was Wilberforce's friend—their families were close and they had been educated together.[44] Way upheld the gentlemanly proprieties and had himself just married into the peerage, wedding Lord Stanley of Alderney's daughter. While Way would be a future archaeologist of note, he was, in Pettigrew's words, resistant to change and so "unpopular among the active members" that it was "desirable to get rid of him".[45] But caution is needed in taking Pettigrew's statements at face value. He could only have been a fair-weather friend for republicans Saull and Lee, because Pettigrew juggled the need for royal patronage with that of sound management.

---

42   Delair 1985. *GM* ns 1 (Apr. 1866): 592–93; "peculiar": *JBAA* 23 (1867): 301.
     Lee was the leading light and first President of the new Numismatic Society, a breakaway from the Antiquaries for specialist ends. On exchange networks and the redistribution of specimens (in another context), see Cornish and Driver 2020; and Heumann, MacKinney, and Buschmann 2022 on the changing concept of "duplicates".
43   *Proceedings of the Society of Antiquaries* 1 (1849): 75–76.
44   As a Cambridge student, Way had befriended Charles Darwin and introduced him to the beetle collecting fad: Burkhardt et al. 1: 58–59, 91; Ashwell 1880–83, 1: 4–6.
45   *Literary Gazette* 1578 (Apr. 1847): 301; also 1572 (Mar. 1847): 196; 1527 (Apr. 1846): 381–82.

He had been surgeon to the Duke of Sussex (the King's brother) and the bibliographer of His Royal Highness's library at Kensington Palace. Pettigrew had also been the Duke's campaign manager for the Royal Society Presidency in 1830 (in opposition to the doyen of physics, John Herschel, part of the "prouder aristocracy of science"[46]), which gained Pettigrew enemies. But he believed that royal "rank would place [the Duke] beyond the operation of any jealousies."[47] Having no problem with royal office, Pettigrew proved himself very unlike Saull. A prickly nature and personal animosities (Pettigrew clearly hated Way) added cross-currents to this politicking. Fair weather ally or not, Pettigrew's relentless debunking in his new book *On Superstitions* (1844) would have been applauded by Saull.[48] Its onslaught on miraculous medical cures, whether from talismans or by tapping the divine through saintly shrine, sat comfortably with Saull's attack on superstition.

Anyway, Pettigrew slated Albert Way as dictatorial, and as contemptuous of those with real "archaeological learning", as shown in Way's mocking of Roach Smith as "this *Liver-puddle Roach!*"[49] Way diplomatically resigned in 1846. The Antiquaries membership continued to plummet, the quality of the papers dropped so as to become a "discredit" to the society, and publications began to run late. Any respect for it was draining away. Not mincing its words, the *Literary Gazette* in 1847 called it "so long a useless (and even worse than useless) body". It did, however, add that it "appears to be on the eve of a revolution for the better".[50] Reforms were expected.

But the Antiquaries were not to be rushed into a "revolution". They did start limiting terms and rotating officers (something demanded but not yet achieved at the Royal Society[51]), so that incumbents could no longer

---

46 Babbage 1832, 381.
47 Pettigrew 1840, 26–27.
48 Pettigrew 1844. As would Pettigrew's exposure of the "horrible" treatment in workhouses of waifs and strays, whom he found malnourished and "rickety", and he publicly complained to Lord John Russell about it (Rosenblatt 1918, 49). A surviving letter shows that Saull was discussing Roman roads with Pettigrew: W. D. Saull to T. J. Pettigrew, 9 Aug. 1852, Beinecke Rare Book and Manuscript Library, Yale University.
49 *Literary Gazette* 1527 (Apr. 1846): 381.
50 *Literary Gazette* 1578 (Apr. 1847): 301.
51 MacLeod 1983, 72. Reformers had demanded a triennial Presidency at the Royal Society, as radicals had demanded triennial parliaments in the country to increase responsiveness.

become entrenched for a quarter of a century. Yet, many noblemen still thought themselves qualified for posts by a pedigree that itself stretched back to the Middle Ages. In 1847, Pettigrew and Saull managed to carry a vote (by a sliver) to send back the new President Lord Mahon's list of nominees for election "for re-consideration", requesting he add those "most active" in the field to the list. But ancestry still bested activity in a Councillor's qualifications.[52]

These minimal compromises, as at the other societies, stopped well short of radical demands. In 1852, Pettigrew, Lee, Saull, and Roach Smith were *still* complaining about the "bad management", meaning the failure to save the antiquities thrown up by London's reconstruction boom and the railway excavations, many of which were destroyed. The middle-class press supported the reformers, echoing grumbles about the Society still being "very ill-managed".[53] And, while the Royal Society in 1847 had (under duress) restricted the number of yearly fellowship entrants and made their admission tougher, to increase exclusivity and raise its scientific prestige, the Antiquaries in 1852 took the opposite tack. To battle the draining membership, the Council halved admission fees. Reformers by now could see the Royal's stock rising again, and argued that the Antiquaries' laxity would tarnish "the character and respectability" of the society. It would open the floodgates rather than restrict the body to dedicated specialists.[54] Pettigrew, Lee, and Saull tried to stall the move but were outvoted. One might have imagined that the radical Saull, who earlier campaigned to have institutions opened up, would have favoured fee reductions. But no, more and more the antiquarian specialist, the Owenite too was now placing meritocracy over democracy.

A Corner of England Revolutionized

In the mapping of progress, images of "archaic" time ... were systematically evoked to identify what was historically new about industrial modernity.

---

52   PSA 1 (1849): 189. They also started a museum, belatedly.
53   *London Weekly Paper and Organ of the Middle Classes*, 5 June 1852, 59. *Conserving* these antiquities was the *sine qua non* of the breakaway British Archaeological Association: *JBAA* 1 (1846), ii.
54   PSA 2 (1853): 258. There followed a spike in fellowship figures in 1852–55, before numbers fell again. MacLeod 1983, 72–74.

The middle class Victorian fixation, with origins, with genesis narratives, with archaeology, skulls, skeletons and fossils—the imperial bric-a-brac of the archaic—was replete with the fetishistic compulsion to collect and exhibit that shaped the *musee imaginaire* of middle class empiricism. The museum—as the modern fetish-house of the archaic—became the exemplary institution for embodying the Victorian narrative of progress.[55]

Saull never totally deserted the Society of Antiquaries. He would continue to talk there on his favourite themes: the progression from British to Roman settlements in Dunstable, the ancient track-ways which became Roman roads; and on ancient Cornish hill-forts betraying the presence of Mediterranean tin traders. And he acted as a conduit for visitors, for instance, introducing the Middle East explorer Major Charles Ker Macdonald's exhibits from Arabia, Palestine, and Egypt, and later passing on translations of runic inscriptions found on a sculptured slab in St Paul's Churchyard.[56]

But his real home now was a new organization, a splinter society forged partly in response to the Antiquaries' intransigence. This was the British Archaeological Association, founded in 1844 by Roach Smith and others with a sympathetic Saull in tow. Its research and preservation agenda made it congenial to the museum owner. Meetings here were more lively, many of them held in Pettigrew's house. Expertise was to be valued and rewarded, even if Antiquaries stalwarts pooh-poohed the upstart Association with its 'specialists' as a fad, a product of "mere fashion".[57] But what totally appealed to Saull was its revolutionary governance. Even the constitutional monarch had been deposed and a democracy established. The President, Vice-Presidents, and the officers were all subject to *annual* election, and every guinea-subscribing 'Associate' had a vote by ballot.[58] If not England revolutionized, then certainly this corner of archaeology had been. It was everything the Tory press feared: "annual elections, annual canvassings, annual ballotings,

---

55   McClintock 1995, 40.
56   *Proceedings of the Society of Antiquaries* 1 (1849): 177, 235; 2 (1853): 91–92, 285, 289; *GM* 39 (Feb. 1853): 186–87. On this runic inscription see also *Proceedings of the Royal Irish Academy* 5 (1853): 351–54. He did present his *Essay on the Connexion Between Astronomical and Geological Phenomena* to the Society in 1853, and the same year he invited members to join him in examining the Castle of Berkhampstead: *PSA* 3 (1856): 42, 99. It showed that ties did remain with the Society.
57   *London Weekly Paper And Organ of the Middle Classes*, 5 June 1852, 59.
58   *JBAA* 2 (1847): 110.

and universal suffrage"—a "monstrous scheme for the right government of a peaceable and scientific profession"![59] Saull was right at home.

The preservation agenda was a priority for Saull and Roach Smith. Of all the influx into the societies, it was seemingly these hands-on men of trade who valued Britain's material heritage most. They were the ones at the forefront of London's salvage archaeology.[60] Preservation was a time-consuming and occasionally soul-destroying job. The odds were often against the survival of fragile objects, given the state of preservational techniques. For instance, by the time Roach Smith and Saull were alerted to an ancient galley raised from the bed of the River Itchen and had applied to the Mayor of Southampton to conserve it, the boat had already crumbled away, leaving nothing but a keel and few timbers.[61] With the Antiquaries uninterested in preservation, and no civic help, and given "the apathy of the government", in contrast to France's mission to preserve "national antiquities", it was often Saull and Roach Smith who had to set up voluntary funding schemes to help protect monuments.[62]

At the fortnightly British Archaeological Association meetings Saull could be seen discussing familiar themes: the City's Roman wall, ancient barrows, and the state of River Thames when aboriginals fished its banks.[63] It was the same at its yearly Congresses, where his talks tracked the social stages through the archaeological strata. At Worcester (1848), he discussed a Roman camp at Malvern; at Manchester (1850), his talk was slated to be on Celtic Cornish antiquities and war chariots (which he doubted that ancient Britons used); and, at Newark (1852), he spoke on the Roman road from Winchester to Old Sarum, and the ancient Celtic earth-works next to it.[64] All were fitted neatly into his over-arching picture of civilizational progression, signals of the stages of "archaic" time to point up London's commercial modernity.

---

59 As the *London Medical Gazette* (29 [15 Oct. 1841]: 119) said of similar institutional demands in the medical colleges.
60 Levine 1986, 13; Hobley 1975.
61 *JBAA* 4 (1849): 382–83.
62 C. R. Smith 1854, 3: Appendix "Antiquarian Excavations on the Site of the Roman Station at Lymne, in Kent".
63 *Literary Gazette* 1622 (Feb. 1848): 138; 1631 (Apr. 1848): 281; *JBAA* 9 (1854): 75.
64 *Berrow's Worcester Journal*, 24 Aug. 1848; *Literary Gazette* 1754 (Aug. 1850): 639; 1758 (Sept. 1850): 710; *GM* 33 (Apr. 1850): 415; 38 (Oct. 1852): 404; *Morning Post*, 26 Aug. 1850, 1; *Times*, 23 Aug. 1852, 8; *JBAA* 8 (1853): 263.

Ancient and Modern Aborigines

Saull's historic phases of social development, pinned to archaeological sites rather than archaic superstitions, might have been expected to appeal to the new Ethnological Society of London. Particularly so, because its founder, the Quaker physician philanthropist and driving force behind the Aborigines' Protection Society, Thomas Hodgkin, had argued specifically in an inaugural address for a study of tumuli.

Hodgkin's Christian understanding of tumuli was very different from Saull's. Hodgkin envisaged a traditional biblical timeframe, which made these earth works the earliest visible remains of Britain's Adamic line. For him, all human types were descended from Adam and Eve, or, more recently, from Noah's descendants.[65] This, as Hodgkin argued, put the onus on ethnologists to focus on the adaptations of these descendants to their local regions—that is, to take, like Saull, an environmental approach to divergence and difference.[66] The two men, despite their religious disparity, were converging in practice. For Hodgkin, the tumuli's creators could not have been very distinct from "present families". Again, like Saull, he saw Britain's "barbarous inhabitants" as similar to today's "uncivilised races". And, as part of his programme to prove the biblical unity of mankind, he suggested that archaeology should look more like ethnology: it should specialize by following the gradations back, say, from Romanized Britons to uncivilized Celts.[67] This was a biblical mirror to Saull's programme, with its rise from "rude" aborigine to civilized Romano-Briton.

In truth, Saull joined the Ethnological Society late, possibly because he had no sympathy with the Christianizing aspect of Quaker philanthropy. But he did round up his aboriginal work here with a paper on 15 March 1848, "Observations on the Aboriginal Tribes of Britain".[68] And yet, despite seeming an obvious placing, the paper was actually ill-suited.

The Ethnologicals were an outgrowth in 1843 of the evangelical Aborigines' Protection Society. This had sought to protect native

---

65 Kenny 2007, 370; Driver 2001, 45.
66 Stocking 1971, 373.
67 Hodgkin 1848, 36–39.
68 *Morning Post*, 13 Mar. 1848, 6.

peoples under Victoria's care, and with protection went a desire to promote "the advancement of uncivilised tribes", which meant saving by Christianizing and civilizing.[69] Even the growing racial warrior, Robert Knox saw it as a case of the wolf taking care of the sheep.[70] A prim evangelicalism and high moral tone invited *Punch*'s cynical claim in 1844 that for these society types "distance is essential to love". Unlike Saull, who had spent his life campaigning in courts and dives for jailed dissidents, the poor, and workhouse indigents, "They have no taste for the destitution of the alley ... but how they glow ... at the misery somewhere in Africa".[71] The offshoot Ethnological Society stood even more distant. It had a harder-edged scientific approach and a smaller share of humanitarianism. As society's xenophobia grew, there was less interest in civilizing savages and more in separating them into ranks. From the Christian heights of Victorian London, the Ethnologicals would use imperial collation from military, merchant, and missionary sources to point up "the distinguishing characteristics" of the human varieties.[72]

Predictably, these well-to-do fellows showed no sympathy for investigating their own ignominious 'aboriginal' roots. And Hodgkin himself had no truck with the idea of "Autochthoni", or aboriginals created in the regions where they are found, truly 'indigenous' peoples, because, to him, they were all Noah's descendants. So Saull's defining his "'Aborigines' [as the] first inhabitants of this or indeed of any other country"[73] might have looked *prima facie* anti-Adamic, even without knowing his monkey-ancestry motive or Bible-exploding timeframe.

Not that it mattered, for the Ethnologicals carried out none of Hodgkin's 'archaeological' plan. They spent their time distinguishing modern ethnic groups, while looking for common linguistic features to trace language separation back to the Tower of Babel. And, by placing emphasis on "national characteristics and cultural groupings", they only

---

69 Driver 2001, 76; Stocking 1971, 369–72; Laidlaw 2007, 138–41; Brantlinger 2003, 3, 35–36, 71; Rainger 1980, 709–10; Kass and Kass 1988, 268–69.
70 Knox's "Lectures on the Races of Men" in *Medical Times*, 8 July 1848, 147.
71 *Punch* 6 (18 May 1844), 240.
72 Kass and Kass 1988, 394–95; Stocking 1987, 243; Rainger 1980, 710–13; Lorimer 1978, 134ff.
73 Saull 1845, 1; Hodgkin 1848, 30; Kass and Kass 1988, 268, 395.

served to strengthen the growing racial awareness of the age.[74] In reality, the *Journal of the Ethnological Society* published no historical papers in its early years. Saull's aboriginal piece in 1848 was itself excluded from the *Journal*, because it was archaeological in fact and historical in scope. And that was irrespective of any ideological stigma, for his aborigine work was driven by an 'evolutionary' and materialist heuristic rather than a Christian Adamic one.

Those who pushed the human story further back received an equally muted reception. The Abbeville antiquarian Boucher de Perthes's discoveries of worked flint, bone knives, and arrows near fossils of mammoths and rhinos led him to propose that ancient humans had lived alongside these extinct giants, indeed that they were butchering them. But when he said as much in the first volume of his *Antiquités Celtiques et Antédiluviennes* (1847), it was dismissed by French geologists as fanciful. It has often been stated that his work made as little impression in Britain.[75] However, he did send his book to Saull's British Archaeological Association in 1849, along with some flint weapons, and their journal reviewed it favourably.[76] They pointed out that these knives and arrow heads were identical to ones found in Celtic tumuli, adding weight to his story. Some geologists at least were also receptive, Mantell particularly, and if he was, undoubtedly Saull was too.[77] It may be significant that Boucher de Perthes was shortly to visit Saull's museum to look over his own Celtic axes.[78]

Given the apathy among the Ethnologicals, Saull had to privately print his rejected sixteen-page aboriginal paper in 1848.[79] It took his bold scheme to its definitive point. Saull now presented a developmental sequence of housing and tooling broken into five theoretical periods, all

---

74   Kenny 2007, 370; Lorimer 1978, 135; Rainger 1980, 703.
75   Stocking 1987, 71; Gamble and Moutsiou 2011, 46; Grayson 1983, 122–31, 172; J. Evans 1949.
76   *JBAA* 5 (1850): 166, 171–72.
77   Mantell 1850. Saull provided John Evans a testimonial for the Society of Antiquaries (MacGregor 2008, 36), from which we might assume that Evans also visited the museum and saw its fossils and flints. Evans went on to study tool use among early humans, contemporaries of the extinct megafauna. His arms-trade links and understanding of gun-flint knapping lead him to become the first, experimental, stone-tool flint knapper (Bulstrode 2016).
78   He visited on 18 September 1851: Perthes 1864.
79   Saull 1848.

pre-Roman. The first was typified by the nomadic or "rude" 'huts' found on the Yorkshire moors and near Whitby and Dunstable. It comprised simple dwellings which were mere depressions in the earth, eight feet round or oblong, with turfed lips which presumably supported branches for walls, and a gap for a door. Evidence of charring inside showed where fires had been. The next, or more "improved" phase, saw uncut stone edges to the depressions and nearby tumuli for interment. Flint knives and hatchets made an appearance. The great stone cromlechs were of this sort, constructed for shelter, not as temples by the Druids. The third period, illustrated by the tumuli of Yorkshire and Wiltshire, was characterized by a new missile technology, arrow heads and spear points, while advances in "civilization" were evident in cist entombing of the dead in foetal positions in barrows, or by cremation, with the burnt human remains placed in unbaked clay urns. Later came the fourth, "pastoral" or settled stage, when wild cattle were corralled in hilltop forts, often of many acres, locally called "Caesar's camps", especially on the Downs of the South East. Finer utensils were now used, including sewing pins carved from antlers; and boar or deer skulls were sometimes placed with the dead.[80] The last period was marked by the arrival of the "Teutons" (the Teutonic, or, as it was already being called in Denmark, the "Iron Age").[81] It began some centuries before the Roman conquest, and, in Saull's, view signalled trade with the more advanced Gauls and Belgae from the Continent. From them, Britons adopted armour and tin and copper coins, sometimes showing crude horse images. Larger hill camps were developed, often near the coast, at Folkestone, Winchester, and Dorchester, where the tumuli contained great ornamented urns as well as cooking utensils and personal adornments.

Saull's diffusionist progression, which saw more advanced tribes sweep in by turn to raise the national stock, moved broadly from Celt to Teuton. This was also the standard progressive sequence portrayed by racial phrenologists. But their transition was conceived differently. For phrenologists, largely fixed mental characters limited the capabilities of the 'lower' races. The Celtic savage, being far beneath the Teuton in capability, was destined to stagnate or die because of his organic

---

80 While the phases were not necessarily criticised by archaeologists, the sites attributed to them sometimes later were, for example, Walford 1883, 2: 494.
81 Rowley-Conwy 2007; Stocking 1987, 72–73.

inferiority.[82] By contrast, Saull saw each invasion offer new scope for improvement to the indigenous tribes. It altered the cultural landscape and encouraged growth. He never doubted the disparity, just as there was between the unschooled wage-slaves and Classics-educated gentry of his own day. But he seems to have envisaged the incursions in a singularly Owenite way, in de-militarized, educational terms. It was more "rational" to see foreigners arriving to trade or mine tin; "in time" they would have "engaged and instructed some of the native inhabitants to assist them", thus passing on esoteric lore and skills and raising them to the same level.[83] Such were also the benefits of the Roman invasion. Saull then completed the sequence in 1851, at the Ipswich meeting of the BAAS, by looking to the final phase, the arrival of the Saxons.[84]

Saull had turned Davy's dream into an archaeological scheme, substantiated by site evidence. But it would be a generation before Darwin's neighbour, the banker and anthropologist Sir John Lubbock, and his father-in-law, the Grenadier Guard and weapons expert Augustus Lane Fox, made this sort of "artefact-based 'philosophy of progress'" acceptable—slotting tools and settlements into chronological sequence—with 'stone age' aboriginal Tasmanians being considered the surviving relics of the oldest period.[85]

However flat Saull's paper fell among the Ethnologicals,[86] the hinterland was another matter. The *Cornwall Royal Gazette* quoted sections with provincial pride, those concerning the first Mediterranean

---

82 Combe 1839, 269–91. On the role of the "Teuton" in growing racial typology, see Horsman 1976, 398–405.
83 *Cornwall Royal Gazette*, 14 July 1848.
84 *Journal of the Ethnological Society of London* 3 (1854): 51. He also read a paper at the Ethnological Society on 17 March 1852 on Saxon ethnology (*Morning Post*, 15 Mar. 1852, 5), which was refereed by Hodgkin (information courtesy of Mrs Beverley Emery at the Royal Archaeological Institute). Saull was on the Ethnological's Council from 1850 until his death, but this was the only paper he submitted during this time. Saull would also argue his chronological sequence of pre-Roman culture at the 1854 BAAS meeting in Liverpool, when discussing Celtic worked flints (*Athenaeum*, 1406 (7 Oct. 1854): 1211: I thank Mick Cooper for first alerting me to this source). By this time (1853–54), Saull was on the Committee of the BAAS's newly established Ethnology and Geography section (*Daily News*, 9 Sept. 1853, 5); *GM* 42 (Dec. 1854): 602; *Journal of the Ethnological Society of London* 4 (1856): 138.
85 Gamble and Moutsiou 2011, 201; Barton 2022.
86 It was widely ignored even in friendly circles. *The Reasoner* 5 (14 June 1848): 44, merely noted its title.

tin and lead miners in the region, traders who had pushed Cornwall ahead of the rest of the country.[87] But the moral of a "superior race of people" pulling the natives up had still greater impact further afield—at the farthest transportable reaches of empire.

Reports of Saull's "interesting" talk were picked up by the *Sydney Morning Herald*. Australia was a dumping ground for penal rejects: mostly poor workers, some convicted of trivial offences. Interest was inevitable here, as the colonists and ticket-of-leave men encroached on the local 'aborigines'. Widely considered "blood–thirsty savages", these indigenous peoples were anything but, said the explorer and magistrate in Murray District, Edward Eyre. He had lived among them for years, and found them only "shy, alarmed, and suspicious" on first contact. While even Eyre talked of the "many brutalising habits that pollute [the aboriginal's] character", the natives still had, in his Anglo-centric view, "an aptness for acquiring instruction" and "the capacity for appreciating the rational enjoyments of life."

This potential squared with Saull's understanding of his aboriginal Britons. To Eyre they were the "poor untutored children of impulse" who needed a hand up. In the racial ranking images of the day, these "children" were placed at the base of the human scale, a "little above the ... brute creation", in Eyre's words.[88] The nomadic Australian was perceived from on high as a throwback, a relict from the infancy of human existence, and the survivor of Saull's first civilizational stage. That aboriginals should have persisted here was not thought so strange by those in the northern hemisphere. Here the Anglo-Saxon was believed to have shot ahead in terms of art, science, and manners, a view reinforced by a faith in the northern superiority of all life-forms.[89]

Jaws of 'marsupials' had been exhumed from Jurassic rocks in Oxford, and had been matched to the newly-discovered Australian numbat.[90] In Britain these ancient marsupials had been replaced by 'higher' placentals.

---

87 *Cornwall Royal Gazette*, 14 July 1848.
88 Eyre 1845, 2: 148, 155–56. This 'childhood' image would become entrenched at the Anthropological Society of London (founded 1863), where C. S. Wake stacked the races by analogy with human growth stages, "infancy, childhood, youth, and manhood"—Australians on the bottom, their development arrested at an infantile stage, up to the 'adult' Europeans (*Journal of the Anthropological Society* 6 (1868) 168; Lorimer 1978, 148).
89 Nelson 1978, 299; Desmond 1982, 103.
90 Desmond 1989, 314.

In the same way, the stone-wielding 'savage' of Saull's phase 1 had been raised in grade by the Celts, Teutons, Romans, and Saxons to become modern mercantile man. Just as there was a contemporary emigration south to Australia (whose population trebled in the 1850s), so in Jurassic times the marsupial colonists had arrived at the palaeontological penal-colony, where they were cut off and stagnated. The same had happened to Saull's stone-using peoples. Waves of progress in the mother country had obliterated this primeval state, but the original primitives making it to the Antipodean backwaters had retained their "rude" condition.

The *Sydney Morning Herald*, the only daily in Australia, was intrigued enough by Saull's speech on Britain's own aboriginal origin to run an 1100-word précis. It exemplified the five stages of civilization, and listed the towns in the mother country where the barrow or tumuli evidence was found.[91] Saull, never a prophet in his own land, was having to look to the penal colonies to gain a hearing.

Skulls

Aboriginal relics, British and imperial, were now finding their place in the museum's lower room. This 'mankind gallery' was filling up, and with the exhibits bearing descriptive labels, and most coming from under Londoners' feet, they were seen as so many relics on the way-stage of "our metropolitan civilisation".[92] Just as provincial and colonial museums privileged local finds, each curator having to "cut his coat according to his cloth",[93] so did Saull, with his metropolitan display running from London Clay crocodiles to an aboriginal skull from Cheapside. Ethnological specimens now merged with antiquities, and, as always, without Saull's voice, it was difficult for many to see the order. The Chartist *Northern Star* got behind the scenes and showed a different side from the genteel tourist guides. It painted a ramshackle picture of typical imperial booty—including a severed head, noted without a hint of surprise or horror, as if to confirm the chasm separating 'it' from

---

91 *Sydney Morning Herald*, 24 Oct. 1848, p. 3.
92 Timbs 1855, 542. That the museum was now well known is indicated by *Sharpe's London Magazine of Entertainment and Instruction* 6 (Jan. 1855): 267.
93 Sheets-Pyenson 1988, 122.

'us'—the lot emphasizing the archaicness of foreign cultures, in time and space.[94] Saull's lower gallery displayed

> two and three pointed spears, made from fish bones, as used by the natives of the South Seas, New Zealand, &c., with a number of rude weapons, dresses, &c., said to be used and worn by the natives of the said clime. Here also may be seen the head of an Indian chief, tattoed, with the hair in its natural state, in an excellent state of preservation; also an Indian canoe and paddles, brought over by the late Captain Cook. Here are also a number of Roman Coins, Skulls, &c, found in the centre of this Great Metropolis, also pieces of fine Roman pavement, found in London, under Allhallows Church during its repairs, as lately as 1843.[95]

But these were never really of interest to Feargus O'Connor's "pestilent publication"[96], as we will see in the next chapter.

The Cheapside skull, found amid the remains of primitive hut-dwellings, with its 'savage' features showing a low moral character, provided Saull's baseline for the rise of mercantile man.[97] His archaic sequence, which would allow the London visitor to re-assess his historical place and future prospects, was now complete: the "connected" fossil series on one floor pointed up to the human archaeological stages on the other. In the racist phrenological parlance of the age, the ancient Londoners were equated with tattooed Maoris and Caribs, the stunted "children" of the living world. This completed the empirical series to show museum visitors how life had risen over unbiblical aeons.

Saull had begun this programme with his simian-hypothesis lectures to the Owenites. With radical euphoria during the Reform Bill optimism, he had promised better things to come, "human perfectibility, and the splendid prospects which are now opening to posterity".[98] His "synoptic series of phases of mental progress" had now established "the principle of the gradual but slow advances of mankind in intellect", thus proving that man was a "progressive being", whatever the current "impediments."[99] Belief in nature's perfectibility remained strong among the dispossessed in Saull's audience. Street poets caught the

---

94 McClintock 1995, 40.
95 *NS*, 31 Oct. 1846, 3.
96 *The Age*, 28 Aug. 1842, 4.
97 *New Monthly Magazine and Literary Journal*, 46 (Feb. 1836): 270.
98 *Crisis* 3 (5 Oct. 1833): 36.
99 Saull 1845, 61.

optimistic Owenite flavour. Saull's hands-on arena inspired secular poems, glorying in a future predicated on the past, and revelling in this direct contact with evolutionary reality:

> Ye that would drink at learning's purest springs,
> Forget your books awhile, and study things;—
> See nature's volumes round you fair outspread,
> Cull'd from her library, too little read;—
> Each line from human pen may err or cheat,
> In her's alone, there cannot be deceit;
> The records of weak man, her youngest born,
> Which he calls truth divine, she laughs to scorn;—
> And points in triumph to each pictured page,
> Replete with monuments of countless age,
> That o'er this quick revolving earth had roll'd,
> Ere ought had come to light of human mould;
> Time was, she seems to say, when thou were not;
> Time will be,—when thy name shall be forgot.
> Though loftier minds, shall surely hold thy place,
> Brightening the features of a nobler race!—
> My bosom deathless,—teeming as tis vast,—
> Shews each new birth more glorious than the last.

These verses were penned in Aldersgate Street museum by an unknown bard, fired by the story Saull told of the fossils. They were equally inspiring at the graveside, as Saull recited them in his funeral eulogy of his old blasphemy partner Gale Jones.[100]

The museum's reach now extended far beyond radical poets. Students, tourists, and the learned elite joined the workers, making it one of the most visited private museums in the capital. With the acclaim, Saull's stock rose. Ten days after the last Owenite Congress ended in chaos, Saull left the old immoral world to join the new learned world.

At a grand meeting of all the scientific clans, gathered under the Lord Mayor's aegis in July 1846, Saull took his place. This once-in-a-lifetime congregation at the Egyptian Hall proclaimed not only the growing importance of science, engineering, and literature but of London as their hub. It shrieked of the city's world status. Here were the new men of literature, science, and art, a veritable Burke's Peerage of the intellectual nobility, cheered by the populace as they entered the

---

100 NMW 8 (12 Sept. 1840): 175; Saull 1838a.

hall. In strolled celebrity leaders of the learned societies and university professors, mingling with the great engineers, architects, military men and manufacturers, sculptors and artists, physicians and explorers, and a hundred others, everyone a 'somebody', with the press scrambling to name them all.

Never, said the *Standard*, had so many "individuals of high literary and scientific attainments" assembled under one roof. The *recherché* banquet was London's act of cultural self-assertion, and no one knew why it had not been done before. Real nobility be damned; here was Saull among the "prouder aristocracy" of intellect, as Charles Babbage had it.[101] Spotted by the hacks, Saull was name-checked as the "Proprietor of the Geological Museum in the City".[102] It left no doubt that the museum had raised his social profile in a way that wine wholesaling never could.

---

101 Babbage 1832, 381. Babbage was himself there.
102 *Standard*, 11 July 1846.

24. Museum and Pantheon for the Masses

What Saull's museum held at this point has to be pieced together serendipitously. Its original core was one of the important early nineteenth-century fossil shell cabinets (Sowerby's), yet we know little of it. Mantell's museum in Brighton, or Roach Smith's in Finsbury, are better known partly because they had printed catalogues. Compilation lists are essential for geology museums because they record data necessary for correlation: the stratum and locale from which the fossils came. It would be surprising if Saull's lacked one, given the radical attacks on the British Museum for its own lamentable cataloguing practices. Here, critics made plain that a catalogue was "the soul" of the collection, and that exhibits without proper classification would lack "any perceptible bond of connexion."[1] The fossil connections, clearly, were something Saull was keen on. But, if Aldersgate Street had a catalogue (perhaps produced by Godfrey, the superintendent), it vanished during the museum's catastrophic breakup after Saull's death.

In lieu of a listing, the contents have had to be construed from scattered sources. That in itself calls for a huge proviso. The results are highly selective, because the literature is obviously skewed. As with fossils themselves, sampling techniques reflect a preservational bias. The press picked high-impact or exotic items to publicize, in order to pique the punters' interest, rather than what was typical. Specimens might be mentioned because they were huge and spectacular, such as the *Iguanodon* dinosaurs or Big Bone Lick mammoths, or giant coal-age tree ferns; or for their beauty, like the pear-shaped sea lilies. These

---

1   MacNeil 2017, 6, 16, 20; McOuat 2001; Knell 2000, 92.

were the show-stoppers.[2] The scientifically significant, new species and such, are known because their details were recorded in monographs. Most notably, the palaeontologist Richard Owen cherry-picked the best Aldersgate Street reptiles for his papers on British fossils. Many of Saull's ichthyosaurs, plesiosaurs, dinosaurs, and more, appeared in these monographs, which were ultimately bound together in Owen's monumental four-volume *History of British Fossil Reptiles* (1849–84).

This is a very low sampling rate. The featured specimens scrape the surface of the 20,000 total and show nothing of the whole. We have no idea of the vast majority of exhibits and know little of their finer arrangement and specific didactic use. Duplicates, found in fellow republican agitator John Lee's collection, might give some clue to the commoner specimens. In that case, *Iguanodon* remains from the Isle of Wight figured large, plus fossils from Saull's native Northamptonshire, an ammonite and fish palate (suggesting a further bias towards this region). There were Tertiary *Pecten* shells from New Jersey, showing that Saull was buying or trading American specimens. Otherwise the samples were the sort that Saull's museum was famous for: the fern fronds of *Pecopteris*, and the perplexing "spotted-stems" (*Stigmaria fucoides*), which in the mid-1840s had finally been shown to be the roots of *Sigillaria* tree ferns, from immediately below the coal seams.[3] The other Lee swaps were nummulites (the sort which Saull exhibited at the Numismatic Society), a small Jurassic sea urchin *Cidaris diadema*, a Cretaceous sponge, and a boat oyster, a common fossil.[4]

The most frequently featured in press reports were those gigantic reptiles that transfixed the Victorians, especially the *Iguanodon*. This is not surprising, given their exposure by Richard Owen. Under his guidance, life-size *Iguanodon* and *Megalosaurus* reconstructions were shortly to be erected in the grounds of the Crystal Palace, when it moved to Sydenham in 1854.[5] Saull's *Iguanodon* too became the cause of the

---

2   For example: A. Booth 1839, 121; *Iguanodon*: G. F. Richardson 1842, 402; largest *Ichthyosaurus platyodon* centrum: Lydekker 1889a, pt. 2: 101–02; the showy ammonite *Ceratites nodosus*: Spath 1934, 477.
3   Confirmation that *Sigillaria* trunks in the coal seams were connected to *Stigmaria* roots, and that these were the same plant, came from Hooker 1848.
4   Delair 1985, catalogue numbers 1332, 1356–62, 2412, 2462, 2599, 2607, 2625, 3388–89.
5   J. A. Secord 2004a.

famous controversy between Mantell and Owen, after Owen had used it to erect the new group, 'dinosaurs'. The other prominent display specimens might have been the tree ferns from the coal measures. Saull had a vested interest in these, with the Parisian expert Adolphe Brongniart having created a new species, *Sigillaria saullii*, from an Aldersgate Street specimen, although whether it took pride of place, we do not know.[6]

How Different was Saull's Museum?

To get some perspective, we might compare Saull's to another museum. For general collections of fossils, there were only three other private museums in London worth speaking of,[7] and only one in the 1840s received press acclaim to rival Saull's. That belonged to the Bishopsgate distiller James Scott Bowerbank.[8] Bowerbank moved his museum to Islington and, in 1846, set it up in a spacious house at 3 Highbury Grove. Eventually he, too, built a dedicated room, forty feet by twenty-eight feet, to take the collection. He was said to have had 100,000 fossils, arranged stratigraphically, and all of them eventually mounted on tablets by his wife Caroline. But here we start to see differences, for the fossils were packed away, as in a modern research museum, in some 400 drawers,

---

6 Two new species were named after Saull, *Sigillaria saulli*: Brongniart 1828 (-1837), 456; Mantell 1851, 32–33; 1854, 129; and *Crocodilus saullii*: Richard Owen 1884 Index to vol. ii, p. vi.

7 According to Morris 1854, iv. Of the other two museums, one belonged to James Baber (1817–1887), an oil-cloth manufacturer in Knightsbridge. It too contained *Iguanodon* vertebrae (Mantell and Melville 1849, 272, 276, 304) and a few British elephant, rhino, and hippo fossils (Mantell 1857–58, 1: 18). *Nautilus baberi* was named after him, so perhaps he specialized in belemnites and ammonites (Morris and Lycett 1850, 10–13, 109; Anon. 1904, 262; Sharpe 1853, 27; Davidson 1854, 89). The museum is scarcely mentioned in the press, making any comparison with Saull's impossible. The other museum rated by Morris was owned by Sowerby's eldest son, James de Carle Sowerby. He carried on collecting after his father's death and continued the *Mineral Conchology of Great Britain* series. This was an identification guide to fossil shells for high-brow enthusiasts. It ran until 1846, in 113 separate parts, often costing 5s apiece (J. B. Macdonald 1974, 389–95; Cleevely 1974, 422; Elliott 1975). Sowerby's museum presumably specialized in shells, for 5,000 were bought by the British Museum for £400 in 1861. The specialism of this little-known museum again militates against a comparison with Saull's.

8 *Civil Engineer* 17 (Feb. 1854): 41–43; G. F. Richardson 1842, 80; A. Booth 1839, 122–23. Timbs (1840, 3: 166) even put Bowerbank's at "the head of private collections", while giving Saull's a bigger write-up, but this was atypical.

rather than being spread out visually, as in Saull's, where they were placed in glass cases or in sequence on open shelves.[9] It was a telling pointer to a deeper proprietorial divergence.

Bowerbank's museum was aimed more at the "Geologist or student of nature".[10] During the London 'season', he reserved Monday evenings for scientific *soirées*, where the geological elite could talk shop. At other times, the public was admitted and Bowerbank, like Saull, was praised for this. But here, too, there was a contrast. Access to Highbury Grove was "by appointment". Advanced "permission" was needed,[11] unlike Saull's open house, which put up no intimidating obstruction for working men and women.

Free and open Saull's museum might have been, but there were still complaints that his Thursday daytime opening was inconvenient. As his aim is the "enlightenment of the masses", chided the *Northern Star*,

> would it not be well for him to throw his museum open one evening during the week, when "the toiler's work is done," but, perhaps, as Mr. Saull is advanced in years, he might think he should be spared this additional gratuitous labour.

The *Northern Star* suggested his superintendent should undertake the task, so that "the benevolent desire of its great and good proprietor would be more surely and effectually accomplished".[12] It had a point, and for a while Saull did opt for a Saturday opening as well. Night time opening also supposed Saull had gas lights installed.[13] If not, the glass-case collection of contoured fossil slabs could hardly have been appreciated.

So the press's bracketing of Saull and Bowerbank belies an instructive difference in the proprietors' attitudes. They were on diverging paths, with different political/professional goals. Both men might have had City trade origins, one a wine importer in Aldersgate, the other a distiller

---

9 On the advantage of Saull's glass case display: *Mining Journal and Commercial Gazette* 1 (7 Nov. 1835): 83; *NS*, 31 Oct. 1846, 3. On Bowerbank: Reeve 1863–64, 2: 133; Timbs 1855, pt. 2: 538; Bowerbank *ODNB*. Mantell's was a hybrid system, part case (he had twenty glass cases, Saull had thirty), part closed drawer: *Lancet*, 29 June 1839, 506–07.
10 A. Booth 1839, 122–23.
11 Williams and Torrens 2016a, 279; Timbs 1840, 3: 166; 1855, pt. 2: 538.
12 *NS*, 31 Oct. 1846, 3.
13 Zorzi 2019, 27.

in Bishopsgate. Even here there were differences. Bowerbank inherited his rectifying distillery business from his father as a going concern. And he could afford to retire and quit any trading ties in 1847 (at the age of 50) to live a more gentrified 'intellectual' life as a fossil specialist. This was unlike Saull, who was a self-made merchant, built his shop up from scratch, and stayed with the trade, and continued to finance freethought, till his dying day.

In terms of exhibits too, Bowerbank's was a much more focussed museum. A visitor could find some of the same fossils in each museum. Take the ancient snake from Sheppey, the *Palaeophis toliapicus* (which in life might have looked like a boa constrictor ten feet long). There were skull fragments in Saull's museum but a better specimen in Bowerbank's.[14] The reason was that Bowerbank *specialized* in fossils found in the sediments containing the snake, the London Clay—Tertiary deposits laid down after the great Age of Reptiles had passed. Here, too, there is an instructive story. Like Saull, Bowerbank exploited the new sewerage and water-supply excavations but for totally different ends. While Saull went to the bottom Roman levels, Bowerbank was descending the shafts below Archway Road because they cut through the London Clay, giving him a unique chance to view the strata foot by foot.[15] He followed the clay outcrop all the way through Kent to the Isle of Sheppey, where his main collection was made. He became an expert on the molluscs, bivalves, and nautiluses from the London Clay, and he made his name in 1840 by monographing the fruits and seeds he found, which resembled those from tropical plants today (showing that Britain was then equatorial).[16] His collection of London Clay fossils was the largest in the world by 1840, and 180,000 of his fruits and seeds eventually passed to the British Museum.[17] Unlike Saull's museum, Bowerbank's was obviously a specialist research hub.

---

14   Richard Owen 1841 [1842], 180; Richard Owen 1850b, 63–65; Mantell 1844, 2: 780.
15   Robinson 2003.
16   Williams and Torrens 2016b; Robinson 2003. It was the same with fossil sponges. Bowerbank began collecting them in the forties, built up the largest collection in the country, and went on to monograph them.
17   By 1879, five thousand of these had been curated and they yielded 154 new species: PP. *An account of the income and expenditure of the British Museum (special trust funds), for the financial year ended 31st March 1879*, LVII.611, 37. C. Tyler, "Memoir of Dr. Bowerbank", in Bowerbank 1864–82, xiv; A. Booth 1839, 122–23; Bowerbank 1840.

And Bowerbank took this interest further. Fossilist *soirées* at his house led to his forming the London Clay Club in 1836 with fellow experts.[18] This vocational dedication was very different from Saull's political motivation.

Being committed, keeping his political nose clean, and spurning trade for science, Bowerbank was respected by the geological gentry. He not only named many new species from the London Clay, but a number were named after him. There was a cowrie shell *Cypraea bowerbankii* and a sea-urchin *Cidaris bowerbankii*, and so forth, even a genus *Bowerbankia*, a bryozoan or simple moss animal. His devotion, research-grade museum, and voluminous publications made him well known. It gained him a place in *Portraits of Men of Eminence*, and a Fellowship of the Royal Society in 1842. His career goals pushed him far from the radical Saull, whose exhibition for working-class instruction had a very dissimilar intent.

Bowerbank's mounted specimens were what fellow experts came to see on a Monday night. The origin of these open Monday sessions again points up how distinct his attitude was from Saull's. Bowerbank was a founder of the Microscopical Society in 1839. Here he tested new recruits to the Society, showing them a pretty slide and weeding out the dilettantes by their response. Protocols were being set up such as would eventually lead to professional approaches to science. He bought the latest microscopes and turned them on fossil fruits, pterodactyl bones, and fossil sponges; in doing so, he was first to show that the flint in chalk was composed of silica from sponges. So many microscopists came to use his Tully-modified achromatic microscope (only the fifth one ever made), that he was forced to set aside Monday night each week for them[19] and, by the forties, the geologists as well. These "scientific" open days, then, were of different complexion from Saull's. They served a distinct function, scientific patronage rather than secular propagandism. Saull welcomed men of science, but, as the *Northern Star* said, the "masses" were his real audience.

---

18 Long, Taylor, Baker, and Cooper 2003, 354. Sowerby illustrated many London Clay specimens in his *Mineral Conchology* (Elliot 1970, 334), fossils that passed to Saull, but he never followed up in the way Bowerbank did.

19 Long, Taylor, Baker, Cooper 2003; Reeve 1863–64, 2: 133; Tyler's memoir in Bowerbank 1864–82; Michael 1895, 10.

The differences even showed in the way satirists spoofed the two men. An innocuous lithograph of "Highbury Grove in 1846" showed Bowerbank's one-storey annex marked "Megatherium Mansion", with jesters standing at the door carrying placards announcing "80,000,000 New Fossil Fruits Just Arrived", and a sign on the house: "Society of ODD FELLOWS every MONDAY NIGHT". Simply innocent badinage, with placard carriers outside offering "Real Fossil Turtle Soup, Pterodactylus Tail D[itt]o."[20] The cartoon was non-threatening, with no dark undercurrent: eccentricity is the worst it implied. How different from the *Penny Satirist's* obsession with Saull's monkey men, with its sinister hint that, for all the absurdity, this was morally pernicious.

Sometimes it was more than a hint. Smith was catering to ever larger readerships now. He had started a new venture, the *Family Herald*. To get issues out with the speed necessary to meet his swelling audience, eager for the latest drama, poetry, and science, Smith initially used machinery to typeset, print, and bind the penny weekly. But only for a year: there was a certain irony to the old socialist union supporter being stymied by the London Union of Compositors, who objected to women working the machinery, so Smith had to revert to manual printing. Still, the venture proved a success, and the *Family Herald* was one of the most successful penny mass-market magazines.[21] But, even here, Smith would not let go of Saull's monkey, typically complaining in 1844 that

> materialism at one time appeared determined to set itself up as a species of religion. Atheism denied the very being of the creative mind, and man's own mind was deemed a mere vapour from the body, which it controlled and animated. Everything was material. Soul, body, and spirit were all so many species of matter; and matter—the dross of existence—was seated on the throne of God himself. With such ideas, down come poetry, imagination, the fine arts, religion, morals; man loses respect for himself. His dignity is compromised, his divinity is denied, his immortality doubted, his divine sonship sneered at. He is merely a logical and philosophical animal—a shaved, and untailed, and cultivated monkey, as Mr. Saull, a materialist and philosophical lecturer, used to describe him, to the amusement of his auditors.[22]

---

20 Robinson 2003; Williams and Torrens 2016a, 281.
21 Blake and Demoor 2009, 213–14; Cox and Mowatt 2014, 8–9; McCalman 1992, 64.
22 *Family Herald* 2 (26 Oct. 1844): 394.

Saull and Bowerbank stood cultural worlds apart. Saull's money and energy were ploughed into dissident causes, and this kept him marginal to gentlemanly geology. His museum, more general, didactic, and exhibitionist, was aimed at artisans, with its shelves of fossils simplifying life's ascent from monad to man in a visible way. Socialist intent meant Saull eschewed Bowerbank's more vocational bent. Only in Highbury Grove did you find a research emphasis on 'museum-quality' fossil fruits and sponges, neatly stowed in drawers, and a bench of microscopes.

One last point will stress how far apart these curators stood in the 1840s. Bowerbank remained focused. He never joined the Antiquaries, Numismatic, or Archaeological Societies, and Saull's goal to push from fossils to savage ascent was never Bowerbank's. By contrast, being integrated into the practising geological community, Bowerbank was instrumental in founding a body dedicated to publishing subscription-only fossil monographs, The Palaeontographical Society (founded 1846). Surprisingly, even though Saull paid his dues to all manner of learned clubs,[23] he was not a subscriber to the Palaeontographical. Yet, this might have seemed closest to his heart. After all, Richard Owen's many monographs on *British Fossil Reptiles* for the Society featured Saull's specimens. Either fossils were losing their appeal, with Celtic and Roman antiquities taking over, or the society was simply too specialized. The Palaeontographical was all technical arcana, by expert fossilists and rock-face collectors. Saull had departed from the clam-shell cognoscenti, men whose arcane knowledge now set them apart. But with the age showing increasing respect for the scientific clerisy, it was Bowerbank who would end up in *Men of Eminence*, whereas the collapse of socialism left Saull in historiographical obscurity.

---

23   Besides those already mentioned, he subscribed to the Ray Society (e.g. *Daily News*, 15 May 1846, 2–3; on whose founding see Gardiner 1993); the Camden Society, set up to publish early historical documents ("Members of the Camden Society ... 1st May, 1847", 14, appended to *Camden Miscellany* 1 [1852]); and was a member (1852–53) and councillor (1853–54) of the new Chronological Institute, established to provide a more exact comparative chronology across cultures: *Transactions of the Chronological Institute* pt. 1 (1852): 39, 65; pt. 2 (1857): 125.

The Eye of the Beholder

Critiques of science ran the gamut among artisan radicals, from distrust because it was in bourgeois hands, to dismissal due to its irrelevance to socioeconomic transformation, while others were ready to co-opt it in an anti-clerical cause. Saull had to connect with all sorts. Then there were the gentlemen dropping in on a Thursday: the geological gentry, Anglican clergy, phrenological enthusiasts, London historians, and Roman antiquarians, all brought their own contexts to bear, to make sense of the exhibits. Whether artisan or bourgeoisie, the visitors came with a bewildering spread of cultural expectations.

Some might not have appreciated Saull's materialist museum at all. Take the sacred socialists from Alcott House, in Burton Street, up the road from the old Owenite institution. They were unlikely comrades. In their "Aesthetic Institution", that refuge "for distressed or curious radicals" reacting to an encroaching materialism, a sentimental judgement of good and evil was substituted for hard-core science as a way of understanding. Acting replaced knowing. Action meant behavioural adjustment, pacifism, celibacy, teetotalism, vegetarianism. And with this physical puritanism came a love of lay-empowering practices: phrenology, hydropathy, mesmerism, and astrology.[24] It was not only the "bourgeoisie's evermore arrogant, elitist, and humanly abstracted utilitarian conception of science"[25] that they feared but the new atheists flexing their muscles. For the sacred socialists, phrenology revealed a deeper "spiritual organization" that made man more than an Owenite rational animal. Their idealism and disdain of science ran counter to Saull's outright materialism. So it is debatable whether these "aesthetical young men with their hair divided down the middle" would have found Saull's monkey-based 'evolution' emancipating or spiritually uplifting.

The same was probably true for a fellow pacifist, the Chartist Thomas Cooper. He was the true impoverished autodidact: an illegitimate dyer's son, lapsed Methodist, and apprentice shoemaker, who had taught himself Latin, Greek, and Hebrew. Cooper was the "Chartist poet" and erstwhile admirer of Feargus O'Connor. He had served his time,

---

24 Latham 1999, 20, 80, 168, 175.
25 Cooter 1984, 202–04.

in Stafford gaol, and used it to write a "Prison Rhyme" in ten books, *The Purgatory of Suicides*, which even an appreciative foe, the *Monthly Christian Spectator*, thought an "incontestably great poem".[26] This man of "immense influence", as Charles Kingsley acknowledged,[27] might have scorned the mysticism of the young fops and their retreat from science. "Arrest science!", he laughed, "You might as well try to put out the sun." Cooper's own wobbly Christianity had finally been knocked over by David Friedrich Strauss's *Life of Jesus*; and, in Cooper's hands, the debates in John Street were shifting from God's existence to the historical veracity of Christ's teachings.[28] The eloquent Cooper, who could recite "Satan's speech from Milton with magnificent effect",[29] remained a deist through the violent years, although he would later regain his Christian faith and go on to attack evolution.[30] But, for the moment, he too lectured Chartists on geology and was happy to upset the "orthodox reckoning of the Mosaic Age of the World". And he saw no "fear for morality if even the New Testament Miracles become generally disregarded and treated as legends", as he said at John Street.[31] Yet even now his respect for science was as much to do with pointing out the "perfections of the divine Mind; for God manifests himself in every object of science". So Saull's religiously liberating explanations would have jarred badly.

Even more opposed to Saull's funny monkey business was his old socialist nemesis 'Shepherd' Smith. He had made no bones about it in every publication since his old *Crisis*-editing days. Smith was another who warmed to the aspirations of socialism just as he warned his readers off a nasty one-sided materialism. "Materialists always attach themselves to the body politic, and sneer at the soul ecclesiastic", ran his leader "Our Double Nature" in the *Family Herald*. It branded Saull, as Smith always did, as an extremist with anti-religious "feelings amounting to abhorrence", who was as bad in his way as the blinkered

---

26 *Monthly Christian Spectator* 2 (Nov. 1852): 669–77; Loose 2014, 42–46, 116–18.
27 Larsen 2004, 48ff; Marsh 1998, 82. The radical hero of Kingsley's *Alton Locke* (1850) was based on Cooper.
28 *Monthly Christian Spectator* 2 (Nov. 1852): 672, 676.
29 W. E. Adams 1903, 1: 170.
30 Thomas Cooper 1878; 1872, ch. 24.
31 *Cooper's Journal* 1 (16 Mar. 1850): 174, 186–87; Goodway 1982, 58. Thomas Cooper ([1842], 11) was another knowledge Chartist who claimed to "popularize Chartism, by delivering familiar and elementary lectures ... on geography, geology, astronomy ...", as if the sciences were integral to the movement.

clerics who "speak contemptuously of the body". Preaching moderation and mutual respect, Smith could nevertheless not help pronouncing the "spiritual power in every country whatsoever is the strongest, the most permanent and enduring", ready to redress any materialist imbalance. This spiritual reserve would be called up if another "revolution like that of '92 [the French Terror] were again to give [the materialists] exclusive possession of power".[32] Readers were left in no doubt that the spiritual element ran deep, while a spirit merchant with a monkey was a shallow entertainer.

Saull had to persuade all sorts. His lecturing over the glass displays, and his question-and-answer sessions, would seek the audience's acquiescence for a new unimpeachable authority outside of Church and Throne, and which worked in the listeners' best interests. His talks re-crafted the social grievances of the downtrodden and related their solution to the new palaeontological science: of planetary changes allowing life to pull itself up unaided, and grant itself inalienable rights. The museum placed the working-class visitor in a new natural position, not at the bottom of the heap, but at the culmination of geological history, by a material process which guaranteed political sovereignty, all of which was expected to transform the artisan's self-perception and bring about the political millennium. This was Owenite self-reformation at work.[33]

Many out-and-outers wanted science 'correctly' interpreted, as in Saull's sense: made to speak as part of their "sociopolitical and socioeconomic struggle and humanist morality". It was to allow workers "to assert their dignity and worth and self-reliance and ... be better equipped to contest obscurantism and social injustice".[34] When it came to readers of Holyoake's *Reasoner* or Robert Cooper's disciples (the current Antichrists of the Christian evangelical press), Saull was preaching to the converted.

We see this in the young Robert Cooper's secular lectures on Moses or the "Origin of Man". These were theatrical stage shows by the early

---

32 *Family Herald* 7 (27 Oct. 1849): 412–13.
33 Lundgren 2013 has shown Francis Galton later using science in his Anthropometric Lab in a similar way to expose "exhibition-goers ... to new perspectives on everyday habits and social practices", in order "to learn how to turn observation back on themselves" and change their own way of life.
34 Cooter 1984, 202–04.

fifties, with a visitor reporting it was like going to "a Drury Lane opera or a gin palace saloon":

> playbills informed us of the nature of the performance—admission was to pit, gallery, and boxes, according to a tariff of charges—persons went round the assembly hawking books and pamphlets—and a professional orchestra diversified the entertainment with stringed instruments and vocal score.[35]

Such twopenny talks trashing Creation or Moses could be hectic, with jostling, evangelical gate-crashing, and heckling. Cooper once even fainted at a particularly fraught event.[36] Given the preponderance of the religious press, we know most about Cooper from these antagonistic sources. New journals like the *Monthly Christian Spectator* (founded 1851), *Bible and the People* (founded 1851), *Defender* (founded 1855) and *Bible Defender* (founded 1856), dutifully shadowed the "Secularists"—Holyoake's neologism was enthusiastically adopted—to expose this "banditti of Freethought".[37] They painted Robert Cooper as "coarse, rude, ludicrous, and outrageous" as he dilated "upon the sublime doctrine of a resurrection from the grave", with the audience "plunged into loud gustos of laughter".[38]

Saull was in the thick of it. He could be found at a Cooper lecture on "The Soul" in the City Road Hall of Science.[39] And it must have been to Saull's taste, with Cooper not only demolishing souls and denying resurrections but more positively referencing "Facts from Anatomy and Physiology in relation to Materialism".[40] Cooper's *Immortality of the Soul*

---

35 *The Association, or Young Men's Magazine* [1855]: 32–33.
36 *Preston Guardian*, 29 May 1852.
37 *Bible and the People* ns 2 (1854): 36.
38 *The Association, or Young Men's Magazine* [1855]: 32–34, painting a derogatory portrait of the "great crowd" at a Sunday meeting in 1852 in an (unnamed) "great Secularist-hall" in London.
39 *Reasoner* 13 (4 Aug. 1852): 128; (25 Aug. 1852): 166.
40 *Reasoner* 13 (27 Oct. 1852): 320; R. Cooper 1853, 57–72 discussing medical aspects of neuro-stimulation, which could be seen to prefigure Henry Maudsley's work, but for decidedly different ends. There is no sign that Cooper was responding to Francis Newman's new edition of *The Soul* (1852). Newman's addition to the dissolvent literature might have augmented the Victorian crisis of faith, but Cooper's was intent on turning it into a catastrophe of faith. Newman's was a 'natural history' of the soul, but Cooper's was a real natural history, with working-class earthiness and recourse to the anatomy of brains, monkeys, and human races.

(1853) was still culled from the old Jacobins, William Lawrence, John Elliotson, and Sir Richard Phillips, but, in places, it began to foreshadow the scientific arguments of respectable bourgeois 'honest doubters', not that the latter would show anything but disdain for these street 'scoffers'.[41] Some of Cooper's language Holyoake considered near the knuckle, but it mirrored a long history of in-your-face rhetoric used by the "vile rabble",[42] smarting at the denial of its rights.

Saull's museum and monkey lectures were suited to angry youngsters like Robert Cooper, uncompromising destructives with a penchant for shock. Cooper positively begged for the ad hominem arguments in the Christian press. A religious interloper at John Street reported on the "feebleness and frothiness" of Cooper's talks. "Effeminate" and "affected" was the *Monthly Christian Spectator*'s summary, sexist imagery used to suggest his (woman's) weakness of mind. To the *Defender*, he was "a little man with spectacles, and a rather well cultivated hirsute appendage, which he wears entirely below his mouth". He "tickles the sillier portion of his audience, with such questions as, In what portion of the human frame does the soul reside?" And then he tells them, "It is these delusions that keep the masses in the mud." A leitmotiv of the reviews was the "miserable audiences" Cooper attracted. They had little capacity to understand, only to be roused—a play on the prevalent view of the hovelled classes as visceral beasts, semi-domesticated animals fit only to be shepherded. His listeners were to be found in filthy dens—as at the Chartist Hall in Newcastle: "up one flight of dirty stairs, from one of the entrances to the Butcher Market", a prejudicial image to stir

---

41   Rectenwald 2013, 2016 ch. 4, discusses 'secularism' as a likely source for the later scientific naturalism, even though Holyoake is not mentioned in Dawson and Lightman 2014. Secularism has been successfully followed into John Chapman's bourgeois circles by Rosemary Ashton (2006). Many of the later arguments for scepticism were being thrashed out at the Utilitarian Society, but in a bitter political context, making the rollover to 'scientific naturalism' far from easy. Aspirational 'honest doubters', notably the young T. H. Huxley, distanced themselves from the street "scoffers". Even though the student Huxley had seen the squalid side of pauper life (indeed his own life had its squalid side, with one brother-in-law addicted to beer and opium, and another jailed for debt [Desmond and Darwin 2021]), and though he claimed that he took "a deep draught of abomination" himself, he was ambitious to climb into scientific society and cautious in his social alignments. Since he saw the key to character in the "temper and tone" of religious views, he despised "those miserable men", who used scepticism for "disturbing the faith of others" (Desmond 1998, 13, 657–58 n. 20).

42   R. Cooper 1853, 79; *Reasoner* 14 (2 Mar. 1853): 137.

the magazine's polite Christian readers. These "lowest of the working classes" were being whipped up, with the "extreme infidel" telling them that "mankind had been befooled, bechurched, and Priest-ridden enough, and that it was now time that they were elevated to that high and glorious position, which by nature they were intended to occupy!"[43]

The *Defender*'s menacing imagery had caught the drift. Cooper was on the verge of starting his own freethought journal, the *London Investigator* (1854), which would run the obligatory "Origin of Man" series as a central pillar. For Saull, that made Cooper a soul mate. Saull attended his lectures and championed the *Investigator*. Even on his deathbed Saull requested that its distribution be boosted. Cooper reciprocated with praise for Saull's evangelizing and called him a merchant who had risen "above the sordid associations a competitive system is calculated to develope".[44] The puffing is not surprising looking at Cooper's derivative post-*Oracle*, post-*Vestiges* dash through the nebular origins of planets and the long geological rise of life leading to the natural "Origin of Man". It was everything Saull had taught for a generation, a gushing of Enlightenment ideals in a "healthy stream of secular knowledge [to] wash into oblivion the dust and mire of superstition" and end "the reign of delusion and slavery". These were Cooper's words, but they could have been Saull's. It was the same exhortation for the worker to read the sermons in stone "as carefully as he has listened to the sermons of the pulpit, and these pious hallucinations will be exploded." Like a piece of Saullist scientism, the series on fossils was written in the same "easy and popular style, to present our readers with the facts of science versus the delusions of superstition. Nothing will so effectually tend to snap the priestly wand."[45]

But, while Saull might have had such secularist sympathizers, there were as many who could not complete the journey with him. The rocks having shown that animals emerged when conditions permitted was one thing, but that "man was developed, as naturally and necessarily", without any "miraculous interposition", was another. Even Cooper's talk stopped at a bland "energy in nature" able "to develope animal

---

43 *Defender* 1 (6 Jan. 1855): 12; (25 Feb. 1855): 119; (17 Mar. 1855): 171; *Monthly Christian Spectator* 2 (Dec. 1852): 718; R. Cooper 1853, 15.
44 *LI* 2 (June 1855): 46.
45 *LI* 1 (May 1854): 26; (June 1854): 41; (July 1854): 58.

forms", including humans. No mention of a monkey ancestry here. On this count, Cooper never got beyond Holbach, positing only that "that matter—'mere matter,' should insensibly develope animal vitality".[46] It showed how heretical Saull had been with his monkey a generation earlier.

Radical Pantheon

[D]isregard the philosophers, the sneerers, and the scoffers ... whilst they are looking with pity on the poor enthusiast, who adores the relic of some pious saint, they themselves are unconsciously actuated by a passion for relics which is, in many respects, less reasonable, less honourable, and less intelligent than his. I have often wondered at the idolatry of the geologist and the antiquarian, and accused them of it. They deny it, as the Catholic does even when caught in the very fact of adoration. They say that they preserve these relics for scientific purposes .... But they are mistaken. It is a real worship; for after having obtained all the scientific information which the relic can communicate, they burn with such desire to be personally possessed of it and to preserve it in their shrine of antiquities...

Saull's nemesis Shepherd Smith, musing on geological idolatry.[47]

If Saull's really was a *radical* museum for the masses, how else did this manifest? What was its most telling difference from conventional museums or vocational meeting-places like Bowerbank's? Moving beyond Saull's use of geology and antiquities to prove human perfectibility for Owenite reasons, or the fossils themselves "as facts much to[o] hard for the parsons",[48] we come to its most unexpected aspect.

Perhaps the best way to see it in its radical light is to follow the *Northern Star*'s reporter as he filed his story in 1846. Recall that Feargus O'Connor's Chartist rag outraged the establishment, who slammed it as politically "detestable, on the ground of sedition. This bad man is like Dante's evil angel, bearing in his hand a two-edged weapon of sin and death." O'Connor was "spreading ... a political and social pestilence" by addressing

---
46   *LI* 1 (June 1854): 41; (July 1854): 57.
47   J. E. Smith 1873 [1848], 1: 310.
48   *NS*, 31 Oct. 1846, 3.

those classes in whose minds disaffection and infidelity are most easily implanted. The chief design of our more licentious writers and speakers is, to deride the Established Church and defame its ministers, and thus weaken man's natural respect for his religion, and, by consequence, his dependence upon his Creator.[49]

The *Northern Star*, then, was an impeccable source. Who better than O'Connor's hack to show us the truly radical aspect, totally missing from the guide books?

Saull had glorified thousands of priceless relics in his altar to 'evolution'. Such sacred objects gave their possessor the power to pontificate, they were the vestiges that connected us with creation, the literal Word of 'evolution'—his direct route to the evolutionary godhead. These prized petrifactions bestowed scientific authority, as much as any saintly bone in a Catholic shrine. But Saull's mausoleum went considerably further when it came to veneration. It was literally a place of radical pilgrimage.

Some exhibits made the museum the ultimate mediating place, where Chartists, socialists, atheists, Christian Freethinkers, and radical millenarians could all find common ground—those who could, and those who could not, accept the monkey-man, or go the "whole orang", as Charles Lyell had it.[50] The museum was more than artefacts pointing towards a radical 'evolution'. These were embedded in a rich cultural environment that tapped a deeper vein of radical emotion. This wider crafting of Saull's display could both draw and unite the radical factions. To them, it was less the *Iguanodons* that were the attraction. Rather, it was Aldersgate Street's real *memento mori*, for the museum was also a shrine—a mausoleum in more than one sense—which made it a proper place of veneration.

The reason lay in one corner of the lower gallery. Saull's warehouseman William Godfrey led the *Northern Star* reporter to it. Here was a closed closet, whose contents "we are sure will much interest many of our readers". On the door Godfrey had written the words:

> Nature stamps all men equal at their birth,
> Virtue alone the difference makes on earth.

---

49 *The Age*, 28 Aug. 1842, 4.
50 K. M. Lyell 1881, 2: 365.

These were instantly recognizable lines from the revolutionary Spencean George Petrie's "noble poem" entitled "Equality". Petrie had penned probably the most celebrated agitational poem of the age (which, the reporter added, had "passed through so many editions, and is yet so much in request"). The lines were Petrie's motto, modified from Voltaire.[51] "Our conductor (Mr. Godfrey) appeared, like most of its readers, to be smitten with that charming work, and from its pages we have imbibed the great truth that 'True Freedom only knows Equality.'"

The Chartist paper then ran a huge extract from "Equality". In fact, well over a third of the review of Saull's museum was a quote from Petrie's poem, suggesting where the real interest lay. Worse things had been said about the King, but the publisher of Petrie's twopenny poem (R. E. Lee) had still been indicted for issuing it[52]:

> Like nature's God, he self-existent reigns,
> And links those rolling suns in golden chains;
> Those suns again their satellites entwine
> With places, pensions, sinecures, and wine;
> The satellites extend the circle more,
> 'Till every idle scamp on Britain's shore
> Obtains a birth among the reckless brood
> Who drink our blood, and eat our flesh for food...

But this den of fossil iniquity did more than celebrate "Equality" from the "poet for all time". It not only had the corpus, but Petrie's corpse as well. For inside the cupboard the reporter was astonished to find Petrie himself, or, at least, his "complete skeleton". Startled by coming face to face with the insurrectionary hero, the reviewer concluded:

> We are sure when the numerous disciples of this truly great poet and veritable democrat, shall learn that the bones of their master are enshrined in this museum, they will at once commence a pilgrimage to the shrine, and while gazing at the dry bones, imagine they hear Petrie's once eloquent lips speak those truthful words, that his pen so copiously indited, and which are sent forth to the world in the poem of "Equality."[53]

---

51 Voltaire, *Eriphile*, act II, scene I (1732).
52 *PMG*, 20 Oct. 1832, 576.
53 *NS*, 31 Oct. 1846, 3. For Chartists reading "Equality" out loud at weekly meetings, see *NS*, 16 Mar. 1844.

It certainly gave Smith's complaint about fossil shrines a new twist. Saull's cathedral actually contained the holy of holies, relics of a heroic sinner.

Quite how Petrie's skeleton arrived here is a mystery, but the roguish Pierre Baume's involvement seems certain. The forty-three-year-old Petrie had died mad in Hanwell asylum in 1836, and tittle-tattle had him driven insane by his wife Mary and Baume—the former for taking up Owen's marriage views too literally, and the latter by his affection for the former. Indeed, there was a whisper on the street that Baume, now co-habiting with Mary, had actually poisoned Petrie.[54] Saull was not alone in stocking his shrine with saintly relics. Baume had wanted William Thompson's skull for his own phrenological lectures. (Where Thompson's skull ended up is an open question.)[55] Curiously, in 1837, Baume offered Saull a body for dissection, which Saull wisely declined, and a skeleton. Even if not Petrie's, it shows that Baume was dispersing anatomical remains.[56] Anyway, Saull's old comrade Petrie—the revolutionary who had once drawn up plans to storm the Tower of London—was now hanging, not in Newgate, but in Saull's closet, the object of real veneration.

The reverent could peek at other radical relics. Phrenological cabinets, which typically stocked the skulls of murderers, madmen, and celebrities to illustrate their cranial anomalies, had accelerated the trend in skull collecting. They help explain Saull's accumulation of human remains, if not his more idolatrous intent. The museum also had the head of that rich, witty patron of freethought, Julian Hibbert.

At least in this case it was semi-legitimate. Hibbert had been a munificent donor to Carlile and Watson, financing their presses. And, with Saull, he had helped set up Carlile in the Rotunda. But he was another to die young, in 1834, only weeks after outraging "public

---

54 Cooter 1984, 211 n. 34; Lovett 1920, 1: 51. Petrie's poisoning was mooted in Petrie [1841], 24–25; Chase 1988, 158 n. 33.
55 Donovan 1876, 202–03; Pankhurst 1991, 130.
56 W. D. Saull to P. Baume, 16 Aug. 1837, Manx Museum, MM 9950 uncatalogued. Roger Cooter kindly supplied a transcription of this letter. Baume's donation of bodies for dissection was not new. He delivered his unmarried sister Charlotte's, who died in childbirth, and her stillborn child's, to London University in 1832, with such speed that he was at first charged with her murder. In fact, she had been a fellow republican, presumably in the Carlile mould, and it had been her wish: R. Richardson 1989, 236.

decency" by declaring his atheism at the Old Bailey.[57] His will stipulated Saull as an executor (along with John Brooks, the radical bookseller, and a coal merchant) and that "there be no funeral". Baume was officially implicated in the body snatching this time, for Hibbert bequeathed him forty guineas that he might "do his best to see that my body or corpse is partly or wholly dissected any where for the benefit of Science & my Skull or head be given to the London Phrenological Society to which I have for many years been a subscriber..."[58]

Hibbert's family was rich on West India pickings. Slavery had bought them civic security among the gentry as sheriffs, members of the judiciary, and Church trustees. Julian's brother was about to buy Bilton Grange in Warwickshire, to signal this social ascent, and have Pugin convert it into what would become one of his masterpieces.[59] Almost in defiance, Julian had led a spartan life, dying in temporary lodgings in Hampstead. His republican and atheist foibles always risked damaging the family's respectability, never mind the Old Bailey ignominy or that he was using Saull and a coalman as trustees. Now the family tried to limit further damage by thwarting his wishes. They had the body removed to a Holborn undertaker, and he was buried at night in Kensal Green Cemetery, attended by close relatives. But it seems that not all of him was in the coffin—by nefarious means (he disguised himself as an undertaker), Baume had managed to extract the head.[60] It seems that a medical school took it to dissect the brain, which is not hard to imagine in the resurrectionist years, when there was a dearth of corpses.[61] How Julian Hibbert's head was subsequently rendered down to a skull can only be conjectured. Anyway, it too ended up in Aldersgate Street.[62]

---

57 *Gauntlet*, 2 Feb. 1834, 824; *DPMC*, 1 (7 Dec. 1833): 356; *Bell's Life in London and Sporting Chronicle*, 2 Feb. 1834.
58 Julian Hibbert, Will, 6 Jan. 1834, Public Record Office, National Archives. An abridged version of the will in the press did not mention Baume: *MC*, 29 Mar. 1834; *The Satirist, and the Censor of the Times*, 30 Mar. 1834; *Patriot*, 2 Apr. 1834.
59 Donington 2014: 204, 224.
60 Wiener 1979; Holyoake 1906, 2: 550–51; McCabe 1908, 1: 294.
61 Wiener 1983, 209. In Manx Museum is another letter from Saull to Baume (undated, MM 9950 uncatalogued, transcription by Roger Cooter) in which Saull responds to Baume's request that he (Baume) be put in touch with a hospital surgeon regarding a donation, although which corpse this relates to is unknown.
62 That it was genuinely Hibbert's skull is suggested by Richard Cull's (1850) craniometric studies, read on 25 April 1849 to the Ethnological Society. This used the skull for measurement. Since Saull was a member, he presumably loaned it.

Saull's museum was part-pantheon. Not only did it enshrine the distant fossilized dead, but more familiar bones made it a place of homage. Here stirring poetry evoked the glory days of the cause, and the lost heroes of the movement drew all under their mediating gaze. It was the stuff of radicals in every sense.

25. Celebrating the Dead

> Materialists always attach themselves to the body politic, and sneer at the soul ecclesiastic. Spiritualists attach themselves to the soul, and speak contemptuously of the body. It was always so. But more so now than ever, for the two are more disunited than they were in former times. The extremes of both sides look upon each other with feelings amounting to abhorrence. They hate each other in life, and are pleased to be separated even in death. The atheists and infidels are beginning, like the Jews, to have their own separate burying ground.
>
> Shepherd Smith on hearing that Saull had bought a funeral plot for his friends.[1]

While the museum was a resting place for Hibbert and Petrie, Saull now set about finding his own seat of repose. Old friends were starting to die, and he was in his mid-sixties. Grave sites were important to radical cults. Freethinkers demanded unconsecrated niches where a last stand might be made, with the interred surrounded by comrades. Cadres were formed even in death. The decline of Owenism and Chartism, if anything, increased the importance of graveside commemorations, to hold steadfast those who remembered and remained.

These veteran freethinkers never had anything like full-blown political funerals. There was rarely that flag-waving community turn out, as in the case of the last Chartist hero Ernest Jones in 1869, with his lavish Manchester send-off, funeral cortège fronted by Peterloo veterans, thousands of spectators, civic celebrities, shuttered shops, and so on.[2] Yet their select cliques employed the same graveside rhetoric of sacrifice, service, and dedication to a just cause under duress—those characteristic elements of martyrdom.

---

1   *Family Herald* 7 (27 Oct. 1849): 412–13.
2   A. Taylor 2003, 31–32.

Saull's French-style obsequies, celebrating a freethinking life in death, were now standard. They emphasized a lifetime lived steadfast in belief, or rather unbelief. Services were almost religious in intent, while totally lacking any theological content. They used the therapeutic value of funerary ritual for cultural ends. But however recalcitrant the orator, the speeches became somewhat stereotyped, being framed in response to a "hostile host culture".[3] Still, that did not stop their equally ritual publication to consolidate the community.

Thus it was that the close-knit group had met in March 1838 over the grave of that veteran Jacobin, John Gale Jones. Saull had probably known Gale Jones for decades and had looked after him (financially) in his declining years. Saull spoke of the "kindred feeling between us": they were fellow Carlileans, Owenites, and Rotundanistas, and "many of our thoughts and sentiments were in unison".[4] Never had there been a more notorious apothecary than Gale Jones, so much so that the conservative *Medical Gazette* ran his obituary as warning "to the younger members of our profession". It told a tale of terrible decline: how a promising apprentice crashed out of the profession during the turmoil of the French Revolutionary years. Fellow students at the Great Windmill Street anatomy school remembered his "great eloquence" and how he looked set to rise "to a high rank" among medical men.[5] Having a golden voice, Gale Jones could tickle the ear, it was said,[6] and that is what made him such a brilliant orator at the London Corresponding Society in the 1790s, when he led the chorus demanding universal suffrage. But the *Medical Gazette* only saw the surgical apostate giving in to the dark side:

> He was now the foremost in attending political meetings; he addressed the populace from the hustings; he travelled as a propagandist of the political faith he had imbibed; and his pen was ever ready to defend the opinions he had embraced; but, alas! these exertions tended not to forward his interests; they only led to prosecutions and imprisonment...

The moral for medical Tories was obvious. Decline and disgrace awaited the radical reprobates. Gale Jones sank into penury, in a small

---

3 Nash 1995a, 167, 179.
4 Saull 1838a.
5 *London Medical Gazette* 22 (19 May 1838): 348–49.
6 *GM* 10 (Aug. 1838): 218–19; *NS*, 17 Mar. 1838.

apothecary shop off Gray's Inn Lane. Here he "suffered privations many and severe".[7] Even Francis Place saw Gale Jones as "a poor emaciated crazy looking creature".[8] This off-quoted statement, though, sits at odds with recollections of Gale Jones retaining "to the last the conversation and manners of a gentleman".[9] But it is undeniable that he declined badly. After the Rotunda years, he retired from politics. His eyesight gave way, his practice never picked up, and activists had to start a relief fund for him. Even though financially supported by Saull, the 68-year-old still died "embittered by poverty", on 4 March 1838.[10]

The *Medical Gazette*'s obituary was typical of the hostile culture. If ever there was an old comrade in need of celebrating, it was Gale Jones. Therefore, while Tories used this tale as a warning, Saull's graveside oration was the reverse. It went beyond the usual levelling sentiment, that all were equal in the "silent tomb", or pointing to the "stern Law of Nature to which Emperors, Kings, and all the magnates of the land must bow". It became a secular celebration. Saull made his Gallic point of rallying the gatherers with "the virtues, the patriotism, or the philanthropy" of their comrade to strengthen their resolve. With no afterlife, the emphasis was shifted back onto the living, with the life of the deceased used for "moral reaffirmation".[11] In honing an uplifting speech, Saull was upturning the *Medical Gazette*. He made a *virtue* of sacrifice. Gale Jones had discovered that most "of the diseases which afflict and desolate humanity, have been the result only of poverty and wretchedness [he was quoting Gale Jones's own words]; that the miserable sufferers wanted not restorative medicines, but actual bread". This had turned him to action, whatever its cost. No doubt "our departed friend" could have "made a market of his splendid talents", in which case

> riches and honours would have been his reward; but no, he was so great a lover of his species, and so devoted to improving the institutions of this country, that he chose to remain in poverty, and even endure distress rather than prove a traitor to his fixed principles.

---

7   *London Medical Gazette* 22 (19 May 1838): 348–49.
8   Miles 1988, 73; Kent 1898, 259; Wallas 1918, 49.
9   *GM* 10 (Aug. 1838): 218–19; *London Medical Gazette* 22 (19 May 1838): 348–49.
10 Parolin 2010, 3; *PMG*, 1 Aug. 1835, 622.
11 Nash 1995a, 162.

It was the medical Tories who were unpatriotic, putting profit before compassion. Gale Jones used his oratorical eloquence for public good at unparalleled personal cost. To the mourners, his was a greater bravery, and Saull drew out the greater moral.[12]

Another old hero reduced to skin and bone was Allen Davenport.[13] In the latter years, he had struggled with failing eyesight to write on freethought, the need for adult education, and poetry, with his epic *Urania* being dedicated to Saull.[14] The seventy-one-year-old died serenely on 30 November 1846, but not before lengthy ministrations by Saull. They had held a benefit ball for the old man, but it was a failure, and so Saull and friends had to chip in to a public subscription. They sold the remaining copies of his new pamphlet, *On the Origin of Man, and Progress of Society* (his history of private property), to help with the funeral costs.[15] Such was the fate of a veteran agrarian polemicist.

Davenport's last request had been that Saull deliver the graveside oration. That was to have been at his interment in Bunhill Fields, because the limited funds would not "carry him further", explained the *Reasoner*.[16] But Davenport had desperately wanted to lie in Kensal Green, where so many reformers were now buried, and had tearfully said so on his deathbed. With no known relatives to cover the additional cost, Saull's fellow mourners got up a subscription, and the funeral was re-located at the last moment. The cortège with its uncovered coffin started off in the Mechanics' Hall of Science and wended its way to Kensal Green.[17] Here the faithful convened, led by Saull, Harney, and Holyoake. The body was

> laid in unconsecrated ground—unconsecrated by the priest, but consecrated by worth—opposite the tomb of 'Publicola.' It was a Reformer's funeral! No mourning clothes were worn, and no ceremony was performed.[18]

---

12 Saull's (1838a) oration made the front page of the *Penny Satirist*, Smith for once not carping.
13 *NS*, 5 Dec. 1846: "wasted to a mere skeleton".
14 *NMW* 4 (11 Aug. 1838): 340; *NS*, 5 Dec. 1846.
15 *UR*, 2 Dec. 1846, 1–2.
16 *Reasoner* 2 (16 Dec. 1846): 18; *NS*, 5 Dec. 1846; *UR*, 2 Dec. 1846, 1–2.
17 *NS*, 12 Dec. 1846; *UR*, 16 Dec. 1846, 5.
18 *Reasoner* 2 (16 Dec. 1846): 18

Only Saull spoke. We do not know what he said, presumably the same uplifting eulogy on the stoical merit of agrarian and freethinking penury. The speech was, said the *Reasoner*, a "brief but impressive discourse on the life, struggling labours, and virtues of the deceased, whom he held up as an example to his hearers."[19]

This funeral at the end of 1846 can only have reminded the ageing Saull of his own mortality—and that Kensal Green, sanctified in its freethought corner, was the place to spend eternity.

The Plot to Bring Freethinkers Together

Two years later, Saull would buy a funeral plot for himself, adjacent to Davenport and Publicola, while the freehold was still available. Like acolytes congregating round the pharaoh, the freethinkers were clustering to ensure the immortality of their doctrines among the surviving faithful. God forbid, was Shepherd Smith's response on hearing of Saull's plot. Smith deplored such "unsocial sectarian antipathy that carries itself even into the grave".[20] Although even he had to admit that the freethinkers had been forced to use "separate burial grounds. The Church has only one service for all, and consecrated ground is attached to its own service and sanctuary".

Kensal Green's growing row of fallen heroes had propagandist value for the activists. But their shrines to unbending unbelief were not so dissimilar to those of religious martyrs. The collective names recalled epic stories of persecution and perseverance. And orations over the dead were designed to be inspirational—these were graves which told a moral tale. And they still do today. There is now an obelisk in Kensal Green with a list of the reformers and freethinkers, raised by a new generation seeking legitimation in history.

Eight months later, and the wisdom of the purchase was clear. Saull himself became seriously ill, enough to alarm Holyoake. He announced in July 1849 that Saull had "been, for some time, in a dangerous state of illness, and his recovery is scarcely expected".[21] A cholera epidemic

---

19 UR, 2 Dec. 1846, 1–2.
20 *Family Herald* 7 (27 Oct. 1849): 412–13.
21 *Reasoner* 7 (18 July 1849): 47. He had evidently been ill all year: he told Richard Owen that "I have been extremely ill since the last week in January": W. D. Saull

was raging in London, and Saull might have succumbed. Whatever the cause, death's door did not open, and *Reasoner* readers were relieved a month later to hear that he was "recovering, contrary to the anticipation of his friends."²² Perhaps because of illness and age, Saull was now turning down jobs. The Chartist Metropolitan Delegate Council wanted him to act as treasurer of the O'Connor fund, to audit the accounts to prove no irregularity, but he had to refuse.²³

Death for a freethinker was not to be feared but celebrated. And the prospect called for a certain defiance, to counter the claim that theirs was a bleak nihilism, with no hope for the present and no faith in the future. How often had they heard that despondency must dominate life's end without the promise of eternal bliss? As the *Defender* put it, "infidelity cannot sustain the infidel in his last hours".²⁴ And it rattled off a whole series of death-bed conversions as proof. Atheist death-bed recantations made good recruiting copy, and good sellers. Atheism puts its dark mark on "everything which makes life bearable". It leaves mankind "bereft of all hope".²⁵ And yet, "How often has the christian's death bed been the brightest scene of his life?"²⁶

To counter this, deaths were equally romanticized by freethinkers. The dying were depicted defiantly clutching the *Age of Reason*, as a Christian would the Bible. Paine's *Age of Reason* "cured me of superstition", ran one verbatim report. "I loved to read its crushing facts ... use them against your opponents, and remember me, who have been martyred into decease by the physical insults ... which furious bigotry has inflicted upon me." Or Holbach's *System of Nature* was exhorted with the final breath, as having liberated the soul in life, or a volume of the *New Moral World*. Once even Mackintosh's *Electrical Theory of the Universe* was praised in an emancipatory last gasp.²⁷ This was the infidel's hallelujah proclamation of salvation.

---

to Richard Owen, 27? Nov. 1849, British Museum (Natural History), Owen Collection, 23: ff. 112–15.
22 *Reasoner* 7 (15 Aug. 1849): 111.
23 NS, 6, 13 Mar. 1852; *Reynolds's Newspaper*, 7 Mar. 1852.
24 *Defender* 2 (10 Nov. 1855): 298–99.
25 *Bible Defender* 1 (16 Feb. 1856): 102.
26 *The General Baptist Magazine* ns 1 (May 1854): 214–18. [Neale] 1848, on the contrasting death bed scenes of Christians and Infidels, went through many editions.
27 *Defender* 2 (10 Nov. 1855): 298–99.

This celebratory aspect helps explain the seeming incongruity of Saull telling the John Street audience at their monthly festivity, amid the speeches, songs, and recitations, that he had bought "two pieces of unconsecrated ground near the grave of Publicola ... for the interment of his family and friends." The "friends" bit struck home, causing Holyoake to "believe that in the event of the death of any such friends as the late Allen Davenport it would be generously offered as their asylum." The word went round that there was now "a burying place for our friends".[28] The wealthy Saull was financing the cause to the last.

The announcement, on 13 November 1848, occurred at the monthly "entertainment" (which Saull chaired) to thank John Street's Directors. One of those directors was the old class warrior who had made his rapprochement with Owenism, Henry Hetherington. The next morning he mused to Holyoake: "Saull has bought a grave, and says he is able to give a friend a lift—there's a chance for us."[29]

And there was. Hetherington was to become the first occupant. Just as Saull started recovering, Hetherington, in August 1849, was struck by the cholera. He had never been one for medicines. It was believed that alcohol exacerbated cholera, and since the fifty-seven-year-old was "almost an absolute teetotaler" he thought he was safe and had refused to call a doctor until too late. Temperance could not protect him.[30]

The two went back a long way—to the old London Mechanics' Institution of the twenties. When Hetherington, the Freethinking Christian, was honing his class analysis, Saull was in league with the "Devil's Chaplain". It seemed that their ways had then parted. As a radical, Hetherington had had little truck with Owenism and his "beatific scenes" of socially-perfected man.[31] The activist fighting in the National Union of the Working Classes, the jailed editor promoting complete suffrage in his pioneering *Poor Man's Guardian*, had no time for sucking up to the rich at social tea parties, in the vain hope that they would voluntarily usher in a co-operative millennium. First must

---

28 *Reasoner* 5 (29 Nov. 1848): 429; 7 (12 Sept. 1849): 164.
29 *Reasoner* 7 (12 Sept. 1849): 164.
30 John Elliotson went so far as to claim that the disease was near "fatal amongst spirit-drinkers": "Health of Towns", *Times*, 2 Nov. 1847; *Reasoner* 7 (29 Aug. 1849): 130; (5 Sept. 1849): 152; (12 Sept. 1849): 162. Hetherington died on 23 Aug. 1849 (Barker n.d. [1938], 61–62, on confusion over the date).
31 Claeys 2002, 175–82.

come "equal rights, and their consequent Equal laws", in short, political redress for the industrious poor.[32]

But Saull had always lent across the aisle to his friend. The two had worked together at the British Association for Promoting Co-operative Knowledge and on the subsequent National Union of the Working Classes and the Metropolitan Political Union. Indeed, the two had battled the stamp duty in lock-step for years while raising fighting funds. Saull the Owenite had never lost sympathy for Hetherington's fight for political justice. Citizen Saull's outstretched hand had always been grasped. Indeed the two had linked arms on so many causes, whether in defence of the Methodist orator Rev. Joseph Rayner Stephens, imprisoned for advising the lock-outs to get their staves, or in support of the condemned Chartist insurrectionary John Frost.[33] Then there were the more conspicuous meeting grounds which cemented this camaraderie—their mutual support for the Mechanics' Hall of Science, or the Chartist Hall of the National Association for Complete Suffrage.[34] Radical and Owenite could always be found together. Saull's fraternal feelings showed when he chaired John Street meetings to get Hetherington out of jail in 1841.[35] The warmth was reciprocated, and the two had come still closer in the years leading up to Hetherington's death.

There was a further reason for that warmth. Historians have generally concentrated on Hetherington the radical firebrand of the *Poor Man's Guardian*. But what the eulogies emphasized was his later migration to Owen's camp. Holyoake, officiating at the funeral, talked of his growing "fervour" for the benevolent Owen's ideas: "they mellowed his manners" and "filled him with hope". Confirmation came from Thomas Cooper, talking on the evening of the funeral. Hetherington's "ever-increasing conviction" that a better character can only be moulded by better institutions led to his eventual "veneration" of Owen.[36] The political failures of the thirties had pushed him towards Saull's camp. By 1843, he could be found at a Harmony Hall *soirée*, singing a song of

---

32 Ibid.; Royle 1998, 52; *PMG*, 14 Jan. 1832. 245–46.
33 *Charter*, 21 Apr. 1839, 200; 15 Dec. 1839, 741; *The Operative*, 21 Apr. 1839; *CGV*, 27 Apr. 1839.
34 *National Association Gazette* 1 (30 July 1842), 243–44; *TS*, 8 Apr. 1834, 6.
35 *NS*, 27 Feb. 1841.
36 *Reasoner* 7 (5 Sept. 1849): 152; (20 Aug. 1849): 132.

his own composition to celebrate Owen's seventy-second birthday.[37] He joined the A1 Branch, attended the Congresses, and, by the time of his death, had been on the Central Board for a few years. He even helped to wind up the failed Harmony experiment.[38]

This was the "mellowed" man in his fifties. To cap it, just before he died, Hetherington produced a "Testament", mostly to show that he remained a freethinker, lest any "gloomy bigot" should try to co-opt him for their death-bed repentance stories. In this he declared his "ardent attachment" to Owen's principles: declaring something that would have shocked his younger self, "I quit this world with a firm conviction that his system is the only true road to human emancipation".[39] It was not surprising, then, that the Owenites should bury him, and Saull's was his tomb of choice. Just as Saull's museum was the resting place of radical atheists Hibbert and Petrie, now Saull's plot was to host Hetherington alongside Davenport in Kensal Green's unconsecrated pantheon.

The Owenites were paying their debt. On the night of Hetherington's death, they convened in John Street and took over the funeral arrangements. No one else had the wherewithal or organizational skill, and they were determined to keep the funeral an in-house affair. It could hardly have been out-sourced to an undertaker with no sympathy and no grasp of the special meaning of the occasion. Press reports said that the arrangements were left to "Mr. Tiffin, of the New Road".[40] Tiffin's revealing credentials were listed in the *Post Office London Directory*:

> Tiffin Charles, bug destroyer to the royal family, upholsterer, undertaker & house decorator (formerly of the Strand & New road), 30 Great Marylebone street.[41]

---

37 *NMW* 11 (27 May 1843): 394; 12 (1 June 1844): 398.
38 Royle 1998, 204–05; *NMW* 13 (5 July 1845): 441.
39 *Reasoner* 7 (5 Sept. 1849): 146; (12 Sept. 1849): 162; see also Nash 1995a, 165–66. It was to counter these conversion stories, and explain the atheist's positive attitude to death during the cholera epidemic, that Holyoake wrote his "Logic of Death" in Hetherington's wake. It reassured the faithful that a virtuous life can ease the pain of death (Goss 1908, xxxviii). It was also a homily on our one-ness with the planet. Man's "home is with the everlasting, and when he sinks, it is into the bosom of nature, the magnificent womb" ("miserable sentimentalism", one critic called this [Anon 1854, 21]). The fact that the penny pamphlet was publicly burned by outraged parsons did not hurt sales, which reached 30,000 by 1860 (Goss 1908, 11).
40 *Reasoner* 7 (12 Sept. 1849): 164.
41 *Post Office London Directory* (1852): 1025.

The joke could be missed, until you realize that Tiffin himself was a John Street director and former investor in Harmony Hall.[42] Thus he was a socialist, and (judging by the entry) a republican. The bugs were bishops and aristocrats, or the royals themselves.

They had all known one another since the London Mechanics' Institution days: Tiffin had started there too.[43] He had then migrated with Saull to Owen's Institution in Gray's Inn Road, and had joined him in setting up the Labour Exchange.[44] Tiffin had also been instrumental in establishing its John Street successor in 1839, where he worked alongside fellow director and friend Hetherington. As a consequence, Tiffin was himself at the party on 13 November 1848 when Saull announced his 'grave' offer to friends.[45] In fact Tiffin was the in-house contractor for 'social funerals'. These were marked by sympathy but also simplicity because Tiffin believed that "extravagance in reference to the dead, whom we cannot benefit, is inconsistent with Socialism, whose end is to produce happiness for the living".[46] And so it was to be for Hetherington's send-off: a moral eulogium for the living.

The result was a Social Funeral on 26 August managed by Hetherington's comrades. Being assured such a good send-off was part of the attraction for poor Owenites.[47] Hetherington's cortège might have been "simple", to meet socialist tenets, but the traditional trappings were important—"mutes" who stood over the coffin, and "pages" to accompany it. An immense crowd with banners turned out, a procession of almost five hundred, walking four abreast, with the women weeping. The pages were followed by John Street officials, all bearing coloured "wands" (long crepe-covered sticks signifying their rank, as was the old fashion). Then came twenty-six carriages, some reported thirty, others

---

42   Holyoake 1906, 2: 599. Tiffin had displayed model houses during the Tytherly planning stage and donated items to Harmony Hall, including a "washing machine": NMW 6 (16 Nov. 1839): 890; 7 (21 Mar. 1840): 1187.
43   Flexner 2014, appendix A, 388.
44   Crisis 1 (21 July 1832): 77. He was Director of the London Co-operative Building Society (founded to establish the John Street Institution), of which Saull was a Trustee: NMW 6 (14 Sept. 1839): 752. To add to his jack-of-all-trades image, Tiffin was a house broker. Saull used him as the letting agent for his eleven–room house, No. 4, Crescent Place, Burton Crescent, which was free to be rented after Owen had left: Reasoner 17 (27 Aug. 1854): 143.
45   Reasoner 5 (22 Nov. 1848): 411.
46   NMW 4 (13 Oct. 1838): 416.
47   Yeo 1971, 102–03.

nearer fifty.[48] Euston and Marylebone Roads were lined with people, and, on the long journey to Kensal Green, passers-by with caps doffed witnessed the immense procession. It was preceded by a hearse covered in silk, on which was emblazoned in silver letters:

WE OUGHT TO ENDEAVOUR TO LEAVE THE WORLD BETTER THAN WE FOUND IT

It was a motto common to freethinkers and socialists, profound yet trite, but clearly a way to unite the brethren on the moral high ground.[49] At the end, the socialist dead could still inspire the survivors.

True to this principle, Hetherington's "Testament" requested that any eulogy should benefit the living by showing his good side to be imitated and defects to be avoided. This allowed for a secular liturgy that combined regret at the loss with moral re-affirmation.[50] As he stipulated that no priest was to "interfere in any way whatever", Holyoake and James Watson did the honours. It went to prove that a freethinker's funeral could be carried off with solemnity and meaning, and in a way that even the religious could find impressive. The poor man's guardian might have expected simplicity, but there was an undisguised splendour in the proceedings. In terms of "imposing effect" nothing like it had "taken place in London for many years."[51] Between one and two thousand mourners thronged Saull's plot as his friend was entombed in it.

No opportunity was lost. The Owenites, adept at pamphleteering, gave away 2000 copies of Hetherington's 'Testament' at the cemetery gates, waiting till after the service so as not to breach etiquette.[52] Death provided an opportunity to repackage life. Some even gave Hetherington's battle a national importance. In his éloge that night at John Street, Thomas Cooper claimed that Hetherington's heroic stand against legalized oppression was more beneficial for mankind than

---

48 *Democratic Review* 1 (Sept. 1849): 155–59; *Reasoner* 7 (12 Sept. 1849): 164; *NS*, 1 Sept. 1849; Thomas Cooper 1849; Barker 1938, 52.
49 It was also William Thompson's death-bed exhortation: *NMW* 12 (23 Sept. 1843): 102–03.
50 *Reasoner* 7 (12 Sept. 1849): 162; Nash 1995a, 162.
51 *Reasoner* 7 (29 Aug. 1849): 129, 144; S. D. Collet 1855, 20; *NS*, 1 Sept. 1849; Goss 1908, xxxvii.
52 *Reasoner* 7 (12 Sept. 1849): 166.

Trafalgar or Waterloo.⁵³ Gone was the image of sedition and scoffing, rather the radical's moral life and "manly" death was being sold to the public. A holier-than-thou Hetherington, fighting the good fight against aristocratic corruption and clerical chicanery, was well marketed. All the eulogies were now added to his dying "Testament" and sold as a twopenny pamphlet.⁵⁴

None of this amused Saull's nemesis. The Rev. James Smith devoted his soothing *Family Herald* editorials to exposing this self-congratulatory atheism. He reassured his readers of the "Duality of Man" and that the material can never extinguish the spiritual. "Mr. Saull, of anti-spiritual notoriety", might isolate his atheists from Christian contamination after death. But speaking of priests, as Hetherington's Testament did, stultifying "the minds of the people by their incomprehensible doctrines, that they may the more effectually fleece the poor deluded sheep", was the language of "disaffection", and as bad as any Christian's. He deplored this "unsocial sectarian antipathy that carries itself even into the grave." Smith was becoming more conventional. That theology was imperfect was true, but so was Saull's science. Could Saull "vouch for the truth ... of his own facts or theories? Would he risk his wealth or his life on the truth of any half-dozen facts, selected at random from one of his lectures?" Saull's public persona as a monkeying wiseacre was still being shaped by Smith's penny literature, lest its gentle readers forget.

With the *Family Herald* selling 125,000 copies a week now, the word about Saull's lop-sided world was still spreading out, but it was not a good word. Smith's warnings were themselves beginning to mirror *Bible Defender* extremism: about the catastrophic consequences rivalling the French Terror should Saull's atheists ever take power. The spectre of a blood-bath hysterically raised the stakes, as families in pantries and parlours were outrageously shown Saull's moral atheism resting, not on hallowed ground, but the horrifying soil soaked in blood from regicide and revolution.⁵⁵ But such melodramatic scare tactics, trotted out in umpteen conservative religious outlets for half a century, were losing their force.

---

53   *Reasoner* 7 (5 Sept. 1849): 145.
54   Holyoake 1849b; Nash 1995a, 166.
55   *Family Herald* 7 (27 Oct. 1849): 412–13. On these leaders being Smith's work, and sales figures: W. A. Smith 1892, vii, 240.

PART IV

1850s

DESTRUCTION

26. Provisions for the Afterlife

His own months-long illness and Hetherington's death forced Saull to face up to the museum's future. He was getting old, and intimations of mortality rekindled his interest in putting his people's museum in a permanent home. Others, too, were chivvying him, knowing the value of an 'evolutionary' display to the freethought cause. Where else, after all, could you find a museum-grade fossil sequence pointing up the self-empowered progress of life? Or the historical artefacts illustrating humanity's rise from savagery, crowned, at the moral apotheosis, by the revered relics of Hibbert and Petrie?

From the first, Saull had wanted to bequeath his museum to the Owenites, to add a hands-on, deep-history experience to their "rational" schooling programme. Typical of Owen's disciples, he never lost his zeal for children's education. Using the fossil display would ease "the infant mind" into naturalistic ways of thinking. By breaking the thrall of parsonism, and the stultified science of Divine Creation used by the gentry to instil docility, it would help the masses challenge their servitude.[1] Implicit in this was that geology, rightly understood, could function ideologically. This distinguished it from that (often cynically) disparaged craft theory taught in mechanics' institutions, which would crank up the output of the workforce, or, in Saull's words, hone the skills to make domestics "better servants".[2] His Enlightenment faith in a liberating materialism never waned. But the aim enlarged over the years, as the museum became targeted more generally at working adults, mostly unschooled themselves, but receptive to the revelations of the new geology. In this respect it was meant to go to Harmony,

---

1 *Crisis* 3 (28 Dec. 1833), 144; (4 Jan. 1834): 150.
2 *Crisis* 2 (1 June 1833): 163; Shapin and Barnes 1976, 1977, 55–56; Johnson 1979.

but with the collapse of that dream, the museum's destination again became uncertain.³

Because his fossil cabinet was intended to benefit the dispossessed, Saull never considered bequeathing it to the aristocrat-controlled British Museum. Crunch time came in the late 1840s, with the collapse of Harmony and Owenism in disarray. It forced Saull to search for a new home. What emerged was a scheme for a working man's hall in London, built from the ground-up, with his museum, now valued at £2,000, lying at its heart. By 1847, Saull was already promising to endow such a public building with a £2,000 cash bequest, "constituting a munificent donation to the cause of science".⁴

Others rallied to the idea. For a couple of years, Holyoake had been floating the notion of a new organizational headquarters, an "Atheon" (a Pantheon without the gods). A "public fraternity" he called it, supporting atheists, republicans, and communists. It was to provide a base for the trades, and share intelligence, hence its projected atheist library, reading room, and "theological museum" of defunct deities. But where Holyoake's museum was to contain *"Blasphemy relics"* from the atheist trials, thus immortalizing them, Saull's promised to substitute real relics to evoke a blasphemous 'evolution'.⁵ One letter-writer in the *Reasoner* immediately saw the link-up. "If Mr. Saul [sic] were to make his Museum the foundation" of an Atheon, what an impetus the cause would receive, "and how much more would it connect the name of Devonshire Saul [sic] with that anti-superstitional progress he has so much at heart, than any posthumous bequest to accidental cultivators of Science?" That was the nub, the posthumous aspect. "Could anything induce that gentleman to make his disposal of his bequest in life, instead of leaving it to the uncertainty of death and the law?" The wisdom in that reflected another salutary event, which Saull knew only too well. Were Saull to die, speculated the correspondent, would it "not be the case of Barber Beaumont repeated?"⁶

John Thomas Barber Beaumont (1774–1841) had made a fortune as managing director of the County Fire Office, one of the largest fire

---

3   Holyoake 1906, 1: 190.
4   *UR*, 15 Sept. 1847, 83.
5   *Movement* 2 (1 Jan. 1845): 4–5; Royle 1974, 88; McCabe 1908, 107.
6   *Reasoner* 3 (6 Oct. 1847): 551.

insurers. The chronic asthmatic was praised as a "poor man's friend". He had set up the first Provident Institution (a savings bank) in the country and, in 1840, just before his death, had ploughed £6,000 into a new Philosophical Institution in Beaumont-square, Mile-End.[7] Like Saull, Beaumont was a fossil collector, and interested in the vegetation of coal fields, in which Saull's museum excelled. His Institution had a museum housing his minerals, and he was still arranging them on the days before his death.[8]

Beaumont's will was written *before* the Institution was built. The building had gone up and looked very grand. But he, evidently, had failed to change his will, which left £13,000 to build it, causing confusion. In 1847, as the *Reasoner* writer was cajoling Saull, the court was *still* trying to decide how, or indeed whether, the sum should be applied.[9] More confusion occurred because Beaumont named Saull as a Trustee, along with John Elliotson (among others). Both however declined, presumably because the will stipulated that the Institution was to cultivate the "principles of natural theology and the wisdom of God". Worse from Saull's perspective, it had a chapel, and was running Sunday lectures to "enforce the great principles of practical religion and morality."[10] These were "introduced by sacred music" in order to give the lecturer's moral and religious "exhortations" solemnity.[11] It was too much for Saull, and the loss of the two Trustees and death of a third further complicated the court's decision, which dragged on.

It was an object lesson, and Saull should have heeded it. Even as Saull became ill in 1849, Holyoake was pleading for the community to build a "unique College and Lecture Hall". Two years had passed, he noted, since the *Reasoner* correspondent had "sought to enlist Mr. Saull's interest, in the disposal of his Museum."[12] Saull's brush with death now gave the project a greater urgency.

---

7   *NMW* 8 (14 Nov. 1840): 313; bank: *MC*, 21 May 1841.
8   *MC*, 21 May 1841; *Proceedings of the Geological Society* 3 (1842): 152–53.
9   *Daily News*, 5 July 1847.
10 *NMW* 8 (14 Nov. 1840): 313; *St. James's Chronicle*, 3 July 1847, 1; *Standard*, 5 July 1847.
11 *Phrenological Journal* 17 (Jan. 1844): 54. For all that, the venue was unsectarian and non-dogmatic, and even F. D. Maurice (1884, 2: 64) in 1851, observed that the 1200–capacity hall was "being used partly for dancing, partly for some infidel lectures, partly for anti-Papal meetings, &c."
12 *Reasoner* 6 (24 Jan. 1849): 57.

As he was recovering a new vista opened up. Another fortune was promised, this time by a wealthy supporter of Owen and Feargus O'Connor. A Harrow gentleman, Charles James Jenkins, had come, like Saull, to see that *"education* was the most important—if not *the* most important—means" of enabling the industrious classes to gain the initiative.[13] Consequently, the bulk of his fortune (£10,000) in September 1849 was willed to trustees tasked with building a new "People's Institute", with another £3000 in the offing.[14] He stipulated a "commodious hall" for lectures, with offices, reading room, and library, somewhere in Central London, to be available as a "place of resort for working men, operatives, and artisans during the evenings, and as a school for the children of such classes during the day-time." He made provisions for schoolmasters, and added the proviso that the "Jenkins' Institution" should be totally non-discriminatory in respect to "country or colour" or "religious or political tenets".[15] It was too good not to attract Saull, who immediately offered to present his 20,000 exhibit "magnificent museum" to the institution, while others pledged a thousand volumes for the library.[16]

But within a year or two, suspicions were raised and it began to seem like another forlorn hope. An impetuous Holyoake started chaffing:

> We could name half a dozen gentlemen of fortune who, for some years past, have publicly avowed their intention of leaving bequests, in some cases to individuals, in others of founding Libraries, Museums, and erecting Public Institutions. We pray those who thus mean well to Free-Thought to profit by the serious failures that from year to year are recorded.

That shaft was aimed at Saull. He rammed the point home. Where is the Beaumont money? "Where is the 'Jenkins Institution?'" he asked in 1851:

> the worthy old gentleman ... died in the certain hope that the "Jenkins Institution" would be a noble and useful monument of his life. The law has stepped in—taken it all away—and not one brick will ever be laid in

---

13   *NS* 13 (13 Apr. 1850): 3; *Reasoner* 7 (26 Sept. 1849): 207. Jenkins died on 7 Sept. 1849 and bequeathed £500 to O'Connor and £200 to Robert Owen: C. J. Jenkins, Will, National Archives, Kew, PROB-11-2101-173, ff. 143-45.
14   *Reasoner* 7 (24 Oct. 1849): 270.
15   *Reasoner* 7 (26 Sept. 1849): 207; (14 Nov. 1849): 318; (5 Dec. 1849): 361-62.
16   *Spirit of the Age* 2 (2 Feb. 1850): 79; *Athenaeum*, 1 Dec. 1849, 1210.

commemoration of his name. Again we say to all those gentlemen who meditate anything munificent for the people—'Do what you intend while you live—nothing can be depended upon that is to come after death.'[17]

Twilight Distractions

The warning was stark. Yet for Saull, running a business, museum, secular and archaeological meets, and filling umpteen offices, the distractions were manifold. There was always something pushing posterity into the background.

All the while he was getting older and, having stared down death, more contemplative. The nostalgia of old age was showing. In the 1850s, he emerged after his illness at his old stamping ground, the City Road Hall of Science, to reminisce about the '20s, Carlile and Taylor, police spies, and state prosecutions, and how, although he escaped prison himself, he took massive financial hits covering the trials.[18] This might have been distant history, yet there were still tangible reminders of those days. Saull's group never forgot the Carlile family. In 1850, Eliza Sharples, "Mrs Carlile" in common law—the fiery "Lady of the Rotunda"—was now following in the "death-wake" of so many herself. As usual, there were tales of penury, of suffering, although with "death staring her hard in the face" she "still adheres most tenaciously to her principles!" And as usual, Saull and his friends contributed to a fund. There were also donations to Carlile's children, which helped his two daughters sail for America in 1852.[19]

The Mechanics' Hall of Science remained one of Saull's favourite haunts. A glimpse of what the venue was like at this time was provided by a Congregationalist minister, who peeked in to view the uninhibited behaviour of "men and women who have altogether thrown off the moral restraints of religion". He bought himself a ticket to a tea party, and insinuated himself among "the enemies of Christianity, at a time when they are the most actively engaged in the prosecution of their godless and debasing enterprise".[20] What resulted was a stinging account sent to

---

17 *Reasoner* 10 (5 Mar. 1851): 354. Holyoake was being hasty. Also, a "Jenkins Secular School" was set up in John Street in 1853.
18 *Reasoner* 8 (20 Feb. 1850): 54; 16 (5 Feb. 1854): 98–99.
19 *Reasoner* 12 (10 Dec. 1851): 64; (21 Apr. 1852): 367; *NS*, 22 Nov. 1851.
20 *British Banner*, 5 Oct. 1853, 705–06.

the big-circulation *British Banner*, a 4*d* Congregationalist weekly, set up in 1847 to fight infidelity and promote Evangelicalism. The paper thrived on such reports, being obsessed with London's "spiritual destitution", growth of socialist halls, and want of anti-infidel preachers to redeem the populace.[21]

The 400-seater hall, ran the report, although not "handsome, is fitted up with a good deal of taste, and lighted by an elegant glass chandelier". A huge platform extended the entire width of the room for the lectures, to the left of which was a small gallery.[22] Up and down the hall walked the "Negative Evangelist," Holyoake, "a dark-haired, lankey [*sic*], amiable-looking young man", resembling "a minister of the Gospel wearing his week-night black stock", chatting to "his loving flock", all seated at tables. It was bad enough that most of the two hundred taking tea were young, showing that the secularists were recruiting a new generation, but the real horror for the *Banner* was that so many were women, long idealized as the angel in the home and moral bedrock of the family.[23] These were lost souls,

> silly creatures, who, having had their minds ... perverted from the right ways of the Lord, vainly expect to be able to cure all the evils of oppression and wrong which exist in the world, by banishing therefrom all that savours of modesty or religion.

The secular halls were in competition with the Congregationalist chapels, and their Sabbath socializing and moral sermonizing clearly irked the interloper. And yet Holyoake's chat about putting parsons on half pay, or thanking the cook instead of God for dinner, and such "mean, grovelling, despicable, and absolutely blasphemous sentiments, found a ready and hearty echo in the breasts of his obtuse, shameless, and godless audience."[24]

---

21 For example, *British Banner*, 24 Oct. 1849, 677; 9 June 1852, 9; 13 Oct. 1852, 691; 11 Jan. 1854, 36; Halévy 1961, 390, on the *Banner*.
22 *British Banner*, 5 Oct. 1853, 705–06.
23 The "Angel in the House" idealisation is covered in B. Taylor 1983, 30; E. Richards 2017, ch. 7; Tosh 1999, 55; Hilton 2006, 363. As Schwartz 2013, 16–17, points out, the idea of women's 'higher' moral nature, emphasized by the religious press, and which led to their civilizing role, inside and outside the home, in education and so on, was shared by the Owenites.
24 *British Banner*, 5 Oct. 1853, 705–06.

Saull, another "negative evangelist", shared Holyoake's sentiments. He was now feeling an anachronism himself among the young. Perhaps, too, it struck onlookers as incongruous that so many secular "tea parties" here, with their pointed abstinence from alcohol, were presided over by a wine merchant. Politics still vied with secularism at City Road, and the last gasps of Chartism here showed that the Hall's radicals retained their teeth.

Saull was among them; his radicalism had never dimmed. This was proved by events at the City Road Hall of Science late in 1851, when Saull joined the agitation over events in Bonner's Fields. This was East London's playground, an area between Spitalfields and Bethnal Green, centred on the three-hundred-acre Victoria Park. Once the site of Bishop Bonner's Palace, now it was used by the Spitalfields artisans, especially in summer when they bathed in the ornamental lake. But the salubrious image belied the jaundiced police view. Problems had started in the late 1840s. The park had opened just before the great Chartist outdoor gatherings in 1848, and stave-wielding demonstrators had made Bonner's Fields synonymous with sedition and riot in the authorities' eyes. In response, the government had banned Chartist assemblies in Bonner's Fields. But still the defiant Chartists continued their "monster meetings". On occasion over five thousand troops, and as many special constables with cutlasses, backed by guns from Woolwich arsenal, had been mobilized to disperse the crowds. The Chartist leaders had been given stiff sentences for inflammatory speeches, while the people had vented their fury by showering the police with stones and smashing up the nearby church, which had sheltered the specials.[25]

The febrile atmosphere continued into the early 1850s. It made Bonner's Fields fertile ground for freethought propagandists, who set up stalls amid the crowds. These outdoor gatherings were a mêlée, part carnival, part "war", as the soap-box orators competed with Christian preachers, who had their own tents for tract distribution and refreshment. As fast as *Reasoners* were sold (eight dozen could go on a Sunday), the freethought posters were torn down. A young Charles Bradlaugh (eventually the first openly atheist Member of Parliament) converted to freethought at this time, and the boy cut his teeth in Bonner's Fields.

---

25   Goodway 1982, 79, 83–88; H. G. Clarke 1851b, 64–65; Hogben n.d., 52.

Soon "scores or hundreds" were milling round his stump.[26] It was to celebrate this recruiting success in Bonner's Fields that the East London freethinkers held a public tea party in the Hall of Science, hosted by Saull, Holyoake, Robert Cooper, and James Watson.[27] But the congratulations were premature. There remained considerable hostility to the atheists in the park. Christian missions were revitalized and tracts with titles such as *Park Visitor and Christian Reasoner* were passed out to counter the threat. Then, in May 1852, *all* assemblages in Bonner's Fields were prohibited, the ban enforced by armed police.[28] The main target was the secularists, which left the Church missions outraged that the Christian baby was being thrown out with the atheist bathwater.

The result was that the "East London Reasoners" retreated back indoors and planned a vast new hall for the swelling crowds. When the Baptists moved out, they took over the dilapidated Morpeth Street Chapel in Bethnal Green. Carpenters among them pulled out the pews and put up a platform. Painters gave it five fresh coats, all working gratis. But the "gas-fitters and paper-hangers require payment in a more vulgar coin" and so they put out a call for help, and that is where Saull and the other donors came in.[29] The funding started, and by January 1853 they opened their 700-capacity "Tower Hamlets Literary Institution", which the East Londoners proudly ranked for size alongside John Street and the City Road Hall of Science.[30]

Saull was still funding secular halls, but not one to house his own museum. He simply had too many irons in the fire. He admitted as much at City Road: "He was now advanced in years", he conceded, "but his interest in the 'good old cause' was undiminished."[31] And as old causes evolved into new, he remained in the fray, and the side-tracking continued.

Nostalgia was again stirred at a new talking shop, the Co-operative League, founded in March 1852 by an alliance of Christian Socialists and

---

26 *Reasoner* 10 (30 Oct. 1850): 35–36; (6 Nov. 1850): 54; Bradlaugh 1891, 6–8; Royle 1974, 210–11.
27 *Reasoner* 11 (12 Nov. 1851): 407.
28 *Reasoner* 12 (4 Feb. 1852): 191; (2 June 1852): 452–53; (9 June 1852): 470.
29 *Reasoner* 13 (15 Dec. 1852): 432.
30 *Reasoner* 14 (12 Jan. 1853): 23. The President was Robert Le Blond, another wealthy businessman and financier of freethought.
31 *Reasoner* 16 (5 Feb. 1854): 97–98.

Owenites to hammer out the direction of the co-operative movement. The League's philosophical bent must have appealed to Saull, who was on board immediately. Another aspect, its educational onus, was dear to his heart, even if the co-op rank-and-file were more concerned with a cash bonus than "eddication".[32] Among the first items for League discussion was the failure of the original Labour Exchange, which Saull, one of those still alive who had been involved, put down to the unequal demand for exchange goods: everyone wanted bread and the daily necessities.[33] But so much had happened in the intervening years. Younger recruits now talked up the latest Co-operative Stores, organized on a buy-and-sell (rather than swap) basis, with profits going to the buyers, after interest had been paid to capital investors.

Saull was still front and centre at Holyoake's London Secular Society. With the shattering of Owenism, this had taken up the slack and taken over the venues: it met at the Literary Institution in John Street. There were also daughter secular societies in many provincial cities within a few years, often just the Owenite branches rebranded.[34] Saull moved with the new men. He became treasurer of the London Secular Society, basically just transferring from his old Owenite duties.[35] And he carried on addressing the new men with little change in tone.

The secular causes were never ending, and securing the museum's permanent home always seemed to be shunted into second place. The astonishing number of campaigns Saull supported in the early fifties simply eclipsed such mundane matters. He was fighting the remaining penny newspaper stamp duty (carrying on Hetherington's campaign),[36] presiding over dinners to Owen, and speaking at the theatrical debates

---

32　J. F. C. Harrison 1961, 105. The League was founded by the Christian socialist Vansittart Neale and old social missionary Lloyd Jones, who were attempting to get the various Co-operative Societies to buy their goods wholesale through one depot, Neale's Central Co-operative Agency in Oxford Street (which opened in 1850): Cole [1944], 110; W. H. Brown 1924, 59–60; Royle 1974, 150; McCabe 1908, 1: 191–92.

33　*Leader* 3 (1852): 509; *The Star of Freedom*, 5 June 1852; *Reynolds's Newspaper*, 1 Aug. 1852. See also Saull's similar talk at the London Working Men's Association: *Journal of Association* (1852): 182.

34　J. F. C. Harrison 1969, 246; Royle 1974, 177.

35　*Reasoner* 16 (15 Jan. 1854) 38; (5 Feb. 1854): 83. Right to the end Saull was putting cash into the Secular Propagandist Fund.

36　*Reasoner* 10 (13 Nov. 1850): 11 (5 Nov. 1851): 383; *Leader* 2 (1851) 1012.

with Christians.³⁷ On other fronts the activity showed no let up. Aldersgate Ward politics went on as usual, as Saull joined other City merchants to petition against the window tax, or to set up a working group to establish a better water supply to the ever-expanding city.³⁸

He was also campaigning to amend the bankruptcy law. The recent removal of imprisonment for £20 debts had led to rise in City fraud, and small creditors wanted the power to seize assets to clear the debt.³⁹ This was especially important to a wine trader like Saull, who dealt extensively with credit. Credit, debt and bankruptcy were all too familiar to him, and sometimes came close to home. His publican nephew had gone bankrupt, perhaps as a result of anti-infidel policing, and in the later years Saull had a bankrupt working as a clerk in his wine depot.⁴⁰

Nor did campaigning on national politics let up. Ever the radical champion, he "poured a little vinegar" into City meetings, which sought Lord John Russell as their Liberal MP—Saull preferred a more radical voice.⁴¹ And, as ever, being one of the "agitators", he incurred "the sneer of pure Whig flunkeyism".⁴² These years, too, saw a revival of middle and working-class alliances. Joseph Hume's old reformers joined anti-corn-law activists and Chartists under Feargus O'Connor in the National Reform Society, which pressed for an extension of the franchise.⁴³ Then there were new lobby groups demanding the ballot.⁴⁴ Officiating left Saull as rushed as ever.

---

37 *Reasoner* 16 (16 Apr. 1854): 261.
38 *Daily News*, 13 Feb. 1850; *MC*, 25 Feb. 1851; *NS*, 1 Mar. 1851.
39 Numerous press reports of Saull taking the small traders' side on the question of debtor and creditor laws, and his role in the Equitable Debtor and Creditor Association, show his extensive engagement with the question, starting in 1845 (e.g., *Examiner*, 29 Mar. 1845) and culminating in 1849 (e.g., *Daily News*, 26 May 1849.)
40 This was Edward T. Tweed: *Evening Star*, 20 Jan. 1858, 4.
41 *Guardian*, 26 May 1852, 359.
42 *The Era*, 11 July 1852; *Daily News*, 25 May 1852; *Times*, 25 May 1852, 8; *MC*, 25 May 1852, 5.
43 This was the Metropolitan Parliamentary and Financial Reform Society (founded 1849), which became the National Reform Association in 1850. The Society's embrace of Chartists proved too much for free-traders such as John Bright, but it appealed to Saull: *Daily News*, 1 Sept. 1849, 1; Jan. 8 1850; 4 Feb. 1851; 12 Mar. 1851; *Express*, 8 Jan. 1850, 3; *Standard*, 8 Jan. 1850, 1; *NS*, 15 Mar. 1851. On this Association: Huch and Ziegler 1985, 150; Maccoby 1935, 315.
44 On Saull at the Ballot Society: *Daily News*, 18 July 1853.

Saull's expanding role in the learned bodies squeezed the last hours out of his free time. Trains now made it possible for him to travel huge distances on the firm's business, and he always took the opportunity to talk locally on the meaning of his fossil sequence or on social progress revealed by nearby antiquities.[45] Added to this, his council duties in the societies led to him criss-crossing the country. He would join the peripatetic British Archaeological Association each August,[46] and visit the British Association for the Advancement of Science meeting the following month.[47] Then, during the season, Saull routinely attended meetings of the Society of Antiquaries in town, and he was on the council of the splinter Numismatic Society from 1851. Add to all of this the Council of the Ethnological Society in 1850, and the Chronological Institute in 1852–54, and one senses that he spent a mint of money on dues, and lost a lot of time on bureaucracy.

Saull's Exhibition and the Great Exhibition

At the Society of Antiquaries, Saull continued to expand on his schema of ancient progress from primeval to pastoral through the Celtic-Roman period. But now there was an innovation: his use of models.[48] At one meeting in 1850, he displayed scaled-down replicas to illustrate the hill forts at the centre of the transition. Four miniatures sat on the table as he talked "On an Ancient Fortified Station, and other Celtic or early British Remains, in Cornwall". They were to illustrate the sophistication of these Cornish forts, with their walls and ditches, and to suggest that, like the local tin-mining dwellings, with stone-walled and ceilinged cells, they were erected by Mediterranean traders. Even here the freethinker protruded. Another model, of Cornish stone circles, were, he thought, where the tribes sat in council, not in religious observance. He was

---

45 *Monthly Notices of the Royal Astronomical Society* 16 (Feb. 1856): 90.
46 He was in Manchester (1850), Derby (1851, on the General Committee), Newark (1852), Chichester (1853), and Chepstow (1854), where again he was on the General Committee.
47 Ipswich (1851), Belfast (1852), Hull (1853, on the Geography and Ethnology Committee), and Liverpool (1854).
48 Despite modelling being a standard way of representing sites three-dimensionally, almost no historiographical studies exist of the archaeological procedure beyond Christopher Evans 2004.

loath to allow any precedence to ancient veneration. Displaying another model, he demurred from those "who were inclined to see in it an object of superstition; and preferred to assign it to another purpose, namely, to the sports and pastimes of the Britons".[49] These models possibly ended up in the museum.

Certainly, miniature models of *saurians* could be found there. A *Mining Manual* reporter commented on seeing them in 1851. Museum modelling had a respectable pedigree. Sowerby had used models of birds and mammals in his museum, and fossil casts were becoming common as exchange items, although not without problems. (Saull's wild millenarian tea-sipping friend Thomas Hawkins even modelled missing bones in his ichthyosaurs and plesiosaurs before selling the 'complete' skeletons to the British Museum, which caused a furore.[50]) The *Mining Manual* simply noted "models of the smaller saurian tribe". Possibly these were the extinct reptiles.[51] If so, they were a didactic tool to give audiences a hands-on experience of Saull's pride of place, the Wealden reptiles. Models could have showed what they were thought to have looked like in life. As such they would have helped illustrate his Owenite sequence, from ruling reptiles through ice-age mammals, and on to perfectible humans—the 'rational' core of his programme for educational regeneration.

If these were reconstructions of extinct saurians, it would have made them the earliest known. Also, they would have represented vastly different values from the commercially-based models of the day. Of the latter, consider the *Iguanodon* and *Megalosaurus* miniatures[52] being sold,

---

49 *PSA* 2 (1853): 91–92. Trips were often multi-purposed, even if they did not involve wine merchandizing, and so we find him presenting fossil sharks' teeth and molluscs from the Suffolk Crag to the Royal Cornwall Geological Society's museum in Penzance at the same time: *Royal Cornwall Gazette*, 4 Oct. 1850, 6; Royal Geological Society of Cornwall, *Thirty-Seventh Annual Report of the Council*, 1850, 25.

50 *MNH* 4 (Jan. 1840): Appendix, 11–44.

51 *Mining Manual and Almanack for 1851*, 136. Given the terminology of the times, the word *saurian* (especially qualified by "smaller") could equally have referred to a living reptile.

52 The sculptor Waterhouse Hawkins (1854) was touting one-inch-to-the-foot models at the Society of Arts in May 1854, where he was encouraged to mass produce them for schools. The marketing opportunity opened up swiftly, because the entrepreneurial geologist James Tennant, who had been making casts of saurian fossils since the 1840s at his shop in the Strand (Moore, Thackray and Morgan 1991, 137), was selling Hawkins' sets of monster miniatures for five guineas. These are listed in the two-page flyer, "Key to a Coloured Lithographic Plate of

scaled-down copies of the full-size monsters that were being constructed in Crystal Palace Park, near Sydenham, just outside London, in 1854. The Park's concrete effigies glorified the Old Immoral World—they turned the past to their own advantage, forced "us to think upwards toward the Creator's past eternity," and "show us His power", said the Methodist *London Quarterly*. And to give further perspective, the reporter looked through the hissing "vein of steam" from the megatherium nostrils to see the reassuring "spire of Penge church" rising in the distance. Tories saw the Palace amazements as a "vast safety-valve" for the multitudes. Even if the exhibits had no educational value for the hordes, they were somehow part of God's inscrutable plan to 'civilize' them. Nor need "Timid politicians" have any fear "of any countenance to socialism" from the experiment: "It is property, in the shape of hard money" which had built this mausoleum of Britain's ancient empire, and a healthy dividend was expected.[53] Speculative capitalism was at root of the venture, the gigantic Crystal Palace saurians being financed by investors who expected a good gate return. The monsters were "Antediluvian" money spinners, and the public paid to see the "prae-Adamite" spectacle. Of course, words like 'antediluvian' and 'preAdamite' were anathema to Saull, part of the grubbing exploitation of the 'Old Immoral World' to be spurned.[54]

The Crystal Palace monsters were to accompany the Great Exhibition, which moved from Hyde Park to Sydenham in 1854. Socialists called for the Exhibition to be opened on a Sunday, and at a reduced rate, so that poorer people could visit it.[55] But why *pay* at all to view the Great Exhibition's geological exhibits, said the *Lancaster Gazette* in a diatribe against the aggrandizing metropolis, when a visitor could see so many for free at Saull's.[56] This was truly free, with unrestricted

---

Waterhouse Hawkins's Restorations of Extinct Animals" (n.d.), my copy of an original owned by Steve Gould. M. Freeman 2004, 21, discusses the sale of strata models before the 1850s.

53 *Westminster Review* 62 (Oct. 1854): 542; *London Quarterly Review* 3 (Oct. 1854): 235, 238; *Quarterly Review* 96 (Mar. 1854): 307; J. A. Secord 2004a, 139; Dawson 2016, 172–208.

54 Their very unfamiliarity led the monsters to be viewed by the public as "Frankensteinic" oddities (*MC*, 2 Jan. 1854, 3), and, as such, they were more freak-show spectacles than educational.

55 *Reasoner* 14 (11 May 1853): 293.

56 *Lancaster Gazette*, 9 Mar. 1850, 4.

access: socialist subsidized free education, for the benefit of the people not the speculator, was on offer. By now, Saull's museum promotion stretched far beyond the artisanal or radical community. There was a vast potential clientele, middle-class hobbyists, students of archaeology or geology, ladies piqued by Roman London beneath their feet, visiting dignitaries, antiquarian gentry, and the "intellectual holyday-maker"[57] from the provinces. All were catered for by the increasing numbers of press listings, and all of the press listed Saull's free museum.[58]

The London guidebooks were becoming an essential part of the tourist's kit. With railways shrinking the country, huge numbers were now visiting London for the first time. In 1850, the *Times* said that "Thirty years ago not one countryman in one hundred had seen the metropolis. There is now scarcely one in the same number who has not spent the day there." But they met a huge, sprawling, smoggy, "strange land", a "Human Awful Wonder",[59] with a gigantic 2.7 million inhabitants—the largest city in the Western world. Its streets swamped foreigners, who were made giddy by the great "waves of people silently surging through the gloom." This was the modern Babylon, "whose extravagant immensity a pedestrian could not encompass in a day's time". It was all "profoundly disturbing and overwhelming", all "turmoil and bustle", and tourists found themselves "lost in a labyrinth".[60]

Hence the profusion of "strangers' guides" and "intellectual guides". They cashed in on the Great Exhibition in 1851, centred on Joseph Paxton's twenty-acre glass and steel building in Hyde Park. This symbol of industrial might, with its vast assemblage of manufactures from around the world, lured six million paying visitors. They included hordes of working people, who so frightened Wellington that he brought in 10,000 troops, fearing the worst.[61] Saull took a co-operative view of the exhibition, as proof of a growing fraternity of feeling between nations. Upturning the press rhetoric, he considered that it laid bare

---

57 *Bell's Life in London and Sporting Chronicle*, 31 Mar. 1850.
58 It was listed in everything from *The Parlour Magazine* to the *Civil Engineer and Mining Manual*; from *Bell's Life in London* to *Reynolds's Newspaper*, and in all the 1851 London guides cashing in on the Exhibition: *Gilbert's Visitor's Guide to London, London As it Is Today, London What to See, London In All Its Glory*; as well as *Black's Guide to London* (1853).
59 J. White 2007, 78.
60 Hogben n.d., 5. This, of course, listed Saull's museum. Tristan 1980, 1–2.
61 J. White 2007, 169.

capitalist greed. It was a "powerful rebuke to those who continued to deprive the men who could produce such results from the power of self-government. Was it not a disgrace that at a time when more wealth was created than at any time in the world's history, that more misery ... should be found among its producers"?[62]

The swathe of city guides capitalizing on the event gave Saull's museum a boost, listing it in places to visit, and highlighting the individual attention offered. ("The proprietor usually explains personally to visitors the various phenomena, and develops some new views on the earth's motion.") "Rich" in specimens, its attractions were the "gigantic" fossils, especially the tree ferns from that distant age of coal which was powering the country's industrial growth.[63] The one benefit they all extolled was the price—it was free, and no questions asked. The cheapest turnstile entry to the Great Exhibition was a shilling, and it was expected to rake in £360,000 net. But even before the first girders from Smethwick Iron Works had arrived,[64] the papers were pointing parsimonious visitors to alternative free venues, particularly Saull's.

The museum was open Thursdays, and Saturdays on some festive occasions.[65] "The favourite, but false idea, that educational institutions are not valued unless they are paid for, stands completely in the way of the poor but zealous student", complained the *Civil Engineer*. The "rank-and-file" are "left to scramble on", it added, unfavourably comparing Britain to France, where state sponsorship meant "There are schools where the greatest professors teach, and the poorest mechanic can enter." How is the new archaeology to spread if students cannot afford the fees asked by the societies, it went on, praising Saull and Bowerbank for opening their cabinets. If only all "Geology and natural history" benefited "from such freedom".[66] Students swelled the visitor numbers,

---

62 NS, 24 May 1851.
63 H. G. Clarke 1851a, 242; 1851b, 143; Gilbert 1851, 139; A. and C. Black 1853, 310.
64 P. Cunningham 1851, xlix.
65 At least that was the case over Christmas 1847: *New Weekly Catholic Magazine*, 26 June 1847, 166; *Morning Post*, 27 Dec. 1847.
66 *Civil Engineer and Architect's Journal*, 17 (Feb. 1854): 42–43. The papers were now separately listing London's free institutions, which included Saull's museum: e.g., *Bell's Life in London and Sporting Chronicle*, 31 Mar. 1850; *Reynolds's Newspaper*, 20 Apr. 1851.

with Saull's collection being credited in text-books,[67] and the mining and civil engineering journals puffing it. It was, said the *Mining Manual and Almanack*, "considered to be the largest private collection of organic remains in the United Kingdom."[68] By 1851, it comprised well over 20,000 exhibits, and Saull would give a walking tour and impromptu lecture on the artefacts if large numbers of students or visitors turned up.

With the students came the professors, a generation older, and with a different take on things. Framed by Saull's 'evolutionary' explanation, the exhibits appealed less to the older party. Consider the stern American Congregationalist and geologist, the Rev. Edward Hitchcock, President of Amherst College in Massachusetts, who was touring the old country in 1850. He was a good surveying geologist, famed for his study of the fossilized footprints left by huge "birds"—or, as it turned out, bipedal reptiles—that once walked in the Connecticut River valley. Hitchcock was equally famous for dealing with threats to biblical orthodoxy, and would return home to see his *Religion of Geology* through the press. The book irked even the *Monthly Christian Spectator*, which regretted that he should have tried to use the strata to deduce not merely God's power, but even our "piety towards God".[69] Nevertheless, it supplied ammunition to fire at the infidel. The *Bible and the People* used it to demonstrate that only Jehovah could replenish life after each geological catastrophe and adapt it to the "improved condition" of the earth's surface at that time.[70] Hitchcock visited the museum and admitted: "Many good things in it, but dirty & not well exposed. The fossils are distributed through the formations in proper order & the collection if put into proper cases & light would be a valuable one." Having presumably suffered a Saull lecture, probably monkey-men and all, Hitchcock was not inclined to be generous. "Mr. Saull seems to me superficial in geology", he concluded in his diary.[71]

---

67 For example, Dixon 1850, 55; Morris 1854, iv; G. F. Richardson 1855, 353, 379, 392.
68 *Mining Manual and Almanack for 1851*, 136; *Civil Engineer and Architect's Journal* 16 (Apr. 1853): 125; 17 (Feb. 1854): 42–43; Timbs 1855, 542.
69 *Monthly Christian Spectator* 2 (June 1852): 379; footprints, Desmond 1982, 129.
70 *Bible and the People* 2 (1852): 447. The freethinkers for their part used the book as a foil: *Reasoner* 16 (25 June 1854): 418; *LI* 2 (Dec. 1855): 136.
71 Herbert, "Edward Hitchcock", 33.

26. Provisions for the Afterlife

The museum was even listed in Paris.[72] Boucher de Perthes turned up to see it at this time. He was no stranger to controversy himself, having raised eyebrows by suggesting his stone tools had been fashioned by humans living alongside extinct mastodons. He did not think much of Saull's Roman and Celtic axes, but he was awed by the upstairs gallery, rich in tropical "plants and fruits coming from English collieries". It was size that mattered in an age of spectacle, and Saull pointed him to a gigantic fossil shark's tooth, which must have come from a fish up to eighty feet long, impressive enough for Boucher de Perthes to record it in his journal.[73]

Each cultural group brought a unique understanding to the artefacts in Aldersgate Street. Distinct interests led to different perceptions, whether it was the *Northern Star* Chartist awed by the skeletal shrine to his fallen heroes, the antiquarian studying the Roman foundations of his world city, or the mine engineer's interest in the swamp-crushed tree ferns lying in the hearth of Britain's coal-powered economy. But to geologists, the focus remained on a two-foot slab of Isle of Wight rock. The rock had been cracked, figuratively speaking, by the prickly Richard Owen, a man now making enemies among a new generation (a "queer fish", the brash young T. H. Huxley called him[74]). Owen had diagnosed the fossil as the five fused sacral vertebrae of *Iguanodon* and used this fusion as the basis of his new order, Dinosauria.[75] But this accelerated a 'proprietary' tussle with Gideon Mantell, who had a vested interest (this was, after all, *Iguanodon Mantelli*, capitalized at the time to bring out the personal importance). Mantell had been the first to illustrate Saull's sacrum (in 1849), having borrowed it to expose more of the fossil, from which he deduced that it had six fused vertebrae.[76] There was no love lost as Owen and Mantell fought over this intellectual property. Saull was caught in the middle, unswervingly faithful to Mantell, but careful to keep Owen onside.[77] One squib fabricated a City trial between Owen

---

72 Duckett 1853–60, 12: 412.
73 Perthes 1863, 416–17.
74 T. H. Huxley to Eliza Scott, 20 May 1851 (letter in the possession of Angela Darwin).
75 Richard Owen 1841 [1842], 130; Torrens 2014, 671.
76 Mantell and Melville 1849, 275; Torrens 1997, 183. In 1851 Mantell had a model of it made, presumably for his own museum (J. A. Cooper 2010, 153).
77 Saull would continue to ask Richard Owen's advice on new fossils, and allow Owen to prepare his specimens. He would send his warehouseman along with

and Mantell, before the Lord Mayor, thus mimicking decades of trials (including Saull's own) of corporation reprobates. Mantell died in 1852, and the squib rather callously suggested that Owen had "worried him to death",[78] although probably an overdose of pain-relieving opium did that. Anyway, with Mantell dead, Owen now had a clear run, and in 1854 he published his own series of plates of Saull's *Iguanodon* sacrum, showing it from all sides, and with only five vertebrae.[79]

The Temple of Free Thought

Whatever the scuffle over the intellectual property, Saull still owned the physical specimen, along with over 20,000 others, and their future remained problematic. While the museum's socialist *raison d'etre* remained paramount, Saull was determined to keep it in a free-to-plebeians, anti-religious institute, where its meaning could remain intact. Any opportunity was seized. With Barber Beaumont scotched, and Jenkins's bequest destined for the courts, Thomas Cooper at John Street suggested they start from scratch. There was a good reason freethinkers had to build or buy an institution of their own. London had very few large halls, and many of those it had were barred to socialists at any price. Others had exorbitant rates, and none could be used on a Sunday.[80] So Cooper now urged them to build their own temple, pitching it in grandiose terms. In a speech, running a diatribe against "Popery" (an intolerance shared by all the sectaries, secular and Protestant alike), he argued for

> a large, if not a splendid building in the metropolis ... one in which all of the intellectual that Catholics adopt should be used—organ, choir, stringed instruments, drums (even) and trumpets—pomp of Handel, sweetness of Haydn, richness of Mozart, sublimity of Beethoven—instruction and eloquence—but none of the painted doll, the petticoated priest, the incense, the smoke, and stench.

---

fossils to be named, thus increasing their intellectual and financial worth: W. D. Saull to Richard Owen, 14 July, 27 July 1851, British Museum (Natural History), Owen Collection, 23: ff. 112–15.
78 Mantell *ODNB*; squib: [Pycroft] 1863.
79 Richard Owen 1854, Tab 3–7.
80 *Reasoner* 13 (22 Dec. 1852): 446; *Reynolds's Newspaper*, 19 Dec. 1852.

Such stirring imagery was guaranteed to fire up the freethinkers. It certainly did Saull, in the audience, for he "sprang up" and promised "£500 to raise such a building".[81]

Actually, there was little new in the rousing approach, Cooper had simply injected his "mad enthusiasm" for Haydn and Handel into the scene. Music drove socialist and secular devotion, and proceedings often began with a social hymn. Many socialist branches had choirs for the purpose. In the old days, Harmony Hall had christened its newly finished rooms with a spectacular concert of social hymns and sacred classics, and John Street had long had an organ and choir, with up to fifty performing on a Sunday evening, led by Cooper.[82] So there was nothing much new in this. Sacred music brought solemnity to reinforce dry sermons,[83] just as it was now used to whip up enthusiasm for the new building drive.

Some shine was taken off the proposal by the fact that many of the sacred lyrics had never been converted. This made them hugely incongruous. No mind that the angels have the best tunes, if the devil can control the lyrics. Complaints that Cooper's performances had not adapted Handel's *Messiah* to socialist ends had their effect, and the John Street choir was finally practising the re-scripted "Liberty the People's

---

81 *Reasoner* 10 (20 Nov. 1850): 80.
82 *NMW* 11 (7 Jan. 1843): 227; 13 (16 Nov. 1844): 165; Thomas Cooper 1872, 110, 315; *LI* 1 (Sept. 1854): 94; 2 (Mar. 1856): 177. Cooper was lecturing on Haydn at John Street at the time. The John Street choir comprised the Apollonic Society. Loose 2014, 63–64 on political hymn singing. Owenites carried their *Social Hymn Book* the way Catholics carried a missal, and, with 155 socially-regenerative hymns, there was one for every occasion. Their ungodly nature was execrable in the eyes of the City of London Mission (Ainslie 1840, 11–12), which singled out the 39th and 57th hymns for blasphemously making nature the creative force:
> Yet *Nature* in her varied forms
>   Applies to local things;
> To men, to beasts, fish, fowl, and worms,
>   As each to nature clings.
> The universe produces all,
>   (As Nature keeps her course),
> Unnumbered beings great and small,
>   By one projectile Force.

83 Royle 1974, 231–32. It continued to add gravity to secular talks right through to T. H. Huxley's 'lay sermon' at St Martin's Hall in 1866 (Barton 2018, 431; Desmond 1998, 344–45).

Messiah, the true Redeemer of our Race" when the call for the new institution surfaced.[84]

In tune with all this, the projected building was to be called, fittingly, the "Temple of Free Thought". Cooper, Saull, and others worked up a blueprint for the "Temple" with its socialist museum, which they planned to inaugurate with all the solemnity of a secular High Mass. Given the contextual shift, Haydn's *Creation* would now be forced to assume a new mantle, as it trumpeted the 'evolutionary' ethos of Saull's museum.

A whirl of activity resulted in a "Metropolitan Building Club" in John Street, set up to raise £10,000 through £1 shares, with Saull as a Trustee and Treasurer. They projected a 3,000-capacity Hall, near Oxford Street, with committee room, library, reading room, and class room for boys and girls, a book depot, and shop—in short, a grander Mechanics' Hall of Science, two decades on.[85]

The John Street Institution was refurbished in 1852, with large gas-chandeliers fitted,[86] but it was a stop-gap measure as the lease was due to expire in 1858. And too many converts meant that the building was bursting. The secularist movement was doing well, so much so that the Bishop of London in 1851 thought that there was now more danger of the spread of "Rationalism" than of "perverts" to Rome (to use the Anglican slang of the day).[87] Saull, in 1852, lamented that for several years the John Street Institution and Mechanics' Hall of Science had had to turn away "Scores often—in some instances, hundreds" on a Sunday because they were packed out, hence the need for a new "Temple", which "would be filled on Sunday evenings ... by an audience eager to listen to the teachings of democracy and freethought".[88]

---

84 *Reasoner* 6 (10 Jan. 1849): 18–22.
85 *Reasoner* 10 1851 (12 Mar. 1851): 371; 13 (21 July 1852): 96, which shows the investment goal being lowered to £5,000.
86 *Reasoner* 13 (11 Aug. 1852): 134.
87 *NS*, 24 May 1851. Although this might have reflected more the fact that debates over the "romanizing tendency of ritualism" (Blomfield *ODNB*) had temporarily died down, rather than "Rationalism" had risen.
88 *Reasoner* 13 (22 Dec. 1852): 446; 12 (3 Mar. 1852): 245; *Reynolds's Newspaper*, 19 Dec. 1852.

A Temple it would be, but not in name. When they actually tried to register it as a "The Temple of Free Thought" under the Friendly Societies Acts, the Attorney-General refused to allow it.[89]

Saull had "grown grey in the service of reform". Now, staring seventy in the face, and contrasting "the gory past" of prosecutions and prison cells with society's growing "smoothness" in the Great Exhibition years, he needed to secure his museum for future generations of working people. He upped the ante, offering "to devote a portion of my property for such a purpose—to double the sum I have already subscribed" if the project took off.[90] The group tried and failed several times to get the company certified under the Friendly Societies Acts, but eventually managed to register it under the Joint Stock Companies Acts, in the name of "The Metropolitan Institution Company".[91] So the title was fixed: it was to be the "Metropolitan Institution".

The Last Astronomical Hurrah

Unfinished business elsewhere was being wrapped up. For twenty years, a staple of Saull's talks to workers had been the relationship between cosmic dynamics and the laws of social progress. He was preaching to the converted, promising heaven to the dispossessed on earth. As one of the last living acolytes of Sir Richard Phillips and Sampson Arnold Mackey, Saull still used planetary perturbations to explain vast-scale geological changes in order to subject planetary prehistory to the deterministic "laws of progress". Phillips had even rejected Newton's gravity for its occult quality. Sir Richard had been in his grave since 1840, and detractors thought his absurd views should have died a lot earlier. Thomas Cooper might be running with the hares, but he could still hunt with the hounds. He too deplored the way Phillips had "sneered" at the Newtonian System, and cynically said he only did it "to put money in his pocket".[92] But Saull remained loyal, right down to

---

[89] *Reasoner* 10 (30 Apr. 1851): 450; 12 (3 Mar. 1852): 245. Cole [1944], 77, on these Acts.
[90] *Star of Freedom*, 2nd ser. 1 (31 June 1852): 5; *NS*, 24 May 1851; *Reasoner* 12 (28 Nov. 1851): 20.
[91] *Reasoner* 12 (3 Mar. 1852): 245; 13 (21 July 1852): 96; 13 (28 July 1852): 112; 13 (4 Aug. 1852): 128.
[92] *Cooper's Journal* 1 (20 Apr. 1850): 249. This was an easy claim to make, when Phillips's print works had churned out huge numbers of compendia for popular

specifics. Phillips had insisted that granite was the oldest bedrock, and Saull would passionately defend that view if it came under attack at the British Association for the Advancement of Science. Here, too, Saull would moot the changing position of the poles, as part of a defence of Phillips's and Mackey's accounts of see-sawing temperatures through geological time.[93]

After twenty years of talking to the unwashed, Saull tried one last time to persuade the impeccably scrubbed. He had one last crack at the gentlemen of the Geological Society. A new paper, read on 3 May 1848, restated his controversial case: that planetary orbits can explain palaeo-environments, so that shifting polar axes can account for a spot on the earth switching from torrid to frigid climes through time. And this, combined with a varying planetary wobble, the precession of the equinoxes, can further explain the movement of oceans from north to south or vice versa, exposing new land or submerging old. In short, the suit of planetary perturbations explained why his museum contained tropical corals, coal-forming tree ferns, and huge reptilian *Iguanodons*, all proving that Britain had at times been much hotter and periodically submerged.[94]

Saull had been touting this astronomical line for twenty years. It remained hopelessly at variance with the elite's programme, which was focussed on empirical work to delineate the successive strata. Their bedrock approach was deemed safer, because it left little room for wanton speculation or the wild social or theological extrapolation which came from the extremist fringes. It shunned anything that might fan the flames of scepticism, transmutation, or discontent. And in 1848, with troops on the streets, Europe ablaze and the Chartists massing, stability seemed doubly important. The Geological Society was tacitly

---

consumption, the flow keeping his financial head just above water; and even easier when Phillips's justified publishing one republican paper with the excuse that "politics were as profitable an article as he could deal in" (*GM* 14 [Aug. 1840]: 212).

93 Saull's paper, "On the Supposed Action of Water in Geological Formations, and the Position of the Poles of the Earth", was read but not printed: *Report of the Twenty-Second Meeting of the British Association for the Advancement of Science; Held at Belfast in September 1852*, (1853): 61; *Athenaeum*, 18 Sept. 1852, 1015. Granite: *Civil Engineer and Architect's Journal*, 15 (Nov. 1852): 383–84.

94 Saull 1853, iv–vi, 19.

policed by its elite. Saull's paper was not published, merely glossed in an abstract.[95]

Saull took matters into his own hands and published it in full in 1853. And priced two shillings, it was clearly not aimed at his penny trash people.[96] Tacked on to this *Essay on the Connexion Between Astronomical and Geological Phenomena* was a giveaway introduction. It made a final plea that science should have a social(ist) meaning. By revealing the planetary laws, it should give secular certainty to life's direction. It was barely coded to bolster an old Enlightenment determinism and rule out a capricious creation. He spelled out the social consequence: this "sublime" astronomical approach, "if properly carried out and expanded in accordance with the universally recognized law of progress, must decidedly be productive of the most beneficial results, by inducing men to conduct themselves in accordance with the bountiful arrangements of nature". The 'is' of science, which revealed "universal harmony", led to the 'ought' of conduct. Why? Because "informed minds adopting these ideas as governing principles" (that is, the geological guardians) will want to allow "participation" in their endeavour to all classes, "and especially to those ... placed in less favourable circumstances". It would lead to a democratizing and secularizing of knowledge to spread social harmony, the Owenite goal.

But the gentlemen had long discarded this Enlightenment non-sequitur. Moreover, any propagandist science designed to steer "our social conduct" in an obscene socialist direction doubly damned itself.[97] Many of the geological knights, like Sir Charles Lyell, had a horror of "mob-rule". The Tory imperialist Sir Roderick Murchison slammed socialism and the assault on ancient aristocratic lineages as "detestable".

---

95 Saull's paper, "An Elucidation of the Successive Changes of Temperature and the Levels of the Oceanic Waters upon the Earth's Surface, in Harmony with Geological Evidences", was left as an abstract in the *Quarterly Journal of the Geological Society* 5 (1849): 7; *Literary Gazette* 1634 (May 1848): 328. Even if the President, Sir Henry de la Beche, did now look wider in his yearly address, it would be to discuss the explanations of axial rotation by the socially and scientifically acceptable banker Sir John Lubbock: *QJGS* 5 (1849) lxxxiv–lxxxix. A later President Edward Forbes (*QJGS* 10 [1854], lxxvi) dismissed Saull's paper outright.

96 *Reasoner* 16 (5 Mar. 1854): 176. It was for sale amid Owenite and dissolvent theological works in Holyoake's 147 Fleet Street shop.

97 Saull 1853, vi–ix.

Saull's dinosaur expert Richard Owen had enlisted in the Honourable Artillery Company, the gentry's volunteer regiment, which supported the police during the 'riots'. The clubbable Tory geologist Edward Forbes also enrolled as a special constable as the Chartists massed on Kennington Common (these specials were singularly hated by Chartists). And the "Government hammerers"—the earth science specialists at the Museum of Economic Geology—took up cutlasses supplied by Scotland Yard.[98] The geological gentry simply stood on opposite sides of the barricades.

Citizen Saull, composing his paper in March 1848, was simultaneously drafting the Finsbury Institute's proclamation on the French revolution, and sending fraternal greetings on the revolutionaries' "glorious accomplishment",[99] while his patron, Robert Owen, was in Paris sharing a platform with the French communists.[100] At the same moment, there was Murchison, geologizing in Italy, outraged by the pistol-toting, tricolour-waving revolutionaries, and declaring that were such rioters to gain the upper hand in Britain "our ruin would be complete".[101] Saull's socialist seeds in the *Essay* were cast on barren ground. Neither the gentry's geology nor its antiquities were about to be purloined to unshackle the masses.

Saull's purview was better suited to Continental radicals (indeed, the *Essay* cited his friend Ami Boué). Saull knew it and accordingly addressed the *Essay* to the geologists of "Europe and America". For all that, there *was* a growing feeling in the British periodicals that planetary perturbations would sooner or later have to be taken account of to explain some of the large-scale geological events. Perhaps it was the liberality of the fifties kicking in. The *New Monthly Magazine* knew that connecting geology with astronomy was "treacherous", and that Saull "boldly ventures into the tabooed field of speculation". But he "had as much right as any one else, sufficiently acquainted with the subject, to enter upon the inquiry, and he appears to have conducted it in a sufficiently close and philosophical spirit".[102] Even the *Gentleman's Magazine* concluded that the carefully marshalled facts meant that "his

---

98 Wilson and Geikie 1861, 433; Desmond 1989, 331–32; Geikie 1875, 2: 87–90; J. A. Secord 2014, 142.
99 *La Voix des Femmes* 7 (27 Mars 1848): 2.
100 *UR*, 19 Apr. 1848, 41; 3 May 1848, 45.
101 Geikie 1875, 2: 87–90.
102 *New Monthly Magazine* 100 (Jan. 1854): 125.

essay will be read with interest", even if the authorities demur.[103] Not, of course, that any respectable reviewer burnt his fingers by touching Saull's inflammatory socialist deductions.

Saull stated that this was to be his last publication, his time being increasingly taken by the 'Metropolitan' to house his museum.[104] The geologians guarding the peace and confronting the Chartists would have been just as unsympathetic to this, an institution promoting irreligious Owenite ideals. The gentry, disagreeing on the social function of science, could give a very different meaning to fossils. When the Museum of Economic Geology was moved from Charing Cross to Piccadilly, it was rebranded the Museum of Practical Geology on its opening by the Prince Consort in 1851. The professors here were required to give evening lectures to local artisans. One of them, Edward Forbes, told his fustian auditors in 1852 that fossils were collected for two reasons, to elucidate the strata, and to help discover coal and minerals. He added that his listeners should collect fossils to make a bit of money by selling them on (tacitly disparaging his audience as the hodmen of geology). Strung together by an expert, these fossils were ultimately a way of "tracing the perfection of the Creator."[105]

The reverse applied to Saull's exhibition. This reflected, in Thomas Cooper's words, Saull's "persevering attachment to the cause of mental and political liberty".[106] Saull tailored his presentation accordingly. He portrayed the self-development of life and society as the liberation of the downtrodden from a conceited Monarch of Creation, whose reflected perfection was mistakenly seen in fossils. With the overthrow came the illegitimization of His policing priests, the state-paid power brokers on earth. Life's self-generating push came from below, where sovereign power lay, not from any Godhead through his supposed agents.

---

103 *GM* 41 (Feb. 1854): 168. Even at the end of the forties more latitude had been evident. *Chambers' Edinburgh Journal* thought that the "question is a promising one, and if steadily pursued, will lead to something more than speculation" (as we might have expected given the *Vestiges* connection): *Chambers's Edinburgh Journal* 253 (Nov. 1848): 297.
104 Saull 1853, ix.
105 *Working Man's Friend* n.s. 1 (28 Feb. 1852): 338–39. In these early years the lectures were not a success, and they were said to have been "of little value" to working men: Ludlow and Jones 1867, 163.
106 *Reasoner* 13 (22 Dec. 1852): 446.

Saull's *Essay* showed the museum still expanding, and it revealed the latest donors. To prove that the Arctic had once been warmer, Saull noted his new acquisitions. Naval vessels returning from their search for Sir John Franklin's missing Arctic expedition brought fossils from these high latitudes. Captain Erasmus Ommanney, in H.M.S. *Assistance*, had been the first to find traces of Franklin's lost crew. That was at Cape Riley. Ommanney returned in October 1851 with a consignment of Cape Riley fossil corals, new tropical species, embedded in what looked like Silurian rock, which went into Saull's museum. Admiral Sir John Ross, the expedition's leader, also used the museum as a repository for his polar corals, illustrating, again, its importance.[107]

With the assemblage at bursting point, Saull and his Metropolitan trustees stepped up a gear to finance their "central citadel". By August 1852, they were taking sixpenny deposits on £1 shares to let the industrious classes buy into the project.[108] In January 1853 the trustees held their first public meeting, and soon after were holding revenue-generating tea parties in John Street. But it was slow going. They had only sold 2,000 shares in July 1854, and, by the rules of the company, they could not start building until half, 2,500 shares, had been taken up. This forced them into the ignominious step of sending a lithographed circular letter to MPs, authors, and other gentlemen to plead for help in shifting shares. Old Joseph Hume, as ever, responded enthusiastically, although apparently without any cash.[109]

The 'industrious' were not picking up the tab, and despondency set in. At a shareholder's meeting on 8 January 1855, Saull, by now tagged with the soubriquet "venerable", and speaking for "the old reformers", made the sad admission that "we could not touch the basis of society", working men. President of the Metropolitan Company, Henry Tyrrell, an expert on Shakespeare and the Devil, talked more starkly of the "apathy of the toilers". Now and then, Saull had a "good round sum brought him to be placed at the bankers", but they had still only sold 2,092 shares and had to hold off from building. They did what they could to keep the

---

[107] Saull 1853, 15. Other donations went to the Museum of Practical Geology. Ommanney: *North British Review* 16 (Feb. 1852): 476–77.
[108] *Reasoner* 13 (11 Aug. 1852): 144.
[109] *Reasoner* 17 (27 Aug. 1854): 138; (2 July 1854): 10; 14 (23 Feb. 1853): 127; (16 Mar. 1853): 175; *Reynolds's Newspaper*, 30 Jan. 1853.

pot boiling, firing off letters to the liberal press and scouting out suitable land. The last word was left to Saull on that cold January day. "He had never deviated from the cause, and, while life lasted, he never would." But life is short, and that was to be his epitaph.[110]

---

110 *Reasoner* 18 (21 Jan. 1855): 38–39. Henry Tyrrell was actually Henry Tyrrell Church, although he dropped the "Church" for obvious reasons. He was just taking over St George's Hall, near the Elephant and Castle.

27. Death and Dissolution

> Without on-going effort the materials which constantly flow into museums can destroy them. Museums are about knowledge and its communication; the natural condition of collections is chaos.
>
> Simon Knell in *Culture of English Geology*, on the cultural binding that prevents the museum's entropic fate.[1]

Saull was the museum's soul. It was his ideological drive that bound its exhibits into a meaningful whole. Without this organized tension based around his talks it had no significance. It would disintegrate, and that is what Holyoake feared most.

Life looked normal at the beginning of 1855, a mundane alternation of professional archaeology and propagandist freethought. At a British Archaeological meeting on 24 January, Saull discussed the Isle of Wight. The island was slowly ceasing to be a rustic backwater, with fishermen's huts and a few lodging houses, although it was yet to see the seaside villas and hordes of holiday trippers.[2] Saull had, he said, been "a constant visitor to that island"—the last time in Spring 1854, typically to examine Wealden dinosaurs and Celtic barrows—and he talked on the Celtic-Roman transition, now pivotal to his museum's existence.[3]

In February, he was donating to the Secular Propagandist Fund and relishing one of Robert Cooper's "bold, unscrupulous, and shameless" attacks on the Bible at the City Road Hall of Science, Saull's stamping ground to the last. He talked to Cooper, praised his *London Investigator* and wished it well.[4] That was on 25 February, and Cooper's lecture, on "Christian Evidences", shows the world turned full circle. This is where

---

1 Knell 2000, xvii.
2 Vitzelly 1893, 131–32; M. Freeman 2004, ch. 1.
3 *JBAA* 11 (1855): 66–67; 12 (1856): 186–87.
4 *LI* 1 (Feb. 1855): 168; 2 (1855): 46; *Young Men's Magazine* (Dec. 1854): 237; *Reasoner* 18 (25 Feb. 1855): 125.

Saull started, a quarter of a century earlier, with the Rev. Robert Taylor's indicted blasphemies at the Christian Evidence Society. Now Saull's life was to end in Strauss's age, with a very different "exegetical knife carving off ... the miracles into legends and myth."[5] Strauss's sensational *Life of Jesus* was becoming part of that new wave of dissolvent literature pushing middle-class England into its own crisis of faith—those "honest doubters" (read bourgeois, non-socially-destabilizing), so frightened of being associated with the "scoffers".[6] As for the "scoffers", those who had taken the brunt, and suffered denigration and incarceration, Strauss, rolling off their presses in three-h'apence parts, was just another weapon in what they considered a class armoury.

Days later, in early March, Saull burst a blood vessel in his lungs. He must have suffered chronic chest pains as he tried to draw breath and spat up blood. Over six weeks, his condition worsened. He remained conscious at the start, and drew up his final Will and Testament on 3 April. He was a born organizer, and a dying one: among his last requests was for a death-bed sojourner to help distribute the *London Investigator*. For these weeks, as he lay dying, he constantly worried over the proposed Metropolitan Institute, which "should be reared without delay" to house his museum.[7] By late April, he was semi-conscious and the doctor held out no hope. He died on Thursday, 26 April 1855, five days shy of his seventy-second birthday.

As befitted a behind-the-scenes activist, a king-maker and facilitator, there was to be no lavish funeral, no cavalcade of flag-waving reformers through the city to Kensal Green, with bands and banners. No fanfare, nor even Holyoake's or Thomas Cooper's panegyrics over the grave to rally the living. It was all very different. We do not even know if the young bloods were there. In fact, the whole funerary episode was not only strange, but it raises more questions than it answers. The end of Saull's life, like the beginning, highlights how little we know.

"His funeral was attended by a number of old and valued friends", the *London Investigator* reported and then added, cryptically, "members

---

5 *LI* 3 (May 1856): 210.
6 A young T. H. Huxley's words: Desmond 1998, 657–58 n. 20; Strauss was already being sold in penny-halfpenny parts in 1843: *OR* 2 (18 Mar. 1843): 112. Larsen 2004, ch. 4; J. H. Brooke 1991, 265.
7 *Reasoner* 19 (22 Apr. 1855): 31; (29 Apr. 1855): 39; (6 May 1855): 47; *LI* 2 (1855): 46.

of a society with which he had been many years connected."⁸ I think this is the City Philosophical Society, founded in 1808 by the "unlettered" silversmith John Tatum, at his house in Dorset Street. The society is well known to historians in its Regency manifestation (largely because Michael Faraday was a member), when it included science-fascinated autodidacts, and, interestingly, the political satirist William Hone. But it was presumed to have died out long before. Not only had it apparently survived in a shadowy form, but it was still in Tatum's house.⁹ Of course it might have been resuscitated, or even infiltrated by Saull's group, or they may have been there from its early days. It seems to have devolved into a select meeting group of old freethinking friends, relics from the heroic age of blasphemous chapels and co-operative start-ups. This is suggested by the *Reasoner*'s report in June 1855 that "At the last meeting of the City Philosophical Society, founded by the late Mr. Saull, Dr. Helsham delivered a biographical sketch [of Saull]."¹⁰ Saull's having 'founded' it (that is, in 1808) seems surprising. But whatever his role, this reclusive set saw him out at the end.

Dr Arthur Helsham was part of this low-profile society, and his elderly group evidently arranged a private secular service and saw their old friend placed in the plot with Hetherington. Saull's oldest living allies—every one nearly a septuagenarian—closed ranks around the grave. (See Appendix 6 for the biographies of this little-known group of activists.) Few were left, and they formed a freemasonry of surviving comrades. The *Reasoner*'s rather unsteady report reveals some of their names:

> Mr. W. D. Saull was interred on Friday last [11 May?], in Kensal Green Cemetery, in *unconsecrated* ground, his grave being situated amid those of 'Publicola,' Hetherington, and Davenport. Mr. Henman spoke at the

---

8   *LI* 2 (1855): 46.
9   Pettigrew (1840 4:10), who attended the society at its Regency height, even thought Tatum was dead by 1840 (he died in 1858). It was reported changing locations in *Journal of Arts and Sciences* 8 (1824): 271–72. The society *was* mentioned a few times in later years, e.g., An American 1839, 178; Mogg 1848, 169, when it was back at Tatum's house. F. James 1992 on Michael Faraday's membership.
10 *Reasoner* 19 (10 June 1855): 87. Saull and Tatum both applied to join the LMI Committee in 1825 (*London Mechanics Register* 2 [3 Sept. 1825]: 312–13), otherwise I can find no connection.

grave. Mr. Prout, Dr. Elsham [Helsham], W. H. Ashurst, jun., and a number of the old friends of Mr. Saull, were present on the occasion.[11]

Obituarists were unexpectedly sympathetic, even if polite society was flummoxed by what it perceived as the incongruities of his life: his kindly heart yet heterodox "politics and religion as well as science". A fellow archaeologist implied that the "excellence of his heart" and "kindness of his nature" trumped all.

> He could differ, aye, and even dispute, but without any feeling of animosity or allowing his temper to be ruffled, and from the peculiarity of some of the opinions he entertained, and considering the manner in which they were occasionally met, this may be regarded as evidence of the benevolence of his disposition and character.

One word that cropped up was "enthusiast", a polite term for 'superficial'. He was uneducated (meaning an autodidact) and, as a result, driven to educate; he was wealthy yet "frugal" in habit, like so many old radicals; a liquor dealer who poured his profits into temperance societies (again supported by so many radicals), but most of all he poured it into his fabulous didactic museum. The sting was extracted from his 'extremism' by considering it the foible of a lovable eccentric, as if he could not really have believed the enormity of what he believed, atheism, evolution, and socialism—no longer crimes, but quirks.

In fact, *what* he believed was avoided altogether in the press obituaries as far too indelicate. Easier to concentrate on the seemingly harmless spectacle of his fossils: as the fusty *Gentleman's Magazine* recalled, "Nothing would more delight this kind but crotchety philosopher than the pleasure of instructing and exhibiting his treasures to the lower classes, and for a long time he was honourably known among geologists as the working man's friend." The obituaries invariably ended on this "valuable", "excellent" and "most instructive museum", his main claim to fame, and pondered its fate.[12] This was to become the burning question.

---

11   *Reasoner* 19 (13 May 1855): 55.
12   The fullest obituary in the popular press appeared in the *GM* 44 (July 1855): 102; abridged versions of this were run in the *Illustrated London News*, *Literary Gazette*, and *Reasoner*. The learned society obituaries were: *JBAA* 12 (1856), 186–87; *Monthly Notices of the Royal Astronomical Society* 16 (Feb. 1856): 90. Foreign obituaries included the untrustworthy Michaud n.d., 38, 47.

All concentrated on the tangibles: his few learned papers, and the professional societies frequented by the clubbable man. None saw (or all carefully avoided) his lifetime of financing blasphemy chapels, Carlilean court cases, Owenite institutions, jailed insurgents, secular education, and anti-Christian propaganda. They missed his major presence behind the political scenes because it was barely visible, and the Robin Hood nature of his brandy trade, which siphoned the rich profits off to the poor. This funding, as his will would show, was to carry on posthumously.

The appreciation was so different in the secularist press, which understood his pump priming and proselytizing. Saull was a rarity, in Robert Cooper's view, a respectable champion of atheism. This self-made autodidact, having penetrated "the middle and commercial" ranks through trade, nevertheless transcended the sordid capitalism of that class, and devoted his profits to educating, defending, and politicizing the poor.[13]

The museum, "one of the sights of London" in Robert Cooper's words,[14] was said by Helsham at the City Philosophical to be left to John Street, and so it was reported by many of the papers.[15] In a sense it was, inasmuch as the organizers of the new Metropolitan Institution were based there. Others saw it going to the Metropolitan Committee, to be held for the new building. Even as the obituaries appeared, the confusion began. Matters were not cleared up by the *Reasoner*'s bowdlerized publication of a small part of Saull's will (never mind its transcription errors). By now quite a lot was at stake, in terms of money and museum, for Saull's worth was proved to be £20,000,[16] and his instructions were far from clear. The will, in short, was confusing. Collating and collecting the various debts and mortgages were the easy part—there was Owen's eleven-room town house in Burton Crescent, while a property in Byfield, Saull's Northamptonshire birth place, went to his younger brother Thomas (as did Saull's share of the wine

---

13 *LI* 2 (1855): 46.
14 *LI* 2 (1855): 46.
15 *Leicester Chronicle*, 21 July 1855; *Preston Guardian*, 14 July 1855; *Reasoner* 19 (10 June 1855): 87.
16 *Illustrated London News*, 30 June 1855, 647; *Reasoner* 19 (9 Dec. 1855): 296.

business).[17] Most importantly, his socialist confrere and solicitor W. H. Ashurst would deal with the auction of Rose Hill mansion and its 62-acre estate, Owen's former house at the centre of the Harmony site.[18] That was the straightforward part of the will.

The Tragic Drama of the Museum's End

It was the wording regarding the museum's fate that was convoluted. In fact Saull's will was an exercise in how to lose a museum. It stipulated:

First, that the museum contents and his scientific books went to Helsham "upon trust to place in or appropriate the same for the use and benefit of the Working Mans Hall or Literary and Scientific Institution John Street Fitzroy Square". It was the second "or" that made it controversial: either the museum was to go to the Metropolitan/Jenkins Institution (yet to be built)—*if* that is what he meant by "Working Mans Hall"—or it was to go to John Street (which was about to lose its lease).

Second, that five hundred pounds was bequeathed to

> Mr. John Whittaker the Secretary of the said Literary and Scientific Institution John Street Fitzroy Square and the committee appointed to act with him under the will of the late Mr. Jenkins of Pinner the interest of such sum of five hundred pounds to be appropriated to the general purposes of the above named Hall and Institution.

So, the John Street Secretary was to get £500 to augment the Jenkins bequest to build the Metropolitan/Jenkins Institution, *if* the "Hall and Institution" referred solely to the projected Metropolitan/Jenkins Institution, and not to the John Street Institution as well!

At this point things get complicated, for the will then stipulated: Third, that if Helsham placed the museum "in the said Working Man's Hall", the interest on the £500 should rather be paid "half yearly to some person acquainted with Geology who may for the time being be the

---

17 Saull, Will, National Archives, Kew, PROB 11/2215. He also owned a counting house with cellarage adapted to the wine trade, also in Burton Crescent (*MC*, 3 Oct. 1848, 1). Three sisters, Caroline, Sarah, and Ann each received small legacies.

18 *Daily News*, 4 Aug. 1856, 8; *MC*, 12 Aug. 1856, 1. Tiffin chaired John Street meetings to liaise with Saull's executors on selling his Rose Hill Estate: *LI* 3 (Nov. 1856): 312; *Reasoner* 21 (21 Sept. 1856): 96; (5 Oct. 1856): 111; (12 Oct. 1856): 115; 22 (22 Mar. 1857): 45; (29 Mar. 1857): 50.

curator of my said Museum". And fourth, that Saull's 500 £1 shares "in the same Literary and scientific Institution" (by which he means the Metropolitan, *not* the John Street Institution) were also to be given to John Whittaker "for the benefit and general purposes of the said Institution".[19]

It was the stuff of legal nightmares, and, although the dying Saull obviously meant it all to go to the Metropolitan Institution Company (of which he had been a Trustee and driving force), his wording was anything but clear. Therefore, the courts were brought in to determine who was to be the recipient of the museum and money. Thus began the weary wrangling of a Chancery suit.

Meanwhile, the museum was cleared out, and in March 1856 the Metropolitan Institution Company made plans to hire a room for it, and settled on the defunct City of London Literary Institution nearby, at 165 Aldersgate Street.[20] This was destined to be its home until the hall could be built. But the expense was heavy, and it had to be paid for by voluntary donations. The trustees at this time said they planned to allow the public in free *"every Sunday"*, explaining that, since the state refused to open public buildings on the Sabbath,

> The working-classes must, therefore, take the initiative, and open all the museums and libraries under their own control on that day. The opening of the Saull Museum will assist the movement now in progress for obtaining a free Sunday for the people.[21]

The museum was being marshalled for the cause even after Saull's death.

It was a huge logistical operation, involving careful dismantling and packaging in hampers, with convoys of carters trundling up Aldersgate Street. Over 20,000 exhibits had been present in 1851, a figure subsequently increased as Saull continued to buy-in, collect on the Isle of Wight, and take in shipments, such as those from captains

---

19 Saull, Will, National Archives, Kew, PROB 11/2215.
20 This was where Richard Owen went to examine the reptile fossils for his continuing monographs: Richard Owen 1859, 22–24. The institution had been part gentleman's club for the City merchants, part scientific institution with a library and museum (Hill 1836, 1: 223–24). It had been wound up in 1852 (Timbs 1855, pt. 2: 459), meaning there was now space for storage there, perhaps in the old museum.
21 *Reasoner* 20 (9 Mar. 1856): 74.

searching for Franklin's expeditions. The *Mining Manual* confirmed that the museum comprised the "largest private collection of organic remains in the United Kingdom".[22] Timbs's *Curiosities of London* at the time of Saull's death saw the "Geological Department" alone "exceeding 20,000 specimens",[23] and that takes no account of the stock piles of British Celtic and Roman ware collected in the 1840s. So the collection being relocated in 1856 was enormous.

It had always been a one-man show—as a London guide book put it, the museum's existence was "due to the perseverance of W. D. Saull".[24] In large part, it was an ideologically-driven endeavour, a shrine to utopian political dreams. The museum was to justify a distinct Carlilean Creator-free rise of life and substantiate the environmentally-driven inevitability of Owen's perfect society. But with Saull dead, this overarching meaning dissipated. He was the last link to this defunct world: Carlile was long dead, Owen was in his dotage—he had even converted to spiritualism in 1853[25] and was to die in 1858, aged eighty-seven. The museum dissolved into so many disparate items. The coherence was shattered, the living relationship Saull forged between the fossil fish and saurians, and aboriginal hut dwellings and Celtic ware, was all lost. The connective stories he told to visitors were now just echoes.

To make matters worse, the old generation, who sympathized with his Carlilean freethought and Owenite goals, and who might have helped preserve the museum intact, were themselves passing away. His younger brother, Thomas Saull, only fifty-three years old, died within months himself, on 1 October 1855.[26] Ashurst expired twelve days later, on 13 October, and Thomas Prout died in 1859. Meanwhile, their sons had grown respectable as solicitors (William Ashurst Jnr. and Robert Helsham), while John Prout, who retained his father's business, was described in Saull's will as a "gentleman". They did not have the same commitment or concern: theirs was a very different professional world.

---

22 *Mining Manual and Almanack for 1851*, 136.
23 Timbs 1855, 542.
24 Gilbert 1851, 139.
25 With socialism withering, a feeble Owen invoked the most sympathetic departed spirits as the new force to re-create the character of man. If his plebeian followers could not do it, then this less fallible agency could usher in the social millennium (Barrow 1986, 19–29).
26 *MC*, 5 Oct. 1855.

Saull's wife Elizabeth would herself die on 22 December 1860, aged seventy-one, the last close family tie with the living museum.[27]

By July 1857, the Committee had chosen a site for the Metropolitan Institution. Many were sad to see John Street go—"there are legends clustered around its platform", mused Robert Cooper wistfully, for the "greatest of our leaders" once graced its "forensic forum." Yet the *London Investigator*, itself on the verge of collapse, hoped to see a "nobler building—one fitted to receive the treasures which the benevolence of Mr. Saull and Mr. Jenkins has endowed it with."[28]

In May 1859, the Court of Chancery decided, correctly,

> that the Metropolitan Institution Company was clearly the Institution meant by Mr. Saul [sic] (although wrongly described in his will) to which he intended to give the 500 shares he held in that company, the legacy of £500, his geological museum and library of scientific books.[29]

And they charged Saull's estate costs to come to that decision. That should have been that, but the court wanted guarantees that an institution *would* be built to receive the museum. And the question remained as to what, meanwhile, "was to be done with the testator's geological collection, which was in 30 glass cases and packed in hampers, and how the costs were to be borne". And so the decision was referred back to the Chief Clerk of the court, and there it remained for a further year. The "weary business" finally terminated five years after Saull's death, in July 1860, when the court decided duty must be paid by the legatees (the Metropolitan Company) "but it would be a pity to raise it by the sale of the collection, which all parties seemed to wish to remain intact". The court decided that the museum should stay packed as it was, until the new institution was built to receive it, while the £1000 in shares and cash be released to the directors.[30]

Building the Metropolitan Institution began at 12 Cleveland Street, Fitzroy Square, apparently a site they were already using for secular

---

27 *Observer*, 31 Dec. 1860, 8; *Reasoner* 26 (6 Jan. 1861): 14. But about her views we know very little, only that she contributed to funds to help the wives of jailed Chartists.
28 *LI* 4 (July 1857): 61–62.
29 *Metropolitan Express*, 13 May 1859, 4; *Reasoner* 24 (12 June 1859): 191.
30 *Times*, 28 June 1860, 10; *Reasoner Gazette*, 8 July 1860, 111; 15 July 1860, 115; *Reasoner* 25 (15 July 1860): 232.

lectures. These were cancelled as the scaffolding went up. By February 1861, the hall walls were thirteen feet high and, by April, the roof was on and joists laid, allowing a public meeting of the shareholders to take place inside for the first time, despite the thicket of scaffolding. The Metropolitan Institution, or Jenkins Institution, or Cleveland Hall as it was coming to be called, could at last be seen in the round, and it looked a "spacious and comely edifice", in fact "the handsomest building the Freethinkers of London have possessed". The ancillary rooms would devolve into "a library, museum, and schools", or such was the idea that Spring. The sale of more shares widened the ownership, and plans were laid for a tea party and festival on 3 June in the finished building, and they began selling shilling tickets for a ball on 10 June. More information was released: it would contain,

> besides the Large Hall, appropriate Committee Rooms, Museum, Library, Reading and Class Rooms for the Tuition of Adults, and school Rooms for the Education of Children of both sexes, and where the industrious classes can assemble to acquire and communicate useful knowledge freed from all sectarian influence and control.

The museum was still on the cards. Thus was the building inaugurated on 3 June, with tea in the school-room, and the 500 guests retreating afterwards to the hall above for the speeches on "unsectarian education, mental freedom, political enfranchisement". Functioning by this point was the "large hall, three school-rooms, coffee-room, library, and several minor rooms, adapted to the wants of the working-classes for committee purposes. There are already a number of children in the schools." The plan was for a "free lending library" and the museum to be added.[31]

They had Saull's books for the library, and, at this point, the fossils should have been unpacked for the museum. It had always been his dream, and it was the wish of the dying man. Yet, despite everyone agreeing with the court that the museum should not be sold, the fossils remained stowed in wine hampers.

Ironically, there had never been a better time for a visual display of fossil evolution and human development. The middle classes had

---

31 *Reasoner* 26 (13 Jan. 1861): 18; (24 Feb. 1861): 127; (3 Mar. 1861): 139; (31 Mar. 1861): 202; (14 Apr. 1861): 228; (28 Apr. 1861): 247; (12 May 1861): 276; (19 May 1861): 288; (26 May 1861): 300; (16 June 1861): 334.

taken a shine to the *Origin of Species* (1859), which Charles Darwin had published with great trepidation—although a 15s cover price (a week's wage for the poorest) signalled that it was not destined for the wrong hands. Punters pushed it into a third edition by 1861, with seven thousand copies in print.[32] That year, T. H. Huxley provocatively stretched its bounds in an *Athenaeum* spat over the similarities of human and ape brains, where an angry Richard Owen accused him of wanting to make man "a transmuted ape".[33] The inflammatory issue, simmering in polite society since the *Vestiges*, had emerged into the open. The public was clamouring to see the evidence for evolution in museums.[34]

Saull's museum at such a moment would have been able to draw fresh reserves for his monkey-ancestry and aboriginal rise. But the exhibits only had meaning in situ with Saull's explanations. Now he was gone and the fossils were hampered. Among freethinkers, the ideological issue of monkey men had spluttered on. That campaigner for life's godless rise, William Chilton, was an arch-"scoffer", and, no less than Huxley, would tactically defend the dignity of man as the "son of an ape", against clerical "rudeness, puerility, and ignorance", but for his own class, in his own time.[35] Chilton's had been a tragically early demise (he died a month after Saull), but human parentage remained a potent anti-clerical weapon in the Halls. Whether the *London Investigator's* obligatory "Origin of Man" series (finished just before Saull's death), or 26-year-old John Watts' "Theological Theories of the Origin of Man" at City Road Hall of Science (delivered and published as the Cleveland Hall was going up in January–March 1861), the warm topic just kept getting warmer.[36] There was no better time for Saull's museum, providing some Owenite spiritualist could summon up Saull's ghost to explain its evolutionary import.

Not that this was impossible. With the decline of socialism had come a rise of the emancipationist spiritualists, even the 'Social Father' himself

---

32 R. B. Freeman 1977, 85. No mind that the *Origin* grew out of a vastly different Malthusian context (Hodge 2009), Saull's palaeontological display would have been just as amenable to Darwin's 'common ancestry' theme in this plebeian venue.
33 Richard Owen 1861, 395.
34 Rev. R. S. Owen 1894, 2: 38–39.
35 Chilton 1854.
36 *Reasoner* 26 (20 Jan. 1861): 47, 48.

would die one. For many marginalized Owenites, now marching to the millennium guided by the spirit powers, contact with past heroes was a feature of *séances*, and Saull's ghost was evidently still hovering overhead. Once it was even summonsed. A co-operator-turned-medium, Dr Jacob Dixon, formerly Secretary of the Labour Exchange,[37] a man who had moved from homoeopathy to mesmerism looking for self-help patient cures, then to spiritualism, called up "Devonshire Saull" by mistake one day at a *séance*. The ethereal Saull was understandably nonplussed according to the medium.[38] But then the arch-materialist's spirit was forever being dragged into uncongenial realms.

John Watts, equally an arch-materialist with an equal distrust of spiritualism, was a new generation secular missionary, but his case shows how much he could have benefited from the museum. He was the son of a Wesleyan preacher, and had learned his preaching techniques well. As a compositor by trade (like Chilton), with type in his hand and words in his head, he was a voluminous reader, and became sub-editor of the *Reasoner*.[39] He threw himself into Darwin's *Origin*. He gave a fair epitome of it in the *National Reformer*, coming to the conclusion that Darwin leaves us "to infer that he includes man,—considered in his corporeal capacity, of course,—amongst the earthly products of 'descent with modification.'"[40]

Strauss had taught the secularists to look at the *evidence* for Gospel statements. The former Oxford tutor Richard Congreve, lecturing on "Positivism" in Cleveland Hall, was stressing "the laws which govern the world"[41] and "the dignity with which [man] submits to them". Put those two approaches together and it explains why Watts' simultaneous lectures in 1861 on the "Origin of Man" in Cleveland Hall was less an attack on Genesis and more a detailing of the proofs of the laws of evolution. And, did he but know it, some of the best fossil proofs were, *as he spoke*, only a few yards away, still packed in W. D. Saull & Co hampers. The audience now craved the "latest intelligence", not a theological

---

37 See Jacob Dixon to Robert Owen Correspondence, Co-operative Heritage Trust Archives, Manchester, ROC/4/23/1–4; *Crisis* 2 (20 June 1833): 196.
38 *Spiritual Magazine* 5 (1 Feb. 1864): 80; Podmore 1907, 2: 610–11.
39 *National Reformer*, 11 Nov. 1866: 305–06.
40 *National Reformer*, 4 Jan. 1862, 6; 18 Jan. 1862, 5–6; on Spencer: 12 Dec. 1861, 2; 28 Dec. 1861, 6–7.
41 *Reasoner* 26 (7 Apr. 1861): 214; (21 Apr. 1861): 238–39.

bash. And Watts was at his best passing on the newest "scientific views on the subject of man's appearance on the earth." He could have put Saull's fossils to advantage, particularly as the two men had shared a goal. Watts, like Saull, had an ideological slant: he sent auditors away with the benevolent Owenite view (far from a Malthusian tooth-and-claw Darwinism), that

> our forming a part in the great whole of animal existence ... instead of conveying the idea of degradation, should induce a better feeling and kinder treatment to those animals it had pleased us to class among the "brute creation".[42]

Watts's twopenny pamphlet on the "Origin of Man", unlike a 15s book, was to change culture, not pretend to stand aloof from it.

The exuberant young Watts, sustained by this positivist air, was "Taking [nature's] facts for our guide". He stressed that aboriginal mankind was a contemporary of extinct cave bears and big cats. This was suggested by Brixham cave finds of human bones gnawed by hyaenas. Then there were the flint knives "mixed with the bones of animals now extinct". He could have pointed to Saull's specimens. And geology, by cataloguing life's rise from the "lowest orders", "polypi, worms" and so on, though the "corals, shell-fish" and eventually fish and reptiles, then "up to man", preserved the sequence "exactly as it must have been had the one been developed from the other." The Cleveland Hall talk was made for Saull's cabinet, which was designed to illustrate just this—that "man, myriads of ages ago, had his origin in the animals now lower in the scale than himself."[43]

There was no denying an audience for lectures highlighting Saull's fossils—and in the very institution which had them secreted away. But the moment was lost. And the audience itself was changing, with the growth of clerks and domestics, who were less concerned with a

---

42 *Reasoner* 26 (27 Jan. 1861): 62.
43 *Reasoner* 26 (17 Feb. 1861): 102–04; (24 Feb. 1861): 119–21; (3 Mar. 1861): 132–34. On the Brixham cave finds in 1858–63, see Riper 1993, ch. 4; Boylan 1978; Grayson 1983, 179–85; Wilson 1996; Bynum 1984. Watts repeated his talks at the City Road Hall of Science on 22 March 1862 (*National Reformer*, 21 Mar. 1863, 8). Watts's potential was never realized. He became ill with consumption in 1863 (aged 29), and died in 1866, aged thirty-two. He was buried in Kensal Green, near Saull, Davenport and Hetherington (*National Reformer*, 3 June 1866, 345; 11 Nov. 1866, 305–06; Royle 1974, 283).

pointed Owenite explanation of evolution.[44] The fossils were, as the *National Standard* had once said, a sealed book without Saull,[45] and with his death the book was being re-sealed. The directors of Cleveland Hall, with different priorities, were losing a sense of its relevance as a whole. Nor was there a willingness to take responsibility.

It was not as if Cleveland Street had lost direction. It remained secularist through the 1860s (evangelicals and spiritualists only got hold of it in the early 1870s). Religious critics continued to damn it till the end of the sixties: "Every cock can crow on his own dunghill", sneered one, "and at Cleveland Hall the Secularists have it all their own way, and are merry at the expense of their opponents. Nor is this all; they often indulge in a style of abuse which sounds even to tolerant ears uncommonly like blasphemy."[46] To Saull's ghost it must have seemed like old times, blasphemy again. But with his demise the space-cluttering exhibits had lost their *raison d'etre*. And without his esoteric understanding, or the paid curator/lecturer he stipulated in his will, they remained a fragmented jumble, all coherence gone.

The Fate of the Fossils

Eight years after Saull's death, in 1863, the directors got rid of the lot. Twenty five years in the making, the haul valued at over £2,000,[47] and no less valuable intellectually, it made no difference to the directors. Nor did they care that Sowerby's historic specimens were included. Without constant curating and reinforcement of their social purpose, collections anyway tend to disintegrate.[48] But this one, boxed, lost from sight, and its moral meaning interred with Saull, was an extreme case. The stowed

---

44 Anon 1904, 322, said that Saull's "money was devoted to carrying on a school, which gradually became little more than a place of evening amusement for the young men and women employed at large shops in the neighbourhood", implying that this caused the Directors to lose interest. Actually, Saull's bequest had gone into the building fund; it was Jenkins's money (£100 a year annuity) that financed the school (*Reasoner* 26 [16 June1861]: 334).
45 *National Standard* 3 (18 Jan. 1834): 44–45.
46 Ritchie 1870, 378.
47 *UR*, 15 Sept. 1847, 83.
48 Jardine, Kowal and Bangham 2019.

exhibits were said to be in a "lamentable state",[49] and taking up space, so they were ditched.

The British Museum was quick off the mark and sequestered the show-stopping exhibits. In fact, the Cleveland Hall Secretary, oblivious to the fossils' real worth, *offered* the British Museum the pick for the bargain price of £30. For this ridiculously-small sum, the museum obtained 201 of the prize specimens, chosen by the keeper of geology, George Waterhouse, undoubtedly guided by Richard Owen, now superintendent of the natural history departments. A quarter were reptiles, fifty fossils, including the *Iguanodon* sacrum made the foundation of Owen's "Dinosauria". Twenty-seven other parts of *Iguanodon* were taken, as well as a cranium of *Crocodilus spenceri*, an ichthyosaur skull, and more. Ten mammal fossils were selected, including four remains of whales. To these were added 45 fish fossils, 69 invertebrates, "the greater portion of which are specimens figured and described in "Sowerby's Mineral Conchology", Waterhouse reported. On top of this were 27 plants—Saull's famous coal-seam fossils, one being the type specimen of *Sigillaria saullii*, we presume.[50] It was daylight robbery of the poor by a state body top-heavy with the country's wealthiest aristocrats. Saull would have been turning in his grave.

Just *how* much of a steal was evident from the market price of fossils. For decades a good ichthyosaur skull could fetch anything from £6 to £25 at Stevens' sales, or a mammoth skull from 12 to 144 guineas. Commercial collectors in Whitby were asking £30 for fossil crocodile skulls.[51] This alone suggests that a single Saull fossil could have been worth the £30 knock-down sum asked for the lot. Knowledgeable collectors got a good price—they could talk up the real value. Mantell had sold his 20,000-object cabinet to the British Museum for £4000 in

---

49   Anon. 1904, 322.
50   British Museum, Central Archive, Trustees Original Papers, Department of Geology, Report respecting Offers for Purchase, 5 Aug. 1863, No. 6607. The sanction for this purchase: Trustees Minutes, 8 Aug. 1863, C10,408; House of Commons, *Finance Accounts I.-VII...1863–4* (28 Apr. 1864): 24–26. Other keepers acquired some of Saull's antiquities in 1863 (Hobson 1903, 109; Walters 1908, 324, 372, 435).
51   Mantell 1846; Knell 2000, 206, 217.

1839.⁵² Roach Smith sold his antiquities, likewise, for £2000 in 1856.⁵³ This casts into relief the paltry sum paid for Saull's choice exhibits bequeathed to guardians ignorant of their worth. The figure seems even more shocking given the price that the British Museum was asking for *casts* of their fossils: £4 10s for an *Ichthyosaurus intermedius* down to 8s for an *Iguanodon* humerus. They were charging, in effect, more than they paid for Saull's original, figured and 'type' specimens.⁵⁴

The Metropolitan managers hived off the duplicate fossils and sold them at auction in June.⁵⁵ Why only the *duplicates* is puzzling.⁵⁶ What happened at this point is an even greater mystery, as is the destination of the remaining fossils, antiquities, ethnographic exhibits, Petrie's skeleton, Hibbert's skull, and the rest.

By all accounts, an unscrupulous con-man carted away seven van loads of remains, as if they were so much bric-a-brac. We do not know whether he paid the managers, or was doing them a favour. Silver-tongued John Calvert, a self-aggrandizing "mining engineer" and "gold prospector"—better known in the mineralogical press as a "blackguard" and "charlatan", and those were the politest things said of the man called "Lying Jack".⁵⁷ Calvert, evidently, cleared the lot out in 1863. The man was a scammer who claimed to have discovered gold in Australia. Even if the near libellous tittle-tattle is colourfully over-inflated, there is a sense in which it helps explain events. It is possible that he not only took Saull's fossils after his death, but conned Saull in life. Calvert's father, a friend of William Blake, indulged his pagan lifestyle to the distress of friends, and son John was probably sympathetic to Saull's

---

52 Cleevely and Chapman 1992, 321–26.
53 PP. *British Museum. An Account of the Income and Expenditure of the British Museum for the Financial Year ended the 31st day of March 1857*, 2.
54 *Synopsis of Contents of British Museum. Sixtieth Edition* (1853): 270.
55 *Express*, 12 June 1863, 1.
56 Although described as "a valuable and interesting Collection of Fossils" by Stevens's sale room, in their auction of 13 June 1863 (*Athenaeum* 1858 [6 June 1863]: 731), these were apparently only duplicates: cf. Cleevely 1983, 255; Chalmers-Hunt 1976, 102. Very little is known about the dismantling of Saull's collection by the uncaring Metropolitan managers, leaving many questions. How did they know which were duplicates? And why, then, did they not auction the valuable originals?
57 My knowledge of Calvert owes much to Mick Cooper, pers. comm.; M. P. Cooper 2006, 85–105; Embrey and Symes 1987, 73; Sherborn 1940, 29; M. A. Taylor 2016, 89; Anon. 1904, 322.

freethought. Given this sympathy and Calvert's puffed-up credentials, Saull put him up for a Geological Society fellowship, unsuccessfully.[58] So Calvert would have known of Saull's valuables. Calvert amassed his own huge "museum" (some 26-million specimens, he claimed unflinchingly!). As a dealer, he profited from buying cheap, and even more from his speculations, taking gold-discovery investors for a ride, so Saull's collection might have seemed an enviable target. Doing the artless managers a favour sounds like Lying Jack's style.

Where is the collection now? This is the strange part. Even before his death in 1897, Calvert started unloading his own hoard on Stevens's auction room, and more went under the hammer after his death.[59] But the bulk, said to be 100,000 shells, fossils, and minerals, gathered cobwebs and dust in a brick building in East London. They supposedly still included the "W. D. Saull coll. ... appropriated from the Metropolitan Inst."[60] The Natural History Museum turned the collection down in 1938. Finally, the trove, now "absolutely filthy with ... London dust and soot", was bought for £2,000 by a New York dealer, Martin Ehrmann, that year, and he had students pack the lot for shipping to America.[61] Expert mineral dealers in New York then processed the collection and brochures were printed, but none mentioned Saull.[62]

And so, for the present, the trail has gone cold. Saull's remaining fossils and antiquities, his ethnographic exhibits and radical relics, all the items that gave his museum its evolutionary coherence and rationalist identity, have disappeared like Arthur in the mist.[63] The whereabouts of Hibbert's head and Petrie's skeleton is unknown. Effectively, the largest private "geology" museum in early Victorian London, possibly in Britain, had vanished. The breakup of the Aldersgate Street museum

---

58   Mick Cooper pers. comm. Calvert (1853, 46) cited Saull in *Gold Rocks of Great Britain*.
59   *Athenaeum* 3652 (23 Oct. 1897): 543; 3690 (16 July 1898): 82.
60   Sherborn 1940, 29.
61   Smith and Smith 1994; cf. Sherborn 1940, 29, who thought the Calvert collection went to Tottenham Castle Museum.
62   Mick Cooper, pers. comm. Some of the collection went to the Smithsonian (*Geological Curator* 3 [June 1982]: 236–37, 242–46), but the provenance of many specimens is unknown.
63   There have been parallel losses to the city. A few years later, Bethnal Green lost the chance to house a fossil museum, when Antonio Brodie's efforts to leave his Pleistocene Mammalia from the Ilford brick pits to the community's East London Museum was thwarted by government indifference (W. Davies 1974, xiv).

prevented the possible re-construction of its meaning for a Darwinian age, and the loss precluded a posthumous celebration of its creator. Any lingering regard for Saull vanished with the museum's dissolution.

Trashing Reputations

"The human race", the *Reasoner* once said in a diatribe against Moses, "has forgotten its own birth" and filled the void with its imagination.[64] The metaphor just as aptly applied to Saull's museum: lost, and its memory erased by a posthumous trashing of Saull's reputation. What sealed Saull's fate finally was his entry in that self-confident *fin de siècle* compendium, the *Dictionary of National Biography*. The *DNB* was a huge exercise of discretion, compression, proportionality, and balance, even if the optimum was not always achieved in its 29,000+ entries. But at least, as Lawrence Goldman said as the superannuated texts were updated and digitized, it was a fair "reflection of late Victorian views of national history".[65] The original intent was to include quirky and offbeat subjects—broadening the dictionary's scope with lesser luminaries whose lives were to highlight the imaginative potential of disparate souls.

Saull probably only squeezed in to the *DNB* because the editors were scouring *Gentleman's Magazine* obituaries.[66] But that disdainful source was problematic. The fogeyish magazine had shuddered at atheism as "an intellectual insult, a social nuisance, a religious pestilence, and a moral curse"—when it dared mention the subject at all. And, believing that socialists were the "only class openly professing infidelity", the *Gentleman's Magazine* took aim at them too, not that it often stooped to such "wretched trash". It loathed the "filth" of socialism and thought that, whatever socialism's benevolent intent to ease the sweated brow, it had to be "leavened by religious impulses and motives".[67] So the *Gentleman's Magazine* was never going to warm to Saull. The *Gentleman's* obituary strained to be fair to the "crochety philosopher", even though his knowledge was "superficial", but it pointed to its own censorious

---

64 *Reasoner* 26 (6 Jan. 1861): 10.
65 Goldman, "Making Histories".
66 Atkinson 2010, 221, 223, 225, 227.
67 *GM* 35 (May 1851): 467–68, 519–23.

review of Saull's *Notitia*, which castigated Saull's belief that the Goddess of Reason's enthronement would be a blessing. The magazine's dismissive stance set the *DNB*'s tone.

Matters were exacerbated by the *DNB*'s choice of obituarist. The *DNB* might have spread its net widely, but "Religion" remained the dominant "field of interest" (a statistic corroborated by the 1995 digitization of the old *DNB*.)[68] The emphasis was on classical learning, clerical piety, good breeding, and scientific merit. The *DNB* offices were in Waterloo Place, off Pall Mall, and the selection of biographers was whittled down to the most reliable *habitués* of London's surrounding clubland. Yet a hack's competence was in proportion to his social and temporal distance from his subject. Therefore, putting a priest with a dual geological calling in charge of a mis-categorized, blasphemous Owenite, dead half a century, who financed freethought and ran a museum for rationalist ends, was spectacularly bad planning. The entry went to the Rev. Professor T. G. Bonney.

Bonney wrote seventy entries for the *DNB*, exclusively on geologists. Unfortunately, Saull, mis-filed as a "Geologist", was parcelled out to him. Bonney was a curate's son who, from his gentrified upbringing to his genteel life at St John's College, Cambridge, was not au fait with the atheist Owenite milieu. His "charmingly written"[69] *Memories* hailed St John's good life, where gastronomy vied with geology. He was an ordained priest and honorary canon of Manchester Cathedral, "a scientific parson, but quite *sans reproche*," said the old agnostic T. H. Huxley,[70] meaning a working petrologist (rock expert) who did not let his cloth intrude. Bonney defended evolution in the religious press, and, in common with his Spencerian age, saw it stretch from crabs to civilization.[71] Nevertheless, he had no truck with Herbert Spencer's view on religion.[72] Bonney insisted that "the earth's history tells its tale of purpose, not of the blind working of physical forces". And the Bible

---

68 Even if the editors dismissed one enthusiast's list of 1400 hymn-writers sent for consideration: Atkinson 2010, 227; Maitland 1906, 367.
69 Rastall 1937.
70 T. H. Huxley to Henrietta Huxley, 5 Feb. 1889, Huxley Archives, Imperial College, London.
71 Bonney 1921, 37; 1891, 23.
72 Gay 1998, 49.

was morally inspiring, even if Genesis was allegorical.[73] Saull, with his virulent distrust of organized religion, was not his ideal subject. Bonney baulked at freethought. He actually penned the Saull entry in 1897 after combatting its latest manifestation, 'agnosticism', in his Boyle Lectures at the Chapel Royal in Whitehall.[74] So even if Canon Bonney had known Saull's views, he would have strained to situate them sympathetically.

In a positivist age, looking for positive scientific attainments—in, say, stratigraphy, fossil classification, or field-work—Bonney could find none in Saull. Bonney simply compressed the *Gentleman's Magazine*'s dismissive snubs of Saull as a crotchety ignoramus. The entry damned with no faint praise at all. Saull "was more enthusiastic than learned". His astronomical explanations of geological events "indicate the peculiarity of his opinions". And his re-publication of Sir Richard Phillips shows him "attacking Newton's theories of gravitation." No one would want to know more, but if they did, they were sent to the *Gentleman's Magazine*.[75] One could never learn of Saull's King-making Carlile benefactions, his financing of the "Devil's Pulpit", or indeed dozens of other radical and co-operative venues, his Labour-Exchange pioneering, or his treasurer's work on so many Owenite and reform causes, national and local, let alone the gigantic enterprise that was his open, didactic, working-man's geology museum, the largest private one in London.

DNB entries were quasi-oracular pronouncements for a century. They were the first and sometimes the only port of call for scholars. So a dismissive entry could dampen research for decades. Yet, sympathy *could* have been achieved. The *DNB* aimed for it: "High-churchmen were to be allotted to high-churchmen", it was said at the start.[76] And they achieved it in ten Holyoake-authored entries: on early deists, co-operators and radicals, notably Carlile and Hetherington. The Holyoake entries were sensitive to context and knowing in their appreciation. Holyoake was by now eighty-years old, and as sharp as ever. In Saull's day, he had moved from political atheism to a more intellectually accommodating secularism, backed by Saull. In these later years, he was on cigar-smoking terms with the new aristocracy of intellect. How different, then,

---

73 *Guardian*, 16 Oct. 1895, 45; Clodd 1902, 186; Bonney 1891, 91.
74 Bonney 1891.
75 Bonney 1897.
76 Maitland 1906, 368.

if he had written Saull's entry: the freethinker he knew well, and the financier who realized his dream of a freethought palace in Cleveland Street—itself still going.[77] But he did not.

The last word must go to the greatest Victorian palaeontologist, Richard Owen. The gaunt, goggle-eyed eighty-year old, just knighted in 1884, was resting on his staggering output of 600 publications. He had been sidelined by the brusque Darwinians to a lonely life in Sheen Lodge, in Richmond Park, a present from Queen Victoria. Forty-three years earlier, on the top floor of Saull's Aldersgate Street museum, Owen had found one of his key fossils, the *Iguanodon* sacrum. It had been the basis for his most enduring creation—the dinosaur—a creature that was acquiring its iconic status thanks to the bone rush in the American West. His life now closing, Owen repaid the debt.

There was some irony to it. The towering figure of his day, Owen had tried Canute-like to stem the transmutationist tide. A devout Anglican, he approached his descriptive work like a religious duty, for his fossil animals, "in the Psalmist's words, 'were telling the glory of God'." With Bonney, he believed the continuous steps from nature to civilization showed "foresight, intention, and successful attainment", and anyone doubting it he called congenitally blind.[78] He had fought tenaciously and occasionally cleverly against the bestial threat of a transmuted-ape inheritance. Yet here he was in 1884 acknowledging an old Owenite freethinker, who had openly dethroned God and made heavenly man a shaved monkey.

Owen cut a forlorn figure, looking for peace and closure in the twilight years. A widower, whose only son was about to commit suicide, he spent his days finishing his magnum opus, *A History of British Fossil Reptiles*, which included a number of Saull's ancient saurians. The four-volume compilation stitched together a long-running series of Palaeontographical Society memoirs, the first from 1849, and it was only now being wrapped up.[79] At the same time, Owen was reconsidering some barely decipherable slabs that had once been in Saull's museum.

---

77 William Morris thought it "a wretched place, once flash and now sordid": Boos, "William Morris's Socialist Diary." It was the home of foreign anarchists in the 1880s, and revamped as a Methodist mission with a food depot for the needy in the 1890s: *Proceedings of the Wesley Historical Society* 35 (May 1986): 141–44.

78 Richard Owen 1860, 314; Gruber and Thackray 1992, 71–74; Rupke 1994a, 210–16.

79 Dawson 2012, 664.

They had come over with the £30 job-lot to the British Museum in 1863. Having spent four years (1880–1884) overseeing the transfer of exhibits to the new Natural History Museum in South Kensington, Owen was re-examining these problematic fossils. The slabs in question contained partial jaws, scutes, and bits of skeleton. Back in 1851, when he first described and illustrated them, he thought this was a young crocodile, of indeterminate species.[80] Thirty-three years later the Grand Old Man of palaeontology finally gave it a name. Tucked away in his newly-added index, Owen called it *Crocodilus Saullii*.[81]

For a moment, Saull's kindly ghost must have smiled. But *Crocodilus Saullii* suffered an ephemeral existence. Almost immediately the name was challenged and dumped.[82]

So *Saullii* disappeared from the record, along with Saull himself. His legacy would have been an Everyman's museum of palaeontological and cultural evolution, had it not been destroyed by uncaring secularists at the onset of the Darwinian age. All that survived was a tattered reputation, hanging in the air like the tail of a Kilkenny cat after being devoured by Victorian orthodoxy. Martyrdom was a popular theme in the French-style obsequies championed by Saull. The fate of this genial socialist facilitator was far more ignominious. He simply vanished.

---

80 Richard Owen 1851, 45, Tab xv.
81 Richard Owen 1849–1884, 2: index vi.
82 A. S. Woodward 1885, 496, 507. Smith Woodward was wrong to suggest Owen had called it *C. Saullii* in 1851; he only did so in 1884. Smith Woodward himself thought Saull's specimen was more likely the newly-named tiny crocodile *Bernissartia*. Buffetaut and Ford (1979) re-examined Saull's slab to confirm that it is *Bernissartia*, a one-metre long crocodile with a short skull and blunt rear teeth for crushing hard-shelled prey.

Appendix 1

The Authorship of "D.". 1826. "Letter From A Friend: On Fossil Exuviae and Planetary Motion".[1]

This letter, published in Richard Carlile's *Republican* in 1826, was signed "D.". It discussed the astronomical causes of the "systematic order" of the geological strata—the orbital wobbles which would finally cause "an entire revolution of the seasons, by bringing the north and south poles eventually into the position originally occupied by the equator". Long-term planetary shifts were used to explain the periodic alternation of "the torrid and frigid zones" during British prehistory. They also accounted for the periodic inundations as the glaciers melted when the poles approached the equator, which explained why the strata alternated between terrestrial and aqueous environments.

The evidence that Saull was the writer can be broken down simply:

1) The letter was "from a Friend", and Saull was Carlile's patron and financier, as shown in the Home Office spy reports.[2]

2) The letter mentions the writer skipping Ogg's lecture on geology on account of the heat. The Plymouth salt refiner George Ogg delivered six lectures on geology at the London Mechanics' Institution, from 31 May to 12 July 1826.[3] Saull was an active member of the London Mechanics' Institution at the time.[4]

---

1 *Republican* 14 (8 Sept. 1826): 265–67.
2 HO 64/11, ff. 197, 446.
3 Flexner 2014, 181, 564; *LMR* 4 (8 July 1826): 163–66.
4 *LMR* 2 (4 June 1825): 91; (3 Sept. 1825): 312–13; (10 Sept. 1825): 327–28, 330; Flexner 2014, app. A, 317, 381.

3) The letter referred to the stock subject of Saull's later "geology" lectures and publications: planetary wobbles, polar approaches to the equator, switching of frigid and torrid zones recorded in the strata, and periodic marine incursions. "D."'s letter was echoing Sampson Arnold Mackey's and Sir Richard Phillips's astronomical explanations. Saull would go on to reprint Phillips's theory of planetary motion and geological periodicity in 1832.[5]

4) The letter was signed "D.", the same as his letter in Appendix 2. Most of the letters in the *Republican* and *Lion* were initialled or pseudonymous, and Saull, still a geological novice, might not have wanted to speak openly yet.

5) There was a remarkable textual and conceptual similarity to the geological part of his signed *Letter to the Vicar of St. Botolph*, written fifteen months later.[6] Parts of the letters map neatly onto one another, use identical language in places, and trace a similar pattern throughout, down to the bullet points. Typical is one such bullet point, given by way of illustration:

"D." "Letter From A Friend"	Saull. *A Letter to the Vicar*
8 September 1826	25 December 1827
A slight examination of a gravel-pit or bed of pebbles presents, in the regular kidney shape of the stones, an evidence of their action upon each other in conjunction with water, from which alone their appearance of that shape could have arisen. This will be further illustrated by a comparison with the shingles on the seashore ...	The obvious fact open to every person's examination on the inspection of a gravel pit. The regular kidney shape of some, and the partially rounded angles of others, are evident proofs of the action of the sea upon them, the effect being exactly the same on the shingles of the sea shore ...

---

5 Saull 1832b.
6 Saull 1828a, 6–7.

Appendix 2

The Authorship of "D.". 1832. *Letter from a Student in the Sciences to a Student of Theology*.[1]

In 1832, a more substantial, sixteen-page, privately printed *Letter from a Student in the Sciences to a Student of Theology* was published, again signed "D.". Although attributed to both Saull and Roland Detrosier in library catalogues, content analysis suggests that Saull was the author.[2]

Saull had attacked theology under his own name in a similar published letter to his vicar on 25 December 1827.[3] Why this one, published on 1 January 1832 (p. 15), was anonymous can be explained by Saull's intervening criminal indictment, when he was charged with facilitating the Rev. Robert Taylor's blasphemies at the "Areopagus". The case was postponed, left hanging over Saull's head. He remained, he said, a "prisoner on bail",[4] expecting a summons at any moment. As one of the "men of property" who financed the blasphemy chapels,[5] he continued to be a police surveillance target. By the time of this "D." letter, Taylor was in Horsemonger Lane gaol. Visiting Taylor in late 1831, Saull found him in solitary confinement, being "treated worse than a thief", because (as the magistracy had it) blasphemy displayed the highest "moral degradation as could be conceived".[6] Saull had no

---

1 "D.". 1832. *Letter from a Student in the Sciences to a Student of Theology*. London: J. Brooks. 16pp.
2 Computational stylometry to determine authorship (Tanghe 2018) cannot be used in Saull's case, because it requires analysis of hundreds of texts, which we do not have.
3 Saull 1828a.
4 Saull 1828c; *Times*, 8 Feb. 1828, 4.
5 HO 64/11, f. 46, Feb. 1828.
6 *Prompter* 1 (13 Aug. 1831): 713; (20 Aug. 1831): 727; (15 Oct. 1831): 860; (29 Oct. 1831): 886; (12 Nov. 1831): 920; Cutner n.d., 29.

illusions about the consequence of controverting theology publicly.[7] He would have seen anonymity as essential in the 1832 New Year's Day letter, if he were to avoid such a fate.

Proof that the letter is Saull's is clear from its content. It is not enough to state that it was printed by his fellow "conspirator" in the Taylor trial, John Brooks, although this is circumstantial evidence that it was Saull's doing.[8]

In the letter, "D." acts the wise old infidel advising the theological novice. "D." indicts rich, tithe-extracted Church livings, explaining that if the theology student just wants a comfortable life, the Church was the corrupt way to go. For Saull, this Church route was the path of social delinquency, as a seminary would not allow the tyro to obtain "real knowledge, to improve yourself, and with yourself your neighbours" (p. 3). All rival knowledge would be blotted out, because *"religion is a despotism*, reigning tyrannically over the human mind" (p. 4). It shuts out those "eternal truths, the knowledge of which entirely ... annihilates your system" (pp. 4–5). "D." starts with the first of these "truths", which disprove the "fabulous narrative" of Genesis. Thus, at the outset, "D." takes up what was to become Saull's stock-in-trade, the geological progression of life. The process had started after millions of years of crustal consolidation, as the first minute organisms changed by "slow and imperceptible gradations". "[S]trange and startling" as it might

---

[7] Saull was no Carlile, who seemed to relish martyrdom, and who declared himself "completely happy" in jail. Saull's prison-escaping faint-hearts, those who are enamoured of "revolutionising" society, but who "don't like vile prisons, where smelling bottles are not allowed, clothes spoiled, and ringlets, a la Absolom, may be most mercilessly cropped", could be written off by later extremists (*Investigator* [1843]: 37, 165). But Saull's trade would have collapsed without his active supervision and, along with it, his power to fund the movement.

[8] Saull and the radical publisher John Brooks were Carlilean comrades. They stood indicted together for backing Taylor; both refused to swear oaths on the Bible; and they worked as executors of Hibbert's will. Brooks was the more self-sacrificing and withheld his church rates, causing his property to be seized. Saull and Brooks worked on Owenite committees, including one in 1832 for the relief of tradesmen, and another in 1833 to assist the Co-operative Congress. They sat on the National Political Union Council together, and were fellow directors in the Educational Friendly Society. Brooks printed the "D." letter amid a plethora of inflammatory literature, which included the *Trial of the Rev. Robert Taylor* (1827) and *Diegesis* (1829), and, at this moment, J. E. Smith's *Antichrist* lectures (1832–33), as they were delivered.

appear to theological minds, the vivifying actions of heat and water produced

> organizations ... superior to any that had preceded them: until, within the last few thousand years, that strange animal called man made his appearance, emerging many steps in advance of the race of the Simians, who had inconsciously been his precursors. (p. 5)

Humans having monkey "precursors" links the letter to Saull. The earliest transcript we have of Saull's geology talks comes from September 1833, twenty-one months later.[9] And its language was a close parallel of "D."'s. Compare the block quote above with the language in Saull's 1833 geology talk. Here he also spoke of a time when

> that most singular of animals, "man," appears, emerging or advancing, perhaps, from some of the simian, the ape or monkey tribe, educed by circumstances over which neither they nor he could have the least control.[10]

Saull's notion of a simian precursor is nearly identical to "D."'s. Both quotes extend Owen's cultural determinism (mankind's character being formed for him, not by him) to an environmental evolutionary drive.

No conscientious theologian, "D." argued, could disregard this revelatory geology. The latter had its own reliquary, the fossils which pointed out life's true ancestry. "D." discussed the tropical animals found fossilized in Britain—notably the startling new saurians "of enormous size and length, even 80 or 100 feet long".[11] Nowadays the northerly latitudes are cold, but in ancient times they must have been hot. That was proved by the palms and fossil fruits, and the coal-age ferns (the speciality of Saull's museum) that spoke of lush, swampy forests. The millennia that had passed since these seams were laid down

---

9   This is our first knowledge of Saull's simian heterodoxy, not necessarily the first time he voiced it. Saull's 1833 lecture was probably a stock one; thus, it was likely the gist of his geology talk at Poland Street on New Year's Day 1832 (the day that "D."'s letter was written). There is circumstantial evidence to support this. J. E. Smith talked of Saull having "an unconquerable tendency" to trace men from monkeys, as if it were old hat (*Crisis* 3 [5 Oct. 1833]: 36). Indeed Smith (1833, 187), in his *Antichrist* lectures on 23 December 1832, had already raised the spectre of "man—sprung from a baboon, as some imagine", and of man as a "shaven monkey", hinting that Saull had by then alerted him.
10 Saull 1833a, 37.
11 This was Saull's friend Gideon Mantell's (1833, 314) estimate.

"would astonish and confound you", he told the novice, "yet we know the same circumstances will occur again and again!" (p. 6) This is the second point which pins the letter's paternity on Saull—the planetary swings which caused ancient environments to repeatedly flush hot and cold. The letter explained the alternating submarine and dry beds characterizing the strata, produced as the "oceanic waters shift their position, in perfect accordance with known astronomical changes". It meant that the "mass of waters, now covering the greater part of the southern hemisphere, will again, in the lapse of rather more than eight thousand years, be prevalent for some thousands of years over this northern hemisphere" (p. 9).

This came straight out of Sir Richard Phillips's work, where orbital fluctuations caused mass oceanic movements. Saull was the chief exponent of Phillips's cyclical theory of inundation and climate change. Only five months later (May 1832), Saull was to send his edited reprint of Phillips's *Essay on the Physico-Astronomical Causes of the Geological Changes on the Earth's Surface* to press. But, in the *Letter*, we see it being made to do anti-theological work first, thrust against the tyro's "sacred books". Phillips's explanations of alternating tropical and frigid zones, based partly on the precession of the equinoxes, had also figured in Saull's 1828 *Letter to the Vicar of St. Botolph*. And both letters pointedly employ an epithet, taken from Paul's epistle to the Thessalonians, "Prove All Things".[12]

"D."'s letter ended on London's latest fashionable profanation: the astrological roots of biblical myth. This is the final aspect—a regurgitation of the Rev. Robert Taylor's solar mythology—which suggests Saull's hand.[13]

Taylor's celestial explanations of biblical myth had figured in Saull's *Letter to the Vicar of St Botolph*. They reappear in "D."'s *Letter*. Saull was a shadowy deacon behind the "Devil's Pulpit", which explains why "D." ended up encapsulating Taylor's Zodiacal explanations "of the twelve Patriarchs, twelve Disciples, twelve Apostles, &c." (p. 12). Starting with

---

12 Saull 1828a, 4; 1832, title page.
13 Rival astro-mythology expositor Sampson Arnold Mackey evidently knew the "D." *Letter* was by Saull. After Saull published his 1832 *Essay*, which supported Phillips's anti-gravitation views, Mackey (1832, 4) complained that Saull's language was unbefitting "a student in the sciences".

the Sun, the "God of Day", "the great I AM; man's First Cause", "D." moved on to discuss how the ancients mapped twelve visionary images onto the celestial orb to represent the twelve months of the year, out of which the patriarchs developed their theology. For instance:

> ARIES, or the RAM, is the sign of March, generally represented with a cross, because of the suns' generally crossing the equator in that month. Hence the whole story of the Passover, the Crucifixion, the Lamb of God which taketh away the sins of the world, (as you have the constellation of the Triangle, the universal emblem of your Christian Trinity, immediately over the head of this mysterious Lamb), &c. [p. 12]

This is a gloss on the *Devil's Pulpit*, which has the Sun crossing the equator in Aries (hence the cross, Crucifixion), making "Aries, the Ram", the "Lamb of God, whose astronomical name, *Yes*, is the root of our *Jesus*, the Lamb of *God*".[14] The theology student heard that the Bible preserved these "ancient mythological fables". "As human history," said "D.", the Bible "is offensive", but "as an allegory of planetary motion ... it is ingenious." "D." was simply précising Taylor. If the novice wanted to follow up on the Hebrew personification of zodiacal myths, "D." pointed him to the "Discourses of the Rev. Robert Taylor" as well as "Dupuis, Volney, Rhegellini [sic]" (p. 13). "D."'s esoteric knowledge of Reghellini's *Freemasonry Considered as the Result of Egyptian, Jewish and Christian Religions* was itself telling. Taylor possessed one of only two copies in the country, and, given that Saull was a friend and visited Taylor's house,[15] he might have seen or read this rare book.

Taken as a whole, the talk of simian antecedents, palaeontological progression, Phillip's geological explanations, Taylor's astro-mythology, and the similar wording to Saull's geological talks point irresistibly to Saull's authorship of the *Letter from a Student in the Sciences to a Student of Theology*.

---

14   He is glossing Taylor's Rotunda lectures, *The Devil's Pulpit*, 11 Mar. 1831, 30; 18 Mar. 1831, 43; 20 Feb. 1831, 159; and passim.
15   HO 64/11, ff. 41–42; *Comet* 1 (23 Dec. 1832): 326.

Appendix 3

Saull's Publications Annotated

Saull published little in professional journals. When he did read a paper to the Geological Society, they refrained from printing it, as being out of step with their stratigraphic norms. But a good deal can be gleaned from reports of speeches in the freethought, radical, Owenite, and popular press, and this approach is better suited to the subject. I have included not only signed articles, monographs, published letters, privately printed, and press pieces, but also reports, if short-hand, or more or less verbatim, of substantive speeches. I doubt this is exhaustive. Further examination of the unstamped periodicals will undoubtedly throw up more publications.

1826. "Letter From A Friend: On Fossil Exuviae and Planetary Motion". *Republican* 14 (8 Sept. 1826): 265-67. [L, signed 'D'. For the attribution, see Appendix 1].

1828. *A Letter to the Vicar of St. Botolph Without Aldersgate, London: By a Parishioner*. London: Printed for the Author, 23 pp. [Privately Printed, L, dated 25 Dec. 1827, addressed to the Rev. W. H. Causton.]

1828. [Memorial Presented to the Common Council of the City of London]. *Lion* 1 (25 Jan. 1828): 115-118. [M, against the charge of blasphemy, dated 12 Jan. 1828. Also in *Morning Chronicle*, 17 Jan. 1828.]

1828. "Oath-Taking". *Examiner*, 9 Nov. 1828: 726-27. [L, to Lord Chief Justice Best, on juries sworn on the Bible, dated 27 Oct. 1828.]

1829. ["At a Dissenters' chapel..."]. *Times*, 26 Feb. 1829: 6. [L, on children signing an anti-Catholic petition, dated 25 Feb. 1829. Also in *Morning Chronicle*, 26 Feb. 1829: 3; and paraphrased in *Examiner*, 1 Mar. 1829.]

[1830. Saull's speech at the Metropolitan Political Union on the removal of John Gale Jones and the Rev. Robert Taylor from the Union, in *Reformer's Register*, Part 1 (July-Sept. 1830), referred to in an advertisement in Letter 3,

A Monitory Letter to Sir Robert Peel, dated 15 Oct. 1830, page 16, in Carpenter 1830-31. Unobtainable, possibly not extant. The *Reformer's Register* is not listed in Wiener 1970.]

1832. *A Letter from a Student in the Sciences to a Student of Theology*. London: J. Brooks. 16 pp. [*L*, signed 'D'. For the attribution, see Appendix 2.]

1832. *Essay on the Physico-Astronomical Causes of the Geological Changes on the Earth's Surface, and of the Changes in Terrestrial Temperature, with Notes. By Sir Richard Phillips. Re-published, with a Preface, by William Devonshire Saull, F.G.S., F.A.S., F.R.A.S.* London: Sherwood, Gilbert, and Piper, iii-viii, 72 pp. [Saull's Preface dated May 1832. Published 29 June 1832 (*True Sun*, 29 June 1832, 3; *Courier*, 29 June 1832, 1); Price 3s 6d.]

1833. "Education of the People—Liberal Offer to Students of Geology". *Mechanics' Magazine* 19 (25 May 1833): 117-118. [*L*, dated 18 May 1833.]

1833. "Lecture of Mr. W. D. Saull in Bristol". *Crisis* 3 (5 Oct. 1833): 37-39. [Report of two-hour lecture on Geology by Saull in Bristol delivered on 23 Sept. 1833; also *Gauntlet*, 1 (22 Sept. 1833): 530-533.]

1836. *An Essay on the Coincidence of Astronomical & Geological Phenomena, Addressed to the Geological Society of France*. London: Printed for the Author, 30 pp. [Privately Printed, dated February 1836.]

1837. "Lecture by Mr. Saull". *New Moral World* 3 (30 Sept. 1837): 397-98. [Report of a lecture on geology and social improvement delivered in the Social Institution, Salford, on 14 Sept. 1837.]

1838. "Progress of Social Reform". *New Moral World* 4 (20 Jan. 1838): 100 [*L*, dated 8 Jan. 1838, Saull, as Treasurer of the new Educational Friendly Society, has £100 cheque, to be used to rent land for the social community.]

1838. "Progress of Social Reform". *New Moral World* 4 (17 Feb. 1838): 131-32. [*L*, dated 3 Feb. 1838, the Treasurer now has £1,000 for the Educational Friendly Society. Also in *Northern Star*, 3 Mar. 1838.]

1838. "Oration". *Penny Satirist*, 17 Mar. 1838, 1. [Oration over the grave of John Gale Jones.]

1838. "Labourers' Friend Society". *New Moral World* 4 (24 Mar. 1838): 174-175. [*L*, informing socialists of the smallholding statistics produced by the Labourers' Friend Society.]

1838. "Educational Friendly Society". *New Moral World* 4 (14 Apr. 1838): 197. [*L*, from Saull and John Henderson (of the Educational Friendly Society Committee), dated 28 Mar. 1838, on obtaining the £1,000 exchequer bills and distribution of interest.]

1838. "Formation of Character". *New Moral World* 4 (23 June 1838): 278-280. [*L*, dated 1 June 1838, on the case for circumstance in character formation, a moral tale of two brothers from his native Northampton, one transported to New South Wales.]

1840. "The Obliquity of the Ecliptic and the Transition of the Poles". A paper read to the Uranian Society on 4 August 1840, with a digest printed in the *Inventors' Advocate*, 3 (1840): 140–41.

1842. [Memorial to the Queen on implementing Owen's Home Colonization plan]. *New Moral World* 10 (19 Feb. 1842): 271. [M, dated 8 Feb. 1842, drawn up at the Institution, 6 Frederick Place, Goswell Road, by Finsbury's inhabitants, calling for the ending of the "appalling distress and deplorable ignorance" of the people by the implementation of socialist measures. Signed by Saull as Chairman.]

1844. "Foundations of the Roman Walls of London". *Archaeologia* 30 (1844): 522-24. [Paper read on 10 Feb.1842 at the Society of Antiquaries of London.]

1845. *Notitia Britanniae; or An Enquiry Concerning the Localities, Habits, Condition, and Progressive Civilization of the Aborigines of Britain; to which is appended, A Brief Retrospect of the result of their Intercourse with the Romans.* London: John Russell Smith, 64 pp. [Introduction dated Jan. 1845, published before 15 Feb., when the *Spectator* review appeared (*Spectator* 18 [15 Feb. 1845]: 162).]

1847. "To the Editor of the Literary Gazette". *Literary Gazette* 1573 (Mar. 1847): 221. [L, dated 9 Mar. 1847, on Wayland Smith's Cave discussed at the Society of Antiquaries and the importance of such ancient British dwelling places.]

1848. "Addresse". *La Voix des Femmes*, 27 Mar. 1848, 2. [Address applauding the "glorious accomplishment" of French citizens, sent to Eugenie Niboyet's "Political and Socialist Journal" advocating the rights of women at the time of the provisional government. It is dated 3 Mar. 1848, from Saull, Gilbert Vale, and Walter Cooper, being the committee of the Finsbury Institution in Goswell Street, signed by Saull, "la tête du comité".]

1848. *Observations on the Aboriginal Tribes of Britain.* London: Effingham Wilson, 16 pp. [Privately Printed, dated 31 Jan. 1848, and read at the Ethnological Society of London on 15 Mar. 1848.]

1853. *An Essay on the Connexion Between Astronomical and Geological Phenomena, Addressed to the Geologists of Europe and America; Including a Paper read before the Geological Society of London, in February 1848, with Notes and Additions.* London: John Russell Smith, x, 40pp. [The *Connexion* is the Preface (pp. iii-ix), while the bulk of the memoir (pp. 11-40), is his paper read to the Geological Society: "An Elucidation of the Successive Changes of Temperature, and the Levels of the Oceanic Waters Upon the Earth's Surface, in Harmony with Geological Evidences", with Table, Diagram, Appendices and Summary. Price 2s.]

Appendix 4

The Major London Lecture Venues of Freethought, Radicalism, and Owenism in the 1830s

Little is known about the Owenite/radical/freethinking venues of the 1830s, and much of what is known comes from asides thrown up by the social excavations of Iorwerth Prothero, Iain McCalman, and Edward Royle. Christina Parolin has studied specific taverns and halls more recently. Older work by historians of science, notably J. N. Hays and Susan Sheets-Pyenson, have looked at so-called 'low' scientific culture and lecturing venues.[1] The study of commercial outlets has been expanded in Aileen Fyfe and Bernard Lightman's *Science in the Marketplace* (2007), but their reach extends generally to conventional journals and venues. Contemporary work on 'popular science' by-and-large sticks with the more respectable theatres and rarely ventures into the radical-blasphemous chapels and socialist halls.

Given this, a preliminary listing of the main venues might be useful. Compiling it was not without its problems. Most of the halls were leased and some very temporarily. Others changed their names more than once, and the names of many were variable, while different activist groups used each venue, sometimes consecutively, at other times concurrently. This is enough to excuse the draft nature of these details. Nor is it an exhaustive list. Tiny social 'halls'—often no more than school rooms, as in Harlington, Middlesex, in 1839—have been ignored (those jumped-up 'institutes', in Holyoake's words, in which "a closet serves for a museum, and the secretary's bed room for a committee room"). Likewise, lecture rooms which were used only fleetingly, or on the outskirts of town,

---

1   Hays 1983; Sheets-Pyenson 1985.

do not figure. Taverns and most coffee shops are also excluded. The focus is on the bigger halls and, then, only in the 1830s, and mostly on Owenite, rather than Chartist, ones, situated within an omnibus ride of London's centre. I have included mutual instruction societies, some of which were sympathetic to Owenism and radicalism, while others, the more expensive ones generally, were neutral. This despite Holyoake's less than sanguine view, that mutuals "never hold long together", their managers being so over-extended.[2] He did, however, exempt some of the best selected here.

Radical Deist/Blasphemous Chapels, Late 1820s to Early 1830s

* Saull financially supported

† Saull lectured in, or entered discussions in

Areopagus, formerly the Presbyterian Salter's Hall Chapel, St Swithin's Lane, Cannon Street. An elegant, double-height hall, with galleries all round, obtained by Saull among others for the Rev. Robert Taylor in November 1826. He moved his religious burlesque here from the Founder's Hall Chapel in Lothbury. Opened January 1827. The base for his "Christian Evidence Society". Saull was on the "Committee".[3] *

Fitch's Chapel, Grub Street, Cripplegate, leased by Saull's group, after Taylor was jailed in 1828, for the former-schoolmaster, the Christian Universalist Rev. Josiah Fitch, who preached in the "very large" chapel. Opened March 1828. It ran till 1829. Saull paid the £100 per annum rent. Audiences could reach 300. Baume and Eliza Macauley spoke here. Saull's group set up the private "Athenaeum" study group in the chapel, which met on Sunday mornings at 11, when they would discuss astronomy, chemistry, and geology.[4] *†

Optimist Chapel, 33 Windmill Street, Finsbury Square (1829–31), opened October 1829 by Pierre Baume. A farthing a week dues. It held deist and anti-Christian meetings in its 500-seater auditorium. After the July Revolution (1830),

---

2 Holyoake 1849c, 8–9.
3 HO 64/11, f. 6; Royle 1979, 468; McCalman 1988, 190; *Republican* 14 (28 July 1826): 73; (10 Nov. 1826): 553; (1 Dec. 1826): 669.
4 HO 64/11 ff. 43, 75, 77–78; *Lion* 1 (29 Feb. 1828): 273; 2 (14 Nov. 1828): 614–19; McCalman 1988, 190, 283 n. 44; Prothero 1979, 260, 263. *Athenaeum* flyer enclosed in W. D. Saull to Robert Owen, n.d., ROC/18/6/1, Co-Operative Heritage Trust Archive, Manchester.

audiences picked up. The police spy reported Saull frequently lecturing here.[5] It had started co-operative meetings by 1830. †

Philadelphian Chapel, 33 Windmill Street, Finsbury Square, City Road. Baume's Optimist Chapel taken over and renamed by James Watson, September 1831–October 1832. The National Union of the Working Classes met here, as well as co-operators and freethinkers. Eliza Sharples and Eliza Macauley spoke at the venue, and Saull talked on "Creation", that is, the 'evolutionary' production of life.[6] †

Borough Chapel, Chapel Court, High Street, Borough (1832–5?). Founded about September 1832,[7] as a successor to the *Philadelphian*. It could hold 800 and was the main NUWC meeting place in 1832-1834. It hosted millenarians, deists, and radicals. Petrie and Davenport spoke here, and J. E. Smith preached his "Antichrist" sermons in 1832–33. Saull lectured here weekly on Fridays when Smith was in residence in 1832–33. †

Bowling Square Chapel, Lower Whitecross Street, Bethnal Green (1833–34). A venue for the NUWC, co-operators, operative boot and shoe makers, Female Society, as well as Watson's and Davenport's "Society for Scientific, Useful and Literary Information", for "youth of both sexes", 1d a week. This was to provide "real Useful Knowledge", with free talks on education, religion, machinery, astronomy (by Davenport), and so on.[8]

Infidel, Radical but mostly Owenite Social Institutions

Albion Hall. Originally the school attached to Albion Chapel, in London Wall. Leased by Saull as a lecture venue for Robert Owen in early 1831.[9]*

Owen's Institutions: These were successively: 1) Burton Street chapel/Burton Rooms, Burton Crescent (1830–1837); 2) Institution of the Industrious Classes, Gray's Inn Road, King's Cross (1831–1833); 3) the Institution, 14 Charlotte Street, Fitzroy Square (1833–1836); 4) 69 Great Queen Street, Lincoln's Inn Fields (Branch A1, 1837–40); 5) Social, Literary, and Scientific

---

5   HO 64/11, ff. 167, 205, 209; Prothero 1979, 260–61; *Lion* 4 (9 Oct. 1829): 459; (28 Oct. 1829): 534–35.
6   HO 64/11, f. 67, also ff. 96, 105, 142; *PMG*, 3 Sept. 1831, 72; *Prompter* 1 (10 Sept. 1831): 782; *Isis* 1 (8 Sept. 1832): 465; *Crisis* 1 (7 July 1832): 68; Prothero 1979, 261–62; McCalman 1988, 198, 202.
7   *Crisis* 1 (8 Sept. 1832): 108; HO 64/11, ff. 150, 170, 177, 188; *Prompter* 1 (27 Aug. 1831): 752; Prothero 1979, 261–62; McCalman 1988, 69, 202; J. E. Smith 1833.
8   *PMG*, 23 Nov. 1833; 30 Nov. 1833; 14 Dec. 1833; 28 Dec. 1833; *Crisis* 3 (28 Dec. 1833): 144; *Gauntlet*, 29 Dec. 1833, 744; 5 Jan. 1834, 762–63; *The Man* 1 (8 Dec. 1833): 171; Prothero 1979, 296–97; McCalman 1987, 331; 1988, 198–99.
9   HO 64/11, f. 237.

Institution (later just Social Institution), 23 John Street, Tottenham Court Road (1840–58).*†

Rotunda, Blackfriars Road (also known as the *Surrey Institution*, originally a scientific institute). When taken over by Carlile in May 1830 it was "a neat dwelling-house" with billiard-rooms, apartments, coffee-room, library, and two theatres. In the smaller theatre, the Rev. Robert Taylor's delivered his theologico-astronomical lectures, and it acquired the nickname the "Devil's Pulpit". Saull was a subscriber, and he talked here often. By 1831, it held anti-theological meetings on Sundays and NUWC political meetings on Mondays. Eliza Sharples took over the running in 1832.[10] It became the first Metropolitan branch (Surrey Branch) of Owen's Institution of the Industrious Classes in 1833. The big theatre could hold two thousand, "comfortably seated". By 1842, it was in the hands of Lambeth Owenite branch 53 and named the *South London Hall of Science*.*†

Theobald's Road Institution, 8, Theobald's Road, Red Lion Square, William Benbow's hall, formerly the *Republican Institution*, was a converted livery stable. The NUWC met in the lower room, the co-operators in the loft above in 1832. The venue could hold almost 2000 and it was nearly full to hear O'Connell speak. The co-operators then moved to Benbow's "Temple of Liberty" in Cumberland Row, King's Cross, later in 1832. Theobald's Road was refurbished and opened as the *Bloomsbury Institution* with a social festival on 11 November 1833. Taylor, supported by Saull, carried on his astro-theological lectures for a short time here in February 1834.[11]†

Western Co-Operative Institute, 59 Poland Street, Oxford Street (1831–32),[12] set up by the First Western Co-Operative Union. Benjamin Warden lectured here, as did Saull on geology; it became the *Western Union Labour Exchange* (1832-33).†

Mechanics' Institution, Circus Street, Marylebone (1833–40s). Owned by the republican democrat John Savage, who ran the Marylebone vestry, and consequently slated as the home of the "Marylebone Savages".[13] Called *Mechanics' Hall of Science* by the *True Sun*. A venue for destructives and socialists, despite being dismissed in James Grant's *Great Metropolis* for its "Radical meetings of some half dozen of the unwashed".

Eastern Institution, 12 Portland Street, Commercial Road East, near Stepney Causeway (1834). Formerly the Portland Assembly rooms.[14] It moved to become:

---

10   Parolin 2010, chs. 6–8; Kurzer 2000; HO 64/11, ff. 445–46, 458, 462; *Prompter* 1 (13 Nov. 1830): 8; *Isis* 1 (3 Mar. 1832): 59–60; *Crisis* 1 (5 Jan. 1833): 174; 2 (30 Mar. 1833): 89; *NMW* 11 (27 Aug. 1842): 74.

11   HO 64/12, ff. 59. 76, 83, 108, 145, 148; *DPMC* 1 (20 July 1833): 200; *Crisis* 3 (9 Nov. 1833): 88.

12   *PMG*, 5, 26 Nov. 1831; and 31 Dec. 1831 for Saull's (earliest?) lecture on geology.

13   *The Age*, 1 May 1836, 141; J. W. Brooke 1839; *Crisis* 3 (28 Sept. 1833): 32; *TS*, 31 Dec. 1835, 4; [James Grant] 1837, 2: 203–04.

14   *Gauntlet*, 2 Mar. 1834, 890; and 16 Mar. 1834, 923, for Robert Owen's founding address; *Crisis* 3 (1 Mar. 1834, 224).

Eastern Social Institution, Curtain Road, near Old Street, Shoreditch (originally Zebulon Chapel) (1836–37).[15] Charles Jenneson was Secretary. Called the *Rational Institution* in 1837.†

Mechanics' Hall of Science, Commercial Place, City Road, Finsbury, James Watson's, opened 7 April 1834 using a bequest from Hibbert, as a venue for Roland Detrosier, who died months later.[16] A 2000-seater barn-like hall, refurbished by Watson and Saull. Freethinking, radical and Owenite talks interspersed with scientific lectures. Long running. At the end of the 1830s, it became Branch 16 of Owen's Universal Community Society of Rational Religionists. In the later 1840s Holyoake based his Utilitarian Society here, and Robert Cooper lectured in the hall in the 1850s. One of Saull's favourite haunts.†

Finsbury (Owenite Branch 16), or the East London Branch of the Association of Rational Socialists, which first met in the Mechanics' Hall of Science, City Road, in 1838-39, then Southwark School-room, Union Street, Borough (1840), finally setting up the *Finsbury Social Institution*, 6, Frederick Place, Goswell Road (founded 1840). This was a compact lecture hall, capacity 300, with coffee room, reading room, and library attached. It became the Finsbury Literary and Mechanics' Institute in 1846.[17]†

Hall of Science, Grosvenor Street, Millbank, 1836. This saw the formation of the City of Westminster Radical Association, under radical/Owenite control. Probably formerly the Owenite *Westminster Rational School and General Scientific Institution* (founded Dec. 1833), at which Carlile, Taylor, and Detrosier guest-lectured. Rebranded the *Westminster Mechanics' Institution* on 10 Jan. 1837. This probably became the *Institution for Instruction and Amusement* in 1839, where Charles Southwell lectured.[18]

Rockingham House, New Kent Road, Elephant and Castle (1838–39). Frederick Hollick lectured here. It held nearly 1,000.[19]

Social Institution, Hunterian Museum, Great Windmill Street, Haymarket (1838–39). It accommodated 400-1,000.[20]

Social Institution, Royal Hill, Greenwich (1839). Joshua Thorne was Secretary. This was commended to social friends as a "change from dirty streets and alleys, to pleasing hills and valleys".[21]

Social Institution, Exeter Street, Sloane Square, Chelsea. Opened by "Miss Reynolds" (Margaret Chappellsmith after her marriage in 1839) on 5 June

---

15 NMW 2 (30 Apr. 1836): 216; Holyoake 1906, 1 139.
16 Crisis 4 (12 Apr. 1834): 5; weekly lectures were listed in the *Penny Mechanic*.
17 NMW 8 (5 Dec. 1840), 368; 9 (27 Feb. 1841): 134; 12 (23 Dec. 1843): 208; *Reasoner* 1 (29 July 1846): 136; *UR*, 8 Sept. 1847, 82; NS, 18 Sept. 1847.
18 TS, 18 Mar. 1836, 2; 26 Mar. 1836, 3; *Crisis* 3 (11 Jan. 1834): 155; *Gauntlet*, 12 Jan. 1834, 776; T. P. Thompson 1843, 4: 365; NMW 6 (16 Nov. 1839): 896.
19 NMW 5 (1 Dec. 1838): 90.
20 NMW 5 (1 Dec. 1838): 90; (15 Dec. 1838): 123–24; (2 Mar. 1839): 303.
21 NMW 5 (18 May 1839): 473.

1839. "As dead as Chelsea" (the old adage) did not apply here, given Mrs Chappellsmith's enlivening lectures.[22]

Kensington Social Institution, 2 High Street, Kensington (1839);[23] then (1840) 1 King Street, Kensington. The Owenite and Chartist T. M. Wheeler was Secretary. Scientific lectures on Monday nights.

Bethnal Green, Trades' Hall, Abbey Street (1839–40).[24] The new Owenite Branch 62 (Secretary George Fleming) used the 1,000-capacity Hall. Charles Southwell and Benjamin Warden lectured here. Also a London Democratic Association stronghold. Police seized arms and made arrests here on 16 January 1840 in the wake of the Newport Chartist uprising.

Social Institution, Lambeth, 28 Mount Street, Westminster Road (1839–42). Charles Southwell lectured here in 1839. This was Owenite branch 53, which re-located to the Rotunda (1842–44) and then a new hall, 5 Charlotte Street, Blackfriars Road, in October 1844.[25]

Mutual Instruction and Related Societies

Allen Davenport reckoned there were nearly fifty mechanics' and mutual instruction institutions in and around London, which saw "young men scarcely out of their teens" discussing and delivering "lectures on ... chemistry, geology, mathematics, and astronomy, with all the gravity, deliberation, and confidence, of old and experienced professors."[26] Only the larger mutual instruction groups are now remembered.

Great Tower Street Mutual Instruction Society, (founded 1836). Cheap (1s a quarter). It had a strong radical and Owenite bent. Allen Davenport lectured here and became President. Classes on mechanics, mathematics and the sciences (including geology);[27] lectures on social and political issues, and the sciences. It was re-founded in the mid-1840s as the City of London Mechanics' Institute at 3 Gould Square, Crutched Friars.†

---

22 *NMW* 5 (1 June 1839): 506; 6 (21 Sept. 1839): 764; (7 Dec. 1839): 940; Frow and Frow 1989, 83.
23 *NMW* 6 (16 Nov. 1839): 893.
24 *NMW* 6 (23 Nov. 1839): 908; 7 (15 Feb. 1840): 1112; (30 Apr. 1840): 1144; (21 Mar. 1840): 1192; (28 Mar. 1840): 1205; Goodway 1982, 33.
25 *NMW* 5 (20 Apr. 1839): 408; 6 (10 Aug. 1839): 665; 13 (12 Oct. 1844): 126; (2 Mar. 1845): 312.
26 Davenport 1845, 74; Rose 2002, 72.
27 *PM* 2 (23 Dec. 1837): 165; 3 (15 Sept. 1838): 200. Davenport: 2 (19 Aug. 1837): 24; (6 Jan. 1838): 184; (7 Apr. 1838): 311; 3 (4 Aug. 1838): 144.

Finsbury Mutual Instruction Society (founded 1836). It met at South Place Chapel, Finsbury, then in 1844 at 66 Bunhill Row. Science, politics and frequent discussion nights. Allen Davenport was an honorary member.[28] The radical MP Thomas Duncombe was President. T. Perronet Thompson, the philosophical radical, Corn Law repealer and supporter of sensible Chartism talked here. It boasted a library, some 350 members in the 1840s, and ran on into the 1850s.†

Society for the Acquisition of Useful Knowledge (founded 1835). 36 Castle Street East, Oxford Market; then from July 1835, 18 Store Street, Bedford Square.[29] 36 Castle Street was formerly the Assembly Rooms, where the NUWC had met, and where Cobbett and the Marylebone radical and Owenite Thomas Macconnel had lectured. It had then been a St Simonian School and a day school for young ladies. J. E. Smith lectured here in later 1834–early 1835, before the Society took over. Saull talked on science here. †

Poplar Literary and Scientific Institution, East India Road, opened January 1837.[30] More staid and expensive (5s a quarter), but could host talks on the "Origin of Mankind". It was still in existence in 1854.

Kentish Town Mutual Improvement Society, 5 Winchester Place. Founded in May 1837. 1s 6d a quarter.[31] Originally the *Kentish Town Mechanics Institution*.

Mutual Instruction Society, 73 Rahere Street, Goswell Road, Finsbury. 1s 6d a quarter. Originally founded in 1834 for carpenters and joiners, but open to the working classes generally in 1838. Self-help evening classes and Wednesday lectures, largely on science (including geology), engineering, architecture, etc.[32]

Lambeth Mutual Instruction Society, North Place, Lambeth (1839). Probably Lambeth Coffee House, 3 North Place, also called Huggett's Coffee House. This was shortly the atheist Charles Southwell's headquarters, and the meeting place of the Chartist National Association.[33]

Bermondsey Mutual Instruction Society, Great George Street (1837). 1s 6d a quarter.[34]

---

28 Davenport 1845, 71; *PM* 1 (21 Jan. 1837): 96; *Reasoner* 1 (22 July 1846): 124.
29 *Shepherd* 1 (28 Feb. 1835): 216; (4 July 1835): 360; *PMG*, 30 May 1835, 550; Roebuck 1835a, 16.
30 *NMW* 3 (4 Feb. 1837): 116; lectures listed in *Penny Mechanic*; *Mechanic and Chemist* 2nd ser. 4 (13 Apr. 1839): 135; Weale 1854, 600.
31 *PM* 1 (6 May 1837): 224 et seq.; Coates 1841, 89.
32 *PM* 2 (24 Mar. 1838): 287.
33 Coates 1841, 89; *NMW* 8 (17 Oct. 1840): 252; 10 (25 June 1842): unpaginated "Advertisements" after p. 424; Royle 1976, 14; *National Association Gazette* 1 (22 Jan. 1842): 25 et seq.
34 *PM* 1 (18 Mar. 1837): 168 et seq.

Appendix 5

Geology Lecturers in Owenite, Radical, and Mutual Instruction Institutions

Besides Saull, many speakers in Owenite or radical venues ran geology lectures under the rubric of attacks on the Deluge and Days of Creation. These stock anti-clerical lectures on the antiquity of the earth and the origin of man were used to underpin Owenite freethought, cultural environmentalism, and natural perfectibility.

At a time when education was rudimentary, resourceful workmen set up self-help groups, sometimes with associated secular schools. Their speakers were ideologically-driven, outspoken, and clever. Too clever, thought the magistrates. The recorder at the trial of Carlile's shop worker, Thomas Ryley Perry, admitted that he had "so much ability as to be a dangerous man to be allowed to be out of Gaol!"[1] Such working men had little free time, and lectures, usually starting at seven or eight in the evening, had to be squeezed in after a day's labour. This night time learning allowed them to master the ideological aspects of geology and, in some cases, get up lecture series themselves, with all the theatrical trappings: dioramas, maps, sections and fossil displays. As the *British Critic* put it, the "scepticism of the day" required geology's "choicest" weapons, and "hundreds of sciolists" were chipping away at "the rock of ages."[2]

Social missionaries were expected to be well-sourced in the science. The Central Board pushed geology to the fore in their rational school curriculum.[3] Undermining the pulpit-props in Genesis; undercutting

---

1   *Republican* 11 (4 Mar. 1825): 288.
2   *British Critic* 1 (Jan. 1827): 200.
3   NMW 9 (13 Feb. 1841): 91; 6 (24 Aug. 1839): 704; (5 Oct. 1839): 789–91.

the notion of death entering the world with the Fall; underlining the natural rise of fossil empires; underscoring a soul-less 'Age of Reptiles' on a planet without men—these were easily explainable in Grub Street venues. On the positive side, geology was expected to prove that perfectibility not depravity maketh man. It contained a promise of better things once the impediments were swept away. The "false" notion of mankind's Fall and the "true" geological bedrock of the coming socialist millennium was a stock-in-trade of the Halls of Science.

Thus, socialist geology, "which the low-minded would wrest to the purposes of infidelity",[4] was widely linked to blasphemy and sedition, and caused a backlash in the evangelical press. As a result, geology itself was often tarred as a "foolish conceit". It is "silly, disgusting, and often injurious", said *Freeman's Journal*.[5] Some of the lecturers listed below were jailed for blasphemy or sedition, and others beaten up by mobs, themselves occasionally fired up by the clergy. Lecturing in a Hall of Science was risky. But the zeal of these science lecturers for Owenism and freethought often reached religious heights, a fervour expressed in Saull's declaration at the Optimist Chapel, that "Materialism was the only true Religion".[6] Many lecturers were young, in their twenties, and most had had conventional religious childhoods. But they had been torn by the injustices of society, its wealth disparity, and a tithe-grubbing established Church. As a result their geology talks invariably bled off into anti-biblical harangues.

The entries are arranged roughly chronologically, according to when the lecturers were active.

London Lecturers

Henry Darwin Rogers (1808–1866). Rogers had been dismissed as a teacher from Dickinson College, Pennsylvania, in 1831 for his reformist views. He wrote for Frances Wright's *Free Enquirer* in New York, and in 1832, only twenty-three, sailed with Robert Dale Owen to England. He moved into Owen's house (4 Crescent Place, owned by Saull) and began a course on geology in December 1832 at Owen's Institution, Gray's Inn Road, occasionally switching venues to Owen's Burton Street chapel and the Rotunda. Rogers' Thursday night

---

4 *The Age*, 8 Jan. 1837, 5.
5 *Freeman's Journal*, 17 July 1839.
6 HO 64/11, f. 167 (22 Nov. 1830).

talks used "rare and beautiful" fossils from Saull's museum[7] and blown-up diagrams of the strata. Saull was simultaneously lecturing here on geology on Tuesdays. Rogers returned to America in 1833. He became professor of geology at the University of Pennsylvania and pioneered state surveys of Pennsylvania and Virginia. From 1857 he was Regius Professor of Natural History at the University of Glasgow.

R. *Penman*. Lectured in 1834 on geology ("Gentlemen 1*d*, Ladies free") at Bowling Square Chapel in Bethnal Green, where Owenite social community classes were convened.[8]

J. *Norman*. Manager of the provision department of the National Equitable Labour Exchange. He worked with Saull on the Dorchester Committee. At Owen's Institution in Charlotte Street, he gave (1835) a "highly important" anti-Mosaic account of "The Antiquity and Duration of the World", which underpinned Owenite rationalism.[9]

Joshua Thorne. A young social missionary who worked out of the Mechanics' Hall of Science, City Road, c. 1838–40. Here and at the Eastern Social Institution, Curtain Road, he lectured on the use of metals and coal formation, and he livened up City Road *soirées* with oxy-hydrogen lights and an electrical machine "to electrify nearly all present at one time".[10] He talked on geology in 1838 at the nearby mutual instruction society in Goswell Road. Thorne spoke on co-operation, geology, and the proofs of T. Simmons Mackintosh's Electrical Theory of the Universe at the Tower Street Mutual Instruction Society, where he was Secretary.

T. *Simmons Mackintosh*. One time Glasgow cotton weaver and a former Carlile shop worker who became a star on the socialist circuit. He incorporated geology extensively into his Electrical Theory of the Universe, which he lectured on at the Mechanics' Hall of Science in City Road in October 1836.[11] He went on to publish an *Inquiry into the Nature of Responsibility* (1840).

[*Thomas Rivers?*] *Mansfield*. At John Street, Mansfield gave a four-lecture and a three-lecture series on geology (1841 and 1843), "fearlessly" exposing the "fallacies" of the reconcilers. They were "illustrated by numerous sections

---

7   *Crisis* 1 (8 Dec. 1832): 159; (15 Dec. 1832): 164; (29 Dec. 1832): 172; (5 Jan. 1833): 174, 176; 2: (12 Jan. 1833): 8; and thereafter weekly to (16 Feb. 1833): 48; *PMG*, 22 Dec. 1832. Gerstner 1994; W. E. Adams 1998; E. S. Rogers 1896, 1: 91–95.
8   *PMG*, 22 Mar. 1834, 56.
9   *NMW* 1 (12 Sept. 1835): 364–66; *TS*, 12 May 1835 1; 20 May 1835, 8; *The Man* 1 (3 Nov. 1833): 134.
10 *NMW* 4 (28 Oct. 1837): 5; 5 (26 Jan. 1839): 224; (13 Apr. 1839): 394–95; (8 June 1839): 520; *Penny Mechanic, and the Chemist* 3 (18 Aug. 1838): 167; (25 Aug. 1838): 176; *MM* 26 (25 Feb. 1837): 405–07; 27 (8 Apr. 1837): 2–4; and throughout this volume (pp. 83–5, 182–84, 291–92, 395–96); *PM* 2 (23 Dec. 1837): 165; (20 Jan. 1838): 200; *Mechanic and Chemist* 2nd ser. 4 (19 Jan. 1839): 24; (18 May 1839): 183; Mackintosh 1846, 333–34.
11 *MM* 26 (8 Oct. 1836); Morus 2011, 78; Holyoake 1875, 1: 373; Anon 1858, 62–63.

and diagrams", all aimed at denying a universal Deluge, while proving the enormous "age of the world, and the antiquity of the human race." Complaints that the lectures were "inimical to revealed religion" led him to discuss critically the origin of Christianity. He also lectured on geology at Branch 53, the Rotunda.[12]

Robert Buchanan (1813–1866). Former Scottish tailor and social missionary, who worked with Saull on the London Tract Society. Buchanan had a "rugged eloquence on the platform"[13] and followed Mansfield in 1842 at the South London Hall of Science with the "Creation and Fall of Man and the Deluge, with reference to Geology and Astronomy". In November and December 1846, he lectured on "Cosmogony, Geology, and Astronomy, considered in conjunction with the recent discoveries of Lord Rosse's Telescope" at John Street. See below (Provincial Owenite Lecturers) for his mission work in Manchester.

[*Thomas?*] *Thomason*. Lectured at the Poplar Institution on 'The People and Geology of Cumberland and Durham.'[14] A Thomas Thomason (probably the same man) was involved with the school at Harmony for a short period.

Thomas Cooper (1805–1892). Former Methodist preacher and shoemaker, who studied the sciences while working with the awl. He became a Chartist (1840) in Leicester, where he taught inter alia geology in the "Shaksperean Rooms". He served two years (1843–45) in Stafford jail for sedition. Thereafter a lecturer in freethought/radical venues, and in the Chartist's National Hall in High Holborn he talked on geology.[15] Cooper remained tantalized by fossils, and after returning to Christianity he printed his later lectures as *Evolution, The Stone Book, and the Mosaic Record of Creation* (1878) to rebut the "godless theorising" of John Tyndall and Herbert Spencer.

Thomas Frost (1821–1908). Croydon-born apprentice in a printing office, who turned to Owenism, Chartism and finally sacred socialism (the rigours of which defeated him). He started his own printing works in 1843 with a satirical paper. In 1842 Hetherington opened his *Free-Thinker's Information for the People* with two Frost articles, which secularized Lyellian palaeontology to illegitimate Moses and the Fall.[16]

John Robinson. Robinson was a member of the London Working Men's Association. From the mid-1840 he ran (as Honorary Secretary) the City of London Mechanics' Institute at 3 Gould Square. The Institute had a museum and laboratory, organized geology classes, hosted a Provident Society, a Society for the Advancement of Secular Education, and a Day school.

---

12   NMW 10 (23 Oct. 1841): 136; (13 Nov. 1841): 160; 11 (6 Aug. 1842): 48; (27 Aug. 1842): 74; (21 Jan. 1843): 243; (4 Feb. 1843): 260.
13   Jay 1903, 3; *NMW* 11 (6 Aug. 1842): 48; (27 Aug. 1842): 74; (22 Oct. 1842): 139; *Reasoner* 1 (25 Nov. 1846): 308.
14   *UR*, 23 Dec. 1846, 8; Royle 1998, 180, 188 n. 133; *NMW* 13 (28 June 1845): 432.
15   *UR*, 12 Jan. 1847, 13; Thomas Cooper 1872, 169.
16   Frost 1880; 1842; *FTI* 1 (n.d. [1842]): 1–16.

Holyoake taught grammar here and dedicated his *Literary Institutions* (1849) to Robinson. Robinson lectured on anatomy, natural philosophy, chemistry, pneumatics, hydrostatics, and the "Geology of the North of England". Robinson and Hedger (see next entry) also gave a joint talk here on geology, illustrated with "shells, minerals, and organic remains," presumably from the Institute's museum.[17]

G. Hedger. He shared a platform with Robinson (above), and spoke on geology at Finsbury Mutual Instruction Society. He also talked on vegetation at Finsbury Social Institution in the Goswell Road, and, like Saull, lectured on "Primaeval Human History".[18]

John Edwards. He talked at the Institute of Political and Social Progress, 1 George Street, Sloane Square, Chelsea. This was set up, in hired rooms, by "self-reliant" mechanics in October 1848. In late 1850, when they had 70 members paying 6*d* a week, it moved to a house, 10a Upper George Street. The mechanics established lectures, classes, a reading room, and a secular school. Thomas Cooper and Holyoake talked here, and Chartist causes were supported. Edwards gave at least two lectures on geology in October 1849.[19]

Later Socialist/Atheist Writers on Transmutation and the Origin of Man

William Chilton (1815–1855). Tin-plate worker's son who became a compositor on the *Bristol Mercury*. An atheist, he seceded from Owenism in 1841 and helped set up the *Oracle of Reason* (1841–43). With Southwell jailed, Chilton took over the "Theory of Regular Gradation" on the seventh number. He ran it in forty-two issues, pressing geological and anatomical texts into transmutatory service. The series covered the gamut from the chemical origin of life to an ape ancestry for man. He took Owenite environmentally-driven evolution to its ultimate expression in an anti-clerical context. Chilton camped in his works to keep the *Oracle* running, but he still found time to become a delegate to the 1842 Birmingham Chartist conference. His evolutionary articles spilled over into the *Movement* and *Reasoner*. In 1851, his "Library of Reason" placed digests of *Vestiges*, Lyell, and Lamarck,

---

17 *Reasoner* 1 (3 June 1846): 16; (15 July 1846): 112; 7 (28 Nov. 1849): 337–38; *UR*, 2 Dec. 1846, 2; 2 Feb. 1847, 19; *NMW* 2 (13 Aug. 1836): 330–31.
18 *Reasoner* 6 (13 June 1849): 383; 7 (17 Oct. 1849): 255; 8 (23 Jan. 1850): 23; *Cooper's Journal*, 26 Jan. 1850, 54.
19 *Reasoner* 5 (29 Nov. 1848): 431; 7 (17 Oct. 1849): 255; (24 Oct. 1849): 271; *Freethinker's Magazine and Review of Theology, Politics, and Literature*, 1 Dec. 1850, 218–19.

alongside Strauss, Hume, and Spinoza. Just before his death, Chilton fired off a defensive letter titled "Man and the Baboon" to the *Reasoner*.[20]

Emma Martin (1812–1851). Bristol-born cooper's daughter, brought up a Baptist. Her feminism developed out of a miserable marriage and she became a socialist, intent on unchaining women from patriarchal bondage. She attended the 1839 Birmingham Congress, and scandalized society by her itinerant lecturing on Owen's marriage system. Appearing demurely on stage, with a sonorous voice, she hit a chord with oppressed women, judging by the huge audiences. In 1844, she softened Chilton's diatribal approach to origins. Familiar with evangelical dialogue tracts, she turned the anti-infidel technique against itself in a pamphlet which had a freethinker quizzing and unsettling a "Theist" on the unaided rise of life. For her, "man" was a "new product of nature's increasing power".[21]

Robert Cooper (1819–1868). With his father a Peterloo veteran, and his Salford Co-Operative schooling (when he studied geology and heard Robert Owen), Cooper was a second-generation atheist-socialist. Fired from his first job for a pamphlet on the Holy Scriptures, he became a social missionary in 1841, lecturing on the "Immortality of the Soul". An invitation to attend lectures at Edinburgh University in 1845 allowed him access to their library.[22] His ensuing *Infidel's Text-Book* (1846) used Sir Richard Phillips's planetary perturbations to establish the globe's antiquity, geology to prove the earth's slow growth, and racial texts to supplant the Adam and Eve story with a pluralist science of discretely originating human types. His theatrical performances were raucous affairs and derided by the anti-infidel press as outrageous.[23] He edited the *London Investigator* (1854–58), and opened it with an eight-part series on the origin of man in 1854.

John Watts (1834–1866). Son of a Bristol tradesman and Wesleyan preacher. Watts, a compositor, became sub-editor of Holyoake's *Reasoner*. He toured the country in 1861-62 lecturing on the "Origin of Man".[24] He was influenced by J. C. Nott and G. R. Gliddon's racist *Types of Mankind* (1854) and Darwin's *Origin of Species* (1859). Watts' Pentateuchal critique took in cave faunas, flint implements and human antiquity as well as higher anatomy and recapitulationist embryology. He reviewed Darwin and Spencer in the *National Reformer*[25] and was talking on evolution at Cleveland Institution[26] while Saull's fossils and flints were in storage there. These talks were repeated at the City Road Hall of Science. Watts became editor of the *National Reformer*

---

20 *Reasoner* 17 (8 Oct. 1854): 225–29; 3 (3 Nov. 1847): 608; Chilton 1854; Chilton ODNB.
21 E. Martin 1844, 6; J. A. Secord 2000, 314; B. Taylor 1983, 64, 70, 130–55.
22 *LI* 2 (May 1855): 28–30.
23 *The Association, or Young Men's Magazine* (1855): 32–34.
24 *Reasoner* 26 (17 Feb. 1861): 102; (24 Feb. 1861): 119; (3 Mar. 1861): 132.
25 *National Reformer*, 12 Dec. 1961, 2; 4 Jan. 1862, 6; 18 Jan. 1862, 5–6.
26 *Reasoner* 26 (27 Jan. 1861): 62; (17 Feb. 1861): 102; (24 Feb. 1861): 119; (3 Mar. 1861): 132; *National Reformer*, 21 Mar. 1863, 8; 11 Nov. 1866, 305–06.

in 1863, only to suffer from consumption and die in 1866, aged thirty-two. He was buried near Saull in Kensal Green.

Provincial Owenite Lecturers

William Hawkes Smith (1786–1840). The Unitarian socialist Hawkes Smith helped found the Birmingham Labour Exchange.[27] His *Birmingham and its Vicinity as a Manufacturing and Commercial District* (1836) dealt extensively with the South Staffordshire coal fields, as did *Birmingham and South Staffordshire: Or, Illustrations of the History, Geology, and Industrial Operations of a Mining District* (1838). He lectured on coal geology at Owenite branches in Worcester and Coventry, as well as at his own Birmingham Mechanics' Institute (where he influenced Holyoake).[28]

Thomas Ryley Perry (1793–1846). Perry, a Leicester druggist, was an atheist, socialist, and Chartist. An erstwhile "itinerant comedian", he and his wife helped run the jailed Carlile's shop, and as a consequence he spent three years in Newgate (1824–27) for selling Palmer's *Principles of Nature*.[29] Moving back to Leicester, he sold unstamped papers, only to be imprisoned again. He campaigned for a reading room for the working classes.[30] As President of the Leicester Owenite branch, based in Market Place, he lectured on geology.[31] The Leicester branch of the Anti-Persecution Union was founded by him. He died in Leicester's Union Workhouse.

Samuel Phillips. Active in the Leamington branch 1838–40. He lectured on geology.[32]

George Connard. A sign painter and social missionary, who delivered a course on geology in Wigan, "illustrated by diagrams painted for the purpose".[33] Despite an idyllic picture in the *New Moral World* of orators delivering "sermons in stones" on sunny field trips,[34] many socialists like Connard suffered privation. Having served time in Lancaster gaol for standing surety on an absconder's debt, Connard had had to remain incarcerated for refusing to take the oath on a Bible.

---

27 *Crisis* 1 (8 Dec. 1832): 157–59.
28 Holyoake 1892, 1: 45–49, 60–61; *NMW* 4 (11 Nov. 1837): 19; (19 May 1838): 237; 7 (25 Apr. 1840): 1264.
29 *Republican* 10. (16 July 1824): 33 et seq; *Newgate Monthly Magazine* (published by Perry et al. 1825 passim); Wiener 1983, 93; Newitt 2016.
30 *DPMC* 1 (10 Aug. 1833): 220; *Gauntlet*, 6 Nov. 1833, 626; 16 Feb. 1834, 861.
31 *NMW* 9 (17 Apr. 1841): 246; *Movement* 1 (8 June 1844): 205; 2 (29 Jan. 1845): 37.
32 *NMW* 5 (9 Mar. 1839): 316.
33 *NMW* 8 (26 Dec. 1840): 414.
34 *NMW* 11 (24 June 1843): 434; 5 (2 Mar. 1839): 304; (6 July 1839): 581; *The Charter*, 15 Sept. 1839, 537; J. F. C. Harrison 1969, 219–30; Holyoake 1906, 1: 212; A. J. Booth 1869, 200.

Robert Buchanan (1813–66). Buchanan was appointed by the Central Board to the Manchester branch. He would stump the surrounding country with his magic lantern, and lectured on everything from communism to priestcraft. He took geology in his stride. He talked "on 'Geology, and the Mosaic Account of World-Making,' with dioramic illustrations" at Manchester[35] and, later, London (see above). The day following his Manchester talk Town Mission preachers and a mob stormed the social institution to stop proceedings.

John Hansom (1790–1878). A weaver-turned-shopkeeper. Hansom was a freethinking radical who supported the hand-loom weavers' struggle, and a socialist who debated Carlile on the formation of character. He was a manager of the Huddersfield Hall of Science in Bath Street (founded 1839, complete in 1840). The town was an Owenite stronghold, and the Hall prospered after many others failed. Hanson delivered a course of twelve lectures here on geology, starting in February 1842.[36]

---

35 *NMW* 10 (14 Aug. 1841): 55–56; J. F. C. Harrison 1969, 220; Jay 1903, 3.
36 *NMW* 10 (11 Mar. 1842): 271; 8 (29 Aug. 1840): 144; Garnett 1972, 192; J. F. C. Harrison 1969, 226; A. Brooke, "Huddersfield Hall of Science."

Appendix 6

Saull's Close Coterie

These are considered to be Saull's closest personal friends—those whom press reports identified standing over his grave in May 1855 to make their farewells.[1]

Edward Henman. Republican, feminist, and Paineite. In 1822, he was with John Gale Jones at the Paine anniversary meeting, endorsing calls that Paine's works "be read by every Tyrant and Bigot". He chaired the Paine celebration at the White Hart Tavern in Bishopsgate Street in 1824, while toasting the "female Republicans" and "The People, the just foundation of power".[2] He supported Carlile's shop, and praised the fortitude of Carlile's wife and jailed sister Mary-Anne Carlile. With Saull, Brooks, and others, he had bankrolled Fitch's take-over of the Grub Street Chapel in 1828. In later years, Henman was chairman of the City of London Mechanics' Institute, where he introduced Holyoake's works.[3]

Thomas Prout (c.1785–1859). An apothecary. His *Annual Register* obituary noted his lifelong advocacy of extreme politics. He financed them through the sale of gout pills, sold with patent medicines in his 229 Strand shop. Whether in *Cleave's Penny Gazette*, Cobbett's *Political Register*, or Owen's *Crisis*, there you could find his pills advertised. With Francis Place and others, he helped return Sir Francis Burdett and Sir J. C. Hobhouse in the Westminster elections. In a *Times* indictment, Saull was called an "intimate friend of Mr. Prout", as if that were enough to condemn him.[4] Prout helped get a repeal of the newspaper stamp duty, and sat on the committee of the London

---

1   *Reasoner* 19 (13 May 1855): 55.
2   *Republican* 5 (22 Feb. 1822): 232; 6 (15 Nov. 1822): 783–84; 9 (6 Feb. 1824): 161; Royle 1976, 26; Prothero 1979, 260, 275, 384 n. 83. Henman's claim (*Republican* 13 [3 Feb. 1826]: 133) that Paine's works had accelerated the Mechanics' Institution movement outraged the *LMR* (3 [4 Feb.1826]: 255), which thought it like comparing "honesty with thieving, or Jaggernaut with the Deity".
3   *UR*, 22 Mar. 1848, 33.
4   *Times*, 23 Jan. 1833, 2; Hobhouse 1819, 30; *Annual Register* (1860): 471–72.

Anti-Corn-Law Association in 1836 with Saull and Ashurst. The three were members of the Radical Club, which approved "The Charter", and the Metropolitan Parliamentary Reform Association, which campaigned for complete suffrage.[5] His son John Prout (d. 1894), who took over the Strand shop, was one of the three Trustees listed in Saull's will.

Dr Arthur Helsham (c.1785–1875), surgeon, of the Mile End Road. Little is known of him, and what there is in the medical press is uninformative, for example, his refusal to bleed or give whisky as a tonic. His donation (with Saull's) to Carlile's children, to enable them to sail to America in 1852, suggests freethought sympathies.[6] He helped deal with Owen's mortgage on Saull's Rose Hill Estate.[7] Saull left his museum in Helsham's hands, and Helsham's son Robert, a solicitor, would later act for Saull's company.

William Henry Ashurst (1792–1855), Saull's friend and solicitor, was ill at the time of Saull's funeral, and represented by his son. He was to die himself months later, on 13 October 1855. He and Hetherington had been Freethinking Christians, though as a solicitor Ashurst was more urbane and he disliked Hetherington's language.[8] A City reformer, he opposed capital punishment and the church rates (which he refused to pay), and acted on deputations with Saull. They worked together against the corn laws and for suffrage.[9] Ashurst was Owen's solicitor, dealing with Harmony, and he helped finance the *New Moral World* and the *Reasoner*.[10] He advised Holyoake's defence in his blasphemy trial of 1842, then bought the *Spirit of the Age* in 1849 and installed Holyoake as editor. It was apparently at Ashurst's suggestion that Holyoake promoted the term 'Secularist'.[11] Ashurst's Muswell Hill home became a radical salon. William Ashurst Jnr was a witness as Saull drafted his final will on 3 April 1855. With no children of his own, Saull fell back on the sons of his best friends, John Prout, Robert Helsham, and William Ashurst Jnr., to see his posthumous interests served.

---

5   Rowe 1970b, document nos. 71a, 99, 126; Prentice 1853, 1: 50; *TS*, 22 Dec. 1836, 1; C. D. Collet 1933, 26.

6   *Reasoner* 12 (21 Apr. 1852): 367; *Lancet*, 26 Oct. 1850, 490; 9 Nov. 1850, 540–41.

7   *Reasoner* 22 (29 Mar. 1857): 50; A. Helsham to Robert Owen, Robert Owen Collection, ROC/8/41, Co-Operative Heritage Trust Archive, Manchester.

8   C. D. Collet 1933, 19; Hetherington 1828; Prothero 1979, 259.

9   *TS*, 13 Aug, 1825, 8; 6 July 1836, 2; 22 Dec. 1836, 1; Rowe 1970b, document nos. 71a, 129.

10   Royle 1974, 91, 93, 154–55; 1998, 79; McCabe 1908, 1: 160; Holyoake 1906, 1: 191; 2: 600; C. D. Collet 1933, 84.

11   Royle 1974, 154–5; Holyoake 1892, 1: 155 and ch. xxxiv; Goss 1908, xxxvi, 67; McCabe 1908, 1: 140, 146; J. F. C. Harrison 1969, 225.

Bibliography

Abrahams, Aleck. 1908. "No. 277 Gray's Inn Road." *Antiquary* 4: 128–34

— 1922. "William Devonshire Saull." *Notes and Queries*. 12 S 11: 230

Adams, W. E. 1903. *Memoirs of a Social Atom*. 2 vols. London: Hutchinson

Adams, Sean Patrick. 1998. "Partners in Geology, Brothers in Frustration: The Antebellum Geological Surveys of Virginia and Pennsylvania." *Virginia Magazine of History and Biography* 106: 5–34

Ainslie, Robert, et al. 1840. *Lectures Against Socialism*. London: Seeley

Åkerberg, Sofia. 2001. *Knowledge and Pleasure at Regent's Park. The Gardens of the Zoological Society of London during the Nineteenth Century*. Umeå: Umeå Universitrts Tryckeri

Akerman, John Yonge. 1847. *An Archaeological Index to Remains of Antiquity of the Celtic, Romano-British, and Anglo-Saxon Periods*. London: J. R. Smith

Alberti, Samuel J. M. M. 2003. "Conversaziones and the Experience of Science in Victorian England." *Journal of Victorian Culture* 8: 208–30, https://doi.org/10.3366/jvc.2003.8.2.208

— 2009. *Nature and Culture. Objects, Disciplines and the Manchester Museum*. Manchester: Manchester University Press, https://doi.org/10.2307/j.ctt1vwmg5x

Albjerg, Victor Lincoln. 1946. *Richard Owen: Scotland 1810, Indiana 1890*. West Lafayette: Purdue University

Allen, David Elliston. 1996. "Tastes and Crazes." In *Cultures of Natural History*. Ed. N. Jardine, J. A. Secord and E. C. Spary, 394–407. Cambridge: Cambridge University Press

— 2009. "Amateurs and Professionals." In *The Cambridge History of Science. 6. The Modern Biological and Earth Sciences*. Ed. Peter J. Bowler and John V. Pickstone, 15–33. Cambridge: Cambridge University Press, https://doi.org/10.1017/CHOL9780521572019.003

Allingham, E. G. 1924. *A Romance of the Rostrum, Being the Business Life of Henry Stevens and the History of Thirty-eight King Street, Together with Some Account of Famous Sales Held There During the Last Hundred Years*. London: H. F. & G. Witherby

Altick, Richard D. 1978. *The Shows of London*. Cambridge, MA: Harvard University Press

Amann, Elizabeth. 2015. *Dandyism in the Age of Revolution*. Chicago: Chicago University Press, https://doi.org/10.7208/chicago/9780226187396.001.0001

An American. 1839. *London in 1838*. New York: Colman

Anderson, Amanda. 1993. *Tainted Souls and Painted Faces. The Rhetoric of Fallenness in Victorian Culture*. Ithaca: Cornell University Press

Anderson, Katharine. 2004. "Almanacs and the Profits of Natural Knowledge." In *Culture and Science in the Nineteenth-Century Media*. Eds. Louise Henson, Geoffrey Cantor, Gowan Dawson, Richard Noakes, Sally Shuttleworth, and Jonathan R. Topham, 97–112. Aldershot: Ashgate, https://doi.org/10.4324/9781315258706-ch-8

— 2018. "Reading and Writing the Scientific Voyage: FitzRoy, Darwin and John Clunies Ross." *BJHS* 51: 369–94, https://doi.org/10.1017/s000708741800050x

Anderson, Gregory. 1976. *Victorian Clerks*. Manchester: Manchester University Press

Anon. 1821. *Report of the Trial of Mrs. Carlile*. 2nd. ed. London: Carlile

Anon. 1822. *Report of the Trial of Mrs. Susannah Wright*. London: Carlile

Anon. 1830. *The Zoological Keepsake; or, Zoology, and the Garden and Museum of the Zoological Society, for the Year 1830*. London: Marsh and Miller

Anon. 1833. *A Catalogue of Books in the Library of the London Mechanics' Institution, No 29, Southampton Buildings, Chancery Lane*. London: Proctor

Anon. 1837. *Report of the proceedings at a public meeting held at the Freemason's Hall, on the 29th of May, 1837, to promote the admission of the public without charge to Westminster Abbey, St. Paul's Cathedral, and all depositories of works of art, of natural history, and objects of historical and literary interest in public edifices, Joseph Hume, Esq., in the chair*. London: Boone

Anon. 1854. *The Logic of Holyoake's "Logic of Death;" or, Why the Atheist Should Fear to Die*. Glasgow: Blackie

Anon. 1858. *Scenes from My Life*. London: Seeley

Anon. 1865. *A catalogue of the valuable and extensive collection of British fossils, formed by Dr. J. S. Bowerbank ... which will be sold at auction by Mr. J. C. Stevens, at the Museum, 20, Highbury Grove, Islington, on Monday, the 27th day of November, 1865 and four following days*. London

Anon. 1904. *The History of the Collections contained in the Natural History Departments of the British Museum. Vol. 1*. London: British Museum (Natural History)

Armytage, W. H. G. 1951. "William Maclure, 1763–1840: A British Interpretation." *Indiana Magazine of History* 47: 1–20

— 1954. "John Minter Morgan, 1782–1854." *Journal of Education* 86: 550–52

— 1858. "John Minter Morgan's Schemes, 1841–1855." *International Review of Social History* 3: 26–42

— 1961. *Heavens Below. Utopian Experiments in England 1560–1960*. London: Routledge and Kegan Paul

Ashton, Rosemary. 2006. *142 Strand. A Radical Address in Victorian London*. London: Chatto & Windus

Ashwell, A. R. 1880–83. *Life of the Right Reverend Samuel Wilberforce, D.D., Lord Bishop of Oxford and Afterwards of Winchester: with Selections from his Diaries and Correspondence*. 3 vols. London: Murray

Atkinson, Juliette. 2010. *Victorian Biography Reconsidered*. Oxford: Oxford University Press, https://doi.org/10.1093/acprof:oso/9780199572137.001.0001

Babbage, Charles. 1832. *On the Economy of Manufactures*. London: Knight

Bakewell, Robert. 1830. "A Visit to the Mantellian Museum at Lewes." *Magazine of Natural History*. 3: 9–17

Bamford, Samuel. 1893. *Passages in the Life of a Radical, and Early Days*. Ed. Henry Dunckley. London: Unwin

Barker, Ambrose G. [1938.] *Henry Hetherington 1792–1849*. London: Pioneer Press

Barnhardt, Terry A. 2005. *Ephraim George Squier and the Development of American Anthropology*. Lincoln: University of Nebraska Press

Barrow, Logie. 1986. *Independent Spirits: Spiritualism and English Plebeians 1850–1910*. London: Routledge and Kegan Paul

Bartholomew, Michael. 1973. "Lyell and Evolution: An Account of Lyell's Response to the Prospect of an Evolutionary Ancestry for Man." *BJHS* 6: 261–303

Bartlett, David W. 1852. *London by Day and Night, Or, Men and Things in the Great Metropolis*. New York: Hurst

Barton, Ruth. 2018. *The X Club. Power and Authority in Victorian Science*. Chicago: University of Chicago Press, https://doi.org/10.7208/chicago/9780226551753.001.0001

— 2022. "The Scientific Reputation(s) of John Lubbock, Darwinian Gentleman." *Studies in History and Philosophy of Science* 95: 185–203, https://doi.org/10.1016/j.shpsa.2022.06.013

Bates, A. W. 2008. "'Indecent and Demoralising Representations': Public Anatomy Museums in Mid-Victorian England." *Medical History* 52: 1–22, https://doi.org/10.1017/s0025727300002039

[Baume, Pierre Joseph]. 1829. *Speech of a Frenchman*. London: Conningham

Baxter, G. R. Wythen. 1841. *The Book of the Bastiles; or, The History of the Working of the New Poor-Law*. 2 vols. London: Stephens

Baylee, Joseph. 1857. *Genesis and Geology; The Holy Word of God Defended from its Assailants*. Liverpool: Holden

—, and Frederick Hollick. 1839. *Substance of the Two Nights' Discussion, in the Social Institution, 69, Great Queen Street, London...on the Genuineness, Authenticity, and Inspiration of the Bible*. London: Cousins

Bayne, Peter. 1871. *The Life and Letters of Hugh Miller*. 2 vols. London: Strahan

Beames, Thomas. 1852. *The Rookeries of London*. 2nd ed. London: Bosworth

Beaven, Alfred B. 1908. *The Aldermen of The City of London*. Vol. 1. London: Eden Fisher

— 1913. *The Aldermen of The City of London*. Vol. 2. London: Eden Fisher

Beer, Gillian. 1985. *Darwin's Plots. Evolutionary Narrative in Darwin, George Eliot and Nineteenth-Century Fiction*. London: Ark

Beer, Max. 1921. *History of British Socialism*. London: Bell

Belcham, John. 1985. *"Orator!" Hunt. Henry Hunt and English Working-Class Radicalism*. Oxford: Clarendon

Bellamy, John. 1811. *The Ophion; or the Theology of the Serpent*. London. Valpy

Bellot, H. Hale. 1929. *University College London 1826–1926*. London: University of London Press

Benbow, William. 1823. *The Crimes of the Clergy: or, The Pillars of Priest-craft Shaken*. London: Benbow

Bennett, Edward Turner. 1829. *The Tower Menagerie: Comprising the Natural History of the Animals contained in that Establishment; with Anecdotes of their Characters and History*. London: Jennings

Bennett, Tony. 1995. *The Birth of the Museum: History, Theory, Politics*. Abingdon: Routledge

Berkowitz, Carin, and Bernard Lightman, eds. 2017. *Science Museums in Transition: Cultures of Display in Nineteenth-Century Britain and America*. Pittsburgh: University of Pittsburgh Press, https://doi.org/10.2307/j.ctt1r6b0c8

Berman, Morris. 1975. "'Hegemony' and the Amateur Tradition in British Science." *Journal of Social History* 8: 30–50

Bickersteth, E. 1843. *The Divine Warning to the Church, at this time, of our Enemies, Dangers, and Duties, and to our Future Prospects: with Information Respecting the Diffusion of Infidelity, Lawlessness, and Popery*. London: Dalton

Biddulph, Thomas Tregenna. 1825. *The Theology of the Early Patriarchs, Illustrated by An Appeal to Subsequent Parts of the Holy Scriptures; in A Series of Letters to A Friend*. 2 vols. Bristol: Short

Black, A. 1955. "Education Before Rochdale. 2 The Owenites and the Halls of Science." *Co-Operative Review* 29: 42–44

Black, Adam, and Charles Black. 1853. *Black's Guide to London and Its Environs*. Edinburgh: A. and C. Black

Blake, Laurel, and Marysa Demoor, eds. 2009. *Dictionary of Nineteenth Century Journalism in Great Britain and Ireland*. Gent: Academia Press

Bloor, David. 1983. "Coleridge's Moral Copula." *Social Studies of Science* 13: 605–19

Boitard, Pierre. 1838. "L'Homme Fossile." *Magasin Universel* t5: 209–40

Bonney, Thomas George. 1891. *Old Truths in Modern Lights. The Boyle Lectures for 1890 with Other Sermons*. London: Percival

— 1897. "Saull, William Devonshire 1784–1855." *Dictionary of National Biography*

— 1921. *Memories of a Long Life*. Cambridge: Metcalfe

Boos, Florence, ed. "William Morris's Socialist Diary." https://www.marxists.org/archive/morris/works/1887/diary/diary.htm#diary

Booth, A. 1839. *The Stranger's Intellectual Guide to London, for 1839–40*. London: Hooper

Booth, Arthur John. 1869. *Robert Owen, the Founder of Socialism in England*. London: Trubner

Bosanquet, S. R. 1843. *Principia: A Series of Essays on the Principles of Evil Manifesting in these Last Times in Religion, Philosophy, and Politics*. London: Burns

Boué, Ami. 1836. *Guide du Géologue-Voyageur*. Tome 2. Paris: Levrault

Bourne, H. R. Fox. 1887. *English Newspapers; Chapters in the History of Journalism*. 2 vols. London: Chatto and Windus

Bowerbank, James Scott. 1840. *History of Fossil Fruits and Seeds of London Clay Pt 1*. London: Van Voorst

— 1864–82. *A Monograph of the British Spongiadae*. Ed. A. M. Norman. 4 vols. London: Ray Society

Bowler, Peter J. 1974. "Evolutionism in the Enlightenment." *HS* 12: 159–83

— 1984. *Evolution. The History of an Idea*. Berkeley: University of California Press

— 2021. *Progress Unchained. Ideas of Evolution, Human History and the Future*. Cambridge: Cambridge University Press, https://doi.org/10.1017/9781108909877.016

Boylan, Patrick J. 1978. "The Controversy of the Moulin-Quignon Jaw: the Role of Hugh Falconer." In *Images of the Earth: Essays in the History of the Environmental Sciences*. Eds. L. Jordanova and Roy Porter, 170–99. BSHS Monographs 1

Bradlaugh, Charles. 1891. *The Autobiography of C. Bradlaugh: A Page of his Life Written in 1873 for the "National Reformer"*. London: Forder

Brady, John H. 1838. *A New Pocket Guide to London and Its Environs*. London: Parker

Brannon, George. [1857.] *The Pleasure Visitor's Companion in Making the Tour of the Isle of Wight, Pointing out the Best Plan for seeing in the Shortest Time every Remarkable Object*. Wootton

Brantlinger, Patrick. 2003. *Dark Vanishings. Discourse on the Extinction of Primitive Races, 1800-1930*. Ithaca: Cornell University Press, https://doi.org/10.7591/9780801468681

Bray, Charles. 1841. *The Philosophy of Necessity*. 2 vols. London: Longman, Orme, Brown, Green, and Longmans

Brayley, Edward Wedlake. 1850. *Topographical History of Surrey*. Vol. 5. London: Willis

Breton, Rob. 2016. "Portraits of the Poor in Early Nineteenth-Century Radical Journalism." *Journal of Victorian Culture* 21: 168–83, https://doi.org/10.1080/13555502.2016.1167766

Brock, Michael. 1973. *The Great Reform Act*. London: Hutchinson

Brock, William H. 1996. *Science for All. Studies in the History of Victorian Science and Education*. Aldershot: Variorum

Broderip, William J. 1835. "Observations on the Habits, &c. of a male Chimpanzee, Troglodytes niger, Geoff., now living in the Menagerie of the Zoological Society of London." *Proceedings of the Zoological Society* 3: 160–69

[—] 1838. "Recreations in Natural History.—No. vi. Monkeys of the Old Continent, &c." *New Monthly Magazine* 52: 88–99

— 1847. *Zoological Recreations*. London: Colburn

Brongniart, Adolphe, 1828 [–1837]. *Histoire des Végétaux Fossiles*. 2 vols. Paris: Dufour et d'Ocagne

Brook, Anthony. 2002. "Gideon Mantell's Inaugural Publication." *Geology Today* 18: 217–19, https://doi.org/10.1046/j.0266-6979.2003.00375.x

Brook, Charles Wortham. 1943. *Carlile and the Surgeons*. Glasgow: Strickland

Brooke, Alan. "Huddersfield Hall of Science", https://undergroundhistories.wordpress.com/huddersfield-hall-of-science

Brooke, James Williamson. 1839. *The Democrats of Marylebone*. London: Cleaver

Brooke, John Hedley. 1979. "The Natural Theology of the Geologists: Some Theological Strata." In *Images of the Earth. Essays in the History of the Environmental Sciences.* Ed. L. J. Jordanova and Roy S. Porter, 39–64. *BSHS* Monographs 1

— 1991. *Science and Religion. Some Historical Perspectives.* Cambridge: Cambridge University Press

Brooks, Simon Vincent. 2009. "Stagoll, Robinson and the Devil's Chaplain", http://vincentbrooksnotes.blogspot.com/2009/02/stagoll-robinson-and-devils-chaplain.html

Brown, John. 1858. *Sixty Years' Gleanings from Life's Harvest. A Genuine Autobiography.* Cambridge: Palmer

Brown, W. Henry. 1924. *A Century of London Co-Operation.* London: Education Committee of the London Co-Operative Society Ltd

Browne, Janet. 1989. "Botany for Gentlemen: Erasmus Darwin and The Loves of the Plants." *Isis* 80: 593–621

— 2001. "Darwin in Caricature: A Study in the Popularisation and Dissemination of Evolution." *Proceedings of the American Philosophical Society* 145: 496–509

Brydon, William. 1988. *Politics, Government and Society in Edinburgh, 1780–1833.* PhD Thesis, University of North Wales

Buchanan, Robert. 1840a. *A Concise History of Modern Priestcraft.* Manchester: Heywood

— 1840b. *Socialism Vindicated.* Manchester: Heywood

Buckland, Adelene. 2013. *Novel Science: Fiction and the Invention of Nineteenth-Century Geology.* Chicago: University of Chicago Press, https://doi.org/10.7208/chicago/9780226923635.001.0001

Buckland, William. 1836. *Geology and Mineralogy Considered with Reference to Natural Theology.* 2 vols. London: Pickering

Buffetaut, Eric. 1987. *A Short History of Vertebrate Palaeontology.* London: Croom Helm

—, and R. L. E. Ford. 1979. "The Crocodilian *Bernissartia* in the Wealden of the Isle of Wight." *Palaeontology* 22: 905–12

Bulstrode, Jenny. 2016. "The Industrial Archaeology of Deep Time." *BJHS* 49: 1–25, https://doi.org/10.1017/s0007087416000017

Burkhardt, Frederick, et al., eds. 1985–2023. *The Correspondence of Charles Darwin.* 30 vols. Cambridge: Cambridge University Press

Butler, Marilyn. 1981. *Romantics, Rebels and Reactions: English Literature and Its Background, 1760–1830.* Oxford: Oxford University Press

Byerley, John. 1831. "An Attempt to Explain the Principal Phenomena of Geology and Physical Geography, by the Precession of the Equinoxes and the Earth's Figure as an oblate Spheroid." *Magazine of Natural History* 4: 308–16

Bynum, W. F. 1984. "Charles Lyell's *Antiquity of Man* and Its Critics." *JHB* 17: 153–87

Byron, Lord. 1822. *Cain; A Mystery*. London: Carlile

Cadbury, Deborah. 2000. *The Dinosaur Hunters. A True Story of the Scientific Rivalry and the Discovery of the Prehistoric World*. London: Fourth Estate

Calvert, John. 1853. *The Gold Rocks of Great Britain and Ireland*. London: Chapman and Hall

Carlile, Richard. 1821. *An Address to Men of Science; Calling Upon Them to Stand Forward and Vindicate the Truth From the Foul Grasp and Persecution of Superstition...* London: Carlile

— 1822. *The Report of the Proceedings of the Court of King's Bench: in the Guildhall, London, on the 12th, 13th, 14th, and 15th days of October: Being the Mock Trials of Richard Carlile, for Alleged Blasphemous Libels, in publishing Thomas Paine's Theological works and Elihu Palmer's Principles of Nature: Before Lord Chief Justice Abbott, and Special Juries*. London: Carlile

— 1832a. "To the Editress of the Isis." *Isis* 1: 341–45

— 1832b. "Progress And Stages Of An Enquiring Mind Working Itself Free From Superstition." *Isis* 1: 369–71

Carnall, Geoffrey. 1953–54. "The Surrey Institution And Its Successor." *Adult Education* 26: 197–208

Carpenter, William. 1830–31. *Political Letters and Pamphlets*. London: Carpenter

—, ed. 1832. *Proceedings of the Third Co-Operative Congress*. London: Strange

Carroll, Victoria. 2007. "'Beyond the Pale of Ordinary Criticism'. Eccentricity and the Fossil Books of Thomas Hawkins." *Isis* 98: 225–65, https://doi.org/10.1086/518187

— 2008. *Science and Eccentricity: Collecting, Writing and Performing Science for Early Nineteenth-Century Audiences*. Pittsburgh: University of Pittsburgh Press

Cash, Derek. 2002. *Access to Museum Culture: the British Museum from 1753 to 1836*. Occasional Paper number 133. British Museum

Chalmers-Hunt, J. M., ed. 1976. *Natural History Auctions, 1700–1972*. London: Sotherby Parke Bernet

Charman, Isobel. 2016. *The Zoo. The Wild and Wonderful Tale of the Founding of London Zoo*. London: Viking

Chase, Malcolm. 1988. "'The People's Farm'. *English Radical Agrarianism 1775–1840*. Oxford: Clarendon Press

— 2000. *Early Trade Unionism. Fraternity, Skill and the Politics of Labour*. Aldershot: Ashgate, https://doi.org/10.4324/9781315257181

Chesney, Kellow. 1972. *The Victorian Underworld*. London: Pelican

Chilton, William. 1842–43. "Theory of Regular Gradation." *Oracle of Reason* 1 (19 Feb. 1842)–2 (11 Nov. 1843), passim

— 1842. "The Cowardice and Dishonesty of Scientific Men." *Oracle of Reason* 1: 193–95

— 1845. "'Vestiges of the Natural History of Creation.' Theory of Regular Gradation." *Movement* 2: 9–12

— 1846. "'Materialism' and the Author of the 'Vestiges'." *Reasoner* 1: 7–8

— 1847a. "Letter from Mr. Chilton. Explanations." *Reasoner* 3: 607–10

— 1847b. "Christian Philosophers.—Dr. Dick and the Rev. A. Burdett." *Reasoner* 3: 373–80

— 1854. "Man and the Baboon." *Reasoner* 8: 225–29, 284–85

Claeys, Gregory. 1987. *Machinery, Money and the Millennium. From Moral Economy to Socialism, 1815–1860*. Princeton: Princeton University Press

— 2000. "The 'Survival of the Fittest' and the Origins of Social Darwinism." *JHI* 61: 223–40, https://doi.org/10.2307/3654026

— 2002. *Citizens and Saints. Politics and Anti-Politics in Early British Socialism*. Cambridge: Cambridge University Press, https://doi.org/10.1017/CBO9780511521324.010

— 2005. *Owenite Socialism. Pamphlets and Correspondence*. 10 vols. London: Routledge

Clark, G. N. 1964–72. *A History of the Royal College of Physicians of London*. 3 vols. Oxford: Clarendon Press

Clark, John Willis, and Thomas McKenny Hughes. 1890. *The Life and Letters of the Reverend Adam Sedgwick*. 2 vols. Cambridge: Cambridge University Press

Clarke, Adam. 1837. *The Holy Bible...with Commentary and Critical Notes*. New ed. vol. 1. New York: Mason

Clarke, H. G. 1851a. *London as it is to-day*. London: Clarke

— 1851b. *London What to See and How to See it*. London: Clarke

Clarke, James Fernandez. 1874. *Autobiographical Recollections of the Medical Profession*. London: Churchill

Clarke, John. 1825. *A Critical Review, of the Life, Character, Miracles, and Resurrection, of Jesus Christ, in a Series of Letters to Dr. Adam Clarke*. London. J. Clarke

Cleevely, R. J. 1974. "The Sowerbys, the *Mineral Conchology*, and their Fossil Collection." *JSBNH* 6: 418–81

— 1983. *World Palaeontological Collections*. London, British Museum (Natural History)

— and Chapman, S. D. 1992. "The Accumulation and Disposal of Gideon Mantell's Fossil Collections and their Role in the History of British Palaeontology." *ANH* 19: 307–64

Clodd, Edward. 1902. *Thomas Henry Huxley*. New York: Dodd Mead

Coates, Thomas. 1841. *Report of the State of Literary, Scientific, and Mechanics' Institution in England*. London: Society for the Diffusion of Useful Knowledge

Cole, G. D. H. [1944]. *A Century of Co-Operation*. London: Allen and Unwin

Collet, Collet Dobson. 1933. *History Of The Taxes On Knowledge*. London: Watts

Collet, Sophia Dobson. 1855. *George Jacob Holyoake and Modern Atheism*. London: Trubner

Combe, George. 1839. "Phrenological Remarks on the Relation between the Natural Talents and Dispositions of Nations, and the Developments of their Brains." In Samuel George Morton. *Crania Americana*, 269–91. Philadelphia: Dobson

Conklin, Lawrence H. 1995. "James Sowerby, His Publications and Collections." *Mineralogical Record* 26: 85–105

Conlin, Jonathan. 2014. *Evolution and the Victorians. Science, Culture and Politics in Darwin's Britain*. London: Bloomsbury, https://doi.org/10.5040/9781474210645

Conway, Moncure Daniel. 1892. *The Life of Thomas Paine: with a History of his Literary, Political and Religious Career in America, France and England*. 2 vols. New York: Putnam

Cooper, Bransby Blake. 1843. *The Life of Sir Astley Cooper, Bart*. 2 vols. London: Parker

Cooper, John A. 2010, ed., *Mantell 1819–1852 Unpublished Journal*, http://www.brighton-hove-rpml.org.uk/HistoryAndCollections/aboutcollections/naturalsciences/Pages/Theunpublishedjournalofgideonmantell.aspx

Cooper, Michael P. 2006. *Robbing the Sparry Garniture. A 200 Year History of British Mineral Dealers 1750–1950*. Tucson: Mineralogical Record

Cooper, Robert. 1846. *The Infidel's Text-Book*. Hull: Johnson

— 1853. *The Immortality of the Soul, Religiously and Philosophically Considered*. London: Watson

Cooper, Thomas, M.D. 1837. *On the Connection between Geology and the Pentateuch*. Boston: Kneeland

Cooper, Thomas. [1842]. *Address to the Jury, by Thomas Cooper, the Leicester Chartist*. Leicester: Warwick

— 1849. *The Life and Character of Henry Hetherington.* London: Watson

— 1872. *The Life of Thomas Cooper. Written by Himself.* London: Hodder and Stoughton

— 1878. *Evolution, the Stone Book, and the Mosaic Record of Creation.* London: Hodder and Stoughton

— 1885. *Thoughts at Fourscore and Earlier.* London: Hodder

Cooter, Roger. 1984. *The Cultural Meaning of Popular Science. Phrenology and the Organization of Consent in Nineteenth-Century Britain.* Cambridge: Cambridge University Press

— 2006. "Pierre Joseph Baume." In Dollin Kelly, ed. *New Manx Worthies.* Douglas: Manx Heritage Foundation.

— and Stephen Pumfrey. 2004. "Separate Spheres and Public Places: Reflections on the History of Science Popularization and Science in Popular Culture." *HS* 32: 237–67, https://doi.org/10.1177/007327539403200301

Cornish and Driver 2020. "'Specimens Distributed': The Circulation of Objects from Kew's Museum of Economic Botany, 1847–1914." *Journal of the History of Collections* 32: 327–40, https://doi.org/10.1093/jhc/fhz008

Corsi, Pietro. 1978. "The Importance of French Transformist Ideas for the Second Volume of Lyell's 'Principles of Geology'." *BJHS* 39: 221–44

— *The Age of Lamarck. Evolutionary Theories in France 1790–1830.* Trans. J. Mandelbaum. Berkeley: University of California Press

— 2005. "Before Darwin: Transformist Concepts in European Natural History." *JHB* 38: 67–83, https://doi.org/10.1007/s10739-004-6510-5

— 2021. "Edinburgh Lamarckians? The Authorship of Three Anonymous Papers (1826–1829)." *JHB* 54:345–74, https://doi.org/10.1007/s10739-021-09646-5

Cowles, Henry M. 2013. "A Victorian Extinction: Alfred Newton and the Evolution of Animal Protection." *BJHS* 46: 695–714, https://doi.org/10.1017/s0007087412000027

Cowtan, Robert. 1872. *Memories of the British Museum.* London: Bentley

Cox, Howard, and Simon Mowatt. 2014. *Revolutions from Grub Street: A History of Magazine Publishing in Britain.* Oxford: Oxford University Press, https://doi.org/10.1093/acprof:oso/9780199601639.001.0001

[Croker, J. W.] 1839–40. "Conduct of Ministers." *Quarterly Review* 65: 282–314

Cruchley, G. F. [1831]. *Cruchley's Picture of London.* London: Cruchley

Cull, Richard. 1850. "Remarks on Three Naloo Negro Skulls." *Journal of the Ethnological Society of London.* 2: 238–45

Cunningham, Hugh. 1980. *Leisure in the Industrial Revolution c.1780–c.1880.* London: Routledge

Cunningham, Peter. 1851. *Modern London; or, London as it is*. London: Murray

Curtis, L. Perry. 1997. *Apes and Angels. The Irishman in Victorian Caricature*. Washington: Smithsonian Institution Press

Cutner, H. n.d. *The Devil's Chaplain. Robert Taylor (1784–1844)*. Pioneer Press

Davenport, Allen. 1845. *The Life, and Literary Pursuits of Allen Davenport*. London: Davenport

D'Archaic, Vicomte. 1847. *Histoire des Progrès de la Géologie de 1834 a 1845*. Paris: Société Géologique de France

Davidson, Thomas. 1854. *A Monograph of British Cretaceous Brachiopoda, Part II*. 8: 55–117. London: Palaeontographical Society

Davies, R. E. 1907. *The Life of Robert Owen, Philanthropist and Social Reformer, an Appreciation*. London: Sutton

Davies, William. 1974. *Catalogue of the Pleistocene Vertebrata, from the Neighbourhood of Ilford, Essex, in the Collection of Sir Antonio Brady*. London: Privately Printed

Davy, Humphry. 1830. *Consolations in Travel, or, The Last Days of a Philosopher*. London: Murray

Dawson, Gowan. 2012. "Paleontology in Parts. Richard Owen, William John Broderip, and the Serialization of Science in Early Victorian Britain." *Isis* 103: 637–67, https://doi.org/10.1086/668961

— 2016. *Show Me the Bone: Reconstructing Prehistoric Monsters in Nineteenth-Century Britain and America*. Chicago: University of Chicago Press, https://doi.org/10.7208/chicago/9780226332871.001.0001

—, and Bernard Lightman, eds. 2014. *Victorian Scientific Naturalism. Community, Identity, Continuity*. Chicago: Chicago University Press, https://doi.org/10.7208/chicago/9780226109640.001.0001

[De Maillet, Benoit]. 1755. *Telliamed*. New ed. La Haye: Gosse

Dean, Dennis R. 1981. "'Through Science to Despair': Geology and the Victorians". *Annals of the New York Academy of Sciences* 360: 111–36

— 1989. "New Light on William Maclure." *AS* 46: 549–74

— 1999. *Gideon Mantell and the Discovery of Dinosaurs*. Cambridge: Cambridge University Press

Dean, Russell. 1995. "Owenism and the Malthusian Population Question, 1815–1835." *History of Political Economy* 27: 579–97

Dear, Pauline Carpenter. 1986. "Richard Owen Invents the Dinosaurs: Cuvierian Paleontological Practice in Britain." Typescript

DeCoursey, Christina. 1997. *The Society of Antiquaries, 1830–1870: Institution, Intellectual Questions, Community, and the Search for the Past*. PhD Thesis, University of Toronto

Delair, Justin B. 1985. "The Fossil Collection of Dr John Lee (1783–1866) of Hartwell." *Geological Curator* 4:69–84

Denman, Thomas. 1828. *An Inaugural Discourse, Pronounced on the Occasion of Opening the Theatre of the City of London Literary and Scientific Institution, in Aldersgate Street, on Friday Evening, April 24th, 1828*. London: Wilson

Desmond, Adrian. 1979. "Designing the Dinosaur: Richard Owen's Response to Robert Edmond Grant." *Isis* 70: 224–34

— 1982. *Archetypes and Ancestors: Palaeontology in Victorian London, 1850–1875*. London: Blond and Briggs

— 1984. "Interpreting the Origin of Mammals: New Approaches to the History of Palaeontology." *Zoological Journal of the Linnean Society*. 82: 7–16

— 1985a. "The Making of Institutional Zoology in London, 1822–1836." *HS* 23: 153–85, 223–50

— 1985b. "Richard Owen's Reaction to Transmutation in the 1830's." *BJHS* 18: 25–50

— 1987. "Artisan Resistance and Evolution in Britain, 1819–1848." *Osiris* 3: 77–110

— 1989. *The Politics of Evolution: Morphology, Medicine and Reform in Radical London*. Chicago: University of Chicago Press

— 1998. *Huxley. From Devil's Disciple to Evolution's High Priest*. London: Penguin

—, and Angela Darwin. 2021. "T. H. Huxley's Turbulent Apprenticeship Years: John Charles Cooke and the John Salt Scandal." *ANH* 48: 215–26, https://doi.org/10.3366/anh.2021.0718

—, and James Moore. 1991. *Darwin*. London: Michael Joseph

—, and James Moore. 2009. *Darwin's Sacred Cause. Race, Slavery and the Quest for Human Origins*. London: Allen Lane

Detrosier, R. 1840 [1829]. *The Benefits of General Knowledge; More Especially, the Sciences of Mineralogy, Geology, Botany, and Entomology, Being An Address Delivered At the Opening of the Banksian Society, Manchester. On Monday, January, 5th, 1829*. London: Cleave

— 1831. *An Address, Delivered to the Members of the New Mechanics' Institution, Manchester, on Friday Evening, March 25, 1831, on the Necessity of an Extension of Moral and Political Instruction Among the Working Classes*. Manchester: reprinted Cleave

Dinwiddy, J. R. 1992. *Radicalism and Reform in Britain, 1780–1850*. London: Hambledon

Dixon, Frederick. 1850. *The Geology and Fossils of the Tertiary and Cretaceous Formations of Sussex*. London: Longman, Brown, Green and Longmans

Dolan, Brian. 1998. "Pedagogy Through Print: James Sowerby, John Mawe and the Problem of Colour in Early Nineteenth-Century Natural History Illustration." *BJHS* 31: 275–304

Donington, Katie. 2014. "Transforming Capital: Slavery, Family, Commerce and the Making of the Hibbert Family." In *Legacies of British Slave-Ownership. Colonial Slavery and the Formation of Victorian Britain*. Eds. Catherine Hall, Nicholas Draper, Keith McClelland, Katie Donington, and Rachel Lang, 203–49. Cambridge: Cambridge University Press, https://doi.org/10.1017/cbo9781139626958.006

Donovan, Daniel. 1876. *Sketches in Carbery, County Cork. Its Antiquities, History, Legends, and Topography*. Dublin: McGlashan and Gill

[D'Oyly, G.] 1819. "Abernethy, Lawrence, &c. on the Theories of Life." *Quarterly Review* 22: 1–34

Driver, Felix. 2001. *Geography Militant. Cultures of Exploration and Empire*. Oxford: Blackwell

Duckett, William. 1853–60. *Dictionnaire de la Conversation et de la Lecture: Inventaire Raisonné des Notions Générales les plus Indispensables à tous, par une Société de Savants et de Gens de Lettres*. 2 ed, Tome 12. Paris: Lévy

Duffy, Kathrinne. 2017. "The Dead Curator: Education and the Rise of Bureaucratic Authority in Natural History Museums, 1870–1915." *Museum History Journal*, 10: 29–49, https://doi.org/10.1080/19369816.2016.1259378

Duffy, Patrick. 2000. *The Skilled Compositor, 1850–1914. An Aristocrat Among Working Men*. London: Routledge, https://doi.org/10.4324/9781315237039

Duncombe, J. 1848. *Sinks of London Laid Open*. London: Duncombe

Durant, John R. 1979. "Scientific Naturalism and Social Reform in the Thought of Alfred Russel Wallace." *BJHS* 12: 31–58

Duthie, Sky. 2019. *The Roots of Reform: Vegetarianism and the British Left, c.1790–1900*. PhD Thesis, University of York

Dyke, Philip J., and E. P. Uphill. 1983. "Major Charles Kerr Macdonald 1806–67." *Journal of Egyptian Archaeology* 69: 165–66

Eagles, Robin. 2000. *Francophilia in English Society 1748–1815*. London: Macmillan, https://doi.org/10.1057/9780230599109

Egan, Pierce. 1821. *Tom and Jerry. Life in London*. London: Chatto and Windus

Elliott, Graham F. 1970. "The Two London Clay Clubs: 1836–1847, and 1923–1940." *JSBNH* 5: 333–39

— 1975. "The Non-Descriptive Palaeontology of the Sowerbys' Mineral Conchology." *Bulletin of the British Museum (Natural History) Historical Series* 4: 389–99

Embrey, P. G., and R. F. Symes. 1987. *Minerals of Cornwall and Devon*. London: British Museum (Natural History)

Emsley, Clive. 2014. *Britain and the French Revolution*. London: Routledge, https://doi.org/10.4324/9781315838984

Endersby, Jim. 2008. *Imperial Nature. Joseph Hooker and the Practices of Victorian Science*. Chicago: University of Chicago Press, https://doi.org/10.7208/chicago/9780226773995.001.0001

Engledue, W. C. 1843. *Cerebral Physiology and Materialism, With the Result of the Application of Animal Magnetism to the Cerebral Organs. An Address Delivered to the Phrenological Association in London, June 20, 1842*. London: Watson

Epps, E. [1875.] *Diary of the Late John Epps*. London: Kent

Epstein, James A. 1994. *Radical Expression. Political Language, Ritual, and Symbol in England, 1790–1850*. New York: Oxford University Press

Evans, Christopher. 2004. "Modelling Monuments and Excavations." In *Models: The Third Dimension of Science*. Eds. Soraya de Chadarevian and Nick Hopwood, 109–37. Stanford: Stanford University Press, https://doi.org/10.1515/9781503618992-008

Evans, Joan. 1949. "Ninety years ago." *Antiquity* 23: 115–25

Evans, Mark. 2010. "The Roles played by Museums, Collections and Collectors in the Early History of Reptile Palaeontology." In *Dinosaurs and Other Extinct Saurians: A Historical Perspective*. Eds R. T. J. Moody, E. Buffetaut, D. Naish and D. M. Martill, 5–29. Geological Society, London, Special Publications 343, https://doi.org/10.1144/sp343.2

Evers, P. 1838. *The Student's Compendium of Comparative Anatomy*. London: Longman

Eyre, Edward John. 1845. *Journals of Expeditions of Discovery into Central Australia*. 2 vols. London: Boone

Faflak, Joel, ed. 2017. *Marking Time. Romanticism and Evolution*. Toronto: University of Toronto Press, https://doi.org/10.3138/9781442699595

Fara, Patricia. 2012. *Erasmus Darwin. Sex, Science, and Serendipity*. Oxford. Oxford University Press

Farley, John. 1972. "The Spontaneous Generation Controversy (1700-1860): The Origin of Parasitic Worms." *JHB* 5: 95-125

— 1977. *The Spontaneous Generation Controversy from Descartes to Oparin*. Baltimore: Johns Hopkins University Press

Faucher, Leon. 1969. *Manchester in 1844. Its Present Condition and Future Prospects*. London: Cass

Fearon, Henry Bradshaw. 1833. *Thoughts on Materialism: And on Religious Festivals, Sabbaths*. London: Longman, Rees, Orme, Brown, Green, and Longman

Fishburn, Matthew. 2020. "The Private Museum of John Septimus Roe, Dispersed in 1842." *ANH* 47: 166–82, https://doi.org/10.3366/anh.2020.0629

Fisher, George Park. 1866. *Life of Benjamin Silliman, M.D., LL.D., Late Professor of Chemistry, Mineralogy, and Geology in Yale College Chiefly from his Manuscript Reminiscences, Diaries, and Correspondence*. 2 vols. New York: Scribner

Fletcher, Alexander. 1815. *The Tendency of Infidelity and Christianity Contrasted*. London: Tew

Flexner, Helen Hudson. 2014. *The London Mechanics' Institution. Social and Cultural Foundations 1823–1830*. PhD Thesis, University College London

Forgan, Sophie. 1994. "The Architecture of Display: Museums, Universities and Objects in Nineteenth-Century Britain." *HS* 32: 139–62

Francis, Frederick John. 1839. *A Brief Survey of Physical and Fossil Geology*. London: Hatchard

Francis, Richard. 2010. *Fruitlands. The Alcott Family and their Search for Utopia*. New Haven: Yale University Press, https://doi.org/10.12987/9780300169447-002

Fraser, Derek. 1979. *Power and Authority in the Victorian City*. Oxford: Blackwell

Freeman, Michael. 2001. "Tracks to a New World: Railway Excavation and the Extension of Geological Knowledge in Mid-Nineteenth-Century Britain." *BJHS* 34: 51–65, https://doi.org/10.1017/s0007087401004277

— 2004. *Victorians and the Prehistoric. Tracks to a Lost World*. New Haven: Yale University Press

Freeman, R. B. 1977. *The Works of Charles Darwin. An Annotated Bibliographical Handlist*. 2nd ed. Folkestone: Dawson

[Frost, Thomas.] 1842. "Mosaic Account of Creation." *Free-Thinker's Information for the People* 1: 1–8

— 1880. *Forty Years' Recollections: Literary and Political*. London: Sampson Low, Marston, Searle, and Rivington

Frow, Ruth, and Edmund Frow. 1989. *Political Women. 1800–1850*. London: Pluto

Fyfe, Aileen. 2004. *Science and Salvation. Evangelical Popular Science Publishing in Victorian Britain*. Chicago: Chicago University Press, https://doi.org/10.7208/chicago/9780226276465.003.0005

— 2005. "Conscientious Workmen or Booksellers' Hacks? The Professional Identities of Science Writers in the Mid-Nineteenth Century." *Isis* 96: 192–223, https://doi.org/10.1086/431532

—, and Bernard Lightman, eds. 2007. *Science in the Marketplace. Nineteenth-Century Sites and Experiences*. Chicago: Chicago University Press, https://doi.org/10.7208/chicago/9780226150024.001.0001

Gamble, Clive, and Theodora Moutsiou. 2011. "The Time Revolution of 1859 and the Stratification of the Primeval Mind." *NRRS* 65: 43–63, https://doi.org/10.1098/rsnr.2010.0099

Garnet, R. G. 1972. *Co-Operation and the Owenite Socialist Communities in Britain, 1825–45*. Manchester: Manchester University Press

Gay, Hannah. 1998. "No 'Heathen's Corner' here: the Failed Campaign to Memorialize Herbert Spencer in Westminster Abbey." *BJHS* 31: 41–54

Geikie, Archibald. 1875. *Life of Sir Roderick I. Murchison*. 2 vols. London: Murray

Genth, F. A. 1854. "On Owenite, a New Mineral." *Proceedings of the Academy of Natural Sciences of Philadelphia* 6: 297–99

George, M. Dorothy. 1952. *Catalogue of Political and Personal Satires Preserved in the Department of Prints and Drawings in the British Museum, Vol. X 1820–1827*. London: British Museum

Gerstner, Patsy. 1994. *Henry Darwin Rogers, 1808–1866*. Tuscaloosa: University of Alabama Press

Gilbert, James. 1851. *Gilbert's Visitor's Guide to London*. London: Gilbert

Gillispie, C. C. 1959. *Genesis and Geology: A Study in the Relations of Scientific Thought, Natural Theology, and Social Opinion in Great Britain, 1790–1850*. New York: Harper

Gleadle, Kathryn. 2003. "'The Age of Physiological Reformers': Rethinking Gender and Domesticity in the Age of Reform." In *Rethinking the Age of Reform: Britain 1780–1850*. Eds Arthur Burns and Joanna Innes, 200–219. Cambridge: Cambridge University Press, https://doi.org/10.1017/cbo9780511550409.009

Godwin, B. 1834. *Lectures on the Atheistic Controversy*. London: Jackson and Walford

Godwin, Joscelyn. 1994. *The Theosophical Enlightenment*. Albany: State University of New York

Godwin, William. 1831. *Thoughts on Man*. London: Wilson

Gold, Meira. 2018. "Ancient Egypt and the Geological Antiquity of Man, 1847–1863." *History of Science* 57: 194–230, https://doi.org/10.1177/0073275318795944

Goldman, Lawrence. "Making Histories: the Oxford Dictionary of National Biography", https://archives.history.ac.uk/makinghistory/resources/articles/ODNB.html

Goldstein, Amanda Jo. 2017. *Sweet Science: Romantic Materialism and the New Logics of Life*. Chicago: University of Chicago Press, https://doi.org/10.7208/chicago/9780226458588.001.0001

Gomme, George Laurence. 1912. *The Making of London*. Oxford: Clarendon

Good, John Mason. 1826. *The Book of Nature*. 3 vols. London: Longman, Rees, Orme, Brown, and Green

Goodfield-Toulmin, June. 1969. "Some Aspects of English Physiology, 1780–1840." *JHB* 2: 283–320

Goodrum, Matthew R. 2002. "Atomism, Atheism, and the Spontaneous Generation of Human Beings: The Debate over a Natural Origin of the First Humans in Seventeenth-Century Britain." *JHI* 63: 207–24, https://doi.org/10.1353/jhi.2002.0011

— 2016. "The Beginnings of Human Palaeontology: Prehistory, Craniometry and the 'Fossil Human Races'." *BJHS* 49: 387–409, https://doi.org/10.1017/s0007087416000674

Goodway, David. 1982. *London Chartism 1838–1848*. Cambridge: Cambridge University Press

Gould, Stephen Jay. 1977. *Ontogeny and Phylogeny*. Cambridge, MA: Belknap Press

Goss, Charles William F. 1908. *A Descriptive Bibliography of the Writings of George Jacob Holyoake With a Brief Sketch of his Life*. London: Crowther & Goodman

[Grant, James]. 1837. *The Great Metropolis*. 2nd ed. 2 vols. London: Saunders and Otley

— 1838. *Sketches in London*. London: Orr

— 1871–72. *The Newspaper Press; its Origin—Progress—and Present Position*. 3 vols. London: Tinsley

Grant, Johnson. 1840. *Sketches in Divinity. Addressed to Candidates for the Ministry*. London: Hatchard

Graves, Algernon. 1906. *The Royal Academy of Arts*. vol. 5. London: Graves

Grayson, Donald K. 1983. *The Establishment of Human Antiquity*. New York: Academic Press

Greaves, J. P. 1827. *Letters on Early Education*. London: Sherwood, Gilbert and Piper

Green, David R. 2010. *Pauper Capital. London and the Poor Law, 1790–1870*. Farnham: Ashgate

Greenwood, Thomas. 1996. *Museums and Art Galleries*. London: Routledge

Grigson, Caroline. 2016. *Menagerie. The History of Exotic Animals in England*. Oxford: Oxford University Press

Grimes, Kyle. 2000. "Spreading the Word: The Circulation of William Hone's 1817 Liturgical Parodies." In *Radicalism and Revolution in Britain, 1775–1848*. Ed. Michael T. Davis, 143–56. London: Macmillan, https://doi.org/10.1057/9780230509382_10

Gruber, Jacob W. and John C. Thackray. 1992. *Richard Owen Commemoration*. London: Natural History Museum

Gunther, A. E. 1978. "John George Children, F.R.S. (1777–1852) of the British Museum. Mineralogist and Reluctant Keeper of Zoology." *Bulletin of the British Museum (Natural History) Historical Series* 6: 75–108

— 1980. *The Founders of Science at the British Museum, 1753–1900*. Suffolk: Halesworth

Hale, Piers J. 2014. *Political Descent. Malthus, Mutualism and the Politics of Evolution in Victorian England*. Chicago: University of Chicago Press, https://doi.org/10.7208/chicago/9780226108520.001.0001

Halévy, Elie. 1950. *The Triumph of Reform 1830–1841*. 2nd ed. revised. London: Benn

— 1961. *Victorian Years*. London: Benn

Harcourt, L. Vernon. 1838. *The Doctrine of the Deluge; Vindicating the Scriptural Account from the Doubts which have recently been cast upon it by Geological Speculations*. 2 vols. London: Longman, Orme, Brown, Green, and Longman

Hardin. Jeff, Ronald L. Numbers, and Ronald A. Binzley. 2018. *The Warfare Between Science and Religion. The Idea That Wouldn't Die*. Baltimore: Johns Hopkins University Press, https://doi.org/10.56021/9781421426181

Hardy, Dennis. 1979. *Alternative Communities in Nineteenth Century England*. London: Longman

Harling, Philip. 1996. *The Waning of 'Old Corruption'. The Politics of Economical Reform in Britain, 1779–1846*. Oxford: Clarendon

Harrison, Brian. 1994. *Drink and the Victorians. The Temperance Question in England 1815–1872*. Keele: Keele University Press

Harrison, J. F. C. 1961. *Learning and Living 1790–1960*. London: Routledge and Kegan Paul

— 1967. "'The Steam Engine of the New Moral World': Owenism and Education, 1817–1829." *Journal of British Studies* 6: 76–98

— 1969. *Robert Owen and the Owenites in Britain and America: The Quest for the New Moral World*. London: Routledge and Kegan Paul

— 1979. *The Second Coming. Popular Millenarianism 1790–1850*. London: Routledge and Kegan Paul

— 1987. "Early Victorian Radicals and the Medical Fringe." In *Medical Fringe and Medical Orthodoxy 1750–1850*. Eds W. F. Bynum and Roy Porter, 198–215. London: Croom Helm

Harrison, Peter. 2015. *The Territories of Science and Religion*. Chicago: University of Chicago Press, https://doi.org/10.7208/chicago/9780226184517.001.0001

Hawkins, B. Waterhouse. 1854. "On Visual Education as Applied to Geology." *Journal of the Society of Arts* 2: 444–49

Hawkins, Thomas. 1834. *Memoirs of Ichthyosauri and Plesiosauri, Extinct Monsters of the Ancient Earth.* London: Relfe and Fletcher

— 1840. *The Book of the Great Sea-Dragons, Ichthyosauri and Plesiosauri, Gedolim Taninim, of Moses. Extinct Monsters of the Ancient Earth.* London: Pickering

Hayes, David. 2001. "'Without Parallel in the Known World': The Chequered Past of 277 Gray's Inn Road." *Camden History Review* 25: 5–9

Hays, J. N. 1983. "The London Lecturing Empire, 1800–50." In *Metropolis and Province: Science in British Culture, 1780–1850.* Ed. I. Inkster and J. Morrell, 91–119. London: Hutchinson

Henderson, Paul. 2015. *James Sowerby: the Enlightenment's Natural Historian.* Kew: Royal Botanic Gardens

Herbert, Robert L., ed. "Edward Hitchcock in Europe, 1850. Unpublished Diary and Notes", https://www.amherst.edu/media/view/383181/original/Edward%20Hitchcock%20in%20Europe%20-%201850%20-%20Unpublished%20Diary%20and%20Notes

Hetherington, Henry. 1828. *Principles and Practice Contrasted; or, A Peep into "The only true Church of God upon earth," Commonly Called Freethinking Christians.* 2nd ed. London: Hetherington

— 1830. *Cure for the Blindness of the People.* London: Hetherington

— [1832]. *Cheap Salvation; Or, An Antidote to Priestcraft.* 2nd ed. London, Hetherington

— 1840. *A Full Report of the Trial of Henry Hetherington.* London: Hetherington

Heumann, Ina, A. G. MacKinney, and R. Buschmann. 2022. "Introduction: The Issue of Duplicates." *BJHS* 55: 257–78, https://doi.org/10.1017/s0007087422000267

Hewitt, Martin. 2014. *The Dawn of the Cheap Press in Victorian Britain. The End of the 'Taxes on Knowledge', 1849–1869.* London: Bloomsbury, https://doi.org/10.14296/RiH/2014/1675

Hibbert, Julian. 1828. *Plutarchus, and Theophrastus, on Superstition; with Various Appendices, and a Life of Plutarchus.* London: Hibbert

Hill, Frederic. 1836. *National Education, its Present State and Prospects.* 2 vols. London: Knight

Hilton, Boyd. 1988. *The Age of Atonement. The Influence of Evangelicalism on Social and Economic Thought 1785–1865.* Oxford: Clarendon

— 2000. "The Politics of Anatomy and an Anatomy of Politics c. 1825–1850." In *History, Religion, and Culture. British Intellectual History 1750–1950.* Eds. Stefan Collini, Richard Whatmore, and Brian Young, 179–97. Cambridge: Cambridge University Press, https://doi.org/10.1017/cbo9780511598487.011

— 2006. *A Mad, Bad, and Dangerous People? England 1783–1846.* Oxford: Clarendon.

Hingley, Richard. 2007. "The Society, its Council, the Membership and Publications, 1830–50." In *Visions of Antiquity. The Society of Antiquaries of London 1707–2007.* Ed. Susan Pearce, 173–97. London: Society of Antiquaries

— 2008. *The Recovery of Roman Britain 1586–1906: A Colony So Fertile.* Oxford: Oxford University Press, https://doi.org/10.1093/oso/9780199237029.003.0010

Hobhouse, John Cam. 1819. *An Authentic Narrative of the Events of the Westminster Election, which Commenced on Saturday, February 13th, and Closed on Wednesday, March 3d, 1819.* London: Stodart

Hobley, Brian. 1975. "Charles Roach Smith (1807–1890). Pioneer Rescue Archaeologist." *London Archaeologist* 2: 328–33

Hobson, R. L. 1903. *Catalogue of the Collection of English Pottery in the Department of British and Mediaeval Antiquities and Ethnography of the British Museum.* London: British Museum

Hodge, M. J. S. 1971. "Lamarck's Science of Living Bodies". *BJHS* 5: 323–52

— 1972. "The Universal Gestation of Nature: Chambers' Vestiges and Explanations." *JHB* 5: 127–51

— 2005. "Against 'Revolution' and 'Evolution'." *JHB* 38: 101–21, https://doi.org/10.1007/s10739-004-6512-3

— 2009. "Capitalist Contexts for Darwinian Theory: Land, Finance, Industry and Empire." *JHB* 42: 399–416, https://doi.org/10.1007/s10739-009-9187-y

Hodgkin, Thomas. 1848. "The Progress of Ethnology." *Journal of the Ethnological Society of London* 1: 27–45

Hogben, John. n.d. *Hogben's Strangers' Guide to London.* London: Hogben

Holbach, Paul-Henri Thiry, Baron d'. [Mirabaud, pseud.]. 1820. *The System of Nature; or, The Laws of the Moral and Physical World.* 3 vols. London: Davison

— [Curé Meslier pseud.] 1826. *Good Sense: or, Natural Ideas Opposed to Ideas that are Supernatural.* London: Carlile

— [Mirabaud, pseud.]. 1834. *The System of Nature: or, The Laws of the Moral and Physical World.* 2 vols. London: Carlile

Hollick, Frederick. 1848. *The Origin of Life: A Popular Treatise on the Philosophy and Physiology of Reproduction.* 20th ed. New York: Nafis and Cornish

Hollis, Patricia. 1970. *The Pauper Press. A Study in Working-Class Radicalism of the 1830s.* Oxford: Oxford University Press

Holyoake, George Jacob. 1842. *The Trial of George Jacob Holyoake, on an Indictment for Blasphemy.* London: Paterson

— [1847]. *Paley Refuted in His Own Words.* 2nd ed. London: Hetherington

— 1849a. *The Life and Character of Richard Carlile*. London: Watson

— 1849b. *The Life and Character of Henry Hetherington*. London: Watson

— 1849c. *Literary Institutions: Their Relation to Public Opinion*. London: Watson

— 1850. *The History of the Last Trial by Jury for Atheism in England: A Fragment of Autobiography*. London: Watson

— 1875. *The History of Co-Operation*. 2 vols. London: Trubner

— 1881. *Life of Joseph Rayner Stephens, Preacher and Political Orator*. London: Williams and Norgate

— 1892. *Sixty Years of an Agitator's Life*. 2 vols. London: Unwin

— 1905. *Bygones Worth Remembering*. 2 vols. London: Unwin

— 1906. *The History of Co-Operation*. 2 vols. Revised. London: Unwin

Hoock, Holger. 2003. "Reforming Culture: National Art Institutions in the Age of Reform." In *Rethinking the Age of Reform: Britain 1780–1850*. Eds. Arthur Burns and Joanna Innes, 254–70. Cambridge: Cambridge University Press, https://doi.org/10.1017/cbo9780511550409.012

Hooker, Joseph. 1848. "On some Peculiarities in the Structure of Stigmaria." *Memoirs of the Geological Survey of Great Britain* 2: 431–39

Horowitz, Alan Stanley. 1986. "Notes on the History of the Paleontological Collection, Department of Geology, Indiana University." *Proceedings of the Indiana Academy of Science* 95: 375–79

Horsman, Reginald. 1976. "Origins of Racial Anglo-Saxonism in Great Britain Before 1850." *JHI* 37: 387–410

Hovell, Mark. 1918. *The Chartist Movement*. Manchester: Manchester University Press

Howlett, E. A., W. J. Kennedy, H. P. Powell and H. S. Torrens. 2017. "New Light on the History of Megalosaurus, the Great Lizard of Stonesfield." *ANH* 44:82–102, https://doi.org/10.3366/anh.2017.0416

Huang, Hsiang-Fu. 2016. "When Urania meets Terpsichore: A Theatrical Turn for Astronomy Lectures in Early Nineteenth-Century Britain." *HS* 54:45–70, https://doi.org/10.1177/0073275315624422

— 2017. "A Shared Arena: The Private Astronomy Lecturing Trade and its Institutional Counterpart in Britain, 1817–1865." *NRRS* 72: 319–41, https://doi.org/10.1098/rsnr.2017.0018

Huch, Ronald K., and Paul R. Ziegler. 1985. *Joseph Hume: The People's M.P.* Philadelphia: American Philosophical Society.

Hudson, J. W. 1851. *The History of Adult Education, in which is Comprised a Full and Complete History of the Mechanics' and Literary Institutions, Athenaeums, Philosophical, Mental and Christian Improvement Societies, Literary Unions,*

Schools of Design, etc., of Great Britain, Ireland, America, etc, etc. London: Longman, Brown. Green & Longmans

Huish, Robert. 1836. *The History of the Private and Political Life of the Late Henry Hunt, Esq. M.P. for Preston.* 2 vols. London: Saunders

Hutchison, Graham. 1835. *A Treatise on the Causes and Principles of Meteorological Phenomena. Also Two Essays; The One on Marsh Fevers; The Other on the System of Equality, Proposed by Mr. Owen of New Lanark, for Ameliorating the Condition of Mankind.* Glasgow: Fullarton

Huzel, James P. 2006. *The Popularization of Malthus in Early Nineteenth Century England. Martineau, Cobbett and the Pauper Press.* Aldershot: Ashgate, https://doi.org/10.4324/9781315237664-102

Inkster, Ian. 1976. "The Social Context of an Educational Movement: A Revisionist Approach to the English Mechanics' Institutes, 1820–1850." *Oxford Review of Education* 2: 277–307

Jacob, Margaret C. 1981. *The Radical Enlightenment: Pantheists, Freemasons and Republicans*, London: George Allen and Unwin

— 2019. *The Secular Enlightenment*. Princeton: Princeton University Press, https://doi.org/10.23943/princeton/9780691161327.001.0001

Jacyna, L. S. 1983a. "Images of John Hunter in the Nineteenth Century." *HS* 21: 85–108

— 1983b. "Immanence or Transcendence: Theories of Life and Organization in Britain, 1790–1835." *Isis* 74:311–29

— 1987. "Medical Science and Moral Science: The Cultural Relations of Physiology in Restoration France." *HS* 25: 111–46

Jaffe, J. A., ed. 2007. "The Affairs of Others: The Diaries of Francis Place 1825–1836." *Camden Fifth Series* 30: 49–352

James, Frank A. J. L. 1992. "Michael Faraday, The City Philosophical Society and The Society of Arts." *Royal Society of Arts Journal.* 140: 192–99

James, K. W. 1986. *"Damned Nonsense!"—The Geological Career of the Third Earl of Enniskillen.* Ulster: Ulster Museum

Janowitz, Anne. 1998. *Lyric and Labour in the Romantic Tradition.* Cambridge: Cambridge University Press

Jardine, Boris, Emma Kowal and Jenny Bangham. 2019. "How Collections End: Objects, Meaning and Loss in Laboratories and Museums." *BJHS: Themes* 4:1–27, https://doi.org/10.1017/bjt.2019.8

Jenkins, Bill. 2019. *Evolution Before Darwin: Theories of the Transmutation of Species in Edinburgh, 1804–1834.* Edinburgh: Edinburgh University Press, https://doi.org/10.3366/edinburgh/9781474445788.001.0001

Johns, Adrian. 2010. *Piracy: The Intellectual Property Wars from Gutenberg to Gates*. Chicago: University of Chicago Press, https://doi.org/10.7208/chicago/9780226401201.001.0001

Johnson, Richard. 1979. "'Really Useful Knowledge': Radical Education and Working-Class Culture, 1790–1848." In *Working-Class Culture. Studies in History and Theory*. Ed. John Clarke, Chas Critcher and Richard Johnson, London: Hutchinson

Jones, Gareth Stedman. 1983. "Rethinking Chartism". In *Languages of Class: Studies in English Working-Class History, 1832–1982*. Ed. G. S. Jones, 90–178. Cambridge: Cambridge University Press

Jones, Greta. 2002. "Alfred Russel Wallace, Robert Owen and the Theory of Natural Selection." *BJHS* 35: 73–96 https://doi.org/10.1017/s0007087401004605

Jones, John Gale. 1819. *Substance of Speeches*. London: Carlile

Jones, R. 1997. "'The Sight of Creatures Strange to our Clime': London Zoo and the Consumption of the Exotic." *Journal of Victorian Culture* 2: 1–26

Karkeek, W. C. 1841a. "The Geological History of the Horse." *Veterinarian* 14: 25–31, 69–78

— 1841b. "The Geological History of the Horse." *Farmer's Magazine* 3: 92–94, 174–77

— ['K' pseud.]. 1841c. "Review — The Naturalist's Library." *Veterinarian* 14:701–708

Kass, Amelie M., and Edward H. Kass. 1988. *Perfecting the World. The Life and Times of Dr. Thomas Hodgkin 1798–1866*. Boston: Harcourt Brace Jovanovich

Keane, Angela. 2006. "Richard Carlile's Working Women: Selling Books, Politics, Sex and *The Republican*." *Literature and History* 15: 20–34, https://doi.org/10.7227/lh.15.2.2

Kenny, Robert. 2007. "From the Curse of Ham to the Curse of Nature: The Influence of Natural Selection on the Debate on Human Unity before the Publication of *The Descent of Man*." *BJHS* 40: 367–88, https://doi.org/10.1017/s0007087407009788

Kent, Clement Boulton Roylance. 1898. *The English Radicals, an Historical Sketch*. London, Longmans, Green

Kidd, Colin. 2016. *The World of Mr. Casaubon*. Cambridge: Cambridge University Press, https://doi.org/10.1017/9781139226646

Kingsley, Charles. 1910. *Alton Locke*. London: Dent

Kirby, William. 1835. *On the Power Wisdom and Goodness of God as Manifested in the Creation of Animals and in their History Habits and Instincts*. 2nd ed. 2 vols. London: Pickering

Klancher, Jon. P. 1987. *The Making of English Reading Audiences, 1790–1832*. Madison: University of Wisconsin Press

Knell, Simon J. 2000. *The Culture of English Geology, 1815–1851. A Science Revealed through its Collecting*. Aldershot: Ashgate

Knight, Charles. 1841. *London*. 6 vols. London: Knight

— 1864. *Passages of a Working Life during Half a Century: with a Prelude of Early Reminiscences*. 3 vols. London: Bradbury and Evans

Knight, David. 2004. *Science and Spirituality. The Volatile Connection*. London: Routledge, https://doi.org/10.4324/9781003416593

—, and Matthew D. Eddy, eds. 2016. *Science and Beliefs. From Natural Philosophy to Natural Science, 1700–1900*. Abingdon: Routledge, https://doi.org/10.4324/9781315243733

Kölbl-Ebert, M., ed. 2009. *Geology and Religion: A History of Harmony and Hostility*. London: Geological Society Special Publication 310, https://doi.org/10.1144/SP310

Kohler, Robert E. 2007. "Finders, Keepers: Collecting Sciences and Collecting Practice." *HS* 45: 428–54, https://doi.org/10.1177/007327530704500403

Konig, Charles. 1814. "On a Fossil Human Skeleton from Guadaloupe." *Philosophical Transactions of the Royal Society of London* 104:107–20.

Ksiazkiewicz, Allison. 2015. "A philosophical Pursuit: Natural Models and the Practical Arts in Establishing the Structure of the Earth." *HS* 53: 125–54. https://doi.org/10.1177/0073275315580956

Kurzer, Frederick. 2000. "A History of the Surrey Institution." *AS* 57: 109–41. https://doi.org/10.1080/000337900296218

Lach, Donald F. 1970. *Asia in the Making of Europe*. 2 vols. Chicago: Chicago University Press

Laidlaw, Zoë. 2007. "Heathens, Slaves and Aborigines: Thomas Hodgkin's Critique of Missions and Anti-slavery." *History Workshop Journal Issue* 64: 133–61, https://doi.org/10.1093/hwj/dbm034

Larcher, P. H. 1844. *Larcher's Notes on Herodotus*. Ed. W. D. Cooley. 2 vols. London: Whittaker

Larsen, Timothy. 2004. *Contested Christianity. The Political and Social Contexts of Victorian Theology*. Waco: Baylor University Press

Latham, J. E. M. 1999. *Search for a New Eden. James Pierrepont Greaves (1777–1842): The Sacred Socialist and his Followers*. Madison: Associated University Presses

Latour, Bruno. 1987. *Science in Action. How to Follow Scientists and Engineers Through Society*. Cambridge MA.: Harvard University Press

Lawrence, William. 1822. *Lectures on Physiology, Zoology, and the Natural History of Man, Delivered at the Royal College of Surgeons*. London: Benbow

— 1840. *Facts versus Fiction! An Essay on the Functions of the Brain*. 2nd ed. London: Watson

Leopold, Richard William. 1940. *Robert Dale Owen A Biography*. Cambridge MA.: Harvard University Press

Levine, Philippa. 1986. *The Amateur and the Professional. Antiquarians, Historians and Archaeologists in Victorian England, 1838–1886*. Cambridge: Cambridge University Press

Lightman, Bernard. 2007. *Victorian Popularizers of Science. Designing Nature for New Audiences*. Chicago: University of Chicago Press, https://doi.org/10.7208/chicago/9780226481173.003.0001

Lindfors, Bernth. 1996. "Hottentot, Bushman, Kaffir: Taxonomic Tendencies in Nineteenth-Century Racial Iconography." *Nordic Journal of African Studies* 5: 1–28

Linton, W. J. 1879. *James Watson. A Memoir of the Days of the Fight for a Free Press in England and of the Agitation for the People's Charter*. Manchester: Heywood

— 1894. *Threescore and Ten years, 1820–1890; Recollections*. New York: Scribner

Litchfield, H. E., ed. 1915. *Emma Darwin, A Century of Family Letters, 1702–1896*. 2 vols. London: John Murray

Lockley, Philip J. 2009. *Millenarian Religion and Radical Politics in Britain 1815–1835: A Study of Southcottians after Southcott*. D.Phil Thesis, University of Oxford

— 2013. *Visionary Religion and Radicalism in Early Industrial England. From Southcott to Socialism*. Oxford: Oxford University Press, https://doi.org/10.1093/acprof:oso/9780199663873.001.0001

Long, S. L., P. D. Taylor, S. Baker, and J. Cooper. 2003. "Some Early Collectors and Collections of Fossil Sponges Represented in the Natural History Museum, London." *Geological Curator* 7: 353–62, https://doi.org/10.55468/gc423

Lonsdale, Henry. 1870. *A Sketch of the Life and Writings of Robert Knox the Anatomist*. London: Macmillan

Loose, Margaret A. 2014. *The Chartist Imaginary. Literary Form in Working-Class Political Theory and Practice*. Columbus: Ohio State University Press

LoPatin, Nancy D. 1999. *Political Unions, Popular Politics and the Great Reform Act of 1832*. Basingstoke: Palgrave Macmillan

Lorimer, Douglas A. 1978. *Colour, Class and the Victorians. English Attitudes to the Negro in the Mid-Nineteenth Century*. Leicester: Leicester University Press

Loudon, I. S. L. 1981. "Origins and Growth of the Dispensary Movement in England." *Bulletin of the History of Medicine* 55: 322–42

Loveless, George. 1838. *The Victims of Whiggery*. London: Cleave

Lovett, William. 1851. *Elementary Anatomy and Physiology*. London: Darton

— 1920. *Life and Struggles of William Lovett in his Pursuit of Bread, Knowledge, and Freedom, with some Short Account of the Different Associations he Belonged to and of the Opinions he Entertained*. 2 vols. New York: Knopf

Lucas, A. M., and P. J. Lucas. 2014. "Natural History 'Collectors': Exploring the Ambiguities." *ANH* 41: 63–74, https://doi.org/10.3366/anh.2014.0210

Ludington, Charles. 2013. *The Politics of Wine in Britain. A New Cultural History*. Basingstoke: Palgrave Macmillan, https://doi.org/10.1057/9780230306226

Ludlow, J. M., and Lloyd Jones 1867. *Progress of the Working Class 1832–1867*. London: Strahan

Lunan, Lyndsay. 2005. *The Fiction of Identity: Hugh Miller and the Working Man's Search for Voice in Nineteenth-century Scottish Literature*. PhD thesis, University of Glasgow

Lundgren, Frans. 2013. "The Politics of Participation: Francis Galton's Anthropometric Laboratory and the Making of Civic Selves." *BJHS* 46: 445–66, https://doi.org/10.1017/s0007087411000859

Lydekker, Richard. 1888. *Catalogue of the Fossil Reptilia and Amphibia in the British Museum (Natural History), Part I Containing the Orders Ornithosauria, Crocodilia, Dinosauria, Squamata, Rhynchocephalia, and Proterosauria*. London: British Museum (Natural History)

— 1889a. *Catalogue of the Fossil Reptilia and Amphibia in the British Museum (Natural History), Part II Containing the Orders Ichthyopterygia and Sauropterygia*. London: British Museum (Natural History)

— 1889b. *Catalogue of the Fossil Reptilia and Amphibia. Part III Containing the Order Chelonia*. London: British Museum (Natural History)

— 1890. *Catalogue of the Fossil Reptilia and Amphibia in the British Museum (Natural History), Part IV Containing the Orders Anomodontia, Ecaudata, Caudata, and Labyrinthodontia; and Supplement*. London: British Museum (Natural History)

Lyell, Charles. 1830–33. *Principles of Geology*. 3 vols. London: Murray

— 1849. *A Second Visit to the United States of North America*. 2 vols. New York: Harper

Lyell, K. M., ed. 1881. *Life Letters and Journals of Sir Charles Lyell, Bart*. 2 vols. London: Murray

Maccoby, S. 1935. *English Radicalism 1832–1852*. London: Allen and Unwin

— 1955. *English Radicalism 1786–1832 From Paine to Cobbett*. London: Allen and Unwin

Macdonald, Jessie Bell. 1974. "The Sowerby Collection in the British Museum (Natural History): a Brief Description of its Holdings and a History of its Acquisition from 1821–1971." *JSBNH* 6: 380–401

Macdonald, Sharon, ed. 1998. "Exhibitions of Power and Powers of Exhibition." In *The Politics of Display. Museums, Science, Culture*. Ed. Sharon Macdonald, 1–21. London: Routledge

Macerone, Francis. [1832]. *Defensive Instructions for the People*. London: J. Smith

— 1837. "On Mackintosh's Electrical Theory of the Universe." *Mechanics' Magazine* 26: 19

— 1848. *Memoirs of the Life and Adventures of Colonel Maceroni*. 2 vols. London: Macrone

MacGregor, Arthur, ed. 2008. *Sir John Evans (1823–1908). Antiquity, Commerce and Natural Science in the Age of Darwin*. Oxford: Oxford University Press

Mackenzie, W. M. 1905. *Hugh Miller. A Critical Study*. London: Hodder and Stoughton

Mackey, Sampson Arnold. 1823. *The Mythological Astronomy of the Ancients; Part the Second: or the Key of Urania, the Wards of which will unlock all the Mysteries of Antiquity*. Norwich: Walker

— 1825. *A New Theory of the Earth and of Planetary Motion; in which is Demonstrated that the Sun is Vicegerent of his own System*. Norwich: Mackey

— 1827 [1822–24]. *The Mythological Astronomy, in Three Parts*. London: Hunt and Clarke

— 1832. *A Lecture on Astronomy, adjusted to its dependent science Geology; in which is shewn the plain and simple cause of the vast abundance of water in the Southern Hemisphere. Given at 91, Dean Street, Soho, Dec. 20, 1832, in consequence of having seen* An Essay On the Astronomical and Physical Causes of Geological Changes, *by Sir Richard Phillips, Edited by W. D. Saull, Aldersgate Street, May, 1832*. London

Mackintosh, T. Simmons. [1840.] *An Inquiry into the Nature of Responsibility*. Birmingham: Guest

— 1846. *The "Electrical Theory" of the Universe. The Elements of Physical and Moral Philosophy*. Boston: Mendum

MacLeod, Roy M. 1970. "Science and the Civil List, 1824–1914." *Technology and Society* 6: 47–55

— 1983. "Whigs and Savants: Reflections on the Reform Movement in the Royal Society, 1830–48." In *Metropolis and Province: Science in British Culture, 1780–1850*. Ed. I. Inkster and J. Morrell, 55–90. London: Hutchinson

Maclure, William. 1838. *Opinions on Various Subjects, Dedicated to the Industrious Producers*. 3 vols. New Harmony, Indiana

MacNeil, Heather. 2017. "Catalogues and the Collecting and Ordering of Knowledge (II): Debates about Cataloguing Practices in the British Museum and the Forebears of the Public Record Office of Great Britain, ca. 1750–1850." *Archivaria* 84:1–35

Maidment, Brian. 2013. *Comedy, Caricature and the Social Order 1820–1850*. Manchester: Manchester University Press

Maitland, Frederic William. 1906. *The Life and Letters of Leslie Stephen*. London: Duckworth

Mamoli Zorzi, Rosella, and Katherine Manthorne, eds. 2019. *From Darkness to Light. Writers in Museums 1798–1898*. Cambridge: Open Book Publishers, https://www.openbookpublishers.com/books/10.11647/obp.0151

Mantell, Gideon. 1831. "The Geological Age of Reptiles." *Edinburgh New Philosophical Journal* 11: 181–85

— 1833. *The Geology of the South-East of England*. London: Longman, Rees, Orme, Brown, Green, and Longman

— 1836. *A Descriptive Catalogue of the Objects of Geology, Natural History, and Antiquity, (Chiefly Discovered in Sussex,) in the Museum, Attached to the Sussex Scientific and Literary Institution at Brighton*. 6th ed. London: Relfe and Fletcher

— 1838. *The Wonders of Geology*. 2 vols. London: Relfe and Fletcher

— 1844. *The Medals of Creation; or, First Lessons in Geology, and in the Study of Organic Remains*. 2 vols. London: Bohn

— 1846. "A Few Notes on the Prices of Fossils." *London Geological Journal* 1: 13–17

— 1847. *Geological Excursions Round the Isle of Wight, and Along the Adjacent Coast of Dorsetshire*. London: Bohn

— 1850. "On the Remains of Man, and Works of Art Imbedded in Rocks and Strata, As Illustrative of the Connexion Between Archaeology and Geology." *Archaeological Journal* 7: 327–30

— 1851. *Petrifactions and their Teachings; Or, A Hand-Book to the Gallery of Organic Remains*. London: Bohn

— 1857-58. *The Wonders of Geology; or, A Familiar Exposition of Geological Phenomena*. Ed. T. Rupert Jones. 7th ed. Revised and Augmented. 2 vols. London: Bohn

—, and A. G. Melville. 1849. "Additional Observations on the Osteology of the Iguanodon and Hylaeosaurus." *Philosophical Transactions of the Royal Society* 139: 271–305

Marsh, Joss. 1998. *Word Crimes. Blasphemy, Culture, and Literature in Nineteenth-Century England*. Chicago: University of Chicago Press

Martin, Emma. [1844.] *First Conversation on the Being of God*. London: Martin

Martin, W. C. L. 1841. *A General Introduction to the Natural History of Mammiferous Animals, with a Particular View of the Physical History of Man, and the More Closely Allied Genera of the Order Quadrumana, or Monkeys*. London: Wright

Matheson, Colin. 1964. "George Brettingham Sowerby the First and his Correspondents." *JSBNH* 4: 214–25, 253–66

Matijasic, Thomas D. 1987. "Science, Religion, and the Fossils at Big Bone Lick." *JHB* 20: 413–21

Mayhew, Henry. 1861–62. *London Labour and the London Poor*. 4 vols. London: Griggin, Bohn

Mcallister, Annmarie. 2013. "Xenophobia on the Streets of London: Punch's Campaign against Italian Organ-Grinders, 1854–1864." In *Fear, Loathing, and Victorian Xenophobia*. Eds. Marlene Tromp, Maria K. Bachman, and Heidi Kaufman, 286–311. Columbus: Ohio State University Press

McCabe, Joseph. 1908. *Life and Letters of George Jacob Holyoake*. 2 vols. London: Watts

— 1920. *Robert Owen*. London: Watts

McCalman, Iain. 1975. *Popular Radicalism and Freethought in Early Nineteenth Century England. A Study of Richard Carlile and his Followers, 1815–32*. MSc. Thesis. Australian National University

— 1984. "Unrespectable Radicalism: Infidels and Pornography in Early Nineteenth Century London." *Past and Present* 104: 74–110

— 1987. "Ultra-Radicalism and Convivial Debating-Clubs in London, 1795–1838." *English Historical Review* 102: 309–33

— 1988. *Radical Underworld. Prophets, Revolutionaries and Pornographers in London, 1795–1840*. Cambridge: Cambridge University Press

— 1992. "Popular Irreligion in Early Victorian England: Infidel Preachers and Radical Theatricality in 1830s London." In *Religion and Irreligion in Victorian Society. Essays in Honor of R. K. Webb*. Eds. R. W. Davis and R. J. Helmstadter, 51–66. London: Routledge

McCartney, Paul J. 1977. *Henry De la Beche: Observations on an Observer*. Cardiff: Friends of the National Museum of Wales

McClintock, Anne. 1995. *Imperial Leather: Race, Gender and Sexuality in the Colonial Contest*. New York: Routledge

McMillan, Nora F., and E. F. Greenwood. 1972. "The Beans of Scarborough; a Family of Naturalists." *JSBNH* 6: 152–61

McOuat, Gordon. 2001. "Cataloguing Power: Delineating 'Competent Naturalists' and the Meaning of Species in the British Museum." *BJHS* 34: 1–28

Mee, Jon. 2016. *Print, Publicity and Radicalism in the 1790s, The Laurel of Liberty*. Cambridge: Cambridge University Press, https://doi.org/10.1017/cbo9781316459935

Michael, A. D. 1895. "The President's Address: The History of the Royal Microscopical Society." *Journal of the Royal Microscopical Society* 1–20

Michaud. [post 1855]. *Biographie Universelle (Michaud) Ancienne et Moderne.* Nouvelle Edition. Tome 38. Paris: Desplaces

Michelotti, Giovanni. 1841. *Saggio Storico dei Rizopodi Caratteristici dei Terreni Supracretacei.* Moderna: Camera

Miles, Dudley. 1988. *Francis Place. The Life of a Remarkable Radical.* Sussex: Harvester

Miller, G. 1826. *Popular Philosophy: or, The Book of Nature Laid Open.* 2 vols. Dunbar: Miller

Miller, Hugh. 1849. *Foot-Prints of the Creator: or, the Asterolepis of Stromness.* London: Johnstone and Hunter

Mitchell, Logan. n.d. [c. 1842]. *The Christian Mythology Unveiled.* London: Cousins

Mogg, E. 1848. *Mogg's New Picture of London.* 11th ed. London: Mogg

Morgan, Lady Sydney. 1862. *Lady Morgan's Memoirs: Autobiography, Diaries and Correspondence.* Ed. W. H. Dixon. 2 vols. London: Allen

Moore, D. T., J. C. Thackray and D. L. Morgan. 1991. "A Short History of the Museum of the Geological Society of London, 1807–1911." *Bulletin of the British Museum (Natural History) Historical Series* 19: 51–160

Moore J. Percy. 1947. "William Maclure—Scientist and Humanitarian." *Proceedings of the American Philosophical Society* 91: 234–49

Moore, James R. 1979. *The Post-Darwinian Controversies. A Study of the Protestant Struggle to come to terms with Darwin in Great Britain and America 1870–1900.* Cambridge: Cambridge University Press

— 1988. "Freethought, Secularism, Agnosticism: The Case of Charles Darwin." In *Religion in Victorian Britain.* Vol. 1. *Traditions.* Ed. Gerald Parsons, 274–319. Manchester: Manchester University Press

— 1990. "Theodicy and Society: The Crisis of the Intelligentsia." In *Victorian Faith in Crisis: Essays on Continuity and Change in Nineteenth-Century Religious Belief.* Ed. Richard J. Helmstadter and Bernard Lightman, 153–86. Stanford: Stanford University Press

— 1997. "Wallace's Malthusian Moment: The Common Context Revisited." In *Victorian Science in Context.* Ed. Bernard Lightman, 290–311. Chicago: University of Chicago Press

Moore, Thomas. 1854. *Life of Lord Byron: With his Letters and Journals.* 6 vols. London: Murray

[Morgan, J. M.] 1834. *Hampden in the Nineteenth Century.* 2 vols. London: Moxon

Morrell, J. B. 1971. "Professors Robison and Playfair, and the Theophobia Gallica: Natural Philosophy, Religion, and Politics in Edinburgh, 1789–1815." *NRRS* 26: 43–63

— 1976. "London Institutions and Lyell's Career, 1820–41." *BJHS* 9: 132–46

— 1985. "Wissenschaft in Worstedopolis: Public Science in Bradford, 1800–1850." *BJHS* 18: 1–23

— 2005. *John Phillips and the Business of Victorian Science*. Aldershot: Ashgate, https://doi.org/10.4324/9781351154888

—, and A. Thackray. 1981. *Gentlemen of Science: Early Years of the British Association for the Advancement of Science*. Oxford: Clarendon Press

— 1984. *Gentlemen of Science. Early Correspondence of the British Association for the Advancement of Science*. Camden Fourth Series, Vol. 30. London: Royal Historical Society

Morris, John. 1841. "Remarks upon the Recent and Fossil Cycadeae." *Annals and Magazine of Natural History* 7: 110–20

— 1854. *A Catalogue of British Fossils: Comprising the Genera and Species hitherto described; With References to their Geological Distribution and to the Localities in which they have been found*. 2nd ed. London: The Author.

—, and John Lycett. 1850. *A Monograph of the Mollusca From the Great Oolite, Chiefly From Minchinhampton and the Coast of Yorkshire. Part I. Univalves*. 4. London: Palaeontographical Society

Morrison, Richard James. [Zadkiel pseud.] 1841. "Philosophy of Geology." *Horoscope* 1: 27–32

Morse, Michael A. 2005. *How the Celts Came to Britain: Druids, Ancient Skulls and the Birth of Archaeology*. Stroud: Tempus

Morus, Iwan Rhys. 1993. "Currents from the Underworld. Electricity and the Technology of Display in Early Victorian England." *Isis* 84:50–69

— 1998. *Frankenstein's Children: Electricity, Exhibition and Experiment in Early-Nineteenth-Century London*. Princeton: Princeton University Press

— 2010a. "Placing Performance." *Isis* 101:775–78, https://doi.org/10.1086/657476

— 2010b. "Worlds of Wonder: Sensation and the Victorian Scientific Performance." *Isis* 101: 806–16, https://doi.org/10.1086/657479

— 2011. *Shocking Bodies. Life, Death & Electricity in Victorian England*. Stroud: The History Press

Moshenska, Gabriel. 2014. "Unrolling Egyptian Mummies in Nineteenth-Century Britain." *BJHS* 47:451–77, https://doi.org/10.1017/s0007087413000423

Moxham, Noah, and Aileen Fyfe. 2022. "Reforms, Referees and the Proceedings, 1820–1850." In *A History of Scientific Journals: Publishing at the Royal Society, 1665–2015*. Eds. Aileen Fyfe, Noah Moxham, Julie McDougall–Waters, and Camilla Mørk Røstvik, 257–95. London: UCL Press, https://doi.org/10.2307/j.ctv2gz3zp1.16

Mudie, Robert. 1836. *London and Londoners: or, A Second Judgment of "Babylon the Great."* London: Colburn

Mullen, Shirley A. 1985. *Organized Freethought: The Religion of Unbelief in Victorian England.* PhD Thesis, University of Minnesota

— 1992. "Keeping the Faith: The Struggle for a Militant Atheist Press, 1839–62." *Victorian Periodicals Review* 25: 150–58

Murphy, Paul Thomas. 1994. *Toward a Working-Class Canon. Literary Criticism in British Working-Class Periodicals, 1816–1858.* Columbus: Ohio State University Press

[Murray, John]. 1831. *The Truth of Revelation, Demonstrated by an Appeal to Existing Monuments, Sculptures, Gems, Coins, and Medals.* London: Longman, Rees, Orme, Brown, & Green

Nares, Edward. 1834. *Man, as Known to us Theologically and Geologically.* London: Rivington

Nash, David S. 1995a. "'Look in Her Face and Lose Thy Dread of Dying': The Ideological Importance of Death to the Secularist Community in Nineteenth-Century Britain." *Journal of Religious History* 19: 158–80

— 1995b. "Unfettered Investigation—The Secularist Press and the Creation of Audience in Victorian England." *Victorian Periodicals Review* 28: 123–35

[Neale, Erskine]. 1848. *The Closing Scene; or, Christianity and Infidelity Contrasted in the Last Hours of Remarkable Persons.* London: Longman, Brown, Green & Longmans

Nelson, G. 1978. "From Candolle to Croizat: Comments on the History of Biogeography." *JHB* 11: 269–305

Neve, M. 1983. "Science in a Commercial City: Bristol, 1820–60." In *Metropolis and Province: Science in British Culture, 1780–1850.* Eds. Ian Inkster and Jack Morrell, 179–204. London: Hutchinson

Newitt, Ned. 2016. "The Who's Who of Radical Leicester." http://www.nednewitt.com/whoswho/P-Q.html

Noel, Baptist Wriothesley. 1835. *The State of the Metropolis Considered, in a Letter to the Right Honorable and Right Reverend the Bishop of London.* 3rd ed. London: Nisbet.

Norman. David B. 1993. "Gideon Mantell's 'Mantell-Piece': The Earliest Well-Preserved Ornithischian Dinosaur." *Modern Geology* 18: 225–45

O'Connor, Ralph. 2008. *The Earth on Show: Fossils and the Poetics of Popular Science, 1802–1856.* Chicago: University of Chicago Press, https://doi.org/10.7208/chicago/9780226616704.001.0001

— 2012. "Victorian Saurians: The Linguistic Prehistory of the Modern Dinosaur." *Journal of Victorian Culture* 17: 492–504, https://doi.org/10.1080/13555502.2012.738896

Offor, George. 1846. *The Triumph of Henry VIII. Over the Usurpations of the Church.* London: Campkin.

Oliver, W. H. 1958. "The Labour Exchange Phase of the Co-Operative Movement." *Oxford Economic Papers*, New Series 10: 355–67

— 1964. "The Consolidated Trades' Union of 1834." *Economic History Review*, New Series 17: 77–95

Ospovat, Dov. 1977. "Lyell's Theory of Climate." *JHB* 10: 317–39

Outram, Dorinda. 1984. *Georges Cuvier. Vocation, Science and Authority in Post-Revolutionary France*. Manchester: Manchester University Press

Owen, Richard. 1835. "On the Osteology of the Chimpanzee and Orang Utan." *Transactions of the Zoological Society of London* 1: 343–79

— 1840. "Report on British Fossil Reptiles." *Report of the Ninth Meeting of the British Association for the Advancement of Science; held at Birmingham in August 1839*. 43–126. London: Murray

— 1841. "A Description of a Portion of the Skeleton of the Cetiosaurus, a Gigantic Extinct Saurian Reptile Occurring in the Oolitic Formations of Different Portions of England." *Proceedings of the Geological Society* 3: 457–62

— 1841 [1842] "Report on British Fossil Reptiles. Pt. 2." *Report of the Eleventh Meeting of The British for the Advancement of Science; held at Plymouth in July 1841*. 60–204. London: Murray

— 1842. "Streptospondylus." *Penny Cyclopaedia* 23: 113–15

— 1846. *A History of British Fossil Mammals and Birds*. London: van Voorst

— 1849–1884. *A History of British Fossil Reptiles*. 4 vols. London: Cassell

— 1850a. "On British Eocene Serpents and the Serpent of the Bible." *Edinburgh New Philosophical Journal* 49: 239–42

— 1850b. *Monograph on Fossil Reptilia of London Clay. Pt 2. Crocodilia, Ophidia*. London: Palaeontographical Society

— 1851. *A Monograph on the Fossil Reptilia of the Cretaceous Formations. Pt I... Chelonia (Lacertilia, &c)*. London: Palaeontographical Society

— 1854. *Monograph on the Fossil Reptilia of the Wealden Formations. Pt 2. Dinosauria*. London: Palaeontographical Society

— 1859. *Supplement (No. 2) to the Monograph on the Fossil Reptilia of the Wealden and Purbeck Formations*. London: Palaeontographical Society

— 1860. *Palaeontology or A Systematic Summary of Extinct Animals and their Geological Relations*. Edinburgh: Adam and Charles Black

— 1861. "The Gorilla and the Negro." *Athenaeum* 1743: 395–96

— 1992. *The Hunterian Lectures in Comparative Anatomy May–June, 1837*. Ed. Phillip R. Sloan. Chicago: University of Chicago Press

Owen, Rev. Richard S. 1894. *The Life of Richard Owen*. 2 vols. London: Murray

Owen, Robert. [1830.] *Lectures on an Entire New State of Society.* London: Brooks

— 1838. *The Marriage System of the New Moral World.* Leeds: Hobson

— 1839. *Robert Owen on Marriage, Religion, & Private Property.* London: Stewart. [1d sheet]

— 1857. *The Life of Robert Owen. Written by Himself.* 2 vols. London: Wilson

—, and Alexander Campbell. 1839 [1829]. *Debate on the Evidences of Christianity.* London: Groombridge

Owen, Robert Dale. 1824. *An Outline of the System of Education at New Lanark.* Glasgow: Wardlaw & Cunninghame

— 1839. *Situations: Lawyers—Clergy—Physicians—Men and Women.* London: Watson

— 1874. *Threading My Way: Twenty-Seven Years of Autobiography.* London: Trubner

Page, Frederick G. 2008. "James Rennie (1787–1867), Author, Naturalist and Lecturer." *ANH* 35: 128–42, https://doi.org/10.3366/e0260954108000120

Palmer, Elihu. 1823. *Principles of Nature; or, A Development of the Moral Causes of Happiness and Misery Among the Human Species.* London: Carlile

Pandora, Katherine. 2017. "The Permissive Precincts of Barnum's and Goodrich's Museums of Miscellaneity: Lessons in Knowing Nature for New Learners." In *Science Museums in Transition: Cultures of Display in Nineteenth-Century Britain and America.* Eds Carin Berkowitz and Bernard Lightman, 36–64. Pittsburgh: University of Pittsburgh Press, https://doi.org/10.2307/j.ctt1r6b0c8.7

Pankhurst, Richard K. P. 1954. "Anna Wheeler: A Pioneer Socialist and Feminist." *Political Quarterly* 25: 132–43

— 1991. *William Thompson (1775–1833) Pioneer Socialist.* London: Pluto.

Paradis, James G. 1997. "Satire and Science in Victorian Culture." In *Victorian Science in Context.* Ed. Bernard Lightman, 143–75. Chicago: University of Chicago Press

Paris, John Ayrton. 1831. *The Life of Sir Humphry Davy.* 2 vols. London: Colburn and Bentley

Parolin, Christina. 2010. *Radical Spaces. Venues of Popular Politics in London, 1790–c. 1845.* Canberra: Australian National University, https://doi.org/10.22459/rs.12.2010

Parsinnen, T. M. and Prothero, I. J. 1977. "The London Tailors' Strike of 1834 and the Collapse of the Grand National Consolidated Trades' Union: A Police Spy's Report." *International Review of Social History* 22: 65–107

Paterson, Thomas. [1843]. *"The Man Paterson." God Versus Paterson. The Extraordinary Bow-Street Police Report.* London: Clarke

Pearce, Susan. 2008. "William Bullock. Collections and Exhibitions at the Egyptian Hall, London, 1816–25." *Journal of the History of Collections* 20: 17–35, https://doi.org/10.1093/jhc/fhm031

Perthes, Jacques Boucher de Crèvecœur de. 1863. *Sous Dix Rois. Souvenirs de 1791 a 1860*. Tome Sixieme. Paris: Jung-Treuttel

Petrie, George. [1841]. *The Works of George Petrie: Comprising Equality, and Other Poems. Select Extracts From the Letters of Agrarius: with A Biographical Memoir of the Author*. 2nd ed. London: Cleave

Pettigrew, Thomas Joseph. 1840. "Thomas Joseph Pettigrew, F.R.S., F.S.A., F.L.S." In *Biographical Memoirs of the Most Celebrated Physicians, Surgeons, etc., etc., who have Contributed to the Advancement of Medical Science*. Ed. T. J. Pettigrew, vol. 4 (9), 1–39. London: Fisher

— 1844. *On Superstitions Connected with the History and Practice of Medicine and Surgery*. London: Churchill

Phillips, John. 1837. *Treatise on Geology*. Vol. 1. London: Longman, Orme, Brown, Green, and Longmans

[Phillips, Richard.] 1812. "True Causes of the Geological Changes in the Earth." *Monthly Magazine* 33: 118–23

— 1821. "On the Changes which take place on the Earth's Surface from its Motion as a Planet." In *The Wonders of the Heavens Displayed in Twenty Lectures*. Ed. R. Phillips, 100–11. London: Phillips

— 1832a. *Essay on the Physico-Astronomical Causes of the Geological Changes on the Earth's Surface, and of the Changes in Terrestrial Temperature, with Notes. By Sir Richard Phillips. Re-published, with a Preface, by William Devonshire Saull, F.G.S., F.A.S., F.R.A.S.* London: Sherwood, Gilbert, and Piper

— 1832b. "Sir J. Byerley's Theory." *Magazine of Natural History* 5: 102–03

— 1835. *A Million of Facts*. New ed. London: Sherwood, Gilbert, and Piper

Place, Francis, 1834. *Improvement of the Working People. Drunkenness—Education*. London: Fox

Podmore, Frank. 1907. *Robert Owen: A Biography*. 2 vols. New York: Appleton.

Porter, Dahlia. 2019. "Catalogues for an entropic collection: losses, gains and disciplinary exhaustion in the Hunterian Museum, Glasgow." *BJHS: Themes* 4:215–43, https://doi.org/10.1017/bjt.2019.15

Porter, George Richardson. 1843. *The Progress of the Nation. Sections V. to VIII*. London: Knight

— 1851. *The Progress of the Nation*. New ed. London: Murray

Porter, Roy S. 1978a. "Philosophy and Politics of a Geologist: G. H. Toulmin (1754–1817)." *JHI* 39: 435–50

— 1978b. "George Hoggart Toulmin's Theory of Man and the Earth in the Light of the Development of British Geology." *AS* 35: 339–52

Powell, Baden. 1834. *A Letter to the Editor of the British Critic, and Quarterly Theological Review*. Oxford: Parker

Powell, B. F. 1837. *Bible of Reason*. 3 vols. London: Hetherington

Prentice, Archibald. 1853. *History of the Anti-Corn-Law League*. 2 vols. London: Cash

Prestwich, G. A. M. 1899. *Life and Letters of Sir Joseph Prestwich*. Edinburgh: Blackwood

Price, John Edward. 1880. *On a Bastion of London Wall, or, Excavations in Camomile Street, Bishopsgate*. London: Nichols

Prothero, I. J. 1979. *Artisans and Politics in Early Nineteenth-Century London. John Gast and his Times*. Folkestone: Dawson

[Pycroft, George.] 1863. *A Report of A SAD CASE, Recently tried before the Lord Mayor, OWEN versus HUXLEY*. London

Quammen, David. 2006. *The Reluctant Mr. Darwin*. New York: Atlas

Qureshi, Sadiah. 2011. *Peoples on Parade. Exhibitions, Empire and Anthropology in Nineteenth-Century Britain*. Chicago: Chicago University Press, https://doi.org/10.7208/chicago/9780226700984.001.0001

— 2014. "Dramas of Development: Exhibitions and Evolution in Victorian Britain." In *Evolution and Victorian Culture*. Eds Bernard Lightman and Bennett Zon, 261–85. Cambridge: Cambridge University Press, https://doi.org/10.1017/cbo9781139236195.011

Rainger, Ronald. 1980. "Philanthropy and Science in the 1830's. The British and Foreign Aborigines' Protection Society." *Man* 15: 702–17

Rastall, R. H. 1937. "Bonney, Thomas George 1833–1923." *Dictionary of National Biography*

Rectenwald, Michael. 2013. "Secularism and the Cultures of Nineteenth-Century Scientific Naturalism." *BJHS* 46: 231–54, https://doi.org/10.1017/s0007087412000738

— 2016. *Nineteenth-Century British Secularism: Science, Religion, and Literature*. London: Palgrave Macmillan, https://doi.org/10.1057/9781137463890

Reeve, Lovell. 1863–64. *Portraits of Men of Eminence in Literature, Science, and Art, with Biographical Memoirs*. 2 vols. London: Reeve

Rennell, Thomas. 1819. *Remarks on Scepticism, Specially as it is Connected with the Subjects of Organization and Life*. London: Rivington

[Rennie, James.] 1830. *Insect Transformations*. London: Knight

[—] 1834. *Alphabet of Natural Theology*. London: Orr and Smith

[—] 1838. *The Natural History of Monkeys, Opossums, and Lemurs*. 2 vols, London: Knight

Richards, Evelleen. 1989. "The 'Moral Anatomy' of Robert Knox: The Interplay between Biological and Social Thought in Victorian Scientific Naturalism." *JHB* 22: 373–436

— 1994. "A Political Anatomy of Monsters, Hopeful and Otherwise. Teratogeny, Transcendentalism, and Evolutionary Theorizing." *Isis* 85: 377–411

— 2017. *Darwin and the Making of Sexual Selection*. Chicago: Chicago University Press, https://doi.org/10.7208/chicago/9780226437064.001.0001

— 2020. *Ideology and Evolution in Nineteenth Century Britain: Embryos, Monsters, and Racial and Gendered Others in the Making of Evolutionary Theory and Culture*. London: Routledge, https://doi.org/10.4324/9780429467042

Richards, Robert J. 1992. *The Meaning of Evolution. The Morphological Construction and Ideological Reconstruction of Darwin's Theory*. Chicago: Chicago University Press

Richardson, George Fleming. 1842. *Geology for Beginners*. London: Hippolyte Bailliere

— 1855. *An Introduction to Geology: and its Associate Sciences Mineralogy, Fossil Botany, and Palaeontology*. London: Bohn

Richardson, Ruth. 1989. *Death. Dissection and the Destitute*. London: Pelican

Rieppel, Lukas. 2012. "Bringing Dinosaurs Back to Life: Exhibiting Prehistory at the American Museum of Natural History." *Isis* 103: 460–90, https://doi.org/10.1086/667969

Riper, A. Bowdoin Van. 1993. *Men Among the Mammoths. Victorian Science and the Discovery of Human Prehistory*. Chicago: Chicago University Press

Ritchie, J. Ewing. 1870. *The Religious Life of London*. London: Tinsley

Ritvo, Harriet. 1987. *The Animal Estate. The English and Other Creatures in the Victorian Age*. Cambridge, MA: Harvard University Press

— 1997. *The Platypus and the Mermaid and Other Figments of the Classifying Imagination*. Cambridge, MA: Harvard University Press

Robinson, Eric. 2003. "Bowerbank: A Forgotten Geologist." *Geology Today* 19: 219–22, https://doi.org/10.1111/j.1365-2451.2004.00436.x

Roe, Shirley. 1983. "John Turberville Needham and the Generation of Living Organisms." *Isis* 74:159–84

Roebuck, J. A. 1835a. "The London Review and the Irish Church Question." In *Pamphlets for the People*. Ed. J. A. Roebuck, vol. 1. London: Charles Ely

— 1835b. "Democracy in America". In *Pamphlets for the People*. Ed. J. A. Roebuck, vol. 1. London: Charles Ely

Rogers, E. S. 1896. *Life and Letters of William Barton Rogers*. 2 vols. Boston: Houghton Mifflin

Rogers, Henry Darwin. 1835. "Report on the Geology of North America, Part I." *Report of the Fourth Meeting of the British Association for the Advancement of Science; Held at Edinburgh in 1834*, 1–66. London: Murray

Rose, Jonathan. 2002. *The Intellectual Life of the British Working Classes*. New Haven: Yale Nota Bene, https://doi.org/10.12987/9780300259827

Rosenblatt, Frank F. 1918. *The Chartist Movement in its Social and Economic Aspects*. PhD Thesis, Columbia University

Rowe, D. J. 1970a. "Class and Political Radicalism in London, 1831–2." *Historical Journal* 13: 31–47

—, ed. 1970b. "London Radicalism 1830–1843. A Selection of the Papers of Francis Place", https://www.british-history.ac.uk/london-record-soc/vol5

Royal College of Surgeons. 1854. *Descriptive Catalogue of Fossil Organic Remains of Reptilia and Pisces Contained in the Museum of the Royal College of Surgeons of England*. London: Taylor and Francis

Royle, Edward. 1974. *Victorian Infidels. The Origins of the British Secularist Movement 1791–1866*. Manchester: Manchester University Press

—, ed. 1976. *The Infidel Tradition from Paine to Bradlaugh*. London: Macmillan

— 1979. "Taylor, Robert (1784–1844)." In *Biographical Dictionary of Modern British Radicals: Volume I: 1770–1830*. Eds Joseph O. Baylen and Norbert J. Gossman, 467–70. Hassocks: Harvester

— 1998. *Robert Owen and the Commencement of the Millennium. A Study of the Harmony Community*. Manchester: Manchester University Press

Rowley-Conwy, Peter. 2007. *From Genesis to Prehistory. The Archaeological Three Age System and its Contested Reception in Denmark, Britain, and Ireland*. Oxford: Oxford University Press, https://doi.org/10.1093/oso/9780199227747.001.0001

Rudwick, Martin J. S. 1985. *The Great Devonian Controversy: The Shaping of Scientific Knowledge among Gentlemanly Specialists*. Chicago: University of Chicago Press

— 2005. *Bursting the Limits of Time. The Reconstruction of Geohistory in the Age of Revolution*. Chicago: University of Chicago Press, https://doi.org/10.7208/chicago/9780226731148.001.0001

— 2008. *Worlds Before Adam. The Reconstruction of Geohistory in the Age of Reform*. Chicago: University of Chicago Press, https://doi.org/10.7208/chicago/9780226731308.003.0001

Rule, John. 1986. *The Labouring Classes in Early Industrial England, 1750–1850*. London: Longman

Rupke, Nicolaas A. 1994a. *Richard Owen Victorian Naturalist.* New Haven: Yale University Press

— 1994b. "C. C. Gillispie's *Genesis and Geology.*" *Isis* 85: 261–70

— 2005. "Neither Creation nor Evolution: The Third Way in Mid-Nineteenth Century Thinking about the Origin of Species." *Annals of the History and Philosophy of Biology* 10: 143–72

Ruse, Michael. 1996. *Monad to Man. The Concept of Progress in Evolutionary Biology.* Cambridge, MA: Harvard University Press

Russell, Lord John, ed., 1853. *Memorials and Correspondence of Charles James Fox.* Vol 1. London: Richard Bentley

[Saull William Devonshire.] 1826. "Letter From A Friend: On Fossil Exuviae and Planetary Motion." *Republican* 14:265–67

— 1828a. *A Letter to the Vicar of St. Botolph Without Aldersgate, London: By a Parishioner.* London: Printed for the Author

— 1828b. [Memorial Presented to the Common Council of the City of London]. *Lion* 1 (25 Jan. 1828):115–118

— 1828c. "Oath-Taking." *Examiner*, 9 Nov. 1828, 726–27

— 1829. ["At a Dissenters' chapel..."] *Times*, 26 Feb. 1829, 6

[—] 1832a. *A Letter from a Student in the Sciences to a Student of Theology.* London: Brooks

— 1832b. "Editor's Preface." In *Essay on the Physico-Astronomical Causes of the Geological Changes on the Earth's Surface, and of the Changes in Terrestrial Temperature, with Notes. By Sir Richard Phillips. Re-published, with a Preface, by William Devonshire Saull, F.G.S., F.A.S., F.R.A.S.* Ed. W. D. Saull, iii–viii. London: Sherwood, Gilbert, and Piper

— 1833a. "Lecture of Mr. W. D. Saull in Bristol." *Crisis* 3: 37–39

— 1833b. "Substance of the Discourse at Bristol." *Gauntlet* 1: 530–33

— 1833c. "Education of the People–Liberal Offer to Students of Geology." *MM* 19: 117–18

— 1836. *An Essay on the Coincidence of Astronomical & Geological Phenomena, Addressed to the Geological Society of France.* London: Printed for the Author

— 1837. "Lecture by Mr. Saull." *New Moral World* 3: 397–98

— 1838a. "Oration." *Penny Satirist* 17 Mar. 1838, 1

— 1838b. "Formation of Character." *New Moral World* 4: 278–80

— 1844. "Foundations of the Roman Walls of London." *Archaeologia* 30: 522–24

— 1845. *Notitia Britanniae; or An Enquiry Concerning the Localities, Habits, Condition, and Progressive Civilization of the Aborigines of Britain; to which is appended, A Brief Retrospect of the result of their Intercourse with the Romans.* London: John Russell Smith

— 1848. *Observations on the Aboriginal Tribes of Britain*. London: Effingham Wilson

— 1853. *An Essay on the Connexion Between Astronomical and Geological Phenomena, Addressed to the Geologists of Europe and America; Including a Paper read before the Geological Society of London, in February 1848, with Notes and Additions.* London: John Russell Smith

Saville, John. 1971. "J. E. Smith and the Owenite Movement, 1833–1834." In *Robert Owen Prophet of the Poor*. Eds Sidney Pollard and John Salt, 115–44. London: Macmillan

Scherren, Henry. 1905. *The Zoological Society of London: a Sketch of its Foundation and Development, and the Story of its Farm, Museum, Gardens, Menagerie and Library*. London: Cassell

Schwartz, Laura. 2013. *Infidel Feminism, Secularism, Religion and Women's Emancipation, England 1830–1914*. Manchester: Manchester University Press, https://doi.org/10.2307/j.ctt1vwmdmp

Scriven, Thomas. 2012. *Activism and the Everyday: The Practices of Radical Working-Class Politics, 1830–1842*. PhD Thesis, University of Manchester

Sebastiani, Silvia. 2022. "Monboddo's 'Ugly Tail': The Question of Evidence in Enlightenment Sciences of Man." *History of European ideas* 48: 45–65, https://doi.org/10.1080/01916599.2021.1950314

Secord, Anne. 1994. "Science in the Pub: Artisan Botanists in Early Nineteenth-Century Lancashire." *HS* 32: 269–315

— 1994. "Corresponding Interests: Artisans and Gentlemen in Nineteenth-Century Natural History." *BJHS* 27: 383–408

— 1996. "Artisan Botany." In *Cultures of Natural History*. Ed. N. Jardine, J. A. Secord and E. C. Spary, 378–93. Cambridge: Cambridge University Press

Secord, James A. 1982. "King of Siluria: Roderick Murchison and the Imperial Theme in Nineteenth-Century British Geology." *Victorian Studies* 25: 413–42

— 1986a. *Controversy in Victorian Geology: The Cambrian-Silurian Dispute*. Princeton: Princeton University Press

— 1986b. "The Geological Survey of Great Britain as a Research School." *HS* 24:223–75

— 1989. "Extraordinary Experiment: Electricity and the Creation of Life in Victorian England." In *The Uses of Experiment: Studies in the Natural Sciences*. Eds. David Gooding, Trevor Pinch, and Simon Schaffer, 337–83. Cambridge: Cambridge University Press

— 1991. "Edinburgh Lamarckians: Robert Jameson and Robert E. Grant." *JHB* 24:1–18

—, ed. 1997. *Charles Lyell Principles of Geology*. London: Penguin

— 2000. *Victorian Sensation: The Extraordinary Publication, Reception, and Secret Authorship of Vestiges of the Natural History of Creation*. Chicago: Chicago University Press, https://doi.org/10.7208/chicago/9780226158259.001.0001

— 2004a. "Monsters at the Crystal Palace." In *Models: The Third Dimension of Science*. Eds. Soraya de Chadarevian and Nick Hopwood, 138–69. Stanford: Stanford University Press, https://doi.org/10.1515/9781503618992

— 2004b. "Knowledge in Transit." *Isis* 95: 654–72, https://doi.org/10.1086/430657

— 2014. *Visions of Science. Books and Readers at the Dawn of the Victorian Age*. Oxford: Oxford University Press, https://doi.org/10.7208/chicago/9780226203317.001.0001

— 2021. "Revolutions in the Head: Darwin, Malthus and Robert M. Young." *BJHS* 54:41–59, https://doi.org/10.1017/s0007087420000631

Sedgwick, Adam. 1833. *A Discourse on the Studies of the University*. Cambridge: Pitt Press

Sera-Shriar, Efram. 2016. *The Making of British Anthropology 1813–1871*. Abingdon: Routledge, https://doi.org/10.4324/9781315654706

Shapin, Steven. 1980. "Social Uses of Science." In *The Ferment of Knowledge: Studies in the Historiography of Eighteenth-Century Science*. Ed. G. S. Rousseau and R. Porter, 93–139. Cambridge: Cambridge University Press

— 1983. "'Nibbling at the Teats of Science': Edinburgh and the Diffusion of Science in the 1830s." In *Metropolis and Province: Science in British Culture, 1780–1850*. Ed. I. Inkster and J. Morrell, 151–78. London: Hutchinson

— 1990. "Science and the Public." In *Companion to the History of Modern Science*. Eds. R. C. Olby, G. N. Cantor, J. R. R. Christie, and M. J. S. Hodge, 990–1007. London: Routledge

— 1994. *A Social History of Truth. Civility and Science in Seventeenth-Century England*. Chicago: Chicago University Press

—, and Barry Barnes. 1976. "Head and Hand: Rhetorical Resources in British Pedagogical Writing, 1770-1850." *Oxford Review of Education* 2: 231–54

—, and Barry Barnes. 1977. "Science, Nature and Control: Interpreting Mechanics' Institutes." *Social Studies of Science* 7: 31–74

Sharpe, Daniel. 1853. *Description of the Fossil Remains of Mollusca found in the Chalk of England. Part 1. Cephalopoda*. London: Palaeontographical Society

Shepherd, Thomas H. 1827. *Metropolitan Improvements of London in the Nineteenth Century*. London: Jones

Sheets-Pyenson, Susan. 1985. "Popular Science Periodicals in Paris and London: The Emergence of a Low Scientific Culture, 1820–1875." *AS* 42: 549–72

— 1988. "How to 'Grow' a Natural History Museum: the Building of Colonial Collections, 1850–1900." *ANH* 15: 121–47

Sherborn, Charles Davies. 1940. *Where is the — Collection?* Cambridge: Cambridge University Press

Shortland, Michael, and Richard Yeo, eds. 1996. *Telling Lives in Science. Essays on Scientific Biography*. Cambridge: Cambridge University Press

Shortt, W. T. P. n.d. *Collectania Curiosa Antiqua Dunmonia; or, An Essay on some Druidical Remains in Devon*. Exeter: Featherstone

Silber, Kate. 1965. *Pestalozzi. The Man and his Work*. London: Routledge and Kegan Paul

Simon, Brian. 1960. *Studies in the History of Education 1780–1870*. London: Lawrence & Wishart

Simons, John. 2012. *The Tiger that Swallowed the Boy. Exotic Animals in Victorian England*. Faringdon: Libri

Sloan, Phillip R. 1997. "Lamarck in Britain: Transforming Lamarck's Transformism." In *Jean-Baptiste Lamarck 1744–1829*. Ed. Goulven Laurent, 667–687. Paris: Comité des Travaux Historiques et Scientifiques

Smith, B. and C. Smith. 1994. "Martin Leo Ehrmann (1904–1972)." *Mineralogical Record* 25: 347–70

Smith, Charles Roach. 1848. *Collectanea Antiqua, Etchings and Notices of Ancient Remains, Illustrative of the Habits, Customs, and History of Past Ages*. Vol. 1. London: J. R. Smith

— 1854. *Catalogue of the Museum of London Antiquities Collected by, and the Property of, Charles Roach Smith*. Privately Printed

— 2015 [1866]. *Retrospections, Social and Archaeological*. 2 vols. Cambridge: Cambridge University Press, https://doi.org/10.1017/cbo9781316155738.004

Smith, James Elishama. 1833. *The Antichrist, or, Christianity Reformed*. London: Cousins

— 1853. *Mercury's Letters on Science*. London: Cousins

— 1873 [1848]. *The Coming Man*. 2 vols. London: Strahan

Smith, John Pye. 1839. *On the Relation Between the Holy Scriptures and Some Parts of Geological Science*. London: Jackson and Walford

Smith, John Thomas. 1839. *The Cries of London: Exhibiting Several of the Itinerant Traders of Antient and Modern Times*. London: Nichols

Smith, Roger. 1972. "Alfred Russel Wallace: Philosophy of Nature and Man." *BJHS* 6: 175–99

Smith, W. Anderson. 1892. *'Shepherd' Smith the Universalist*. London: Simpson Low, Marston

Solomonescu, Yasmin. 2014. *John Thelwall and the Materialist Imagination*. Basingstoke: Palgrave Macmillan, https://doi.org/10.1057/9781137426147

[Somerville, Alexander]. 1848. *The Autobiography of a Working Man*. London: Gilpin.

Sopwith, Thomas. 1843. *Account of the Museum of Economic Geology, and Mining Records Office*. London: Murray

Southwell, Charles. [1840]. *An Essay on Marriage; Addressed to the Lord Bishop of Exeter*. London: Roe

— 1841–42. "Theory of Regular Gradation." *Oracle of Reason* 1 ([6 Nov. 1841]–8 Jan. 1842), passim

— 1842. *The Trial of Charles Southwell, (Editor of "The Oracle of Reason") for Blasphemy*. London: Hetherington

— 1850. *The Confessions of a Free-Thinker*. London: Southwell

Spary, Emma. 2000. *Utopia's Garden: French Natural History; from Old Regime to Revolution*. Chicago: University of Chicago Press, https://doi.org/10.7208/chicago/9780226768700.001.0001

Spath, L. F. 1934. *Catalogue of the Fossil Cephalopoda in the British Museum (Natural History). Part IV. The Ammonoidea of the Trias*. London: British Museum (Natural History)

Spencer, Howard. 2009. "Atkins, John (c.1754–1838), of Halstead Place, nr. Sevenoaks, Kent", http://www.historyofparliamentonline.org/volume/1820-1832/member/atkins-john-1754-1838

Sprigge, S. Squire. 1897. *The Life and Times of Thomas Wakley*. London: Longmans Green

Stafford, Robert A. 1989. *Scientist of Empire: Sir Roderick Murchison, Scientific Exploration and Victorian Imperialism*. Cambridge: Cambridge University Press

Stenhouse, John. 2005. "Imperialism, Atheism, and Race: Charles Southwell, Old Corruption, and the Maori." *Journal of British Studies* 44:754–74, https://doi.org/10.1086/431940

Stocking, George W. 1971. "What's in a Name? The Origins of the Royal Anthropological Institute (1837–71)." *Man* 6: 369–90

— 1987. *Victorian Anthropology*. New York: Free Press

Stack, David. 1999. "William Lovett and the National Association for the Political and Social Improvement of the People." *Historical Journal* 42: 1027–50

Stange, Douglas Charles. 1984. *British Unitarians against American Slavery, 1833–65*. Rutherford: Associated University Presses

Stott, Rebecca. 2012. *Darwin's Ghosts. In Search of the First Evolutionists*. London: Bloomsbury

Strasser, Bruno J. 2012. "Collecting Nature: Practices, Styles, and Narratives." *Osiris* 27: 303–40, https://doi.org/10.1086/667832

"Student in Realities". 1836–37. *Serious Thoughts, Generated by Perusing Lord Brougham's Discourse of Natural Theology*. 4 pts. London: Brooks

Swenson, Astrid. 2013. *The Rise of Heritage. Preserving the Past in France, Germany and England, 1789–1914*. Cambridge: Cambridge University Press, https://doi.org/10.1017/cbo9781139026574

Swinney, Geoffrey N. 2010. "Robert Jameson (1774–1854) and the Concept of a Public Museum." *ANH* 37: 235–45, https://doi.org/10.3366/anh.2010.0006

Tanghe, Koen B., and Mike Kestemont. 2018. "Edinburgh and the Birth of British Evolutionism: A Peek behind a Veil of Anonymity." *BioScience* 68:585–92, https://doi.org/10.1093/biosci/biy049

Taquet, Philippe, and Kevin Padian. 2004. "The Earliest Known Restoration of a Pterosaur and the Philosophical Origins of Cuvier's *Ossemens Fossiles*." *Comptes Rendus Palevol* 3:157–7, https://doi.org/10.1016/j.crpv.2004.02.002

Taylor, Anthony. 2003. "Radical Funerals, Burial Customs and Political Commemoration: The Death and Posthumous Life of Ernest Jones." *Humanities Research* 10: 29–39, https://doi.org/10.22459/hr.x.02.2003.04

Taylor, Barbara. 1983. *Eve and the New Jerusalem: Socialism and Feminism in the Nineteenth Century*. London: Virago

Taylor, Charles. 2007. *A Secular Age*. Cambridge, MA: Harvard University Press, https://doi.org/10.2307/j.ctvxrpz54

Taylor, Michael A. 1989. "Thomas Hawkins FGS, 22 July 1810–15 October 1889." *Geological Curator* 5: 112–14

— 1994. "The Plesiosaur's Birthplace: the Bristol Institution and its Contribution to Vertebrate Palaeontology." *Zoological Journal of the Linnean Society* 112: 179–196

— 2016. "'Where is the Damned Collection?' Charles Davies Sherborn's Listing of Named Natural Science Collections and its Successors." *ZooKeys* 550: 83–106, https://doi.org/10.3897/zookeys.550.10073

— and Hugh S. Torrens. 1986. "Saleswoman to a New Science: Mary Anning and the Fossil Fish *Squalaraja* from the Lias of Lyme Regis." *Proceedings of the Dorset Natural History and Archaeological Society* 108: 135–48

— and L. I. Anderson. 2017. "The Museums of a Local, National and Supranational Hero: Hugh Miller's Collections over the decades." *Geological Curator* 10: 285–368, https://doi.org/10.55468/gc242

Taylor, Robert. 1828a. *Trial of the Reverend Robert Taylor...Upon a Charge of Blasphemy*. London: Carlile

— 1828b. *Syntagma of the Evidences of the Christian Religion*. Boston: Mendum

— 1831. *The Devil's Pulpit.* Nos. 1–23. London: [Carlile]

— [Talasiphron pseud.] 1833. *The Philalethean.* London: Benbow

Temkin, O. 1977. "Basic Science, Medicine, and the Romantic Era." In *The Double Face of Janus and Other Essays in the History of Medicine.* Ed. O. Temkin, 345–72. Baltimore: Johns Hopkins University Press

Tennant, James. 1858. *Catalogue of Fossils, Found in the British Isles, Forming the Private Collection of James Tennant.* London: Tennant

Thelwall, Mrs [Cecil Boyle]. 1837. *The Life of John Thelwall.* 2 vols. London: Macrone

Thiemeyer, Thomas. 2015. "Work, Specimen, Witness: How Different Perspectives on Museum Objects Alter the Way They Are Perceived and the Values Attributed to Them," *Museum and Society* 13: 396–412, https://doi.org/10.29311/mas.v13i3.338

Thompson, Dorothy. 1984. *The Chartists. Popular Politics in the Industrial Revolution.* New York: Pantheon

Thompson, E. P. 1980. *The Making of the English Working Class.* London: Gollancz

Thompson, T. Perronet. 1843. *Exercises, Political and Others.* 2d ed. 6 vols. London: Wilson

Thompson, William. 1824. *An Inquiry into the Principles of the Distribution of Wealth.* London: Longman, Hurst, Rees, Orme, Brown, and Green

— 1826a. "To the Members and Managers of the Mechanics' Institutions in Britain and Ireland". *Co-Operative Magazine and Monthly Herald* 1: 22–27, 43–48

—, 1826b. "Physical Argument for the *Equal* Cultivation of all the Useful Faculties or Capabilities, Mental or Physical, of Men and Women." *Co-Operative Magazine and Monthly Herald* 1: 250–58. Farnham: Ashgate

Timbs, John. 1840. *The Literary World. A Journal of Popular Information and Entertainment.* Vol. 3. London: Timbs

— 1855. *Curiosities of London.* London, Bogue

— 1866. *Club Life of London, with Anecdotes of the Clubs, Coffee-houses and Taverns of the Metropolis during the 17th, 18th and 19th Centuries.* 2 vols. London: Bentley

Topham, Jonathan. 1992. "Science and Popular Education in the 1830s: the Role of the *Bridgewater Treatises.*" *BJHS* 25:397–430

— 1998. "Beyond the 'Common Context'. The Production and Reading of the Bridgewater Treatises." *Isis* 89: 233–62

— 2005. "John Limbird, Thomas Byerley, and the Production of Cheap Periodicals in the 1820s." *Book History* 8: 75–106, https://doi.org/10.1353/bh.2005.0012

— 2007. "Publishing 'Popular Science' in Early Nineteenth-Century Britain." In *Science in the Marketplace. Nineteenth-Century Sites and Experiences*. Eds Aileen Fyfe and Bernard Lightman, 135–60. Chicago: Chicago University Press, https://doi.org/10.7208/chicago/9780226150024.001.0001

— 2009a. "Introduction." *Isis* 100: 310–18, https://doi.org/10.1086/599551

— 2009b. "Rethinking the History of Science Popularization/Popular Science." In *Popularizing Science and Technology in the European Periphery, 1800–2000*. Eds. Faidra Papanelopoulou, Agustí Nieto-Galan, and Enrique Perdigeuro, 1–20, https://doi.org/10.4324/9781315601472

— 2022. *Reading the Book of Nature. How Eight Best Sellers Reconnected Christianity and the Sciences on the Eve of the Victorian Age*. Chicago: Chicago University Press, https://doi.org/10.7208/chicago/9780226820804.003.0001

Torrens, Hugh. 1992. "When did the Dinosaur get its Name?" *New Scientist* 134: 40–44

— 1995. "Mary Anning (1799–1847) of Lyme; 'The Greatest Fossilist the World ever Knew'." *BJHS* 28: 257–84

— 1997. "Politics and Paleontology: Richard Owen and the Invention of Dinosaurs." In *The Complete Dinosaur*. Eds. James O. Farlow and M. K. Brett-Surman, 175–90. Bloomington: Indiana University Press

— 1998. "Geology and the Natural Sciences: Some Contributions to Archaeology in Britain 1780–1850." In *The Study of the Past in the Victorian Age*. Ed. Vanessa Brand, 35–60. Oxford: Oxbow

— 2000. "New light on William Maclure's Early Geology and Travels." *Abstracts with Programs - Geological Society of America* 32: 87

— 2014. "The Isle of Wight and its Crucial Role in the 'Invention' of Dinosaurs." *Biological Journal of the Linnean Society* 113: 664–76, https://doi.org/10.1111/bij.12341

Tosh, John. 1999. *A Man's Place. Masculinity and the Middle-Class Home in Victorian England*. New Haven: Yale University Press

Toulmin, George Hoggart. 1790. *The Eternity of the Universe*. London: Johnson

— 1854 [1824]. *Antiquity and Duration of the World*. Boston: Mendum

Tresise, Geoffrey R. 1989. "Chirotherium - The First Finds at Storeton quarry, Cheshire, and the role of the Liverpool Natural History Society." *Geological Curator* 5:135–51

— 2003. "Chirotherium and the Quarry Men: The 1838 Discoveries at Storeton Quarry, Cheshire, U.K." *Ichnos* 10: 77–90, https://doi.org/10.1080/10420940390257897

Treuherz, Nick. 2016. "The Diffusion and Impact of Baron d'Holbach's Texts in Great Britain, 1765–1800." In *Radical Voices, Radical Ways: Articulating and Disseminating Radicalism in Seventeenth- and Eighteenth-Century Britain*. Eds

Laurent Curelly and Nigel Smith, 125–48. Manchester: Manchester University Press, https://doi.org/10.7228/manchester/9781526106193.001.0001

Tristan, Flora. 1980. *Flora Tristan's London Journal. A Survey of London Life in the 1830s*. Trans. Dennis Palmer and Giselle Pincetl. London: Prior

Tromp, Marlene, ed. 2008. *Victorian Freaks. The Social Context of Freakery in Britain*. Columbus: Ohio State University Press

Turner, Colin. 1980. *Politics in Mechanics' Institutes, 1820–1830: A Study in Conflict*. PhD Thesis, Leicester University

Turner, F. M. 1978. "The Victorian Conflict between Science and Religion: A Professional Dimension." *Isis* 69: 356–76

Tylecote, Mabel. 1957. *The Mechanics' Institutes of Lancashire and Yorkshire Before 1851*. Manchester: Manchester University Press

Underhill, Paul. 1993. "Alternative Views of Science in Intra-Professional Conflict: General Practitioners and the Medical and Surgical Elite 1815–58." *Journal of Historical Sociology* 5:322–50

Urry, Amelia. 2021. "Alfred Newton's Second-hand Histories of Extinction: Hearsay, Gossip, Misapprehension." *ANH* 48: 244–62, https://doi.org/10.3366/anh.2021.0720

Vadillo, Mónica A. Walker. 2013. "Apes in Medieval Art", http://mad.hypotheses.org/172

Vallance, Edward. 2016. "'The Insane Enthusiasm of the Time': Remembering the Regicides in Eighteenth- and Nineteenth-Century Britain and North America." In *Radical Voices, Radical Ways: Articulating and Disseminating Radicalism in Seventeenth- and Eighteenth-Century Britain*. Eds Laurent Curelly and Nigel Smith, 229–50. Manchester: Manchester University Press, https://doi.org/10.7228/manchester/9781526106193.001.0001

van Wyhe, John, and Peter C. Kjærgaard. 2015. "Going the Whole Orang: Darwin, Wallace and the Natural History of Orangutans." *Studies in History and Philosophy of Biological and Biomedical Sciences* 30: 1–11, https://doi.org/10.1016/j.shpsc.2015.02.006

Ville, Simon, Claire Wright and Jude Philp. 2020. "Macleay's Choice: Transacting the Natural History Trade in the Nineteenth Century." *JHB* 53:345–75, https://doi.org/10.1007/s10739-020-09610-9

Vizetelly, Henry. 1893. *Glances Back Through Seventy Years. Autobiographical and Other Reminiscences*. 2 vols. London: Kegan Paul

Volney, C. F. 1819. *The Ruins: or a Survey of the Revolutions of Empires*. London: Davison

Waddington, I. 1984. *The Medical Profession in the Industrial Revolution*. Dublin: Gill and Macmillan

Wait, D. G. 1811. *A Defence of a Critique on the Hebrew Word "Nachash,"...in which it is Proved, from the Hebrew Text, and the Oriental Languages, that a Serpent, not an Ape, deceived Eve*. London: Valpy

Walford, Edward. 1883. *Greater London: a Narrative of its History, its People, and its Places*. 2 vols. London: Cassell

Walker, Martyn. 2013. "'For the Last Many Years in England Everybody has been Educating the People, but they have Forgotten to find them any Books': The Mechanics' Institutes Library Movement and its Contribution to Working-Class Adult Education during the Nineteenth Century." *Library and Information History* 29: 272–86, https://doi.org/10.1179/1758348913z.00000000048

Wallace, Alfred Russel. 1905. *My Life. A Record of Events and Opinions*. 2 vols. London: Chapman & Hall

Wallace, Mark Coleman. 2007. *Scottish Freemasonry 1725–1810: Progress, Power, and Politics*. PhD Thesis, University of St Andrews

Wallas, Graham. 1918. *The Life of Francis Place, 1771–1854*. London: Allen

Wallbank, Adrian J. 2012. *Dialogue, Didacticism and the Genres of Dispute: Literary Dialogues in an Age of Revolution*. London: Routledge, https://doi.org/10.4324/9781315655239

Walters, Henry Beauchamp. 1908. *Catalogue of the Roman Pottery in the Departments of Antiquities, British Museum*. London: British Museum

Warwick, Alexandra. 2017. "Ruined Paradise: Geology and the Emergence of Archaeology." *Nineteenth-Century Contexts* 29: 49–62, https://doi.org/10.1080/08905495.2017.1252485

[Watts, William.] 1830. *The Yahoo; A Satirical Rhapsody*. New York: Simpson

Weale, John. 1854. *The Pictorial Handbook of London*. London: Bohn

Webb, Sidney, and Beatrice Webb. 1920. *The History of Trade Unionism*. Rev. ed. London: Longmans, Green

Weinstein, Benjamin. 2011. *Liberalism and Local Government in Early Victorian London*. Woodbridge: Boydell Press, https://doi.org/10.1017/upo9781846158490.008

Weisser, Henry. 1975. *British Working-Class Movements in Europe 1815–48*. Manchester: Manchester University Press

Welch, Charles. 1896. *Modern History of the City of London; a Record of Municipal and Social Progress, from 1760 to the Present Day*. London: Blades, East & Blades

Wennerbom, Alan John. 1999. *Charles Lyell and Gideon Mantell, 1821–1852: Their Quest for Elite Status in English Geology*. PhD Thesis, University of Sydney

Wheeler, Alwyne. 1997. "Zoological Collections in the Early British Museum: the Zoological Society's Museum." *ANH* 24: 89–126

[Whewell, William.] 1832. "Lyell's Geology." *Quarterly Review* 47: 103–32

White, George W. 1977. "William Maclure's Maps of the Geology of the United States." *JSBNH* 8: 266–69

White, Jerry. 2007. *London in the Nineteenth Century. 'A Human Awful Wonder of God'*. London: Cape

Wickwar, William H. 1928. *The Struggle for the Freedom of the Press 1819–1832*. London: Allen and Unwin

Wiener, Joel H. 1969. *The War of the Unstamped. The Movement to Repeal the British Newspaper Tax, 1830–1836*. Ithaca: Cornell University Press

— 1970. *A Descriptive Finding List of Unstamped British Periodicals 1830–1838*. London: The Bibliographical Society

— 1979. "Julian Hibbert." In *Biographical Dictionary of Modern British Radicals*. Eds. J. O. Baylen and H. J. Gossman, 1: 221–22. Hassocks: Harvester

— 1983. *Radicalism and Freethought in Nineteenth-Century Britain. The Life of Richard Carlile*. Westport: Greenwood

— 1989. *William Lovett*. Manchester: Manchester University Press

Williams, Eric. 1994. *Capitalism and Slavery*. Chapel Hill: University of North Carolina Press

Williams, John. [Publicola pseud.] 1840. *Life of Publicola. First Series*. London: Cunningham

Williams, Gwyn A. 1965. *Rowland Detrosier. A Working-Class Infidel 1800–34*. York: St Anthony's Press

Williams, R. B., and Hugh S. Torrens. 2016a. "No. 3 Highbury Grove, Islington: the private geological museum of James Scott Bowerbank (1797–1877)." *ANH* 43:278–84, https://doi.org/10.3366/anh.2016.0383

— 2016b. "A History of the Fossil Fruits and Seeds of the London Clay (1840): A Historical and Bibliographical Account of James Scott Bowerbank's Unfinished Monograph." *ANH* 43:255–77, https://doi.org/10.3366/anh.2016.0382

Wilson, George, and Archibald Geikie. 1861. *Memoir of Edward Forbes, F. R. S.* Cambridge and London: Macmillan

Wilson, Leonard G. 1996. "Brixham Cave and Sir Charles Lyell's ... *Antiquity of Man*: the Roots of Hugh Falconer's Attack on Lyell." *ANH* 23: 79–97

Winchell, N. H. 1890. "A Sketch of Richard Owen." *American Geologist* 6: 135–45

Winter, Alison. 1998. *Mesmerized. Powers of Mind in Victorian Britain*. Chicago: University of Chicago Press

Woodward, Arthur Smith. 1885. "On the Literature and Nomenclature of British Fossil Crocodilia." *Geological Magazine* NS Dec. 3, 2: 496–510

— 1901. *Catalogue of Fossil Fishes in the British Museum (Natural History)*. 4 vols. London: British Museum (Natural History)

Woodward, Horace B. 1908. *The History of the Geological Society of London*. London: Longman Green

Wright, G. N. 1837. *The Life and Reign of William the Fourth*. 2 vols. London: Fisher

Wright, Thomas, ed. 1845. *The Archaeological Album; Or, Museum of National Antiquities*. London: Chapman and Hall

Yanni, Carla. 1999. *Nature's Museums. Victorian Science and the Architecture of Display*. London: Athlone

Yeo, Eileen. 1971. "Robert Owen and Radical Culture." In *Robert Owen: Prophet of the Poor*. Eds Sidney Pollard and John Salt, 88–103. London: Macmillan

Yolton, John W. 1983. *Thinking Matter. Materialism in Eighteenth-Century Britain*. London: Blackwell

Youatt, William. 1836. "Contributions to Comparative Pathology. No. V." *Veterinarian* 9: 271–82

Young, G. M. 1960. *Portrait of an Age*. 2nd ed. Oxford: Oxford University Press

Zimmerman, Virginia. 2008. *Excavating Victorians*. Albany: State University of New York. https://doi.org/10.1086/597725

Index

abattoir 305
Abbeville 461
Abbott, Charles 118–119
aborigines 384, 419, 429–430, 435, 443, 459–461, 464
Aborigines' Protection Society 429, 459
Abrahams, Aleck 23
Adam, Admiral 333
Address to Men of Science (Carlile) 64, 90
Admiral Rodney pub 60, 348, 409
admission fees and policies 185–188, 241, 517
Aesthetic Institution 477
Agassiz, Louis 39
Age 20, 318–319, 344, 394
Age of Reason (Paine) 78, 80, 304, 324, 494
Age of Reptiles 33, 57, 202, 473, 574
Agitator, and Political Anatomist 162, 289
agnosticism 323, 549–550
agrarian radicalism 53, 81, 163, 425, 427, 492–493
Agricultural Employment Institution 426
Akerman, John Yonge 430, 445
Albion 20, 264–265
Albion Chapel 166, 567
Albion Hall 166, 176, 299, 567
Alcock, T. 316–317
Alcott, A. Bronson 305
Alcott House 326, 477
aldermen 65–66, 119–121, 125, 127, 192, 314, 318–319
Aldersgate 1, 16, 31, 34, 36–37, 40, 59–60, 64, 129, 179, 316
Alphabet of Geology 398

Alton Locke (Kingsley) 478
Amherst College 518
ammonites 323, 347, 354, 427, 432, 470–471, 518
amphorae 420, 422, 424, 427, 446
An Account of the Regular Gradation in Man (White) 367
Anatomy Bill 192
anatomy museums 35
anatomy schools 67, 192, 318, 352, 448
Ancient and Honourable Lumber Troop 314
Animal and Vegetable Physiology (Roget) 380
animate, living, atoms 105–106, 111–112, 205–206
Anning, Mary 41–42, 182
Anthropological Society of London 464
anthropology 441
anti-Catholic, oath 52, 121, 135, 254, 316, 451, 483
Antichrist (J. E. Smith) 101, 227, 229, 556–557
anti-corn-law 297, 359, 408, 512
Anti-Corn-Law League 72, 298
anti-cruelty petition 304
anti-gravitation 38, 152, 155, 189, 274, 523, 550, 558
anti-Malthusian 163 9, 163, 169, 193, 296, 346
Anti-Persecution Union 388, 579
Antiquités Celtiques et Antédiluviennes (Perthes) 461
Antiquity and Duration of the World (Toulmin) 88, 90, 92, 257
anti-semitism 64, 78, 85, 310, 367, 392–393
anti-smoking 454

ape/monkey, as human ancestor 1, 3–6, 13–14, 21, 38–39, 49, 52, 55–57, 199, 211, 213–215, 217–220, 223–224, 227, 232–237, 240–241, 243, 246, 267, 271, 283, 285, 288, 290–291, 346–347, 351, 365, 371–372, 396, 418, 441, 460, 466, 475, 477, 479, 481, 483–484, 518, 541, 551, 557, 559, 577
apes 241–244
Apollonic Society 521
Apology for the Bible (Watson) 64, 128
Appeal of One Half the Human Race (Thompson) 171
Arcana of Science and Art 97
Archaeological Institute of Great Britain 436
Archbishop of Canterbury 53, 145, 315, 377, 452
Archbishop of Paris 315
Archbishop of York 180, 436
Archives of Natural History 35
Areopagus 117, 120–121, 124–127, 170, 555, 566
Argus 20, 344, 389
Artisans and Politics (Prothero) 18
Ashmolean museum 420
Ashton 385
Ashurst, William Henry 317, 403, 408–409, 534, 536, 538, 582
Ashurst, William jnr 408, 534, 538, 582
Assistance, H.M.S. 528
Association for Removing the Causes of Ignorance 164
Association of All Classes of All Nations 321, 337, 357
astrology 38, 142–144, 229, 249–250, 477, 558
Astronomical Society of London 68, 148, 454
astronomy 8, 23, 25, 38, 48, 64, 73, 84, 88, 90, 101, 107, 128, 134, 139, 142, 144, 147–148, 152–154, 189–190, 194, 202, 227–230, 248, 250, 265, 273, 277, 296, 339, 341, 346, 350, 387, 426, 457, 478, 524–526, 550, 553–554, 558–559, 566–568, 570, 576
atheism 2, 4–6, 14, 19, 22–23, 29, 37, 49, 53–54, 59, 64, 68–69, 82, 85–86, 90, 104–105, 110–111, 129, 203–204, 218, 225, 230, 234–236, 238, 245–246, 261–264, 272, 286–287, 290, 292, 305, 319, 325–326, 352, 355, 360, 362–363, 366, 368–371, 376, 383, 387–391, 393–397, 405–406, 413, 438, 475, 477, 484, 487, 489, 494, 497, 500, 504, 509–510, 534–535, 548–550, 571, 577–579, 626
Atheist and Republican 363
Atheistical Society 413
Athenaeum 435, 541
Athenaeum, Saull's Cripplegate Chapel group 133–134, 566
Atheon 504
Atkins, John 121, 125–126
Atkinson, Henry 263
Atlas 20, 57, 213
Attorney-General 78, 118, 523
Augero, F. A. 127, 130
autochthons 110, 210, 460, 523

Babbage, Charles 468
Baber, James 40, 471
baboon, in dog pits 240–241
Babylonians 88, 142, 153
Bakewell, Robert 262
Ball, G. M. 81
Ballot Society 512
Balmoral Castle 377
Bamford, Samuel 123
Bank of England 124, 227
bankruptcy 62–63, 348, 402, 512
Banksian Society 102
Baptists 65, 263, 271, 351, 403, 411, 510, 578
Barmby, Goodwyn 346, 439
Barnes, Barry 25
barrows 418, 423, 431, 435, 445, 458, 462, 465, 531

Bartholomew Fair 239
Barton, F. B. 440
basal granite rocks 249, 296, 524
Baume, Charlotte 135, 137, 486
Baume, Julian Hibbert 137
Baume, Pierre 135–137, 151, 169–170, 243, 254, 258, 306, 326, 486–487, 566–567
Baylee, Joseph 327–328
Beagle, H.M.S. 21, 113
Bean, William 40, 184–185, 188
beards 232, 326, 360–361
Beaumont, Thomas Barber 504–506, 520
Beche, Henry De la 180, 187, 372, 377, 525
Beethoven, L. van 520
belemnites 471
Belfast 513
Belfast News-Letter 214
Bell, John 216, 291
Bellona, goddess 388, 445
Belzoni, Giovanni 143
Benbow, William 4, 98, 100, 123, 197, 239, 259–260, 568
Ben Jonson pub 148
Bennett, Tony 32
Bentham, Jeremy 272
Berard, Chevalier de 287
Berkhampstead, Castle 457
Bermondsey Mutual Instruction Society 571
Bernissartia, crocodile 552
Best, Lord Chief Justice 561
Best, William Draper 129
Bethnal Green 509–510, 547, 567, 570, 575
Bible and the People 480, 518
Bible Defender 480, 500
Bible of Reason (Powell) 94
Big Bone Lick 279, 469
Biographie Universelle (Michaud) 251
Birkbeck, George 64, 113, 179, 220
Birmingham 93, 245, 274, 286, 327, 329–331, 337, 345, 357, 361–363, 371, 386, 388, 406, 410, 414, 577–579
Birmingham Mechanics' Institute 327, 330, 579
Birth of the Museum (Bennett) 32
Bishop of Exeter 330, 343, 358
Bishop of London 53, 522
Bishop of Oxford 451
Bishop's Court 357
Blake, William 546
Blasphemer 363
blasphemy 4–9, 13–16, 18–19, 22–23, 26, 29, 35, 38, 43, 45, 53–55, 59, 66, 72–73, 77–79, 81–82, 84, 93, 95–97, 100–101, 105, 115, 117–123, 125, 127–129, 136, 138, 142, 167, 176, 183–184, 194, 203, 206, 217–219, 223, 225–226, 228–229, 238, 244, 246, 257, 264, 269, 285, 294, 303–304, 312, 315, 318, 341, 343–344, 350–352, 364, 368, 376, 386, 388, 390–392, 394–395, 408, 410, 429, 431, 467, 504, 508, 521, 532–533, 535, 544, 549, 555, 561, 565, 574, 582
blasphemy chapels 5, 7–8, 18, 93, 257, 535, 555
Blond, Robert le 510
blood sports 304
Bloomsbury Institution 568
bodies donated to science 137, 352, 486, 568
Boitard, Pierre 371
Bonner's Fields 509–510
Bonney, T. G. 549–551
book clubs 47 25, 47, 257
Book of the Bastiles (Baxter) 295
Booth, Abraham 178, 354, 421
Borough Chapel 225–230, 267, 567
Bory de St Vincent, J.-B. 111
Botanic Garden (E. Darwin) 114
Boué, Ami 248, 526
Boulogne 331
Bourbons 287
bourgeois influx into learned bodies 443, 447, 458

Bowerbank, Caroline 471
Bowerbankia 474
Bowerbank, James Scott 40, 68, 452, 471–476, 483, 517
Bowling Square Chapel 567, 575
Bow Street Police Station 80, 100, 167, 311, 313, 393
Boyle Lectures 550
Bracklesham Bay 42, 550
Bradford 112, 204, 326
Bradlaugh, Charles 509
Bridewell and Bethlem hospitals 100
Bridgewater, Earl of 53, 374
Bridgewater Treatises 53–54, 322, 380
Bright, John 512
Brighton 40, 179, 469
Brighton Herald 44
Brindley, John 57, 302
Bristol 15, 176, 201–202, 207, 209, 212–213, 215, 233, 238, 267–268, 271, 342, 357, 359, 361, 363, 368, 371, 385–386, 578
Bristol Institution 207
Bristol Job Nott 207–208
Bristol Mercury 362, 368, 577
British Archaeological Association 434, 443, 454, 456–458, 461, 476, 513
British Association for Promoting Co-operative Knowledge, BAPCK 48 26, 48, 163–164, 168, 176, 228, 426, 496
British Association for the Advancement of Science 164, 210, 281, 285, 288, 291, 329, 331, 378, 435, 463, 513, 524
British Banner 508
British Critic 573
British Legion 313
British Museum 5, 36, 40, 54–55, 108, 111, 182, 184, 186–187, 287, 377, 420, 434, 446–447, 452, 469, 471, 473, 504, 514, 545, 552
Brixham cave 543
Brodie, Antonio 547

Brongniart, Adolphe 36, 102, 248, 250–251, 323, 471
Brooke, John Hedley 165
Brookes's Club 113, 544
Brooks, John 126, 151, 165, 257, 487, 555–556, 581
Brougham, Henry 119, 128, 214, 291, 344, 349
Brownian motion 111
Brown, Robert 111
Brushfield, Thomas 126
bryozoan 474
Buchanan, Robert 341, 350, 357–358, 409, 576, 580
Buckingham Palace 43, 377
Buckland, William 53–54, 56–57, 108, 322, 365, 372, 374–376, 387, 435
Bugbrook 347
Bull Ring riot, Birmingham 329, 331
Bunhill Fields cemetery 253, 492
Bunyan, John 154
Burdett, Francis 581
Burke and Hare, resurrectionists 89
Burke, Luke 414, 440–442
Bursting the Limits of Time (Rudwick) 94
Burton Street/Crescent 62, 321–324, 477, 498, 535–536, 567, 574
Byerley, John 154
Byfield 347, 535
Byron, Lord 97–99, 102–103

Caesar 419, 432–433
Cain (Byron) 97–100
Calvert, John 546–547
Cambrian strata 375
Cambridge University 4–5, 14, 19, 37, 54, 82, 122, 169–170, 180, 214, 315, 364, 376, 410, 448, 454, 549
Camden Society 476
Campbell, Alexander 343, 350
Carbonari revolutionaries 68
Carlile, Jane 19, 80, 118, 160–161, 581
Carlile, Mary-Anne 80, 581

Carlile, Richard 163 13–14, 18–19, 27, 29, 32, 37, 64, 66, 72–73, 77–88, 90–113, 115, 118–119, 122–125, 127–128, 131–134, 137–140, 142, 144, 146–147, 151, 153, 159–163, 165, 176, 179, 183, 194, 197, 201, 205, 207, 217–218, 225, 228, 237–238, 245, 257–258, 260–261, 271–272, 274, 286, 304–306, 314, 340, 350–352, 360, 367, 373, 383, 389, 393, 406, 409, 423, 426, 430, 437, 486, 490, 507, 535, 538, 550, 553, 556, 568–569, 573, 575, 579–582
Caroline, Queen 378
Carpenter, W. B. 372
Castle, J. 62, 278
Castlereagh, Lord 139
Catholic emancipation 118, 135, 316, 451
Cato Street conspiracy 81, 118, 166, 426
cave fossils 97, 108, 210, 371, 417, 445, 543
Cecilian Society 166
Celt 251, 418–420, 423, 430, 433, 439, 444, 458–459, 461–463, 465, 476, 513, 519, 531, 538
Central Co-operative Agency 252, 511
Ceratites nodosus 323
Cerebral Physiology and Materialism (Engledue) 262
Cetiosaurus 33, 281, 286
Chaldeans 127, 142, 145
Chalmers, Thomas 108
Chambers' Edinburgh Journal 363, 527
Chambers, Robert 363, 372, 415
Champion Commercial Hotel 278
Chapel Royal 550
Chapman, John 412, 420, 481
Chappellsmith, Margaret 389, 569
Charles I 2, 65, 140, 258, 287, 447, 453
Charlotte Street 61, 199, 201–202, 224, 230, 233, 267, 270, 272, 301, 303, 413, 567, 575
Charter-Socialists 333

Chartism 33, 35, 37, 47–48, 66–67, 204, 243, 292, 298, 302–303, 307, 328–329, 331–334, 340, 348, 359, 362, 385–386, 413–414, 416, 427, 440, 465, 477–478, 481, 483–485, 489, 494, 496, 509, 512, 519, 525–527, 539, 566, 570–571, 576–577, 579
Chase, Malcolm 163
Cheap Magazine 95
Cheapside 423, 432, 465–466
Chelsea Social Institution 569
Cheltenham Mechanics' Institution 386
chemistry 48, 99, 134, 195, 202, 273, 339, 566, 570, 577
Chepstow 513, 570
Chester Assizes 313
Chichester 513
Chilton, William 49, 64, 359, 361–362, 370, 372–376, 379–384, 386, 395–396, 407, 411, 415, 437, 541–542, 577–578
chimpanzee 213, 241, 243, 285, 383
Chipping Warden 432
cholera 246, 260, 333, 493, 495, 497
Christ Church, Oxford 54
Christian Beacon 69
Christian Evidence Society 117, 122, 532, 566
Christian Israelites 224
Christianity as law of the land 19, 79, 118, 129, 138, 368, 393
Christian Lady's Magazine 390
Christian Mythology Unveiled (Mitchell) 205
Christian Observer 104, 108
Christian Socialism 413, 440, 510
Christian Times 69
Chronological Institute 454, 476, 513
Church and State Gazette 385
Church Examiner 315
Church Magazine 344
Church Missionary Society 348
Church of England Magazine 13, 57, 202, 268

church rates 19, 72, 126, 151, 315, 408, 556, 582
Church taxes, annual 19, 79, 142, 375, 379, 387
Cidaris diadema 470
circulating libraries 47, 263
City Chapel, Grub Street 5, 18, 132–134, 566
City Hall 423, 570
City of London Corporation Reform Association 317
City of London Institution 63
City of London Literary Institution 166, 537
City of London Mechanics' Institute 442, 570, 576, 581
City of London Mission 521
City of Westminster Radical Association 569
City Philosophical Society 533
Civil Engineer 188, 517
civil registration of marriage 451
Civil War, English 452–453
Clarke, Adam 61, 244–246
Clarke, John 245
Cleave, John 163 93, 151, 163, 177, 197, 217, 257, 290, 312, 315, 353, 581
Cleave's Penny Gazette 290, 581
clergy, London earnings 315, 374
clerks 166, 339, 349, 543
Cleveland Institution 539–545, 551, 578
coal fields, providence of 38, 178, 376
Cobbett, William 61, 118, 138, 149, 420, 445, 448, 571, 581
coffee houses 123, 139, 257–258, 305, 319, 394, 413–414, 566, 571
Coldbath-fields jail 258
Cole, Lord 39, 43, 287
College of Physicians 448
College of Surgeons 34, 99–100, 270, 281, 286, 448
Collyer, Bengo 124
Combe, George 328

Comet 79
Comité des Arts et Monuments 434
Common Council, City of London 65–66, 120–121, 254, 314, 316, 318, 561
Community Friendly Society, John Street 324
compositors 49 8, 19, 49, 64, 92, 257, 361–362, 382, 475, 542, 577–578
Confessions of a Free-Thinker (Southwell) 301, 367
Congregationalists 5, 124, 258, 507–508, 518
Congreve, Richard 542
Connard, George 579
Connection between Geology and the Pentateuch (T. Cooper) 262
Consolations in Travel (Davy) 218–219, 418
Constitution of England (Lolme) 179
Cook, James 466
Cooley, W. D. 431
Cooper, Astley 182
Co-operative Associations of Britain and Ireland 195
Co-Operative Building Society 307, 338, 498
co-operative congresses 194–196, 329–330, 337, 340, 357, 390, 401–402, 404, 428, 443, 467, 497, 556, 578
Co-operative League 510
co-operative schools 16, 61, 191, 202, 212, 270, 272–273, 325, 369, 419, 569, 573, 578
Cooper, Dr T. 262
Cooper, Robert 59–60, 368–369, 412, 440–441, 479–482, 510, 531, 535, 539, 569, 578
Cooper, Thomas 262, 440, 477–478, 496, 499, 520–523, 527, 532, 576–577
Cooper, Walter 429, 440, 563
Cooter, Roger 9, 26, 137, 262
copyright 97, 99–100
corals 40, 209, 249, 269, 524, 528, 543

Cornwall 426, 432–433, 464, 513–514
Corporation and Test Bill 118, 451
corporation reform 314, 317, 408
Corsi, Pietro 20, 110
Cosmopolite 105
Courier 330, 353
Court Gazette 279
Court of Chancery 100, 539
Court of King's Bench 119 117
Cousins, B, D. 93, 227, 289, 292
craniometry 423–424, 487
Crawford, William 314
Crimes of the Clergy (Benbow) 239
Crisis 8, 196, 199, 201, 205, 211, 215, 220, 224, 227, 233–235, 289–290, 306, 321, 430, 437, 478, 581
crocodiles 54, 89, 178, 286, 465, 471, 545, 552
Crocodilus Saullii 552
Crocodilus spenceri 545
cromlechs 425, 436, 462, 545
Crosse, Andrew 275
Cross's Menagerie 240
Crown and Anchor Tavern 47, 123, 193, 241, 310
Crown and Rolls Room 123
Cruikshank, G. 240
Crystal Palace 470, 515
Cull, Richard 487
cultural levels, pre-Roman 418, 432, 435, 437
Cultural Meaning of Popular Science (Cooter) 26
cultural relativism 324, 367, 419
Culture of English Geology (Knell) 18, 531
Curiosities of London (Timbs) 446, 538
Cuvier, Georges 43, 94–96, 98–99, 101–109, 153, 159, 173, 194, 285

Dante, A. 483
Danton, G. J. 395
Darwin, Charles 1–5, 14, 21, 39, 100, 113, 362, 382, 442, 454, 463, 541–543, 548, 552, 578

Darwin, Erasmus 112–114, 170, 218
Davenport, Allen 305, 426–427, 492–493, 495, 497, 533, 543, 567, 570–571
Daventry 432–433
Davison, Thomas 84, 91, 118
Davy, Humphry 218–220, 377, 418, 439, 449, 463
Dawson, George 414
Dean of Hereford 180
Dean of York 395
death-bed affirmations 412, 494, 497, 499, 532
death before Adam 103–104, 110, 202–203, 364
Defender 480–482, 494
Defensive Instructions for the People (Macerone) 259
Deist 80, 86, 112, 566
Democratic Review 113
Derby 513
design argument 85, 92, 101, 106, 112, 115, 207, 214, 264, 322, 380–381
Detrosier, Rowland 102, 264–265, 304, 555, 569
Devil's Pulpit (R. Taylor) 559
Dickens, Charles 216
Dictionary of National Biography 22, 548
Diderot, Denis 7
Diegesis (R. Taylor) 259, 556
Diluvium 417, 422
dinosaurs 2, 34, 36, 282, 285, 469–471, 519, 526, 531, 545, 551
discussions after lectures 46 46, 338, 349, 410, 430
disestablishment 15, 28, 38, 151, 180, 204, 208, 254, 313, 375, 409
dissection, human 137, 328, 339, 352, 486
Dissenters 19, 38, 71, 123, 126, 143, 315, 409, 448, 451
Dixon, Jacob 542
Dixon, Samuel 121, 126
Doctrine of the Deluge (Harcourt) 436

Dorchester Committee 309, 311, 313, 575
Dorchester gaol 78, 80, 92, 218, 561
Drift 417
Dr Johnson's House 314
Druids 419, 462
Dugdale, William 98, 392
Duke of Bedford 160 160
Duke of Newcastle 42
Duke of Sussex 455
Duncan, Jonathan 410
Duncombe, Thomas 297, 571
Dunham, Samuel Astley 435
Dunstable 435, 444, 457, 462
Dupuis, Charles François 144–145, 154, 559
Duthie, Sky 150

Eagles, Robin 176
Eamonson, James 93
Earl of Aberdeen 451
Earth on Show (O'Connor) 31
Eastern Institution 568
Eastern Social Institution 569, 575
East India Company 143, 450
East London Branch of the Association of Rational Socialists 201, 569
East London Democratic Association 427
East London Museum 547
economic depression 52, 297, 329–330, 337, 340, 371, 379–380, 385, 387–388
Edinburgh 20, 88–89, 93, 95, 203, 243, 248, 288, 327, 333, 369, 385, 578
Educational Friendly Society 325, 556, 562
education, children's 16, 26, 47, 52, 61–62, 148, 164, 168, 172–173, 183, 191, 202, 212, 230, 254, 270–273, 275, 277, 325–326, 328, 330, 332, 337, 341, 345, 369, 396, 398, 400–401, 404, 415, 419, 429, 503, 506–507, 514, 517, 540, 544, 571, 573, 576–578

Edwards, John 577
Edwards, Milne 111
Egerton, Philip 39, 43, 287
Egypt, Egyptian 109, 143–145, 367, 370, 446, 457, 467
Egyptian Hall 143, 467
Ehrmann, Martin 547
Electrical Theory of the Universe (Mackintosh) 274–275, 494, 575
electricity 25, 47, 152, 273–275, 322, 328, 339, 494, 575
Elementary Anatomy and Physiology (Lovett) 328
elephants, fossil 89, 210, 252, 284, 354, 471
elephants, fossil 283
Eliot, George 436
Elliotson, John 263, 328, 481, 495, 505
Emerson, R. W. 415
Engledue, William 262–263
Epps, John 4, 428, 439
Equality, poem 71, 227–228, 427, 485
Equitable Debtor and Creditor Association 512
Equitable Labour Exchange, bazaar 1, 16, 25–27, 37, 43, 66, 81, 133, 185, 196–198, 223, 230, 233–235, 258, 272, 289, 292, 299, 306, 324, 425–426, 437, 498, 511, 542
Era 188
Essay on the Coincidence of Astronomical & Geological Phenomena (Saull) 249
Essay on the Connexion Between Astronomical and Geological Phenomena (Saull) 457, 525
Essay on the Physico-Astronomical Causes of the Geological Changes on the Earth's Surface (Phillips ed. Saull) 154, 209, 558
eternalism, planetary, human 14, 30, 73, 85–88, 90, 93–95, 101, 114–115, 159, 172, 206
Ethnological Journal 441–442
Ethnological Society of London 459–460, 463, 487, 513

ethnology 414, 420, 440, 459, 463
Evangelical Magazine 300
Evans, Arthur 439
Evans, George de Lacy 313
Evans, John 461
Eve and the New Jerusalem (B. Taylor) 352
Evolution, The Stone Book, and the Mosaic Record of Creation (T. Cooper) 576
evolution, the word and concept 31, 160, 170, 172, 207, 210–212, 282, 285, 296, 366, 370, 372, 382–383
Examiner 129
exams, in socialist science 62, 273
expertise, scientific 452, 457
extinction 87, 89, 95, 101, 103–104, 108, 181, 205, 248, 275, 364, 371–372, 417, 461, 514, 519, 543
Eyre, Edward 464

Facts versus Fiction: An Essay on the Functions of the Brain (Lawrence) 262
Family Herald 21, 224, 475, 478, 500
Faraday, Michael 533
Farringdon ward 317–318
Fearon, Henry Bradshaw 317–318
ferns, fossil 36, 69, 102, 209, 247, 250–251, 323, 354, 454, 469–471, 517, 519, 524, 557
Finch, John 306, 400, 402, 404
Finsbury Literary and Mechanics' Institute 439, 569
Finsbury Mutual Instruction Society 293, 426, 571, 577
Finsbury Social Institution 386, 414, 425, 427, 429, 569, 577
First Bristol Co-operative Society 201
First Western Co-Operative Union 568
fish fossils 39, 252, 287, 347, 366, 470, 514, 519, 538, 545
Fitch, Josiah 5, 30, 133–134, 160, 566, 581
Fleet Prison 254

Fleming, G. A. 399, 439, 570
Fletcher, Alexander 132–133, 166
Flexner, Helen 177
flint, worked 108, 251, 417, 461–463, 519, 543, 578
Flood, biblical 96–97, 102, 104, 134, 190, 322, 350, 417, 436, 573, 576
Flourens, Pierre 369
foetal recapitulation of fossil ascent 31, 269
Forbes, Edward 37, 525–527
fossils, price of 63, 132, 182, 545–546
Founder's Hall Chapel 124, 566
Fourier, F. C. M. 340
Fox, Augustus Lane 439, 463
France, French connection 26, 36, 62, 102, 111, 151, 173, 176, 180, 188, 213, 247, 249–255, 277–278, 295, 303, 306, 310, 315, 331, 429, 434–435, 444, 453, 458, 461, 490, 517, 526, 563
Francis, F. J. 108
Francophilia in English Society (Eagles) 176
Frankenstein 114, 515
Frankenstein's Children (Morus) 274
Frankland, Edward 404
Franklin, John 528, 538
Franklin's Miscellany 292
Fraser's Magazine 301, 318, 351
freak shows 236, 243, 515
free admission to public buildings campaign 186
Free Enquirer 574
Freeman's Journal 574
Freeman, William 126
Freemasonry Considered as the Result of Egyptian, Jewish and Christian Religions (Reghellini) 144, 559
Freemasons 88–89, 559
Free-Thinker's Information for the People 261, 363–366, 382, 576
Freethinking Christians 26, 28, 165, 317–318, 408, 495, 582
French Protestant Church 424

French Revolution, Terror 7, 26, 53, 90, 114, 122, 169, 175, 177, 252, 295, 344, 425, 453, 479, 490, 500
Friendly Societies 252, 523
Friends of the People 89
Frost, John 67, 303, 331, 496
Frost, Thomas 364–366, 576
Fruitlands 326
fruits, fossil 40, 68, 252, 473–476, 519, 557
Fyfe, Aileen 24, 565

Galileo 52, 100, 262
Galileo and the Inquisition 262
galley 458
gallows 77, 146
Galton, Francis 479
Garcimartin, Colonel 287
Gardeners' Gazette 279, 282
Garden of Eden deciphered 144
Gardner, Richard 348
Gaselee, Stephen 129
Gast, John 166
gatherers, category 41, 43
Gauntlet 162, 201, 238
General Dispensary 63–64, 179
General Post Office 1, 64, 180–181, 415
General Practitioners 4, 179, 262, 448
Genesis and Geology (Gillispie) 7
Genesis and Geology; The Holy Word of God Defended (Baylee) 328
Gentleman, occupation 376
Gentleman's Magazine 424, 438, 453, 527, 534, 548, 550
Geological Society 43, 68, 148–149, 179, 185, 188–190, 195, 215, 245, 248–250, 279, 288, 378, 421, 454, 524–525, 547, 561, 563
geology, agitators' use of 6, 8–9, 13, 15, 20, 24–26, 33, 38, 45, 48–52, 56–57, 61–62, 64, 67, 71, 73, 90, 94, 96, 101, 103–104, 106–108, 114, 128, 130, 134, 147, 153–154, 173, 177–178, 193–195, 201–204, 206, 208, 217, 226, 235, 248, 254, 267, 270–272, 275, 277, 288, 292–293, 296–298, 303–304, 321–322, 330, 341, 345, 350, 354, 357, 364–366, 371–372, 379, 390, 398, 411, 418, 421, 425, 437, 445, 478–479, 482–483, 503, 523, 553, 556, 573
Geology and Mineralogy (Buckland) 53–54, 322, 374
George Inn, Oakham 131
Gibraltar 88
Gibson, Thomas Field 279
Gillispie, C. C. 7
Giltspur Street prison 238
glass cases, museum 277, 472, 539
Gliddon, G. R. 578
Globe 312, 413
Globe, Fleet Street 123
Gloucester 319, 388, 444
Gloucester Mechanics' Institute 319
Godfrey, William 416, 469, 484–485
Godmanchester 444
Godwin, Benjamin 204–205
Godwin, Joscelyn 88
Godwin, William 272, 340, 398, 424
Goldman, Lawrence 548
Goldsmith's Hall 421
Gomme, Lawrence 439
Good Sense (Holbach) 84, 87
Goswell Road 414, 425–427, 429, 563, 569, 571, 575, 577
Gould, S. J. 269
gradation, merging of species 31, 107, 155, 207, 210, 215, 269, 282–283, 382, 437, 442, 556
Grand Lodge of Operative Gardeners 81
Grand National Consolidated Trades Union 32, 289
Grant, James 568
Grant, Johnson 268
Grant, Robert Edmond 285, 372, 380–381
Gray, H. 98
Gray's Inn Road 37, 51, 194–195, 224, 267, 498, 567, 574

Great Exhibition 513, 515–517, 523
Great Metropolis (Grant) 568
Great Queen Street 337–338, 567
Great Tower Street Mutual Instruction Society 46 46, 201, 267, 274, 293, 427, 570, 575
Great Windmill Street anatomy school 490
Greaves, James Pierrepont 71, 305, 325, 328
Green, David 294
Greenwich Social Institution 569
Grey, Earl 192
Griffin, printer 100
Grimwood, Charles 125
Grote, George 314
Guadaloupe skeleton 108
Guest, James 93, 245, 274
Guildhall 66, 108, 119, 127, 129, 149, 192–193, 314, 317, 408
Guizot, F. P. G. 434
Guy's and St Thomas's hospitals 114, 122, 429
Gwynne, George 414

Hale, Matthew 368
Hall, Abel 81, 125–126
halls of science 8, 16, 37, 45–46, 63, 81, 177, 201, 264–265, 267, 272, 274, 288, 298, 302, 306, 323, 327, 332, 338, 340–345, 347–350, 352–353, 357–358, 361, 368, 385, 394, 399, 403, 405–406, 410, 412, 427–428, 437, 440, 480, 492, 496, 507, 509–510, 522, 531, 541, 543, 568–569, 574–576, 578, 580
Halse, William H. 423
Handel, G. F. 341, 520–521
Hanger, John 126
Hansom, John 580
Hansom, Joseph 398
Hanwell asylum 486
Harcourt, Leveson Vernon 436
Hardy, Thomas 253
Harmer, James 318–319

Harmony 16, 23, 386, 397–406, 408, 412, 418, 424, 496, 498, 503, 521, 536, 576, 582
Harney, Julian 113, 204, 333–334, 367, 427, 492
Harrison, J. F. C. 33, 333
Haslam, Charles Junius 364
Hawes, Benjamin 297
Hawkins, Benjamin Waterhouse 514
Hawkins, Thomas 35, 40, 189–190, 514
Haydn, F. J. 341, 520–522
Hays, J. N. 565
Hebrews, seers 143
Hedger, G. 577
Helsham, Arthur 161, 533–536, 582
Helsham, Robert 538, 582
Henderson, John 562
Henman, Edward 132, 161, 533, 581
Herald to the Trades' Advocate 15, 177
Hercules Tavern 151
Herschel, John 455
Hetherington, Henry 160, 163 2, 17, 28, 32–33, 37, 42, 67, 79, 81, 93–94, 146, 151, 160–161, 163–164, 167, 175–178, 183, 190–191, 198–199, 207, 217, 244, 253, 257, 259, 261, 305–306, 312, 317, 325, 330, 332–333, 363–366, 368, 379, 382, 387–388, 393, 395, 408, 495–500, 503, 511, 533, 543, 550, 576, 582
Heywood, Abel 93
Hibbert, Julian 23, 81–82, 129, 139–140, 146–148, 183, 225, 258, 261, 263–264, 304, 379, 486–487, 489, 497, 503, 546–547, 556, 569
Hibbert, Julian 140, 146
Highbury Grove 471–472, 475–476
Hill, James 53
Hill, Rowland 408
hippopotamus fossils 42, 210, 283, 288, 345, 471
Histoire des Végétaux Fossiles (Brongniart) 251
Histoire Naturelle des Animaux sans Vertèbres (Lamarck) 172

History of British Fossil Reptiles
  (Richard Owen) 285, 470, 476,
  551
History of Greece (Grote) 314
Hitchcock, Edward 35, 518
Hobhouse, J. C. 581
Hobson, Joshua 93, 245
Hodgkin, Thomas 429, 459–460, 463
Hogarth, William 314
Holbach, Paul-Henri Thiry Baron d'
  83–85, 87–92, 106, 112, 114, 206,
  227, 257, 261, 271, 296, 317, 398,
  407, 442, 483, 494
Holborn 47, 317, 332, 413, 487, 576
Holland 303
Hollick, Frederick 327–329, 363, 569
Holyoake, Eleanor 388
Holyoake, George Jacob 23, 30, 52,
  66–67, 305, 321, 327, 329–330, 340,
  342, 348, 357, 362, 368, 386–389,
  392, 395–397, 402, 405–410, 412,
  414–415, 427, 436, 439–441, 447,
  479–481, 487, 492–493, 495–497,
  499, 504–511, 525, 531–532, 550,
  565–566, 569, 577–579, 581–582
Holyoake, Madeline 388
Holyoake, Mazzini Truelove 340
Holywell Street 392–394, 399, 411
home colonies 363, 386, 392, 408, 563
Home Colonization Society 408
Home Office 19, 59, 138, 142, 164,
  229, 259, 311, 392, 553
homoeopathy 428, 542
Hone, William 78, 100, 118, 533
Honourable Artillery Company 35,
  526
Hooper, George 410
Horne, Richard Henry 54
Horoscope 250
Horsemonger Lane gaol 146, 228, 555
hospital consultants 91, 262, 448
House of Commons 175, 310
House of Lords 36, 192, 266, 304, 317,
  330, 343, 358
Hull 412, 513

Hulley, Dr 344
human fossils 108, 210, 248, 422, 543
Hume, David 396, 578
Hume, Joseph 148, 186, 192, 297, 512,
  528
Hunterian Museum, College of
  Surgeons 34, 281, 377, 416
Hunter, John 100
Hunt, Henry 17, 32, 118, 253
Hunt, Leigh 220
Hutchinson, Alexander 386
Hutton, James 89
Huxley, T. H. 481, 519, 521, 532, 541,
  549
Huzzars 17, 99
hyaena fossils 97, 210, 283, 543
Hyde Park 515–516
hydropathy 477

Ichthyosaurus, ichthyosaurs 94, 102,
  182, 189–190, 281, 323, 364, 470,
  514, 545–546
idolatry, fossil 52, 346, 483, 486
Iguanodon 2, 33–34, 57, 97, 279–282,
  284, 286, 323, 354, 364, 454,
  469–471, 484, 514, 519–520, 524,
  545–546, 551
Illustrated London News 534
Immortality of the Soul (R. Cooper)
  480
Independents 124, 295
infidelity, the word 79
Infidel's Text-Book (R. Cooper) 412,
  578
Information for the People 363, 372
Inglis, Robert 451–453
inheritance of acquired
  characteristics 172
Inquiry into the Nature of Responsibility
  (Mackintosh) 575
*Inquiry into the Principles of the
  Distribution of Wealth* (Thompson)
  171
insects produced by electricity 275
Insect Transformations (Rennie) 112
Instauration of Trial by Jury 129, 252

Institute of Political and Social Progress, Chelsea 577
Institution for Instruction and Amusement, Millbank 569
Institution of the Industrious Classes 61, 194–195, 224, 266–267, 277, 567–568
introduction, for museum entry 185, 188, 268
Introduction to Geology (Bakewell) 262
Investigator! 396
Ipswich 463, 513
Ireton, Henry 258
Irish Disturbances Suppression Bill 193
Iron Age 417, 462
Irving, Edward 234
Isis 105, 196
Isle of Sheppey 286, 473
Isle of Wight 33–34, 42–43, 278–281, 286, 470, 519, 531, 537
Islington 68, 137, 305, 471

Jacco Maccacco 240
Jacob, Margaret 89
Jenkins, Charles James 506–507, 520, 536, 539–540, 544
Jenneson, Charles 428, 569
Jenneson, Mary 428–429
Jerry the Satyr 240
Jew, as untrusted other 393
Jew Book, anti-semitic slur 64, 78, 367, 392, 411
John Bull 239, 304, 389, 392–395
Johnson, B. 98
John Street Institution 93, 202, 267, 323–324, 338–340, 345, 349, 358–360, 364, 368–369, 382, 386, 399, 403, 405, 412, 437, 442, 478, 481, 495–499, 507, 510–511, 520–522, 528, 535–537, 539, 568, 575–576
Jones, Ernest 489
Jones, Gareth Stedman 67, 168

Jones, John Gale 81, 123, 139–140, 148, 151, 167, 183, 225, 253, 258, 304, 467, 490–492, 561–562, 581
Jones, John Gale 140
Jones, Lloyd 326, 511
Jones, Thomas Rupert 35
Journal of the Ethnological Society 461
July Revolution 1830 45, 136, 138, 151, 173, 176, 252, 566
Jurassic 209, 464, 470
jury service 129

Kant, I. 396
Kemble, Francis 314
Kennington Common 311, 340, 359, 526
Kensal Green Cemetery 2, 352, 487, 492–493, 497, 499, 532–533, 543, 579
Kensington Palace 455
Kensington Social Institution 570
Kentish Town Chapel 268
Kentish Town Mechanics Institution 571
Kentish Town Mutual Improvement Society 571
King's Bench 78, 117–119, 126, 130
King's College, London 112
Kingsland and Newington Co-operative Society 185, 197
Kingsley, Charles 478
King, Thomas 372
King, William IV 65, 140–141, 287, 316
Kinross 333
Kirkdale cave 97
Knell, Simon 18, 24, 40–41, 184, 531
Knowledge Chartists 47–48, 332, 478
Knox, Robert 89, 367, 441, 460

Labourers' Friend Society 425, 562
labour notes 27, 196–198, 230, 272
lachrymatories 420, 422, 424, 562
Lady's Magazine 353
Lady's Newspaper 288

Lamarck, Jean-Baptiste 20, 55–56, 101, 104, 110, 115, 172, 213–219, 243, 248, 353, 372, 383, 577
Lambeth 35, 151, 181–182, 359–360, 371, 568, 570–571
Lambeth Mutual Instruction Society 571
Lambeth Social Institution 570
Lancashire mills 385
Lancaster Gazette 515
Lancet 4, 100, 179, 309, 448, 515
land nationalization 231, 324
Landseer, E. H. 240
Lane, Charles 71, 305, 325–326
Laplace, Pierre-Simon 173
last rites, radical 252
La Voix des Femmes 429, 563
Lawrence, William 91, 99–102, 114, 127, 170, 217, 223–224, 257, 262, 318, 352, 438, 481
Lectures on an Entire New State of Society (Robert Owen) 299
Lectures on Physiology, Zoology, and the Natural History of Man (Lawrence) 4, 99, 170, 217, 257, 262, 318, 352
Lectures on the Atheistic Controversy (B. Godwin) 204
Lectures on the Marriages of the Priesthood in the Old Immoral World (Robert Owen) 300
Leeds 88, 93, 202, 296, 345–346, 353
Lee, John 40, 447, 453–454, 456, 470
Lee, R. E. 485
Leiningen, Countess of 290
Letter from a Student in the Sciences to a Student of Theology (Saull) 194, 211, 555
Letters on the Laws of Man's Nature (Atkinson and H. Martineau) 263
Letters to Dr. Adam Clarke (J. Clarke) 245
Letters to the Clergy (Haslam) 364
Letter to the Vicar of St. Botolph (Saull) 165, 179, 554, 558
Lewes, G. H. 409
Liberal Registration Association 315

Liége caves 371
Life of Jesus (Strauss) 478, 532
Lightman, Bernard 25, 565
Lincoln's Inn 146
Lion 79, 94, 102, 106–107, 131, 133, 554
literacy of radical readers 8
Literary Gazette 22, 444, 447, 451, 455, 534, 563
Literary Institutions (Holyoake) 577
Liverpool 160, 274, 327, 343, 463, 513
Liverpool Royal Institution Museum 186
Locke, J. 396
Lockley, Philip 224
Lolme, Louis de 179
London and Birmingham railway 286
London Anti-Corn-Law Association 297, 581
London City Mission 299
London Clay 286, 465, 473–474
London Clay Club 474
London Co-operative Building Society 307, 338, 498
London Co-operative Trading Fund Association 154, 163
London Corresponding Society 114, 129, 253, 490
London Democratic Association 570
London Investigator 59, 482, 531–532, 539, 541, 578
London Labour and the London Poor (Mayhew) 302
London Mechanics' Institution 47, 51, 67, 93, 97, 113, 139, 161, 179, 220, 257, 325, 495, 498, 533, 553
London Oriental Institution 143
London Phrenological Society 423, 487
London Quarterly 515
London Secular Society 511
London Tract Society 358, 576
London Union of Compositors 475
London University 143, 263, 314, 321, 380, 486
London vestries 65, 294–295, 345

London Wall 166, 567
London Weekly Review 102
London Working Men's Association 273, 511, 576
Long Acre 127
Long Buckby 348, 409, 432
Lord Chief Justice 118, 122, 129, 303, 561
Lothbury 124, 566
Loveless, George 312
Lovett, William 47–48, 70, 151–152, 163, 168, 178, 192, 197, 217, 258, 260–261, 300, 328–329, 331–333, 413, 428
low/high concept in society and nature 7, 95, 115, 282, 372, 376
Lubbock, John 463, 525
Ludington, Charles 62
Lyell, Charles 2, 53–57, 68, 91, 115, 172, 213–215, 219, 243, 287, 365–366, 372, 375, 377, 383, 391, 417, 484, 525, 576–577
Lyme Regis 41, 182, 323

Macauley, Eliza 133, 566–567
Maccoby, Simon 96
Macconnel, Thomas 571
Macdonald, Charles Ker 457
Macerone, Francis 259, 275
Mackay, Charles 439
Mackey, Sampson Arnold 87–88, 144–145, 153–155, 174, 190, 250, 523–524, 554, 558
Mackintosh, T. Symmonds 152, 165, 273–275, 323, 494, 575
MacLeod, Roy 449
Maclure, William 50, 71, 217
Magasin Universel 371
Magazine of Natural History 112
Mahon, Lord 456
Maiden Lane 433
Maidstone quarry 281, 284
Maillet, Benoit de 109–110
Making of London (Gomme) 302, 439
Malthus, Thomas 9, 73, 162–163, 169, 193, 294, 296, 346, 442, 541, 543

Malvern 458
mammoths 283, 461, 469, 545
Man, As Known to Us Theologically and Geologically (Nares) 69, 214
Manchester 41, 93, 99, 102, 186, 210, 264, 274, 313, 340–342, 344–345, 347, 350, 357, 364, 385–386, 403, 458, 489, 513, 549, 580
Manchester Natural History Society museum 186
Manchester Zoological Society 344
Mandeville, B. de 437
mandrills 240–241
Mansfield, Thomas Rivers 350, 575–576
Mansion House 125, 148
Mantell, Gideon 33–35, 40, 179, 185, 189–190, 202, 250, 279–282, 284–286, 323, 417, 420, 461, 469, 471–472, 519–520, 545, 557
Man, The 25
Maoris 367, 429, 466
Marat, J. P. 204, 333, 427
Marchant, Thomas 401, 403
Marquess of Northampton 449
marriage, Owenite 171, 299–304, 328–329, 340, 343, 351, 359–360, 390, 394, 486, 578
Marriage System of the New Moral World (Robert Owen) 301
Marseilles 176, 252, 405
Marsh, Joss 29, 360
marsupial fossil 364, 464–465
Martineau, Harriet 263
Martineau, James 414
Martin, Emma 351–352, 389–390, 411–412, 578
Marx, Karl 16
Marylebone 26, 571
Marylebone, Western, and Richmond Literary and Scientific Institutions 108
Mason, Josiah 410
Mason's College 410
mastodons 279, 323, 354, 364, 519

materialism 2, 4, 7, 21, 25–26, 28–30, 37, 65, 68–69, 72, 83–85, 87, 89, 91, 99, 105–106, 110–112, 114–115, 131, 133, 136–137, 147, 163, 178, 183, 189–190, 202, 211, 213, 218, 221, 228, 232, 234, 237–238, 246, 258–259, 263, 282, 285, 288, 290–292, 297, 304–305, 317–318, 324–326, 330, 339, 354, 360, 373, 377, 379, 383, 411, 438, 461, 475, 477–480, 489, 503, 542, 574
Maudsley, Henry 480
Maurice, F. D. 505
Mayhew, Henry 302
Mayor of London 65, 120–121, 125, 148, 318, 467, 520
M'Callum, Sandy 42
McCabe, Joseph 23
McCalman, Iain 18, 83, 133, 162, 229, 565
Mechanics' Hall of Science, City Road 81, 264, 266, 274, 298, 332, 340, 348, 352, 360, 394, 406, 410, 412, 427, 440, 480, 492, 496, 507, 509–510, 522, 531, 541, 543, 569, 575, 578
Mechanics' Hall of Science, Manchester 202, 264
Mechanics' Hall of Science, Marylebone 81, 177, 568
Mechanics' Magazine 53, 185, 274, 277, 293
Medical Gazette 352, 490–491
Medical Times 262
Megalosaurus 282, 286, 470, 514
Memories (Bonney) 549
mental materialism 318
merchant, status 67
mermaid 109
mesmerism 25, 232, 263, 423, 477, 542
meteorology 250, 265
Methodists 244–246, 300, 313, 360, 403, 477, 496, 515, 542, 551, 576, 578
Metropolitan Anti-Corn-Law-Association 298

Metropolitan Association for the Repeal of the New Poor Law 295
Metropolitan Building Club 522
Metropolitan Churches Committee 345
Metropolitan Institution 523, 535, 537, 539–540
Metropolitan Parliamentary and Financial Reform Society 512, 523
Metropolitan Parliamentary Reform Association 314, 408, 410, 428, 582
Metropolitan Political Union 48 48, 67, 127, 151–152, 162, 325, 496, 561
Metropolitan slip-decorated ware 446
Michaud 251, 534
Michelotti, Giovanni 250
microscopes 474, 476
Microscopical Society 474
Militia Laws 260
Millbank 272, 569
millenarians 4, 6, 33, 107, 189–190, 205–206, 220–221, 224, 226–228, 230, 233–234, 236, 238, 289, 389, 484, 514, 567
millennium, political 1, 32–33, 37, 44, 185, 191, 205, 212, 224, 230–231, 235, 365, 386, 407, 479, 495, 538, 542, 574
Miller, George 95
Miller, Hugh 69, 188
Miller, James 369
Million of Facts (Phillips) 151, 277
Mill, James 171
Milton, John 50, 189–190, 193, 206, 478
Mineral Conchology of Great Britain (Sowerby) 36, 182, 471, 474, 545
Mining Journal 181, 277–278
Mining Manual 514, 518, 538
miracles 139, 207, 232, 341, 363, 380, 532
Mississippi delta deposits 391
Mitchell, Logan 205
models 71, 181, 513–514
molluscs 40, 473, 514

Monboddo, Lord 218, 220
monkey fossils 366, 383
monkeys 211, 239–241, 243–244
Montemolin, Conde de 287
Monthly Christian Spectator 478, 480–481, 518
Monthly Repository 54
Moore, James 52, 324
Moorgate 166
Morgan, John Minter 173, 437
Morning Chronicle 135, 310, 439
Morning Herald 265, 291
Morning Journal 135
Morning Post 280–281, 311, 354, 392
Morpeth Street Chapel 510
Morrison, Richard James, 'Zadkiel' 250
Morris, William 551
mortarium 446
Morus, Iwan 274–275
Moses 51, 53, 55, 190, 322, 327, 370, 391, 441, 479–480, 548, 576
Movement 395–396, 411, 577
Mozart, W. A. 520
Muggletonian 152
Muir, Thomas 252
mummy-unrolling 143
municipal reform 65, 317
Murchison, Roderick Impey 42, 374–375, 377, 526
Murray, John, publisher 97
Murray, John, scriptural literalist 108
Muséum d'Histoire Naturelle 102, 172, 247
Museum of Economic Geology 187, 526–527
museum, Saull's 1–3, 8, 10, 13, 16–18, 22–24, 26–28, 30, 32–40, 43–45, 49, 56–57, 61–62, 64, 68, 70–72, 147, 150–152, 175–177, 179–184, 187–188, 191, 193–195, 202, 210, 215, 228, 247–249, 251, 254, 267–268, 271–272, 277–282, 285–287, 289, 292–293, 297–298, 310, 323, 330–332, 346, 353–354, 399, 405, 415–416, 418, 420–425, 427, 432–434, 444, 446, 452, 461, 465–473, 476–477, 479, 481, 483–486, 488–489, 497, 503–506, 510–511, 514, 516–520, 522–524, 527–528, 531–532, 534–542, 547–552, 557, 575, 582
music in Halls of Science 166, 197, 265, 300, 322, 338, 340–341, 349, 505, 520–521
mutual instruction 45–46, 201, 267, 274, 293, 426–427, 566, 570, 575, 577
My Life (Wallace) 323
Mythological Astronomy (Mackey) 87–88, 144, 153, 174, 190, 250

Nares, Edward 69, 214
Nashville University 51
National Assembly, French 82
National Association, Chartist 47, 328, 332, 413, 428, 496, 571, 576
National Community Friendly Society 357
National Land and Building Association 405
National Political Union, NPU 47–48 47–48, 67, 127, 161, 191–193, 253, 264, 405, 428, 556
National Reformer 542, 578
National Reform Society 512
National Standard 17, 70, 188, 267–269, 544
National Union of the Working Classes, NUWC 26, 37, 48, 138–139, 151, 161, 168–169, 171, 173, 176, 183, 191–192, 198, 207, 225, 227, 255, 258–260, 495–496, 567–568, 571
Natural History Museum 547, 552
Natural Theology (Chalmers) 108
Natural Theology (Paley) 388
Neale, Vansittart 511
Newark 458, 466, 513
Newcastle 481
Newgate prison 80, 96, 111, 118, 139, 218, 245, 486, 579

New Harmony 50, 195–196, 217
New Jersey 470
New Lanark 164, 173, 248
Newman, Francis 480
New Monthly Magazine 526
New Moral World 48–49, 54, 57, 72, 105, 263, 274, 293, 321, 325–326, 329, 337, 347, 358, 362, 387–389, 394, 399, 401, 404, 408, 415, 430, 437, 494, 579, 582
New Poor Law 163, 193, 293–295, 313, 339, 378
Newport Chartist uprising 302–303, 570
Newport, George 372
New Religion; or, Religion Founded on the Immutable Laws of the Universe (Robert Owen) 165
Newspaper Press Directory 20
newspapers, London's 3, 19
newspaper stamp duty 160 19, 72, 151, 160, 164, 167–168, 229–230, 314, 319, 408, 496, 511, 581
news vendors 37, 44, 47, 168, 176, 192, 207, 228, 230, 260, 288, 389, 395
Newton, Isaac 6, 152, 274, 523, 550
New York 195, 264, 327–328, 547, 574
New Zealand 367, 429, 466
Niboyet, Eugenie 429, 563
Norman, J. 575
Northampton 60, 294, 343, 347–348, 409, 422, 432–433, 470, 535
Northern Star 33, 331, 416, 465, 472, 474, 483–484, 519, 562
Norwich Museum 186
Notes and Queries 23
Notitia Britanniae (Saull) 430–431, 433–435, 437–438, 444, 549
Nott, J. C. 442, 578
numbat 464
Numismatic Society 331, 422, 443, 445–447, 454, 470, 476, 513
nummulites 422, 470

Oakham prison 129–131, 146

oath swearing 126, 129, 254, 316, 407, 556, 579
O'Brien, Bronterre 260, 303
obsequies, French style 253, 490, 552
O'Connell, Daniel 193, 265–266, 568
O'Connor, Feargus 303, 332, 466, 477, 483–484, 494, 506, 512
O'Connor, Ralph 31, 50
Odd Fellow 395
Offor, George 65
Ogg, George 51, 97, 553
Old Bailey 331–332
Old Bailey court 100, 487
Oldham 36, 250, 274, 385, 403
Ommanney, Erasmus 528
On Superstitions (Pettigrew) 455
On the Origin of Man, and Progress of Society (Davenport) 427, 492
Operatives' Literary Association 257–258
Optimist 136
Optimist Chapel 5, 30, 135–136, 161, 166–167, 169, 176, 192, 258, 426, 566–567, 574
Oracle of Reason 37, 49, 362–363, 366–368, 370, 373, 375–376, 380–386, 388–389, 391–393, 395–396, 430, 482, 577
orang-utan 55, 214, 217–218, 223, 241–244, 484
ordinands, armed against infidelity 57, 204, 267
organ grinders 240
Origin of All Religious Worship (Dupuis) 144
Origin of Life (Hollick) 327–328
Origin of Species (C. Darwin) 1, 541, 578
Ornithocephalus 95
Owen, David Dale 50–51, 193, 195
Owenite, mineral 50
Owen, Richard, palaeontologist 2, 26, 34–36, 43, 270, 281–282, 285–286, 331, 363, 372, 374, 377, 416, 470–471, 476, 493, 519–520, 525–526, 537, 541, 545, 551–552

Owen, Richard, Robert Owen's son 51
Owen, Robert Dale 49, 51, 193, 196, 304, 306, 574
Owen, Robert, Owenites 13–14, 16–18, 23–28, 30, 32–33, 35–37, 39, 43–54, 57, 59–64, 66, 68–69, 71–72, 81, 107, 137, 149–150, 152, 154, 162, 164–173, 176, 178–179, 189, 191, 194–199, 201–206, 208–212, 217, 223–224, 227–228, 230–231, 233–236, 238, 240, 243, 245, 247–248, 252–254, 257–259, 261, 263, 265–275, 277, 287–289, 291, 293–295, 298–303, 305–306, 313, 315, 317, 321–333, 337, 339–346, 349, 351, 353, 355, 357–369, 371–372, 379, 382–383, 386, 388–391, 394–397, 399–410, 416, 418–420, 423–424, 426–431, 437–444, 446, 454, 456, 463, 466–467, 477, 479, 483, 486, 489–490, 495–499, 503–504, 506, 508, 511, 514, 521, 525–527, 535–538, 541, 543–544, 549–550, 552, 556–557, 561, 565–571, 573–582
Oxford Dictionary of National Biography 2
Oxford Street 201, 311, 511, 522, 568
Oxford University 15, 19, 37, 53–54, 56, 69, 204, 214, 281, 285, 315, 322, 363, 365, 376, 410, 420, 432, 448, 542
oyster fossil 96, 470

pabulum 209, 295–297, 346, 354
Paine, Tom 29, 64, 78, 80, 83, 92–93, 132, 139, 150, 161, 252, 261, 272, 304, 324, 340, 412, 453, 494, 581
Palaeontographical Society 476, 551
palaeontology, the word 31
Palaeophis 286, 473
Palaeotherium 364
Paley Refuted in His Own Words (Holyoake) 388, 415
Paley, William 380–381, 388

Palmer, Elihu 80, 86–87, 91–92, 105–106, 258, 261, 579
palm fossils 195, 268–269, 323, 557
Paradis, James 21
Pare, William 331, 403
Paris 36, 55, 62, 94–95, 97, 102, 111, 135, 173, 247–248, 250, 254–255, 313, 315, 317, 415, 429, 434, 471, 519, 526
Park Visitor and Christian Reasoner 510
Parochial Libraries, Church 344
Parolin, Christina 257, 565
Paterson, Thomas 93, 392–394, 396
Patriot 69, 214, 295, 436
Paxton, Joseph 516
Pecopteris 251, 470
Pecten 470
Peel, Robert 281, 374, 448, 450
Pegge, Samuel 435
Penman, R. 575
Penny Magazine 167
Penny Mechanic 46 46, 293, 353, 569, 571
penny postage 408
Penny Satirist 21, 224, 290–291, 346, 391, 395, 408, 475, 492
perfectibility 49
perfectibility, in society and nature 14, 25, 31, 33, 49, 65, 73, 106–107, 134, 154, 159–160, 165, 168–170, 172–174, 184, 188, 194, 196, 198–199, 203–204, 208, 211–213, 215, 231–233, 237, 243, 269–270, 324–325, 363, 365, 370, 372, 418, 437, 444, 466, 483, 495, 514, 538, 573–574
Perry, Thomas Ryley 573, 579
Perthes, Boucher de 35, 251, 461, 519
Pestalozzians 50, 53, 71, 325, 431
Peterloo massacre 17, 99, 118, 368, 426, 489, 578
Petrie, George 27, 71, 163 227–228, 300, 427, 485–486, 489, 497, 503, 546–547, 567
Petrie, Mary 486
Pettigrew, Thomas 451, 454–457, 533

Philadelphian Chapel 5, 59, 149, 258–260, 263, 567
Philippa, Queen 398
Phillips, John 372
Phillips, Richard 14, 31, 35, 108, 147, 150–155, 165, 189–190, 194, 209–211, 248–249, 275, 277, 282, 295–296, 304, 372, 481, 523–524, 550, 554, 558, 578
Phillips, Samuel 579
Phoenicians 419, 433, 505
Phrenological Association 262–263
phrenology 25, 47, 67, 137, 250, 262, 292, 322, 367, 371, 423–424, 428, 439, 462, 466, 477, 486
Piccadilly Hall 420
Pickering, William 322
Pictet, Charles 173
piracy 4, 78, 84, 87, 96–100, 107, 170, 217, 257, 262
Place, Francis 113, 191, 305, 491, 581
planetary causes of hot/cold, marine/terrestrial geological periodicity 78, 88, 107, 134, 144, 152–155, 165, 174, 179, 189–190, 194, 209–210, 212, 248–250, 296, 346, 354, 523–526, 553–554, 558, 563, 578
Plesiosaurus, plesiosaurs 94, 102, 180, 281, 364, 470, 514
Police Bill 316
Political Justice (Godwin) 398
Political Register 149, 448, 581
Politics of Wine (Ludington) 62
Poor Man's Guardian 2, 8–9 28, 32, 42, 105, 160, 164, 167–168, 175, 177–178, 198, 227, 252, 259–261, 321, 333, 495–496
Poplar Literary and Scientific Institution 571, 576
Popular Philosophy (Miller) 95
pornography printers 4, 18, 98, 392
Porter, Roy 90
Portraits of Men of Eminence (Reeve) 474, 476
Portsoken Ward 314
positivism 22, 436, 542–543, 550

Post Office London Directory 60, 497
Powell, Benjamin 94
pre-Adamite people 98, 189, 284
pre-human conditioning of character 49 49, 515
Presbyterian Review 50
Prestwich. Joseph 43
Prichard, J. C. 440–441
Primary era 94
primeval archaeology 386, 418, 421, 445
Prince Albert 377, 527
Prince of Wales 113
Principles of Geology (Lyell) 53, 55, 115, 213, 365, 383
Principles of Nature (Palmer) 80, 86, 105, 258, 579
printing press 107, 128, 167, 261, 358, 404, 414, 475
professions 41, 377
Prompter 146
Prothero, I. J. 18, 24, 51, 72, 163, 565
Prout, John 538, 582
Prout, Thomas 161, 264, 534, 538, 581–582
Provident Institution 505
Prussia 303
pterodactyl 95, 474–475
public debates 30, 46, 57, 206, 326, 338
public libraries 150, 415, 537
pubs 25, 60, 63, 131, 148, 257, 305–306, 319, 348, 409, 512
Pugin, A. W. 487
Pumfrey, Stephen 9
Pummell, John 130, 139, 228
Punch 21, 449, 460
Purgatory of Suicides (T. Cooper) 478
Purnell, Samuel 125

Quakers 4, 129, 398, 428, 459
Quarterly Journal of Education 148
Quarterly Review 99, 108, 152, 302
Queen Mab (Shelley) 87
Queenwood 398, 404, 406

racism, racial ranking 218, 301, 310, 352, 367, 369, 423, 431, 441–442, 462, 464, 466, 578
Radical Club 428, 582
radicalism and millenarianism 224
radicalism vs co-operation 198–199, 330, 363, 365, 495
Radical Reform Association 127, 151, 162, 582
Radical Underworld (McCalman) 18
Rationale of Religious Enquiry (J. Martineau) 414
Rational Institution 201, 267, 274, 569
Rational Society 358, 403 399–402, 439
Rational Tract Society 127, 358
Ray Society 476
reading rooms, libraries 93, 137, 171, 179, 207, 257, 263, 332, 338, 369, 394, 398, 413–414, 425, 504, 506, 522, 537, 539–540, 568–569, 571, 577–579
Reasoner 110, 406–407, 409, 411, 415, 447, 479, 492–494, 504–505, 509, 533–535, 542, 548, 577–578, 582
Recherches sur les Ossemens Fossiles de Quadrupèdes (Cuvier) 95–96
Reform Bill 15, 19, 24, 36, 48, 59, 66–67, 118, 141, 148, 161–162, 175, 191–193, 212, 216, 309, 466, 568
refugees 68, 192
Regent's Park 241, 377, 450
Reghellini, Macon, de Schio 144, 559
Religion of Geology (Hitchcock) 518
religious tracts 107, 358–359
Rennie, James 112
Reports of the British Association for the Advancement of Science 288
Republican 77–80, 85, 99, 101–102, 107, 109, 112, 153, 245, 553–554
Republican Institution 568
Revelation 6, 13, 57, 69, 107, 136, 202, 224, 226, 234, 268, 343, 351, 368, 380
Revell, Henry 191
Revolt of the Bees (Morgan) 173, 437

rhinoceros fossils 210, 283, 354, 461, 471
Rights of Man (Paine) 150
River Itchen 458
Robespierre, Maximilien 252, 395
Robinson, Harriet 126
Robinson, John 576–577
Rochdale 273
Rockingham House 569
Roebuck, J. A. 187
Rogers, George 67, 331, 333
Rogers, Henry Darwin 51, 67, 195, 321, 574
Roget, P. M. 380–381, 449
Romans, Roman ware 31, 251, 419–422, 424–425, 429–435, 439, 444–446, 455, 457–459, 462–463, 465–466, 473, 476–477, 513, 516, 519, 531, 538
Roman wall 421, 424, 458
Ronge, Madame 305
Roome, John 126
Rose Hill 397–401, 403–404, 408, 536, 582
Rosse, Lord 576
Ross, John 528
Rotunda 26, 81, 101, 137–141, 144–148, 161, 176, 183–184, 191, 207, 219, 224–225, 228–229, 239, 277, 350, 394, 414, 486, 490–491, 507, 568, 570, 574, 576, 582
Rousseau, Jean Jacques 301, 369, 414, 439
Royal Academy 378
Royal Cornwall Geological Society 514
Royal Mint 287
Royal Pavilion Theatre 309
Royal Polytechnic Institution 279
Royal Society 53, 218, 279, 378, 447, 449, 455–456, 474
Royal Victoria Theatre 309
Royle, Edward 18, 301, 565
Rudwick, Martin 94

Ruins, Or a Survey of the Revolutions of Empires (Volney) 83, 93, 139, 144, 206, 261
runic inscriptions 68, 457
Ruskin, John 375, 445
Russell, Lord John 455, 512
Rutland 130
Ryall, M. Q. 396

sacred socialists 25, 54, 305, 325, 328, 343, 477, 576
Salford 274, 293, 296, 342, 578
Salford co-operative school 61, 369, 578
Salter's Hall Chapel 5, 124–126, 129–130, 566
salvage archaeology, preservation 420–421, 435, 447, 457–458
Samian pottery 422, 424–425, 433, 446
Sanatorium, New Road 64
satire 20–21, 23, 100, 122, 239, 289, 475, 533, 576
Saul, Jane 22
Saull, Ann 536
Saull, Caroline 536
Saull, Elizabeth 303, 539
Saull, John 60, 348, 409
Saull, Sarah 536
Saull, Thomas 60, 62, 535, 538
Saull, William 348
Saull, William Devonshire. *See* specific entries
Savage, Charles 410
Savage, John 177, 568
Saxons, Saxon ware 422, 444, 463–465
Scales, Christopher 125
Scales, Michael 314
Scarborough 40, 184–185
Scarlett, James 118, 120
Schelling, F. W. J. von 106
School of Free Discussion 271
Science in the Marketplace (Fyfe and Lightman) 25, 565
Scotland Yard 526

scripturalist responses 57, 79, 107, 480
sea lilies 69, 354, 469
Secondary era 94–95
Secord, Anne 25, 41–42, 257
Secord, James 9, 14, 55, 91, 219, 373, 410, 412
secularism 23, 29–30, 37, 323, 405–409, 415, 441, 480–482, 508–510, 522, 535, 542, 544, 550, 552, 582
Secular Propagandist Fund 511, 531
Sedgwick, Adam 169–170, 214, 375–376, 390, 395
Seditious Meetings Act 65, 531
Select Committee on British Museum 184, 447, 452
Senior Scholars' Institute 345
Serres, Marcel de 248
Shakespeare, W. 241, 360, 367, 528
Shapin, Steven 8, 25, 67, 96, 373
shark fossil 252, 514, 519
Sharples, Eliza 160, 183, 196, 219–220, 239, 507, 567–568
Sheets-Pyenson, Susan 565
Shelley, Percy Bysshe 87, 93
Sheridan, R. B. 113
shipwrights' union 166
Shortt, W. T. P. 432
Sigillaria, S. saulli 36, 250–251, 323, 470–471, 545
Silliman, Benjamin 262
Silurian System 374
skulls, ancient 352, 371, 420–421, 423, 457, 465–466
Slap at the Church 315
Smith, Charles Roach 421, 433–435, 443, 447, 451, 455–458, 469, 546
Smith, J. 100
Smith, James Elishama 21, 33, 52, 101, 205–206, 220–221, 223–238, 244, 246, 289–293, 319, 346, 354–355, 367, 390, 475, 478–479, 483, 486, 489, 492–493, 500, 556–557, 567, 571
Smith, John Pye 130, 436

Smith, Joseph 342
Smithsonian Institution 547
Smith, William Hawkes 579
snake fossils 286, 473
Social History of Truth (Shapin) 96
social hymns 341, 521
Société Géologique de France 248–249, 254, 331
Society for Obtaining Free Admission to Public Monuments and Works of Art 186
Society for Scientific, Useful and Literary Information 567
Society for the Acquisition of Useful Knowledge 201, 267, 571
Society for the Diffusion of Useful Knowledge 25, 112
Society for the Extinction of Ecclesiastical Abuses 38, 180
Society for the Suppression of Vice 80, 97, 145, 392
Society of Antiquaries 68, 245, 378, 421, 424, 443, 445, 447, 449–458, 461, 476, 513
Society of Arts 514
Somerset House 378
Somers Town 68
Soul (Newman) 480
Southampton 444, 458
Southcott, Joanna 33, 224, 226, 228, 232, 361
Southern Metropolitan Political Union 151, 191, 444
South Place Chapel 426, 571
Southwell, Charles 49 49, 64, 69, 79, 301, 340, 359–363, 366–372, 380, 386–387, 396, 413, 441, 569–571, 577
Sowerby, George Brettingham 22, 181–182
Sowerby, James 22, 35–36, 62, 181–185, 251, 469, 471, 474, 514, 544–545
Sowerby, James de Carle 471
Spencer, Edward 421, 443
Spencer, Herbert 549, 576, 578

Spence, Thomas 81, 86, 425–427, 485
spies 14, 18–19, 30, 78, 80–82, 93, 100, 121, 125–127, 129–133, 135–136, 138–141, 143, 145–148, 160–161, 164–167, 183, 192, 197, 216, 225, 227–229, 253, 258–259, 263, 292, 507, 553, 567
spinners 17
Spinoza, B. 396, 578
Spirit of the Age 409, 582
spiritualists 305, 538, 541–542, 544
Spitalfields 279, 309, 318, 509
sponge fossils 40, 454, 470, 473–474, 476
spontaneous generation of individuals and ecosystems 110–111, 210, 217, 249
Stack, David 47 47
Staffordshire Gazette 343
St Aidan's Theological College 327
Standard 193, 287, 351, 389, 392–394, 468
Stanley, Lord 450, 454
Star 182–183
Star in the East 53
St Botolph Without Aldersgate 64, 79
Stephens, Joseph Rayner 313, 496
Stevens, Joseph 145
Stevens's auction room 182, 279, 547
St George's Hall 327, 529
St Giles 67, 302
Stigmaria 470
St Martin's Hall 521
Stockport 273
St Paul's Cathedral 186, 457
Stranger's Intellectual Guide to London for 1839-40 179, 354, 421
Stratford 284
stratigraphical arrangements 17, 182, 184, 249, 277–278, 471
Strauss, David Friedrich 478, 532, 542, 578
Streptospondylus 286
St Simonians 345, 571
St Thomas's Hospital 114 122, 352

Student in Realities 272
Suffolk Crag 514
Sunday Times 133
Surrey Institution 568
Surrey Zoological Gardens 240
Sussex Literary and Scientific Institution 185
Swift, Jonathan 50, 122
Sydney Morning Herald 464–465
Syntagma (R. Taylor) 130
System of Nature (Holbach) 83–85, 91, 112, 130, 261, 494

Tailors' Union 81
Tasmanians 463
Tatum, John 533
Taylor, Barbara 300, 352
Taylor, Robert 4–5, 14, 29, 31, 37, 59, 64, 66, 79, 81, 93, 117–133, 135, 137–147, 149, 151, 154, 160, 162, 170, 176, 225, 228–229, 258, 287, 367, 389–390, 426, 507, 532, 546, 555–556, 558–559, 561, 566, 568–569
tea parties 46, 306, 359, 365, 389, 495, 507, 509–510, 528, 540
teetotalism 16, 25, 137, 305–306, 400, 453, 477, 495
Telliamed (De Maillet) 109–110, 112, 218
temperance 82, 257, 305–307, 454, 495, 534
Tennant, James 40, 514
Tenterden, Lord 118–119, 129
Tertiary beds 44, 94–95, 215, 250, 470, 473
Test and Corporation Acts 451
Teutons 462–463, 465
Thackeray, William Makepeace 17, 70, 188, 268
Thelwall, John 114, 216, 253
Theobald's Road Institution 197, 228, 253, 568
Theory of the Earth (Hutton) 89
Theosophical Enlightenment (J. Godwin) 88

Thief 196
Thistlewood, Arthur 118, 425
Thomason, Thomas 576
Thompson, E. P. 29, 83
Thompson, George 124
Thompson, T. Perronet 571
Thompson, William 171–172, 299–300, 304, 306, 486, 499
Thorne, Joshua 569, 575
Three Age System 417
Tiffin, Charles 497–498, 536
Tilgate Forest 97, 280
Timbs, John 446, 471, 538
Times 2, 23, 54, 59, 65, 118, 148–150, 204, 287, 291, 318–319, 374, 387, 393–394, 453, 516, 581
tithes 6, 15, 19, 28, 38, 56, 72, 79, 151, 192, 204, 208, 212, 229, 231, 293, 315, 375, 379, 381, 556, 574
Todd, R. B. 381
Tolpuddle Martyrs 4, 68, 261, 287, 309, 312
Tommy, chimpanzee 242–243
Topham, Jonathan 53–54
Torrens, Hugh 2, 33, 40–42, 279
Tottenham Castle Museum 547
Toulmin, George Hoggart 88–96, 105, 134, 159, 172, 257
Tower Hamlets 124, 148
Tower Hamlets Literary Institution 510
Tower of London 227, 486
tract distribution 47, 107, 261, 329, 333, 358–359, 369, 396, 408, 509–510
trial by jury 129, 167, 252
Trial of the Rev. Robert Taylor (R. Taylor) 556
Truelove, Edward 340, 412–413
True Sun 20, 42, 56, 67, 149, 215–217, 247, 291, 305, 568
Truth of Revelation (Murray) 108
Tsar 192, 375
tumuli 420, 431, 459, 461–462, 465
Tyndall, John 404, 576

Types of Mankind (Nott and Gliddon) 578
type specimens 1, 35–36, 181–182, 545–546
Tyrrell, Henry 528
Tytherly 386, 397, 498, 545

Unitarians 26, 54, 82, 400, 403, 414, 439, 448, 579
United Trades' Loan Fund 199
Universal Community Society of Rational Religionists 357–358, 569
University College London 137, 285, 401
University of Cambridge. *See* Cambridge University
University of Glasgow 575
University of Pennsylvania 51, 575
Uranian Society 148, 563
Urania poem 426, 492
U.S. Geological Survey 50
Utilitarian Record 415
Utilitarian Society 348, 409–410, 412, 442, 481, 569

Vale, Gilbert 563
vegetarianism 25, 82, 150, 210, 304–305, 325–326, 477
Vestiges of the Natural History of Creation (Chambers) 14, 21, 39, 395, 410–415, 482, 527, 541, 577
Victim's Fund 167
Victims of Whiggery (Loveless) 312
Victorian Infidels (Royle) 18
Victorian Sensation (Secord) 14
Victoria, Queen 242, 290, 429, 460, 551
Vincent, Henry 243
Visions of Science (Secord) 10
vitalism 99–100, 114
vitrified forts 100, 444
Volney, Constantin François de 83–84, 87, 92–93, 111, 139, 144, 154, 206, 217, 261, 317, 340, 398, 559

Voltaire, F. M. A. de 7, 54, 93, 204, 340, 412, 439, 485

Wake, C. S. 464
Wakley, Thomas 4, 67, 179, 297, 309–311, 334, 427, 439, 445, 448, 451
Wallace, Alfred Russel 169, 323–324
Warden, Benjamin 26, 306, 568, 570
ward meetings 64, 512
wardmotes 65
Waterhouse, G. R. 420, 545
Watson, Bishop 64, 128
Watson, Dr James 86
Watson, James, obstetrician 85, 110, 113
Watson, James, printer 163 81, 86, 93, 151, 163, 176–178, 192, 197, 216, 245, 257–264, 274, 312, 332, 340, 388, 398, 406, 426–427, 486, 499, 510, 567, 569
Watts, John 306, 541–543, 578
Way, Albert 454–455
Wayland Smith's Cave 445, 563
Wealden beds 278, 281, 514, 531
wealth, connections of scientific elite 54, 56, 373–380, 382, 448, 525, 545
weavers 83, 309, 385, 580
Wedderburn, Robert 218
Weekly Dispatch 318–319
Weisser, Henry 176
Welch, Georgina 329
Wellington, Duke of 27, 42, 138, 140–141, 193, 287, 451, 516
Western Co-Operative Institute 26, 108, 147, 194, 201, 267, 568
Western Union Labour Exchange 568
West India docks 82
West-London Central Anti-Enclosure Association 405
Westminster Abbey 2, 186, 453
Westminster Mechanics' Institution 272, 569
Westminster pit 240
Westminster Rational School 272–273, 569

Westminster Review 337
whale fossils 36, 545
Whalley, Robert 210
Wheeler, Anna 171, 229
Wheeler, T. M. 570
Whitby 431–432, 462, 545
Whitby Literary and Philosophical Society 431
Whitechapel day school 271, 429, 581
White, Charles 367
White Conduit House 309, 312
White Hart Tavern 581
Whittaker, John 403, 536–537
Wiener, Joel 82, 139, 423
Wilberforce, Samuel 451, 454
Williams, John, 'Publicola' 319 492–493, 495, 533
window tax 512
wine trade 63, 201
Witness 69
Wollaston Medal 44
Wollstonecraft, Mary 171
women, as activists 79, 167, 177, 184, 303, 339, 342, 351–353, 389–390, 394, 399, 428–429, 507–508, 578
women, in audiences 1, 38, 93, 133, 142, 171, 177, 184, 207, 277, 353, 389–390, 394, 472, 507

Wonders of Geology (Mantell) 33
Woodward, Arthur Smith 552
Worcester 458
Word Crimes (Marsh) 29
workhouses 163, 193, 293–296, 313, 402, 455, 460
Working Men's Association 273, 348, 511, 576
Wright, Frances 264, 574
Wright, Susannah 118
Wright, Thomas 434
Wroe, John 224

Yahoo 306
Yeo, Eileen 300
Yorke, Charles 139
Yorkshire ancient huts 432, 444, 462
Young, George 431–432, 439

Zebulon Chapel 569
Zetetic Society 440, 569
Zoological Society of London 185, 239, 241–242, 250, 377–378, 447, 449
Zoonomia (E. Darwin) 113

About the Team

Alessandra Tosi was the managing editor for this book. Rose Cook provided editorial assistance.

Jennifer Moriarty proofread this manuscript.

Jeevanjot Kaur Nagpal designed the cover. The cover was produced in InDesign using the Fontin font.

Cameron Craig typeset the book in InDesign and produced the paperback and hardback editions. The main text font is Tex Gyre Pagella. The heading font is Californian FB.

Cameron also produced the PDF and HTML editions. The conversion was performed with open-source software and other tools freely available on our GitHub page at https://github.com/OpenBookPublishers.

Jeremy Bowman created the EPUB.

This book was peer-reviewed by Bernard Lightman, professor of humanities at York University, Canada, and two anonymous referees. Experts in their field, these readers give their time freely to help ensure the academic rigour of our books. We are grateful for their generous and invaluable contributions.

This book need not end here...

Share

All our books — including the one you have just read — are free to access online so that students, researchers and members of the public who can't afford a printed edition will have access to the same ideas. This title will be accessed online by hundreds of readers each month across the globe: why not share the link so that someone you know is one of them?

This book and additional content is available at:
https://doi.org/10.11647/OBP.0393

Donate

Open Book Publishers is an award-winning, scholar-led, not-for-profit press making knowledge freely available one book at a time. We don't charge authors to publish with us: instead, our work is supported by our library members and by donations from people who believe that research shouldn't be locked behind paywalls.

Why not join them in freeing knowledge by supporting us:
https://www.openbookpublishers.com/support-us

Follow @OpenBookPublish

Read more at the Open Book Publishers BLOG

You may also be interested in:

The Scientific Revolution Revisited
Mikuláš Teich

https://doi.org/10.11647/obp.0054

From Darkness to Light
Writers in Museums 1798-1898
Rosella Mamoli Zorzi and Katherine Manthorne (eds)

https://doi.org/10.11647/obp.0151

Cultural Heritage Ethics
Between Theory and Practice
Sandis Constantine (ed.)

https://doi.org/10.11647/obp.0047

www.ingramcontent.com/pod-product-compliance
Lightning Source LLC
Chambersburg PA
CBHW052052300426
44117CB00013B/2094